...IER ENGINEERING AND TECHNOLOGY

...ve Editor: D. V. Rosato
...: R. B. Akin, H. F. Mark, J. J. Scavuzzo,
...S. S. Stivala, L. J. Zukor

...ONAL VOLUMES IN PREPARATION

POLYM

Executi
Editors

SYNTH
edited b

FILAM
and des
D. V. R

REINF
edited

ENVIR
(2 volu
edited

FUND
COMP
edited

CHEM
(3 volu
edited

THE S
(2 volu
edited

PLAS
edited

EPOX
edited

POLY
edited

ADDIT

The Science and Technology of Polymer Films

Edited by
ORVILLE J. SWEETING

Yale University and Quinnipiac College

VOLUME II

Wiley—Interscience

a Division of John Wiley & Sons, Inc.
New York · London · Sydney · Toronto

CHEMISTRY

6128-0276 ✓

Copyright ©1971, by John Wiley & Sons, Inc.

Library of Congress Catalog Card Number: 67-13963

ISBN 0 471 83894 2

Printed in the United States of America

10 9 8 7 6 5 4 3 2 1

For Mildred and Thelma

PREFACE

Since the conception of this volume on self-supporting films took place more than five years ago, many changes have occurred in the packaging field. Some of the chapters written then have been revised at least twice and last-minute changes have been made in the page proofs.

Despite this effort we shall probably fail to present a completely current picture of film materials in 1969, which was our goal. I wish to thank all of the contributors for their patience in writing the chapters initially and for their willingness to correct the data currently. The task has been a difficult one for all concerned to bring all of the chapters of the manuscript to print simultaneously.

This volume includes as Chapter 1 the treatment of barrier properties that was initially planned for Volume I. Otherwise the present Volume II attempts to present a thorough discussion of all films that are important in the packaging business today, excepting laminates and other types of composite. The purpose and application of coatings, as well as extrusion coating and laminating, were included in Volume I.

As costs of distribution rise and inflation grows worse, the cost-conscious manufacturer or distributor will pay increasing attention to the role played by packaging films in getting products to the consumer in durable, protective, pilferage-resistant, attractive units that preserve the qualities of the contents until use. It is the hope of the authors that this book will play an important part in making available a convenient reference to the various self-supporting films that are available today. Comparison of properties from film to film can now readily be made in a single convenient reference.

As before, I welcome comments and criticism. The favorable reception accorded Volume I has been gratifying, and it is hoped that with the publication of this volume each of the two will enhance the value of the other to the reader. They must be read together in order to derive the greatest benefit in practice.

ORVILLE J. SWEETING

New Haven, Connecticut
March 1970

vii

AUTHORS

GARTH H. BEAVER
Designed Polymers Research, The Dow Chemical Company, Midland, Michigan

HARRIS J. BIXLER
Amicon Corporation, Lexington, Massachusetts

HARVEY A. BROWN
3M Company, St. Paul, Minnesota

LAWRENCE D. BURKINSHAW
General Electric Company, Pittsfield, Massachusetts

R. L. BUTLER
The Dow Chemical Company, Midland, Michigan

DAVID W. CAIRD
General Electric Company, Pittsfield, Massachusetts

GEORGE H. CRAWFORD
3M Company, St. Paul, Minnesota

DONALD ESAROVE
Tremco Manufacturing Company, Cleveland, Ohio

CARL J. HEFFELFINGER
Film Department, E. I. du Pont de Nemours and Company, Inc., Circleville, Ohio

HARRY A. KAHN
Easton, Connecticut

KENNETH L. KNOX
Film Department, E. I. du Pont de Nemours and Company, Inc., Circleville, Ohio

HERBERT K. LIVINGSTON
Department of Chemistry, Wayne State University, Detroit, Michigan

WALTER R. PAVELCHEK
Film Operations, American Viscose Division, FMC Corporation, Marcus Hook, Pennsylvania

JULES PINSKY
Monsanto Company Packaging Division, Hartford, Connecticut

HUGH W. RICHARDS
Polymer Technology Division, Eastman Kodak Company, Rochester, New York

CHARLES S. SCHOLLENBERGER
The B. F. Goodrich Company Research Center, Brecksville, Ohio

H. H. SINEATH
Film Operations, American Viscose Division, FMC Corporation, Marcus Hook, Pennsylvania

ORVILLE J. SWEETING
Yale University and Quinnipiac College, New Haven, Connecticut

A. T. WIDIGER
The Dow Chemical Company, Midland, Michigan

CONTENTS

xii CONTENTS

*The Science and Technology
of Polymer Films*

VOLUME II

CHAPTER 1

BARRIER PROPERTIES OF POLYMER FILMS

HARRIS J. BIXLER

Amicon Corporation, Lexington, Massachusetts

and

ORVILLE J. SWEETING†

Yale University, New Haven, Connecticut

† Formerly Associate Director of Research, Olin Film Division, Olin Corp., New Haven, Connecticut. Present address: Quinnipiac College, New Haven 06518.

I. Introduction

Polymer films are generally employed as barriers to the free transmission of gases, vapors, liquids, ions, and other substances across phase boundaries. Classically, such films have been selected to serve this general function when, for instance, a paint or varnish is used to protect metals against corrosion, paper is coated to render it moisture-resistant, or a freestanding polymer film is used to package foodstuffs. In the not-too-distant past these applications were looked upon simply as a means of isolating some object from its surroundings, be it a metal chemical reactor vessel containing a corrosive solution, a cellulose fiber immersed in water, or a package of baked goods on a grocer's shelf.

Through the pioneering work of Daynes [1], Barrer [2], and van Amerongen [3], it became evident that no polymeric film is completely impervious to its surroundings, but it was deduced that temporary protection, at least, could be afforded. The detailed understanding of the barrier properties of polymer films, however, has matured rapidly over the last decade. Today more sophisticated utilization is being made, or is about to be made, of polymers in film form in which not only their over-all *impermeability* (or permeability) is important but also their ability to discriminate is employed in the transmission of molecular species on the basis of size, shape, polarity, charge, or other factors. Otherwise stated, advantage is being taken of the *permselectivity* of polymer films.

Examples of these latter applications include controlled-atmosphere packaging of produce, the desalination of seawater by reverse osmosis, the desalination of brackish water by electrodialysis, the separation and purification of a polymer in solution by ultrafiltration, the life-sustaining purification of the blood of uremics by hemodialysis, and the preparation of moisture "breathable" plastic films as leather substitutes. The term *membrane* has been applied to the polymer film used in many of these examples, and its use has generally connoted small volume applications of polymers of little consequence to the chemical industry. It is our conviction, however, that the membrane aspect of barrier film technology is no longer inconsequential from a commercial standpoint and the principles outlined are applicable to this emerging field.

This chapter has been designed to serve several purposes. First, the physical chemistry of molecular transport kinetics in polymer films is discussed with an eye to relating polymer structure to transport parameters and providing logical methods of correlating and extrapolating permeability data. The authors belong to the school that attempts to keep mathematical modeling and physical modeling in step with each other in the analysis of complex physical phenomena, and although the value of the phenomenological approach espoused by Katchalsky, Curran, and Kedem [4, 5] and by Lakshminarayanaiah [6] is appreciated this approach is not yet a well-functioning part of their kit of tools.

Also, methods of experimentally measuring the gas permeability parameters of polymer films are covered. The emphasis in the experimental area is on simplicity of measurement, on the one hand, and accuracy and reliability on the other. No effort has been made to include a description of all the devices reported in the literature for permeability measurement, but instead only those instruments that are readily available commercially or those methods that have stood the test of multiple laboratory usage are described. Data taken from the technical literature and our own files have been tabulated for reference (Appendix Tables 1—3). Several such tabulations already exist in the literature [7–9] and when possible have been upgraded by more accurate values and more complete characterization of the films. Unfortunately, lack of information in the latter area is still the bane of the barrier technologist's existence, and a strong word of caution must be introduced on applying data recorded under a generic polymer name to all materials carrying that same name.

Last, it is intended in this chapter to stimulate thinking along the lines of the tailorability of polymers with respect to permeability and permselectivity. The chemical composition, morphology, and molecular topology of polymer films strongly influence transport properties, and the key to more imaginative uses of existing polymers, in film form, and to the development of new commercially significant polymers for films will in part be in the hands of those engineers and scientists who understand and can apply concepts of the interrelated structure and properties to the solution of barrier problems.

A word of apology might be offered for the type of coverage sought in this chapter. No attempt has been made to make this an in-depth treatise. Nor has an effort been made to include a complete bibliography on the subject. Several excellent reviews already exist for this purpose [2, 7, 8, 10]. Instead, we have drawn heavily on our own experience in this field in an effort to relate to the reader those aspects of the subject that seem understandable and exciting.

II. The Permeability of Polymer Films to Permanent Gases

A homogeneous polymer film, free of pinholes, cracks, and other macroscopic defects, presents a barrier to gas molecules made up of polymer chains in a fairly densely packed configuration but in which, nonetheless, chain segments are undergoing thermal motion (Volume I, Chapter 2). The penetration of this barrier by gas molecules involves short-range interactions with these moving chain segments, and it is therefore not unlike the penetration of a liquid film by a gas. In fact, viewing the polymer participating in the transport process as a liquid of high viscosity and low volatility is usually helpful from the standpoint of developing physical models and will seldom lead to erroneous conclusions. It should be emphasized that this liquid model is applicable, even in a qualitative sense, only because of the short-range nature of gas-polymer interactions, and it fails completely when longer range interactions are involved, as, for instance, in the case of macro-mechanical deformation processes.

With this liquid model in mind it is a fairly simple matter to reconcile the general permeability of polymers to gases in terms of classical solution-diffusion equilibria and kinetics. When a liquid or polymer film is subjected to differential partial pressure of a gas across it, a flux of the gas will be observed emanating from the surface of the film in contact with the lower gas pressure. If the boundary pressures are maintained constant, the flux of gas through the film will eventually become constant. The magnitude of the steady-state flux will be observed to increase as the pressure difference across the film is increased, and it will decrease as the thickness of the film is increased (as long as the polymer morphology is not thickness-dependent). These observations can be formalized into the following relationship,†

$$J = -\bar{P}\frac{\Delta p}{L} \tag{1}$$

† An unbelievable variety of units is still used in (1). The American Society for Testing and Materials has recently adopted the *barrer* (equal to 10^{-10} ml (STP)-cm/sec-cm²-cm Hg) as the standard unit for the permeability coefficient, but it has not yet received widespread use. The most frequently used practical units for the permeability coefficient are ml (STP)-mil/100 in.²-24 hr at 1 atm and 25°C. A table of conversion factors is given in Table 1 for several of the more frequently used sets of units.

where J is the flux in quantity of gas per unit film area per unit time, Δp is the external pressure difference across the film, L is the film thickness, and $\bar{P}$ is a constant of proportionality known either as the permeability coefficient or simply the permeability. The minus sign appears in this equation because the quantity $\Delta p/L$ is negative when the flux is positive.

As (1) is written, it applies equally well to the viscous or laminar flow of gases through porous media. The absence of discernible pores in dense homogeneous polymer films and the failure of the flux to follow the fully formulated Poiseuille or Knudsen flow equations eliminate these mechanisms from detailed consideration. If flaws are present in the film, however, these modes of transport can reach dominant proportions. Pore flow is not considered with respect to gas transport in polymers, although it is of major importance in other membrane transport processes.

The nondependence of $\bar{P}$ on thickness must be carefully checked before it can be assumed. Besides variations in polymer morphology with thickness, other factors can lead to the failure of (1). Many polymer films of commerce are coated for various reasons, ranging from improving printability to reducing solvent sensitivity. If the coating is much less permeable than the substrate and the thickness of the substrate is varied, a dependence of the apparent $\bar{P}$ on L will be observed. Coated papers definitely fall into this category. In these cases the *permeance* Q is most often reported as the practical barrier property.

$$J = Q \, \Delta p \tag{1a}$$

Following is a brief analysis of the permeability coefficient.

Graham [11] was the first to suggest in print that the physical process leading to (1) involves the dissolution of the permeating species in the film at the high-pressure surface, diffusion of the dissolved gas through the film along a concentration gradient, and re-evaporation of the gas from the low-pressure surface of the film; this view is still regarded as the most generally valid physical model for the process.

Fick [12] is credited with the mathematical formulation of the physical process occurring within the film, and von Wroblewski [13] was the first to apply Fick's laws of diffusion to the permeation process

TABLE 1
Conversion Factors
Transmission Conversion Factors

To obtain	Multiply		
	$\dfrac{\text{g}}{24\ \text{hr-m}^2}$ by	$\dfrac{\text{g}}{24\ \text{hr-100 in.}^2}$ by	$\dfrac{\text{grains}}{\text{hr-ft}^2}$ by
$\dfrac{\text{g}}{24\ \text{hr-m}^2}$	1	15.5	16.7
$\dfrac{\text{g}}{24\ \text{hr-100 in.}^2}$	6.45×10^{-2}	1	1.08
$\dfrac{\text{grains}}{\text{hr-ft}^2}$	5.97×10^{-2}	0.926	1

Permeability Conversion Factors

To obtain	Multiply						
	barrer by	$\dfrac{\text{ml-mm}}{\text{cm}^2\text{-sec-cm Hg}}$ by	$\dfrac{\text{ml-mm}}{\text{cm}^2\text{-sec-atm}}$ by	$\dfrac{\text{ml-mm}}{\text{cm}^2\text{-24 hr-atm}}$ by	$\dfrac{\text{ml-mil}}{\text{cm}^2\text{-24 hr-atm}}$ by	$\dfrac{\text{ml-mil}}{100\text{ in.}^2\text{-24 hr-atm}}$ by	$\dfrac{\text{in.}^3\text{-mil}}{100\text{ in.}^2\text{-24 hr-atm}}$ by
barrer	1	10^9	1.32×10^7	1.52×10^2	3.88	6.00×10^{-3}	9.80×10^{-2}
$\dfrac{\text{ml-mm}}{\text{cm}^2\text{-sec-cm Hg}}$	10^{-9}	1	1.32×10^{-2}	1.52×10^{-7}	3.88×10^{-9}	6.00×10^{-12}	9.80×10^{-11}
$\dfrac{\text{ml-mm}}{\text{cm}^2\text{-sec-atm}}$	7.60×10^{-8}	76.0	1	1.16×10^{-5}	2.94×10^{-7}	4.56×10^{-10}	7.47×10^{-9}
$\dfrac{\text{ml-mm}}{\text{cm}^2\text{-24 hr-atm}}$	6.57×10^{-3}	6.57×10^6	8.64×10^4	1	2.54×10^{-2}	3.90×10^{-5}	6.45×10^{-4}
$\dfrac{\text{ml-mil}}{\text{cm}^2\text{-24 hr-atm}}$	0.258	2.58×10^8	3.40×10^6	39.4	1	1.55×10^{-3}	2.54×10^{-2}
$\dfrac{\text{ml-mil}}{100\text{ in.}^2\text{-24 hr-atm}}$	167	1.67×10^{11}	2.19×10^9	2.54×10^4	6.45×10^2	1	16.4
$\dfrac{\text{in.}^3\text{-mil}}{100\text{ in.}^2\text{-24 hr-atm}}$	10.2	1.02×10^{10}	1.34×10^8	1.6×10^3	0.394	6.10×10^{-2}	1

in polymers.   For the steady-state referred to above Fick's first law of diffusion† is assumed to apply *within the film*, namely,

$$J = - D \frac{dc}{dx},$$ (2)

where D is the diffusivity of the gas in the polymer and dc/dx is the local concentration gradient of the gas in the x direction.   (Uniaxial diffusion or diffusion normal to the film surface is of prime interest in thin polymer films where the gradients $\delta c/\delta y$ and $\delta c/\delta z$ vanish.)

At this point the concept of a permanent gas must be introduced in order to integrate (2) over the film thickness.   Permanent gases are those gases that at room temperature are above their critical temperatures and thus are not readily condensable.   They tend to be sparingly soluble in polymers: generally no more than one molecule of gas per 100 mer units of the polymer at atmospheric pressure.   At this low level of solubility they have a negligible plasticizing effect on the polymer.   At any reasonable temperature and pressure the bulk physical properties of the polymer are not detectably influenced by the presence of the gas.   Furthermore, such permanent gases do not chemically associate with the polymer or with each other.   Gases that fall in this category are the air gases (except for water vapor in many polymer systems), the rare gases, and certain hydrocarbons of low molecular weight such as the C_1, C_2, and C_3 alkanes.

For these permanent gases diffusing in a polymer at constant temperature the diffusivity is usually constant, independent of concentration. Therefore, when a steady-state flux is achieved, dc/dx must also be constant.   Equation 2 can then be integrated across the film thickness to give

$$J = - D \frac{\Delta c}{L}$$ (3)

Combination of (1) and (3) yields

$$\bar{P} = D \frac{\Delta c}{\Delta p}.$$ (4)

† This relationship is strictly applicable only to diffusion in a binary system consisting of the polymer film and a single gas in which the flux J is relative to the mass-average velocity of the system.   As long as the polymer is in solid film form and the concentration of gas in the film is low this flux is also equivalent to the flux relative to stationary coordinates assumed in developing (2).

Thus the permeability coefficient is dependent on the gas diffusivity and a phase-boundary coefficient relating gas concentration *inside* the polymer to gas pressure outside the polymer. (D is often called the diffusion coefficient.)

Drawing again on the analogy between polymers and simple liquids, the solubility of a sparingly soluble gas in a polymer usually obeys Henry's law, that is,

$$C = S \cdot p, \tag{5}$$

where S is the Bunsen coefficient expressing the volumetric gas solubility (at 0°C and 1 atm) per unit volume of polymer per unit pressure. In many polymers S is found to be essentially independent of pressure at constant temperature for permanent gases.

Assuming that thermodynamic equilibrium can be maintained at the film surfaces under conditions in which transport is occurring, (4) and (5) can be combined to yield

$$\bar{P} = D \cdot S. \tag{6}$$

Therefore the permeability coefficient in the ideal case of permanent gas transport through polymers is a constant, independent of the pressure driving force.

Table 2 shows typical values of these constants determined at different

TABLE 2

Effect of Pressure on Gas Transport Parameters at 30°C

Polymer	Gas	Upstream pressure (cm Hg)	$\bar{P}$ (barrer)	D (cm²/sec × 10⁷)	S [cc (STP)/cc polymer- cm Hg × 10⁴]
Pliofilm NO	CO_2	6.2	0.18	0.0052	350
		30.1	0.17	0.0052	350
		60.1	0.17	0.0052	350
Nylon 6	CO_2	7.6	0.15	0.017	88
		25.5	0.15	0.017	88
		48.8	0.14	0.017	82
		72.7	0.14	0.017	82

pressures in two different polymers for carbon dioxide, a permanent gas under these conditions [7]. It can be seen from this table that $\bar{P}$, D, and S are independent of pressure within experimental error.

Equation 6 is the fundamental modeling equation for gases permeating polymers. It relates the practical barrier coefficient of flux (the permeability coefficient) to the phase-boundary equilibrium constant (the solubility constant) and the mass-transfer coefficient within the polymer (the diffusivity). Since S is a thermodynamic property and D is a kinetic property of the system, it is not surprising that attempts to correlate $\bar{P}$ with a single property of the gas and a single property of the polymer have been quite unsuccessful. Wherever possible S and D should be examined separately for potential correlations with physical properties of the gas and polymer and $\bar{P}$ should be estimated from such correlations.

So far attention has been focused on a binary system consisting of one gaseous component. In many instances of permanent gas permeation through polymers multicomponent gas mixtures are encountered. Very little work has been done on the analysis of mixture data, and usually a gross permeability coefficient for a mixture is obtained without analyzing compositional changes across the film. In those cases in which sufficient data exist the process would appear to involve negligible coupling between the flows of the various gaseous species. This means that (1) and (3) apply to each species in a mixture, with the partial pressure and concentration of individual species being employed as the driving forces.

Before we turn to the examination of permanent gas transport data, we shall discuss methods of measuring the parameters D, S, and $\bar{P}$.

III. Measurement of the Barrier Properties of Polymer Films to Permanent Gases

The three parameters in (6) are amenable to separate and independent measurement. The permeability coefficient can be obtained by detecting the steady-state flux of gas across a film. This is usually accomplished in a manometric, volumetric, or isostatic device [14].

In the volumetric technique gas at constant superatmospheric pressure is kept in contact with one face of the film. Downstream of the film is a calibrated capillary tube open to the atmosphere but containing a small slug of liquid of low surface tension which thoroughly wets the capillary.

Excellent wetting of the capillary is required to prevent contact-angle hysteresis, and a liquid with low surface tension is employed to encourage wetting and to minimize Δp across the slug if hysteresis should occur. The displacement of the slug of liquid with time is measured and the permeability coefficient is calculated from the displacement volume (corrected to standard conditions of temperature and pressure) per unit time, the film area, the film thickness, and the pressure difference across the film. This simple technique reached a high state of development and refinement under Park at The Dow Chemical Company [15, 16].

The American Society for Testing and Materials has recently subjected the volumetric instrument shown in Figure 1 to careful round-robin testing, and with suitable precautions this instrument can be used with good accuracy ($\pm 5\%$) for the determination of permeability coefficients at least as low as 0.01 barrer (*ASTM D 1434, Volumetric Method*). The particular unit shown was originally designed in research at the Tonawanda Laboratories of the Linde Company [17] and is currently being manufactured by Custom Scientific Instruments of Kearny, New Jersey. It provides for good support and sealing of the film and exposes a polymer disk $3\frac{5}{8}$ in. in diameter to test. The latter is large enough to eliminate problems of small-scale inhomogeneities in films and yet is small enough to allow experimental polymer films to be tested. The unit is designed to be partially submerged in a thermostatted water bath for temperature control; for measurements at the low end of the permeability scale, however, total submersion of the unit (including the capillary) in a well-controlled liquid or air bath should be used. The volume of gas between the downstream face of the film and the liquid slug in the capillary is large enough to make the unit a very sensitive air thermometer if it is not kept totally isothermal.

An instrument of similar design which can employ a bank of four permeation cells is known as the Aminco-Goodrich Gas Permeability Apparatus and is manufactured by the American Instrument Co., of Silver Spring, Maryland. This unit will also meet the revised *ASTM D 1434 Method V*. A device of the volumetric displacement type which is adaptable to operation at higher pressure than the Linde cell and in which the upstream and downstream pressure can be independently varied is described by Li and Henly [18].

In the manometric technique of measuring permeability coefficients a relatively high pressure gradient is maintained across the membrane while the increase in pressure downstream of the membrane is measured

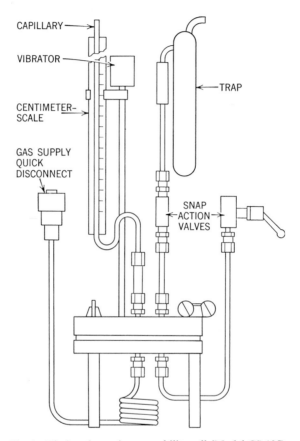

Fig. 1   Linde volumetric permeability cell (Model CS-135).

as a function of time and the upstream pressure is held constant.   A small-bore mercury manometer is used to measure the downstream pressure so that the change in gas receiving volume with time is negligible.   Under these conditions the pressure gradient across the film is reasonably constant and a steady permeation rate can be achieved.

A manometric unit that complies with the requirements of *ASTM D 1434–66 Manometric Method* is shown schematically in Figure 2. This unit was developed by the Plastics Technical Service group at The Dow Chemical Company and is sold by Custom Scientific Instruments

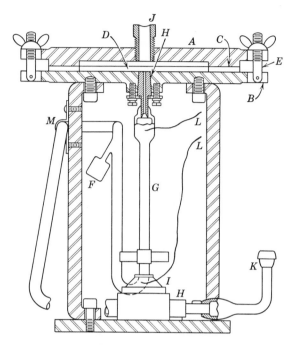

Fig. 2   Dow manometric permeability cell (*ASTM D 1434–58*): *A*, upper plate; *B*, lower plate; *C*, rubber gasket; *D*, porous filter paper supporting sample film; *E*, swivel bolts; *F*, Hg storage reservoir; *G*, calibrated portion of instrument; *H*, Kovar seals; *I*, Demi-G valve; *J*, gas supply tube; *K*, to vacuum; *L*, wires to recorder; *M*, glass supporting clip.

of Kearny, New Jersey.   In a 1963 modification of the unit a platinum resistance wire is mounted in the downstream capillary manometer to allow continuous recording of pressure as a function of time (Figure 3).

The volumetric and manometric methods suffer from the disadvantage that a hydrostatic pressure gradient is maintained across the film during measurement, whereas under most use conditions polymer films are subjected to partial pressure gradients with no net hydrostatic gradient across the film.   For films of very low modulus this can make an appreciable difference in the permeability even when the film is adequately supported, as in the Linde cell.   If the film were not adequately supported when under a pressure gradient, erroneous

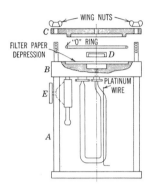

Fig. 3 Gas transmission cell (*ASTM D 1434–66*): *A*, supporting legs; *B*, lower plate; *C*, upper plate; *D*, adapter; *E*, vacuum valve.

results would be obtained even for polymers of quite high modulus, as has been demonstrated [19].

Furthermore, the volumetric method is always subject to insidious leaks and the introduction of clamping defects at the film edges. To overcome these shortcomings isostatic techniques have been employed by numerous researchers over the years. The method, until recently, was limited to research work because it required expensive and cumbersome methods of gas analysis to detect the change in gas composition downstream of the film.

Recently two commercial units have been made available and are receiving considerable attention. Both employ thermal conductivity cells to sense differences in gas composition. A schematic of the measuring chamber in the unit sold by Hans Sickinger Company, Bloomfield Hills, Michigan, is shown in Figure 4 [20]. The measuring procedure consists of filling chambers *J* and *K* with a reference gas and chambers *H* and *L* with the test gas—all at constant total pressure. The change in composition of the gas in chamber *J* due to permeation through the test film results in a change in thermal conductivity of the gas in this chamber. By bucking this change in thermal conductivity against the thermal conductivity in the reference chamber (*K*) on a bridge circuit an accurate measurement of the change in composition with time is obtained. Chamber *L* simply provides thermal symmetry in the system when absolute permeability measurements are being made; however, the "reference film" can be replaced with a second test film to facilitate making relative permeability measurements between two different samples. This unit is now undergoing evaluation by an ASTM task force. An instrument operating on the same principle has recently become available from Incentive Research and Development, AB., Stockholm, Sweden.

Gas analysis by chromatography has also been used as a sensing technique in developing an isostatic permeameter [21].

A word of caution should be introduced to inexperienced workers in

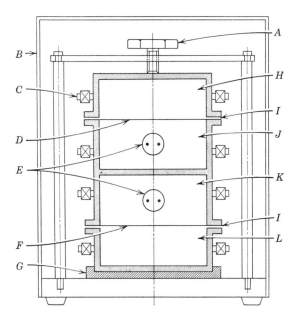

Fig. 4 Isostatic permeability cell: *A*, center clamp; *B*, cover; *C*, valve; *D*, film under test; *E*, thermoconductor cells; *F*, reference film; *G*, insulation; *H*, upper gas chamber; *I*, airtight joints; *J*, measuring chamber; *K*, reference chamber; *L*, lower gas chamber.

this field who are undertaking the measurement of permeability coefficients. Satisfactory results can be obtained only if steady-state changes in volume displacement pressure rise, or thermal conductivity change, are obtained with the above methods. If thick films of relatively impermeable polymers are being tested, it is difficult to discern by visual observations of the data when the steady state is reached. Equations 10 and 11 can be used to help in deciding how long it will take to reach a steady state if a crude estimate can be made of the diffusion constant for the gas-polymer system under investigation. To illustrate that the time to reach steady state is not inconsequential it would take about one hour with oxygen in 1-mil Saran, two hours for carbon dioxide in 1-mil Mylar, and six hours for carbon dioxide in 1-mil Saran.

In general only permeability coefficients are measured and reported by most of the workers in this field. This reflects the fact that $\bar{P}$ is the parameter of most direct utility in barrier development. For developing correlation and extrapolating techniques, however, as well as for using transport data to gain additional insight into polymer structure, the solubility and diffusion constants must be obtained.

The solubility of gases in polymers is fairly difficult to measure. Gravimetric methods generally cannot be used because of the low weight gains encountered.

The technique with which we are most familiar is that developed by Michaels and Parker [22]. This volumetric or static sorption method has been used with good results for about 10 years in the Department of Chemical Engineering at the Massachusetts Institute of Technology. A schematic of the system is shown in Figure 5. Its operation consists

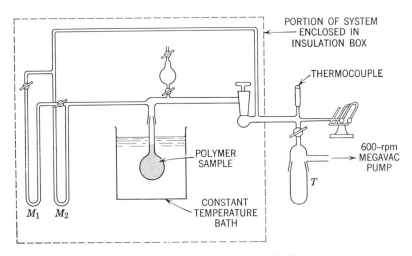

Fig. 5   Static sorption apparatus for low-solubility constants.

of charging a polymer in pellet, fiber, or film form to the sample chamber. (When a film is used, it is wrapped in a helical coil with small nonsorbing spacers such as wires or glass rods inserted between individual loops of the coils to maintain gas access to all of the film surface.) The key to the success of the method relies on getting a high polymer-to-system volume ratio (preferably greater than 0.5). The polymer is degassed under vacuum and then allowed to sorb the test gas at constant pressure. By the use of thin film or fine fibers the time to achieve equilibrium sorption is greatly diminished. After the system has come to equilibrium the void volume is quickly evacuated to a pressure of 0.1 mm Hg or lower with a high-capacity vacuum pump. The system is then isolated from the pump and desorption equilibrium is established. By suitable design and manipulation the aforementioned pumpdown can be accomplished without removing appreciable sorbed gas from the polymer.†

The solubility constant is calculated by a material balance, which in its simplest form is given by

$$Sp_1 V_p = p_2(V_v' + SV_p), \tag{7}$$

where p_1 is the initial sorption equilibrium pressure, p_2 is the final desorption equilibrium pressure, V_p is the volume of polymer in the system, and V_v' is the void volume of the system adjusted to standard conditions. If the isotherm is nonlinear, (7) contains two different values and S cannot be determined uniquely. It is usually possible, however, to reduce p_1 to a level at which the isotherm becomes linear and S can be determined. A known value of S in the linear isotherm region having been established, values of S in the nonlinear region can be obtained.

A somewhat simpler procedure can be employed for measuring solubility constants when S is greater than 0.5 ml (STP)/cc polymer-atm. This consists of charging a known volume of gas at known pressure to an initially degassed polymer sample and determining the equilibrium sorption pressure. A typical apparatus is shown in Figure 6. Again a simple material balance is used for determining S,

$$p_1 V = p_2(V_v' + SV_p), \tag{8}$$

where p_1 is the pressure of the gas in the inlet bulb, V is the bulb volume (corrected to STP), p_2 is the final equilibrium pressure, V_v' is the void volume of the sample system (corrected to STP and including the inlet bulb), and V_p is the polymer volume. As with the preceding apparatus isothermal conditions must be maintained through the system. This

---

† This technique can be used to obtain gas-diffusion constants in polymers by determining the pressure-time history for the unsteady-state desorption process. It is particularly advantageous when extremely low diffusion constants are encountered, being superior to the time-lag apparatus discussed below. Knowing the diffusion constant, we can allow for the amount of gas lost during pumpdown as a correction to (7).

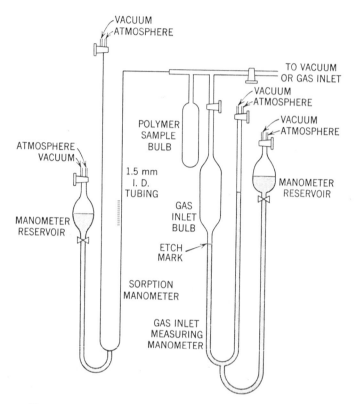

Fig. 6   Static sorption apparatus for high-solubility constants.

is usually achieved by mounting the entire apparatus in a thermostated air chamber.

The most sophisticated tool available for the study of transmission properties of polymer films is the time-lag apparatus developed and refined by Barrer [23, 24].   This is also a manometric technique of measurement.   This instrument takes advantage of Fick's second law of diffusion for the determination of diffusivities of gases in polymers by operating in the nonsteady state.   It also yields permeability co-efficients by achieving a quasi-steady state.   Therefore in theory all three transport parameters can be obtained with the method—two directly and the third by calculation from (6).

The time-lag method consists of degassing a film and then suddenly subjecting one face of the film to a constant, elevated gas pressure. The pressure downstream of the film is measured as a function of time, but conditions are adjusted so that a high pressure-ratio is maintained across the film throughout the experiment. A typical time-lag apparatus is shown in Figure 7.

The above experimental conditions satisfy Fick's second law of diffusion,

$$\frac{\delta c}{\delta t} = D \frac{\delta^2 c}{\delta x^2}.$$ (9)

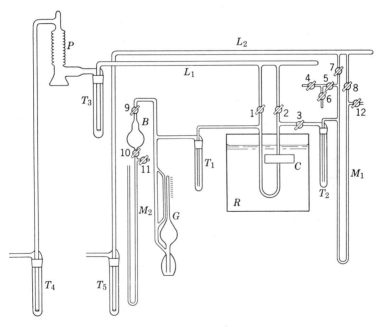

Fig. 7 Time-lag apparatus: L_1, high vacuum line; L_2, roughing vacuum line; P, oil diffusion pump, using Octoil S; B, bulb of known volume for determination of system volume; G, McLeod gage; C, permeability cell; M_1, mercury manometer for adjusting test gas pressure; M_2, mercury manometer for use in determining system volume in conjunction with B; T_4, T_5, cold traps leading to mechanical vacuum pumps; T_3, cold trap for oil diffusion pump; T_1, T_2, cold traps for condensable gas collection; R, constant temperature bath; 1 to 12, high vacuum stopcocks.

For the boundary conditions, applying Henry's law

$$
\begin{aligned}
t &= 0 \quad \text{all } x \quad c = 0, \\
t &> 0 \quad x = 0 \quad c = Sp_1, \\
t &> 0 \quad x = L \quad c = 0.
\end{aligned}
$$

Equation 9 can thus be solved to yield a value of $\delta c/\delta x$ at the downstream face of the film (L). Since the amount of gas entering the pressure measuring chamber is given by

$$
-\int_0^t D\left(\frac{\delta c}{\delta x}\right)_{x=L} dt,
$$

the increase in measured pressure with time can be predicted.

In theory the diffusion constant could be obtained by curve fitting the pressure-time data in the unsteady-state region. However, measuring pressures accurately at the beginning of the experiment is difficult. Fortunately an extension of the above mathematics yields a simple graphical extrapolating procedure for obtaining diffusivities with surprisingly good accuracy. Figure 8 shows a typical pressure-time curve for a time-lag run. The quasi-steady state referred to above is represented by the region in which the pressure rises linearly with time. If this portion of the curve is extrapolated to the time axis, the intercept θ_L, or the time lag, can be shown to be related to the diffusivity by

$$
D = \frac{L^2}{6\theta_L}. \tag{10}
$$

The reader is referred to any of several sources for a detailed mathematical development of the time-lag procedure [2, 25, 26].

It is frequently difficult to discern by eye exactly which data points should be included in the steady-state extrapolation, but the mathematics also predicts that the quasi-steady state should be reached for all practical purposes when

$$
t_{ss} = 3\theta_L. \tag{11}
$$

A little trial and error with a straight edge, or a simple computer program, therefore can be used to make the extrapolation reasonably precise.

A question has been raised from time to time about the validity of assuming that sorption equilibrium is maintained at the downstream

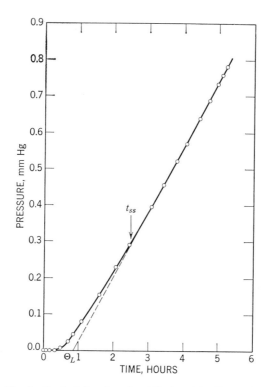

Fig. 8 Typical time-lag plot, CO_2 in Mylar [8, p. 614].

face of the film when pressures in the submicron Hg pressure range are employed. The authors have estimated that the rate of arrival of gas molecules at the membrane surface from the gas phase always exceeds the rate of "evaporation" from the surface by at least three orders of magnitude under the most unfavorable experimental conditions. Thus the assumption of equilibrium being maintained throughout the experiment from time zero appears to be valid.

The permeability coefficient is determined from the rate of pressure rise in the steady-state region, the volume of the gas receiving system (corrected to standard conditions), the film area, the film thickness, and the upstream pressure. The solubility constant is then calculated by using (6). We have shown that good agreement is obtained between solubility constants determined in this manner and those obtained by the direct method for permanent gases in natural rubber and a variety of polyethylenes [27, 28].   Cases of divergence between the two types of measurement have been reported for heterogeneous materials containing, for instance, voids, but these cases are exceptional.

A variety of pressure-sensing devices has been used on the low-pressure side of the film.   The McLeod gage has certainly stood the test of time and when used carefully and maintained in a pristine clean state it is virtually infallible.   A question is often asked how condensable a gas must be before problems in accuracy of measurement are encountered with a McLeod operating at room temperature.   Bixler, Michaels, and Parker [29] have shown that no appreciable error is encountered with a series of gases whose critical temperatures are above room temperature.

Simple-to-operate, direct-reading, electronic pressure sensors of satisfactory reliability have become available in recent years for low-pressure measurement.   We have experienced good results with the Decker Pressure Sensor made by The Decker Corporation of Bala-Cynwyd, Pennsylvania.   This instrument employs a stretched-diaphragm differential pressure sensor whose displacement is electronically detected by a capacitance change.   It is recommended that a unit in which the diaphragm is welded to the holder to eliminate all polymeric gasketing material in the sensing chamber be obtained for high-precision work.   Several laboratories are now using the MKS Baratron made by MKS Instruments of Burlington, Massachusetts.   This unit operates on the same general principle as the Decker unit but is a more sophisticated (and expensive) instrument.

The cell design in the time-lag apparatus is quite critical.   One design that has proved quite satisfactory in our laboratory is shown in Figure 9.   Note that the unit employs no flat gasket or "O" ring seals. A vanishingly thin coating of fluorocarbon stopcock grease on the mating metal surfaces affords a good vacuum seal without contaminating the film.   At one time it was felt that making and breaking blown glass seals was the only reliable method of changing films in the system, but several years' experience with high-vacuum quick couplings of the compressed "O" ring variety has proved them to be sufficiently leak-tight for time-lag work.

As a final note on the time-lag technique, two useful working relationships are presented.   First, when making repeated measurements on

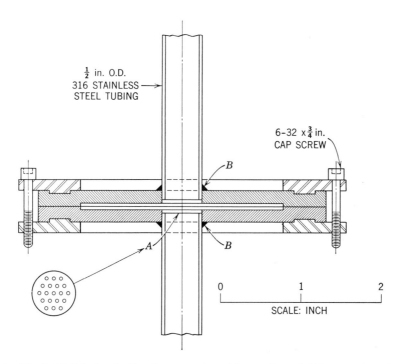

Fig. 9   Cell design for time-lag apparatus: A is a perforated disk approximately 0.495 in. in diameter and 0.025 in. thick. The disk has 12 to 20 holes, each approximately 0.025 in. in diameter. The disk fits moderately tight above tubing in a hole in the bottom plate, flush with the surface. For high-vacuum use silver solder is applied at B.

the same film, it is desirable to know how long the film must be degassed between runs.   A minimum safe pumpdown period has been calculated from the unsteady-state mathematics to be $6\theta_L$, and this has been borne out in practice.   It is also desirable to know under what operating conditions (10) will be reasonably accurate; that is, when does the boundary condition of "zero pressure" downstream of the film break down.   The ratios of true to apparent values of D, S, and $\bar{P}$ as a function of system parameters is shown in Figure 10, from the work of Paul and DiBenedetto [30].   As can be seen from this plot, the errors become significant when η exceeds 0.1.   In practice, this restraint is rarely exceeded with standard time-lag equipment, but it can be exceeded

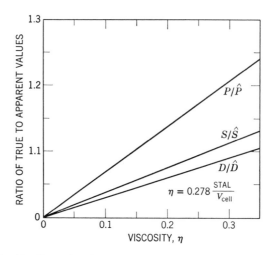

Fig. 10   Correction factors for time lag transport parameters.

when one attempts to use the Linde or Dow cells for time-lag measurement in which the downstream cell volume is quite small.

For more extensive information on methods of measurement the reviews of van Amerongen [3], Rogers [8], and Stannett and Yasuda [31], and the bibliographies by Weiner and associates of the Institute of Paper Chemistry [32] are recommended. For a definitive discussion of the reliability of the volumetric, isostatic, and manometric techniques of permeability measurement the work of Taylor, Karel, and Proctor [33] should be consulted. More recently excellent agreement between the manometric and volumetric techniques has been obtained in an ASTM round-robin evaluation.

IV. Analysis of the Transport of Permanent Gases through Polymer Films

Having considered the experimental and theoretical groundwork for categorizing the transport of small molecules through polymer films, we can now turn to a more detailed analysis of the thermodynamics and kinetics of the permeation process. Solubility constants and diffusivities are examined separately in order to show that the effects of temperature and gas and polymer properties on these transport parameters can be correlated reasonably well. The analysis to be presented places emphasis on helping the reader improve his ability to predict the gas permeability of modified or totally new polymer systems.

A. The Solubility Constant

The liquidlike model for a polymer can be used to heuristic advantage in analyzing gas solubility data. This approach must be tempered by several restraints, however. First, the model is usually valid only above the glass transition temperature of polymers, that is, in the rubbery state. The presence of a well-formed crystalline phase within the polymer will also modify the degree of liquidlike behavior. In the rubbery state the thermal motion of the polymer chain segments is generally sufficient to prevent the localization of the free volume into static or slow-moving holes. This state is molecularly similar to that of simple organic liquids. It is not inconceivable that polymers containing bulky side chains or bulky backbone units could appear rubbery from all classical methods of measuring the glass-to-rubber transition but still be in a low enough entropy state to fall short of responding as liquidlike toward gas solubility. With these limitations in mind, however, it is still possible to unify the data surprisingly well with the liquid model.

The classical work of Hildebrand [34] on the solubility of gases in simple liquids provides the most direct key to suitable correlation. In its most advanced state Jolley and Hildebrand [35] have shown that the natural logarithm of the solubility constant of permanent gases in an organic liquid is a linear function of the Lennard-Jones potential parameter $\varepsilon/\overline{k}$ for the gases. Michaels and Bixler [36] have shown that the form of the expression for polymers, which makes use of the Flory-Huggins thermodynamic expression for the chemical potential of the dissolved gas, μ_{dg}, should be

$$\mu_{dg} = RT\left[\ln(1 - \bar{v}_p) + \left(1 - \frac{1}{m}\bar{v}_p + \chi\bar{v}_p{}^2\right)\right],$$

where $\bar{v}_p$ is the volume-fraction of polymer, m is approximately the number of $-CH_2-$ groups per amorphous chain segment, and χ is the mixing parameter relating to the heat of dilution (R and T have the usual significance). The solubilities encountered were all so low that

$\bar{v}_p \sim 1.0$. The probable minimum value of m is about 20 for the poly-crystalline polymers studied, as determined from the vapor sorption measurements of Rogers, Stannett, and Szwarc [37].

Michaels and Bixler have applied Hildebrand's correlation to data on the solubility of gases in several rubbery amorphous polymers [36] and the resulting linear regression, as shown in Figure 11, is quite good.

By using the Flory-Huggins expression for the chemical potential of the gas dissolved in the polymer Michaels and Bixler were further able to show that the form of correlation in Figure 11 at 25°C should be given by

$$\ln S^* \overline{V}_g = 0.026 \frac{\varepsilon}{k} - (1 + \chi), \tag{12}$$

where $\overline{V}_g$ is the partial molal volume of the dissolved gas and χ is the mixing parameter related to the heat of dilution of the polymer by the gas; S^* is the solubility constant in a hypothetical completely amorphous polymer and is related to $1 - v_p$ as follows:

$$S^* = 22,400 \frac{1 - v_p}{\overline{V}_g}$$

In constructing Figure 11, as a first approximation, it has been assumed that $\ln \overline{V}_g$ and χ are proportional to $\varepsilon/\bar{k}$ (there is no method for reliably estimating $\overline{V}_g$ and χ commensurate with the precision of S^* or ε/k). The success of the correlation is a good indication of the validity of this assumption.

Independent measurements of $\overline{V}_g$ and χ have not been made for gases in polymers, although the values of $\overline{V}_g$ for gases dissolved in chemically similar organic liquids [38] and values of χ for gases dissolving in elastomers have been calculated [39].

Furthermore, Kwei and Arnheim have shown that $\overline{V}_g$ is eliminated from (12) if it is assumed that the dissolved gas occupies free volume within the polymer [40]. This isosteric model for the mixing process may be more accurate than the lattice model used to develop (12), and this may account for the good correlation of $\ln S^*$ versus $\varepsilon/\bar{k}$ without regard for $\overline{V}_g$.

The available data all indicate that S and $\overline{V}_g$ are generally propor-tional to one another. The slopes of the regression lines in Figure 11 agree remarkably well with the coefficient multiplying $\varepsilon/\bar{k}$ in (12), which indicates that χ is essentially constant for a variety of gases dissolving

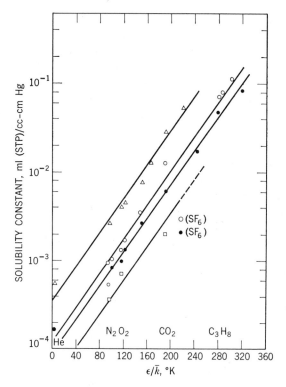

Fig. 11 Correlation of solubility of several gases in rubbery, amorphous polymers: △ silicone rubber, ○ natural rubber, □ styrene/acrylonitrile rubber (39% acrylonitrile), ● amorphous polyethylene. $T = 25°C$. (Two unique points for SF_6 are shown, far from related points for other gases.)

in a single polymer. The displacement of the lines for different polymers, however, indicates that χ is quite sensitive to the polymer in which the gases are dissolving.

The Lennard-Jones force constant enters (12) as an empirical parameter for predicting the enthalpy of condensation of the gas, and thus it is the degree of gas condensability that is the key factor in determining the level of permanent gas solubility in a polymer. Other guides of condensability such as normal boiling temperature or critical temperature would therefore be expected to be equally reliable thermodynamic correlating parameters. Barrer [2] and van Amerongen [3] have indeed demonstrated good correlation of gas solubility constants in rubbers based on T_b and T_c. Stern, Mullhaupt, and Gareis [41] have also shown good correlation of gas solubility with critical temperature and pressure, following the principle of corresponding states (and leading to a linear relationship between log S and $(T_c/T)^2$). The authors' preference for the potential field constant stems primarily from a desire to focus attention on molecular interaction as the key to understanding the subtleties of gas solubilities in polymers. It also does a better job of bringing the solubility constants of helium and hydrogen onto the regression lines by compensating for quantum mechanical effects associated with these two molecules. Even with this refinement, however, solubility data for certain gases that otherwise display all of the characteristics of permanent gases noted earlier fail to correlate. Sulfur hexafluoride, for instance, displays an unusually low solubility constant in comparison to its $\varepsilon/\bar{k}$ (or T_c). The exceptionally large size of this molecule may play a role in this lack of correlation and the rather cavalier treatment of V_g in (12) may not be satisfactory in this case.

In addition to correlating the solubility constants of several gases in a single polymer, it would be desirable to have a method of correlating the solubility constants of a single gas in several rubbery, amorphous polymers, that is, a method of predicting the displacement of the lines in Figure 11. A totally satisfactory technique for this type of cross correlation has not yet evolved. Some guidelines can be laid down, however. The displacement of the regression lines for the four polymers included in Figure 11 should be related to the magnitudes of the mixing parameter χ. The mixing parameter for these dilute solutions is given approximately by

$$\chi \sim \frac{\overline{V}_g(\delta_p - \delta_g)^2}{RT}, \qquad (13)$$

where δ_p and δ_g are the Hildebrand solubility parameters of the polymer and gas, respectively.

It will be recalled that the solubility parameter is the square root of the cohesive energy density and thus is a measure of the force field intensity around a gas molecule and a polymer chain segment. In general, solubility parameters for permanent gases are much lower in magnitude than are the apparent solubility parameters of polymers; so χ is primarily dependent on δ_p, as indicated earlier. In light of these considerations it might be expected that the solubility constant of a given gas in a

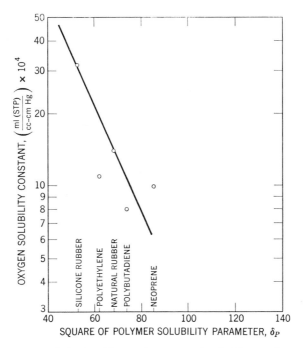

Fig. 12   Oxygen solubility in rubbery polymers correlated with the polymer solubility parameter.

variety of rubbery amorphous polymers would correlate reasonably well with δ_p [42]. This correlation is tested for oxygen in Figure 12: the form of the variables used in this plot has been inferred from (12) and (13). The correlation is at best qualitative but the solubility constants for oxygen decrease with increasing δ_p as expected from (12) and (13). Therefore this procedure can be used for an order-of-magnitude comparison of gas solubility constants of relatively nonpolar gases in various polymers at constant temperature.

For a polar gas such as carbon dioxide the solubility constants *increase* with increasing δ_p. This increase is not predicted on the basis of tabulated thermodynamic parameters for the systems, but intuitively the increase might be expected for polar gases such as carbon dioxide, sulfur dioxide, ammonia, and water in which strong polymer-gas interactions can manifest themselves. In general, it is observed experimentally that the solubility constants of these very polar gases increase with increasing polymer polarity, whereas the opposite is true for more nearly ideal gases.

In summary, the correlation shown in Figure 11 can generally be used to estimate solubility constants for gases at room temperature when $\varepsilon/\bar{k}$ is less than $320°K$ ($T_c < 250°C$). From a single measured gas solubility constant in a particular rubbery amorphous polymer an estimate of other gas solubility constants in that polymer can be made. Extrapolation from one polymer to another can be made via a plot of the type shown in Figure 12 for the more noble permanent gases. For the very polar but still permanent gases a satisfactory method of extrapolation does not yet exist.

1. Temperature Dependence of Solubility Constants in Rubbery Amorphous Polymers

The temperature dependence of oxygen-solubility data in four rubbery amorphous polymers (Figure 13) follows an Arrhenius relationship of the form

$$S = S_0 \, e^{-\Delta H/RT}, \tag{14}$$

where S_0 is an extrapolated constant obtained by letting $1/T$ go to zero and ΔH is, under ideal conditions, the enthalpy of the sorption process. The ideal conditions that must prevail for the experimentally determined ΔH to be equal to the thermodynamic enthalpy of the solution process are (a) that the polymer undergo no phase transitions or other morphological changes within the experimental temperature interval and (b) that the gas behave as an ideal gas in this same temperature interval. In general ΔH values reported in the literature for rubbery amorphous polymers were obtained under circumstances in which the above conditions apply with reasonable vigor. Nevertheless, it is usually not advisable to extrapolate solubility data more than $50°C$ outside the range over which ΔH was determined because of possible phase transitions, morphological changes, etc. Obviously, volumetric gas solubility constants should be corrected for the temperature dependence of the density of the polymer if ΔH values are not to include a contribution from the coefficient of thermal expansion of the polymer. Over small temperature intervals, however, this effect is usually small.

Correlation of the temperature dependence of gas solubility in poly-

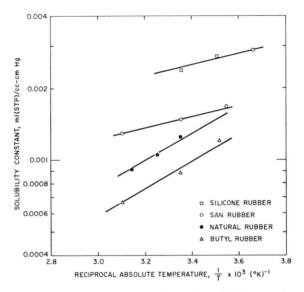

Fig. 13 Correlation of solubility constants for carbon dioxide at various temperatures in four amorphous rubbery polymers.

mers has usually focused on ΔH in order to extrapolate experimental data obtained at one temperature to some other temperature. Little attention has been given to the tabulation of correlation of S_0 values in (14), since this parameter is of specious physical significance and its magnitude is difficult to obtain with precision.

Michaels and Bixler have followed a procedure similar to that leading to (12) in order to develop a predictive thermodynamic relationship for ΔH [36]. At 25°C this relationship is given by

$$\Delta H = 0.59\chi - \frac{0.0156\varepsilon}{\overline{k}}. \tag{15}$$

The second term on the right side of (15) represents the estimated enthalpy of condensing the gas at 25°C; the first term is the enthalpy of mixing the condensed gas with the polymer.

Again, assuming that χ is primarily dependent on the polymer cohesive energy density and not that of the gas, ΔH should be a linear function of $\varepsilon/\overline{k}$. Figure 14 shows a plot of ΔH as a function of $\varepsilon/\overline{k}$ for the four rubbers being used for illustration and the predicted linear correlation

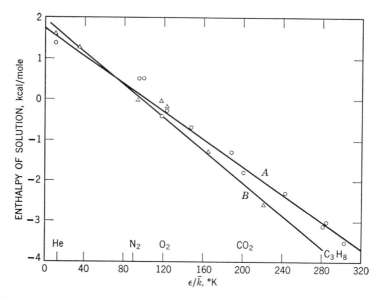

Fig. 14 Enthalpies of solution for various gases in elastomers: *A*, ○, natural rubber; *B*, △, silicone rubber.

seems to be reasonably accurate. As in the case of the solubility constants the displacement of the lines in Figure 14 for the different polymers can be correlated against χ and thus again δ_p.

It will be noted that the partial molal volume of the gas dissolved in the polymer does not enter into (15), whereas it did in (12). This may explain why the enthalpy data for sulfur hexafluoride correlate well in Figure 14, whereas at the same time the isothermal solubility data for this gas correlated poorly in Figure 11.

So far gas solubilities in rubbery amorphous polymers parallel very closely those in simple organic liquids. We consider next the consequence of a microcrystalline phase being present in the polymer. Since many commercial polymer films are polycrystalline (Volume I, Chapter 7), this is an important aspect of barrier film technology.

2. Polycrystalline Polymers

Most flexibly backboned polymers with any significant degree of stereoregularity will undergo partial crystallization on being cooled from the melt. For the more highly crystalline polymers such as linear polyethylene the volume fraction of crystalline polymer may exceed 0.8, although in commercial films it is generally in the 0.4 to 0.6 range. The presence of this crystalline phase has a marked effect on gas barrier properties.

According to the more recent views of Keller, Geil, Keith, and others (Volume I, Chapter 7), the crystallites in these polymers are lathlike structures consisting of ordered segments of chains that fold back and forth on themselves. The crystallites are further aligned into bundles known as spherulites. Spherulites per se do not influence gas transmission in polymers, but it is shown below that their presence exerts a secondary effect. This picture of the morphology of polycrystalline polymers diverges considerably from the older fringed micelle view (Volume I, Chapter 2). From the standpoint of gas transmission the newer morphological model is far more consistent with experimental observations—especially when the diffusion of gases in these polymers is considered.

Except for helium, and perhaps hydrogen, individual polymer crystallites are impenetrable by gas molecules, or at least their permeability relative to the noncrystalline portion is negligible. Therefore the solution of gases in microcrystalline polymers occurs to a significant extent only in the noncrystalline regions. Van Amerongen [3] was the first to investigate carefully gas transmission in a microcrystalline polymer and its amorphous counterpart. He studied a gutta percha (*trans*-polyisoprene), which at room temperature was about 50% crystalline. The crystalline melting point of this polymer was about 50°C and above this temperature range the polymer was essentially amorphous. The maximum crystalline melting point of gutta percha is 74°C, but this point is generally not observed in regularly prepared samples. Table 3

TABLE 3

Gas Solubilities in *trans*-Polyisoprene
S in ml (STP)/cc-atm

Morphology	$T(°C)$	Oxygen	Nitrogen
Crystalline	25	0.067	0.033
Essentially amorphous	50	0.106	0.057

shows the solubility of oxygen and nitrogen in crystalline and amorphous gutta percha. Since the true enthalpies of solution for oxygen and nitrogen in polyisoprene are quite small, the changes in solubility between these two temperatures reflect primarily an increase in the noncrystalline fraction of the polymer available for gas dissolution.

The advent of low- and high-density polyethylenes has led to increased interest in the study of gas solubility in microcrystalline polymers [36, 43–45]. This work led to the conclusion that above the glass transition temperature (a) gases are soluble only in the noncrystalline portions of the polymer, (b) gas molecules at least as large as sulfur hexafluoride have access to all noncrystalline regions of the polymer, and (c) the extrapolated gas solubilities per unit volume of amorphous polymer (completely amorphous polyethylene cannot be prepared) fall in line with the values for completely amorphous polymers of comparable cohesive energy density or solubility parameter.

Points (a) and (b) are illustrated in Figure 15 in which the solubilities of a number of gases in polyethylenes of different crystallinities have been plotted as a function of the volume fraction of noncrystalline polymer (determined by density measurements) [36]. A linear correlation between these parameters which leads to the following relationship can be seen:

$$S = S^* \alpha, \qquad (16)$$

where S^* is the solubility constant for the amorphous polymer ($\alpha = 1$) and α is the volume fraction of amorphous polymer. Point (c) is illustrated in Figure 11 in which the extrapolated solubilities of gases in amorphous polyethylene fall in line with those of other amorphous polymers when its solubility parameter is taken into consideration.

These observations indicate that a microcrystalline rubbery polymer such as polyethylene can be treated as a simple two-phase mixture of crystalline and noncrystalline polymers. Permanent gases dissolve in the noncrystalline polymers. Permanent gases dissolve in the noncrystalline regions much as they would in a completely amorphous polymer of similar chemical composition. This model has also been found to apply to polyethylene terephthalate above its glass transition temperature [46]. Polyethylene and polyethylene terephthalate are relatively simple microcrystalline polymers with only one crystalline phase configuration.

Isotactic polypropylene has a more complex crystalline morphology

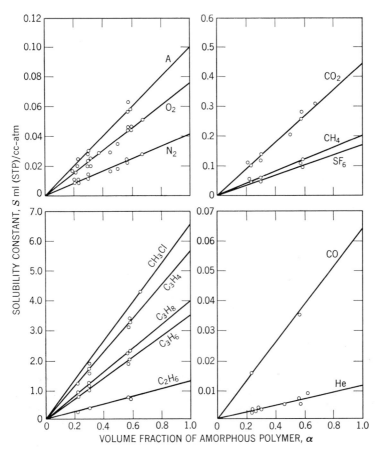

Fig. 15   Solubility constants for gases in polyethylenes as a function of the volume-
fraction of amorphous polymer α.

and gas solubility data do not respond to accurate correlation by this
simple two-phase model [47, 48].   This polymer appears to consist of
a well-defined monoclinic crystalline phase in which gas is not soluble
and a smectic or "liquid crystal" phase in which gas is partly soluble.
Solubility in the latter phase, however, is less than it is in atactic completely amorphous polypropylene.   Solubility in the smectic phase
also depends on the size of the gas molecule as well as its molecular

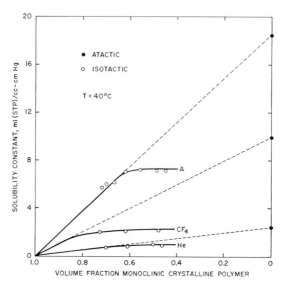

Fig. 16   Variation of solubility constant with differing amounts of monoclinic
crystallinity in polypropylenes.

force parameter.   Figure 16 shows the solubility of helium, argon, and
carbon tetrafluoride in polypropylene as a function of the volume fraction of monoclinic crystalline content.   Apparently divergence from
the simple two-phase model for gas solubility can be expected for any
polycrystalline polymer in which multiple crystallite forms are evident.

Complications can develop even in a simple polycrystalline polymer
such as polyethylene.   The spherulite morphology of polyethylene is
established upon nucleation from the melt.   Annealing of a quenched
linear polyethylene film results in alteration of the crystallites residing
within these preformed spherulites.   The ribbonlike crystallites tend
to thicken on annealing, a condition that introduces defects or voids
within the spherulites.   These defects are caused by the inability of
the preformed spherulites to accommodate readily to the changes in
crystallite morphology.   Small gas molecules such as helium fill these
voids, and the gas solubility is therefore enhanced over what would be
predicted from the observed crystallinity of the polymer [49].   The
solubility of helium in quenched and subsequently annealed polymer
can be as much as 50% greater than indicated by the results in Figure

15, which apply only to polymers cooled directly from the melt. For larger gas molecules the influence of these defects is negligible.

Polybutene-1 exhibits another morphological phenomenon that can confound interpretation of gas solubility data [50]. On cooling from the melt this polymer forms a low-density crystalline phase which slowly changes to a crystalline phase of higher density. The transformation introduces voids into the polymer which slowly disappear with time. Therefore gas solubility goes through a maximum as a function of time, reflecting the formation and disappearance of voids.

For permanent gases there is no evidence that the crosslinking effects of crystallites influence solubility constants. These gases are so sparingly soluble in polymers that negligible expansion of the polymer is required to accommodate the gas. The crosslinking effect becomes important for the more soluble condensable vapors and liquids considered in a later section.

The temperature-dependence of the solubility constants of gases in microcrystalline polymers can generally be correlated by (14), but ΔH will contain a contribution due to any melting of crystallites within the

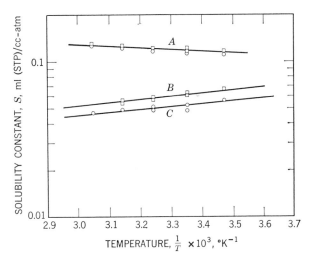

Fig. 17 Variation of the solubility constant of methane in polyethylenes with temperature: A, Alathon-14, $\alpha = 0.57$; B, Grex, $\alpha = 0.30$; C, Grex, $\alpha = 0.23$. Points represent O, time-lag and □, equilibrium solubility values.

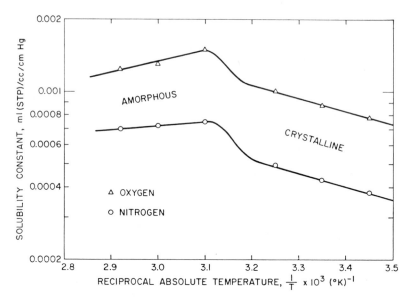

Fig. 18   Solubility of gases in crystalline and amorphous gutta percha.

TABLE 4

Apparent Heats of Solution

Gas[a]	ΔH(kcal/mole)		
	Grex	Alathon 14	Hydropol
He	1.4	2.5	2.4
N_2	0.5	1.9	1.8
CO	0.5	1.7	1.8
O_2	−0.4	0.6	0.5
A	−0.3	0.8	0.7
CH_4	−0.7	0.5	0.4
CO_2	−1.3	0.1	−0.1
SF_6	−1.8	−0.4	−0.8
C_2H_6	−2.3	−1.5	−1.8
C_3H_6	−3.1	−2.1	−2.0
C_3H_8	−3.0	−2.1	−2.1
C_3H_4	−3.5	−2.6	−2.2

[a] Minimum purity 99%; except for ethane and allene 95%.

temperature range under investigation. Melting of crystallites occurs to some degree for all polycrystalline polymers even far removed from the maximum crystalline melting point. As an example the apparent values of ΔH for methane are less negative in branched than in linear polyethylenes, since the former undergoes more crystalline melting in the vicinity of room temperature than the latter. This is illustrated graphically in Figure 17 [36]. For Grex, a linear polyethylene, ΔH is -0.7 kcal/mole, and for Alathon 14, a branched polyethylene, ΔH is $+0.5$ kcal/mole. The 1.2 kcal difference is a reflection of the endothermic crystallite fusion taking place in the branched polyethylene. At the maximum crystalline melting point there is, of course, a step change in gas solubility. This is illustrated in Figure 18 for oxygen and nitrogen in gutta percha.

Apparent heats of solution for a variety of gases are shown in Table 4 [36].

3. Glassy Polymers

There are two characteristic changes in the molecular or morphologic behavior of polymers in going from the rubbery viscoelastic state to the glassy brittle state, which strongly influence gas transport.

1. A reduction in polymer chain-segment mobility that results from the loss of certain translational and rotational modes of molecular motion.

2. The relative immobilization and partial coalescence of some portion of the polymer-free volume into polymer-free or polymer-lean voids capable of imbibing gas.

Gas diffusivities in glassy polymers are influenced by both (1) and (2), whereas gas solubilities are primarily influenced by (2). Many film-forming polymers such as the cellulosics, the polyesters, the polyamides, polyvinyl acetate, polyvinyl chloride, and polystyrene are glassy at use temperature, and an understanding of gas transport in the glassy state is therefore essential for the barrier technologist.

The first clear indication that the solution process for gases in glassy polymers was different from the solution process in rubbery polymers came from an examination of the enthalpies of solution above and below the glass transition temperature in polyvinyl acetate [51], although anomalies had previously been observed in the diffusion of organic vapors into glassy polyvinyl acetate [52] and glassy polystyrene [53]. Meares found that the enthalpy of solution was considerably more negative for permanent gases below T_g (ca. 29°C) than above T_g. Table 5 lists values of ΔH from his work [51].

TABLE 5

Effect of Glass Transition on Enthalpies of
Solution in Polyvinyl Acetate

Gas	ΔH(kcal/mole)	
	Above T_g	Below T_g
Helium	2.11	−1.00
Hydrogen	2.47	−1.42
Neon	1.05	−4.62
Oxygen	−1.10	−6.26
Argon	−1.88	−3.70

Similar results have been observed by Michaels in polyethylene terephthalate [46], and by Norton in Lexan polycarbonate [54]. These results have led to the postulation of a dual sorption mechanism below the glass transition temperature by which the sorbed gas is partly dissolved in the conventional manner in the polymer and partly sorbed in the immobilized microvoids created by transition to the glassy state. Since the gas in the microvoids is probably physically adsorbed to the polymer, this portion of the sorption process should be exothermic. The results in Table 5, therefore, reflect the importance of the exothermic void-filling process below T_g in relation to the conventional dissolution process.

Michaels, Vieth, and co-workers have extensively studied the sorption process in glassy polymers [46, 55, 56]. Figure 19 shows the effect of pressure on the solubilities of several permanent gases in polystyrene at room temperature. These nonlinear isotherms have been analyzed on the basis of the following relationship

$$C = C_D + C_H = S_D p + \frac{C_H' bp}{1 + bp}, \tag{17}$$

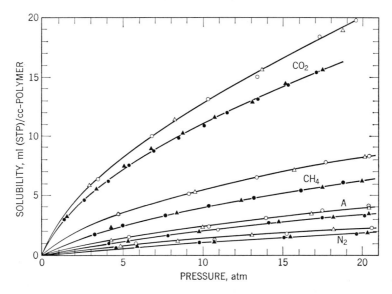

Fig. 19 Solubility of CO_2, CH_4, A, and N_2 in polystyrene at 25°C: ●▲ biaxially oriented film; ○△, cast annealed film; ●○, first runs; ▲△ second runs.

where C is the solubility [cc(STP)/cc total polymer], C_D is the solubility due to conventional dissolution, C_H is the solubility due to microvoid filling, S_D is the solubility constant associated with the conventional dissolution process, C_H' is a void-filling contant [cc(STP)/cc total polymer], b is a void-affinity constant (atm^{-1}), and p is pressure (atm). It will be observed that the microvoid-filling process has been modeled on the basis of a Langmuir adsorption isotherm and that Henry's law has again been used to model the conventional dissolution process. The curves in Figure 19 are analyzed by obtaining S_D from the slope at high pressures where C_H is negligible; C_H is then obtained by difference at each pressure, and a plot of p/C_H as a function of p is constructed. C_H' and b are then obtained from the slope and intercept of the latter plot. These investigations found that the values of S_D obtained in this manner fell in line with the correlation for rubbery amorphous polymers presented in Figure 11 and also the correlation presented in Figure 12, when the solubility parameter of polystyrene (9.1) was taken into consideration. They also found that the void-affinity constants correlated with the Lennard-Jones force constants for the gases and the void-filling constants correlated with the reciprocal of the cross-sectional area of the gas molecules, as would be predicted from physical sorption of gases into microvoids.

The importance of this void-filling process might best be illustrated by examining C_D and C_H in several polymers at 1 atm gas pressure. These solubilities are shown in Table 6.

TABLE 6

Dual Sorption in Glassy Polymers at 25°C and 1.0 Atm Pressure

Glassy polymer	Gas	C_D [ml(STP)/cc polymer]	C_H [ml(STP)/cc polymer]
Polyethylene terephthalate, unoriented, "amorphous"	CO_2	0.38	1.62
Polyethylene terephthalate, unoriented, crystalline[a]	CO_2	0.22	1.30
Polyethylene terephthalate, oriented, crystalline[b]	CO_2	0.37	1.93
Polystyrene, unoriented, atactic	CO_2	0.65	2.1
Polystyrene, unoriented, atactic	CH_4	0.16	0.72
Polystyrene, unoriented, atactic	A	0.065	0.19
Polystyrene, unoriented, atactic	N_2	0.025	0.11
Polystyrene, oriented, atactic[c]	CO_2	0.57	1.7
Polystyrene, oriented, atactic[c]	CH_4	0.13	0.44
Polystyrene, oriented, atactic[c]	A	0.06	0.16
Polystyrene, oriented, atactic[c]	N_2	0.02	0.10

[a] 5-mil Mylar, 43% crystalline.
[b] 1-mil Mylar, 46% crystalline.
[c] 7.5- mil Trycite, amorphous.

The results in Table 6 show that void filling makes the major contribution to gas solubility, even for gases as inert as nitrogen. Furthermore, biaxial orientation influences the sorption process. In an amorphous polymer orientation appears to reduce the microvoid content, probably by bringing about closer chain alignment. In a crystalline polymer orientation also alters the crystalline morphology which appears to introduce more microvoids per unit volume of non-crystalline material and counteracts the effect of chain alignment. These results clearly indicate that the presence of microvoids must be considered in analyzing gas transport in glassy polymers.

Even though there is now a preponderance of data to support the void-filling sorption mechanism, some studies of sorption and diffusion in glassy polymers have shown no change in the apparent enthalpy of

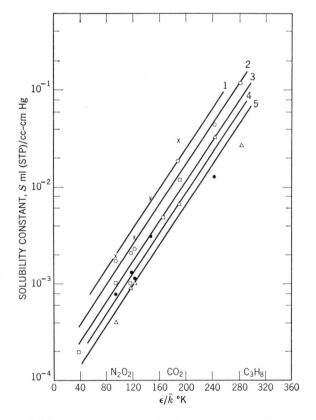

Fig. 20   Solubility constants of several glassy polymers at 25°C and pressures below 1 atm: 1, ×, biaxially oriented polystyrene [56]; 2, ○, ethylcellulose, [61]; 3, □, polyethyl methacrylate, [58]; 4, ●, "amorphous" polyethylene terephthalate, [58]; 5, △, nitrocellulose, [61].

solution at the glass transition temperature for small molecules, but the change is observed for larger molecules [57]. In the case of polyethyl methacrylate gas molecules as large as SF_6 showed no change in the apparent enthalpy at the glass transition temperature [58]. It now seems that nonlinear isotherms at elevated pressures may be a more sensitive test of the dual-sorption mechanism.

In spite of the complications introduced by the void-filling process, total gas solubilities in glassy polymers can usually be correlated empirically by Henry's law up to atmospheric pressure with suitable accuracy for most barrier work. The resulting Henry's law solubility constants can be correlated further in the manner shown in Figure 11. Figure 20 is a plot like Figure 11 for permanent gases in several glassy polymers.

4. Miscellaneous Factors that Influence Gas Solubility

We have shown that permanent gas solubility in polymers is strongly influenced by (a) the chemical composition of the polymer, (b) the crystallinity of the polymer, and (c) the presence of microvoids in the glassy state. Most other factors influencing gas solubility can be analyzed in terms of these considerations, although the interrelations can become quite complex.

In commercial film practice plasticizers, diluents, and fillers are frequently added to the polymer. If the plasticizer or diluent converts the polymer from the glassy to the rubbery state, a reduction in gas solubility resulting from elimination of the void-filling sorption process can be expected. Such a reduction was observed by Barrer, Mallinder, and Wong [58], when comparing the solubility of hydrogen and neon in unplasticized polyvinyl chloride to the solubilities in polyvinyl chloride containing 40% tricresyl phosphate. At 35°C (below T_g of the former and above T_g of the latter), the gases were 20 to 30% less soluble in the plasticized film. A larger relative reduction in gas solubilities is reported by Takeda and Yamaguchi [59] for films plasticized by dioctyl phthalate, but their extensive solubility data have unfortunately been found to contain numerous errors; these results must therefore be considered inconclusive. The reason that greater changes in the gas solubility were not observed by Barrer may reflect a compensating effect of the added plasticizer or the cohesive energy density or solubility parameter of the system. So-called fugitive plasticizers introduce another complication by creating a dispersed liquid phase into which gases can dissolve separately from the polymer.

Fillers are usually dense solids in which gases are quite insoluble. Therefore their presence can be expected to reduce gas solubilities in much the same way they are reduced by polymer crystallites. If the fillers are poorly dispersed in the polymer, however, aggregates that contain unfilled voids will be present and an increase in apparent solubility will generally be observed. Van Amerongen [60] has shown that a marked increase in apparent gas solubility can also occur when a filler such as carbon black is incorporated in rubber. The apparent gas solubilities correlated very well with the BET surface area of the carbon blacks, indicating that the internal surface area of highly structured carbon blacks apparently does not become coated with rubber even when excellent reinforcing action is achieved. The adverse effects of poor dispersion and the internal porosity of consolidated porous fillers are frequently overlooked when fillers are selected as a means of reducing polymer permeability.

Crosslinking of polymers can affect gas solubilities in various ways. Chemical crosslinking can alter the chemical composition, with attendant changes in solubility. If crosslinking converts a rubbery polymer to a glassy polymer, gas solubilities due to void formation will increase. Crystalline polymers are sometimes mildly crosslinked to improve mechanical properties or to impart heat shrinkability. This may be done chemically or by radiation [61, 62].

If this treatment reduces the crystallinity of the final polymer film, an increase in gas solubility can be expected. Unless crosslinking is carried out in an inert gas or in a vacuum, oxidation can be expected to accompany crosslinking. For some gases such as carbon dioxide, water vapor, ammonia, and others this oxidation will increase solubilities, whereas for highly polar gases the opposite effect will occur.

In summary, the solubilities of permanent gases in rubbery polymers can be reasonably well understood in terms of a model that treats the polymer thermodynamically like a liquid. This model generally accounts for the variation in solubilities from gas to gas and polymer to polymer as well as the variations with temperature. Readily condensable gases are more soluble in any polymer than less condensable gases. The solubility of a highly polar gas such as carbon dioxide usually increases with increasing solubility parameter of the polymer, whereas the opposite is true for the less polar gases. Crystalline polymers contain at least one phase in which gases are insoluble, although the presence of smectic phases, multiple crystalline phases, and microvoids can complicate the analysis. Glassy polymers contain a high concentration of immobilized microvoids, and permanent gas solubilities in these polymers are best understood in terms of a dual sorption mechanism that involves conventional dissolution in the polymer plus void filling.

B. THE DIFFUSION CONSTANT

1. Diffusion in Polycrystalline Polymers

The most extensive studies of morphology of polycrystalline substances with relation to permeability have been performed with polyethylene, both as-cast and after irradiation [62]. It became apparent quite early that study of gross permeability was inadequate unless a separation of the effects of solubility and diffusion could be made.

The early work [36, 43] indicated that the process of dissolution in polyethylene is confined to what was called the noncrystalline regions of the polymer. The volume fraction of this amorphous material was determined from density measurements, assuming that the amorphous and crystalline polymers have characteristic densities. Except for a slight effect on the density of the amorphous phase, the mode of polymerization was found not to alter the solubility of a gas per unit volume of amorphous polymer. Molecular weight and method of sample preparation were found also to exert no unusual influence on the thermodynamics of gas dissolution in the amorphous polymer.

This simple two-phase model for solubility held over a fairly wide temperature range (5 to 55°C) as long as the effect of temperature on the amorphous volume fraction (16) and the thermodynamic mixing process of a gas with a liquid were considered. The relationship

$$S = S_0 e^{-\Delta H/RT}, \tag{14}$$

where S_0 is a constant, ΔH is the apparent heat of solution, R is the gas constant, and T is the absolute temperature, applied in all systems studied. (ΔH is not simply the heat of solution of the gas with the amorphous polymer but must include a contribution from any crystalline melting that occurs in the temperature range of investigation.)

The diffusion process in polyethylene indicates that this polymer behaves like a dispersion of highly anisometric impenetrable crystallites in penetrable amorphous polymer. These crystallites impede the flow of gas by constricting the available passageways for flow. Since diffusion of gases is an activated process in polymers, that is,

$$D = D_0 \, e^{-E_D/RT}, \tag{18}$$

where D_0 is a constant and E_D is the apparent activation energy, the crystallites also impede the flow of gas by altering the activation process through their crosslinking action. The effect of crosslinking on the activation process could either be to alter the size of the activation zone and thus the entropy of activation or directly increase the activation energy by chain restriction. To account for these two impedance factors the following expression has evolved:

$$D = \frac{D^*}{\tau\beta} \tag{19}$$

where D^* is the diffusion constant in amorphous polyethylene, τ is the geometric impedance factor accounting for pore constriction, and β is the chain immobilization factor accounting for the crosslinking action of crystallites. Natural rubber has been assumed to be a completely amorphous homolog of polyethylene which gives experimental values of D^* so that values of τ and β can be determined indirectly.

Values of τ have been related to the degree of anisotropy of crystallites and to the mode of polymer synthesis. Values of β have been related to the size of the gas molecule and the volume fraction of amorphous polymer.

Table 7 summarizes the results obtained for the diffusion and permeability constants and activation energies of helium, nitrogen, methane, and propane in irradiated and nonirradiated samples of polyethylene [62]. It is clear that there are two major effects of irradiation on the diffusion constants in polyethylene. Irradiation reduces the diffusion constants and the percent reduction increases with increasing molecular size of the diffusing gas. The energies of activation for diffusion show only a slight increase with radiation. No appreciable size-dependency for the diffusing molecule is observed. (The reported reduction in the activation energy for helium is doubtful because of the difficulty in obtaining reasonable time lags over a significant temperature range.)

TABLE 7

Values of D, $\bar{P}$, E_D, and $E_{\bar{P}}$ at 25°C

Gas	Film[a] sample	$D \times 10^7$ (cm²/sec)	$\bar{P} \times 10^7$ [ml(STP)/ cm-sec-atm]	E_D (kcal/g-mole)	$E_{\bar{P}}$ (kcal/g-mole)
He	1	77.0	0.43	5.6	7.4
He	2	54.0	0.44	4.7	7.7
N_2	1	2.90	0.078	9.7	10.6
N_2	2, 3	1.90	0.070	10.5	11.5
CH_4	4	1.80	0.280	11.3	10.9
CH_4	3	0.95	0.190	11.5	11.5
C_3H_8	4	0.26	0.69	15.1	11.7
C_3H_8	3	0.117	0.39	15.3	12.6

[a] Films are Du Pont Alathon 3, a high-pressure polyethylene. Samples 1 and 4 had $d = 0.9183$ and 0.9154 g/cc, respectively. Sample 3 is sample 4 after Co^{60} irradiation (10^8 r at 1.34×10^6 r/hr) $d = 0.9311$ g/cc. Sample 2 is sample 1 after irradiation (same dose), $d = 0.9268$ g/cc. All samples $\alpha = 0.55$.

These results parallel the results of Barrer and Skirrow [63] for lightly sulfur-crosslinked natural rubber in comparison with the work of Michaels and Bixler [28] on unvulcanized natural rubber. The diffusion results for nitrogen, methane, and propane in a rubber containing 2.7 % sulfur compared with unvulcanized natural rubber are almost identical to a similar comparison between the results in irradiated and unirradiated Alathon 3 [28]. Ascribing the effect of irradiation on the diffusion process to crosslinking therefore seems reasonable.

From (18) it can be seen that D_0 or E_D, or both, must change for the diffusion constant to be lower in the irradiated polymer at constant temperature. Using the model of Brandt [64] in which the activation energy for diffusion is the energy required for symmetrical separation of two chain segments, Michaels and Bixler have shown that to a first approximation in polyethylene

$$E_D \sim ld_p \left[d - \frac{\phi^{1/2}}{2} \right] \delta_p^2, \qquad (20)$$

where l is the length of a chain segment involved in a diffusion step, ϕ is the free volume per unit length of a CH_2 group, d_p is the chain diameter,

d is the diffusion diameter of the gas molecule, and $\delta_p{}^2$ is the cohesive energy density.

From the theory of absolute reaction rates [65], D_0 is given by

$$D_0 = e\lambda^2 \frac{\bar{k}T}{h} e^{\Delta S^*/R}, \tag{21}$$

where λ is the length of a successful diffusion step, $\bar{k}$ is Boltzmann's constant, h is Planck's constant, and ΔS^* is the entropy of activation.

If it is assumed that D_0 is unaltered by irradiation, the experimental values of D in Table 7 would indicate that E_D should be 0.2 kcal/g-mole higher in the irradiated film for helium. This incremental increase in E_D should be progressively higher for the larger gas molecules. An incremental increase of 0.5 kcal/g-mole would be required for propane. These incremental increases in E_D fall within the precision limits of the data, and therefore it is impossible to ascertain conclusively whether the experimental values of E_D reflect these changes. An increase in E_D is consistent with (20), however, since an increase in the cohesive energy density and decrease in the amorphous free volume have been argued in conjunction with the solubility data.

In general, when an activated process involves the participation of a large number of degrees of freedom to achieve the activated state, there is a proportional increase in the activation entropy with increasing activation energy. Qualitatively, this indicates that the larger the activation energy, the larger the number of degrees of freedom over which this energy is most likely to distribute itself. This relationship is usually observed for gases diffusing in amorphous polymers. The parameters E_D, λ, and ΔS^* are, however, mean values for an Avogadro number of diffusion steps. For individual diffusion steps there is undoubtedly a spectrum of energies, entropies, and diffusion-step lengths that gives rise to the mean experimental values. In a completely amorphous matrix of macromolecules this spectrum may be fairly broad. If the amorphous polymer chain segments are gradually restricted in their mobility via crystallization or crosslinking, or both, the first change to be observed with respect to the diffusion process may be the loss of availability of some low-energy diffusion sites. There may be no significant change in the measured value of E_D accompanying this process. Depending on the activation energy distributions, a significant reduction may occur in the probability that gas molecules will find regions in the polymer where conditions exist that are favorable for a diffusion step.   This would result in a reduction in the activation entropy without a compensating change in the activation energy.

If the restraints on the polymer chains are heterogeneously distributed on a molecular scale, the above argument seems even more reasonable. Such a heterogeneous distribution would more conclusively remove certain regions in the amorphous polymer from probable participation in a successful diffusion step.   Mild irradiation probably induced this small-scale heterogeneity.   In fact, the low carbonyl concentrations encountered in this work may lead to a similar heterogeneity in favorable diffusion zones.   Thus in the irradiated films it is believed that there is no significant reduction in E_D but a reduction in D_0, which causes the reduced apparent diffusion constants.   It appears that a similar condition may exist in lightly crosslinked rubber vulcanizates [63].

This argument may be better illustrated by considering what would happen if the crosslinking in irradiated polyethylene was so heterogeneously distributed that small fragments of highly crosslinked polymer were formed.   Assume that the fragments are impenetrable to gas molecules and that these fragments are widely dispersed, compared with the mean activation zone size.   Clearly, under these conditions E_D should not be affected, but the diffusion constants would be reduced by the loss of available sites in the polymer for diffusion and a probable increase in mean diffusion path length through the polymer.   Although this degree of heterogeneity is not expected in most polycrystalline films, it is this type of reduction in diffusion constants that is being considered.

In conclusion it may be noted that a high crosslink density should result in an appreciable increase in both E_D and ΔS^*.   If the average distance between crosslinks is less than the average zone size in the absence of crosslinks, a greater energy of activation must be distributed over a large number of degrees of freedom to give rise to a successful diffusion step.   The latter condition results from the fact that isolated segments of the polymer chain can no longer move independently, and any chain motion must disturb a large number of units in the network polymer.   This situation has been observed in highly crosslinked rubber vulcanizates [63].

The diffusion process for unirradiated films should be considered in reference to the microporous model proposed to account for the effects of crystallinity on the diffusion process in the amorphous phase of polyethylene.   The interpretation of the diffusion data in the unirradiated films is fairly straightforward with respect to this model.   It is

assumed that τ is the impedance offered to a helium molecule in travers-
ing the amorphous phase due simply to the irregular diffusion path
imposed by impenetrable crystallites. If it is further assumed that in a
50% crystalline film the mean distance between crystallites is larger than
the size of the activation zone; β is essentially equal to 1.0. From these
arguments β is related to the probability that crystallites in polyethylene
will significantly alter the spectrum of E_D and ΔS^* values that would be
encountered in the absence of crystallites. A value of β equal to 1.0
indicates that no significant change is expected to occur for the small
helium molecule. Therefore τ for helium is given by the ratio of D in
completely amorphous polyethylene to the experimental value in
Alathon 3.

From the experimental value of D in unvulcanized natural rubber for
the former ($D^* = 2.16 \times 10^{-5}$ cm^2/sec), a value of τ equal to 2.8 is
obtained. This is in excellent agreement with the value predicted from
the relationship of Michaels and Bixler [28] for branched polyethylenes:
$\tau = \alpha^{-1.88}$.

The definitions of τ and β are seen to be highly arbitrary; although
thus defined, τ does represent the minimum impedance that would be
encountered by a gas molecule of any size due only to the geometry of
the amorphous regions as a result of the presence of crystallites. The
utility of τ rests in its ability to give some insight into the shape of
crystallites in polyethylene; for instance, the mean axial ratio of the
crystalline lamellae (assumed to be disks) in the unirradiated samples
is about 20/1, as determined from τ. Crystallites of this shape are
consistent with light-scattering and electron microscopy data.

Values of β calculated in the above manner have been correlated in
polyethylenes by the relationship

$$\ln \beta = \gamma \left(d - \frac{\phi^{1/2}}{2} \right)^2 , \qquad (22)$$

where γ is a constant expected to be dependent on α and the size and
shape of crystallites. Analytical dependence of γ on these quantities
has not been fully determined. Values of β obtained from the unirradiated films are correlated in accordance with (22) in Figure 21,
from which γ was found to equal 0.045.

According to (20), a linear correlation of E_D with $[d - (\phi^{1/2}/2)]$
should be obtained. It should be borne in mind that the values of E_D
for a branched polyethylene will contain a contribution from crystalline

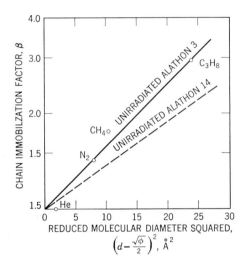

Fig. 21   Effect of size of gas molecules on the chain immobilization factor.

melting as it affects the parameters τ and β.   An energy contribution of this type is not implicit in the model of (20).   For polyethylene $\phi^{1/2}/2$ equals 0.9 Å in the unirradiated films and values of d are included in Table 8 [62].

Application of the micropore model to irradiated films leads to some interesting conclusions.   It appears futile to attempt to differentiate between τ and β in the irradiated films.   The crystallinity appears to be unaltered by irradiation; therefore it is tempting to say that τ is again

TABLE 8

Gas Parameters

Gas	$\varepsilon/\bar{k}$ (°K)	d (Å)
He	10	2.2
N_2	94	3.7
CH_4	148	4.1
C_3H_8	284	5.8

equal to 2.8. In other words, a helium molecule diffusing through the irradiated polymer would, on the average, traverse a micropore of amorphous polymer geometrically similar to a pore in the unirradiated polymer. A moment's reflection indicates that this assumption is probably invalid. One properly positioned crosslink could effectively block a region between two crystallites to helium flow that would otherwise be readily accessible to this small gas molecule. Thus crosslinking resulting from irradiation could alter the geometric impedance factor τ.

In spite of the fact that τ and β cannot be logically separated to retain the significance attributed to them in the microporous model, the product $\tau\beta$ can be calculated. A small correction of the values of D^*, allowing for the fact that the amorphous phase is denser after irradiation, has been neglected. The analogy between diffusion constants in natural rubber and amorphous polyethylene is not sufficiently exact to warrant attempting such a correction.

The interpretation of the effect of crosslinking on β in the irradiated films should be similar to the interpretation ascribed to β in a microcrystalline polymer. The chain mobility in irradiated polyethylene should be reduced by both crystallites and chemical crosslinking. Since τ has been assumed to be independent of gas molecule size, the product $\tau\beta$ in the irradiated films should have the same size-dependence as β alone. Figure 22 is a plot of $\ln \tau\beta$ as a function of $[d - (\phi^{1/2}/2)]^2$ and a correlation is observed. Because of the densification of the amorphous phase, a slightly reduced value of $\phi^{1/2}/2$ (0.8 Å) has been used. The slope of this line is 0.071, indicating the higher gas-size dependency as already noted. It appears from the correlation in Figure 22 that τ cannot be appreciably less than 4.0, indicating a significant increase upon irradiation. It appears that crosslinking has converted some regions in the amorphous polymer that were previously merely a tight fit for helium into regions that are now impassable with any reasonable activation energy.

It has already been noted that the experimental values of E_D are not appreciably affected by irradiation. With respect to the microporous model this would indicate that no significant crystalline melting is occurring in the temperature range of the investigation over and above that which was occurring in the unirradiated films. This is additional support for the argument that the values of ΔH in the irradiated polymer are more positive as a result of changes in chemical composition of the film and not from additional crystalline melting.

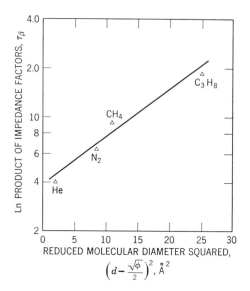

Fig. 22   Effect of size of gas molecules on the product of impedance factors, τ and β, for irradiated Alathon 3.

Chmutov found an even more marked reduction in the diffusion constant of water in irradiated polyethylene [66]. At 25°C, D was found to be 0.6×10^{-7} cm^2/sec. A comparison of this value with a value of 2.9×10^{-7} cm^2/sec reported by Stannett and others [7] for a similar unirradiated polymer indicates a fivefold reduction for an irradiation dose at 10^8 r. This seems like an excessive reduction, but without a knowledge of the exact thermal histories of the films of Chmutov and Stannett the comparison cannot be precise. Since Chmutov's films may have suffered from a gradient in chemical composition, caution should be exercised in attaching undue significance to his apparent values of D. The effect of a gradient in carbonyl content would probably be much more serious than in the case of nonpolar gas transmission.

In summary, the apparent solubility constants are higher by an equivalent fraction for all nonpolar gases consistent with the chemical compositional changes occurring upon irradiation in air. The apparent diffusion constants are lower after irradiation, the fractional decrease being greater, the larger the molecules. The suggested model for gas transmission in polyethylene predicts this behavior, consistent with reduced chain mobility brought about by crosslinking in the amorphous phase. Although all aspects appear to be internally consistent, the possible effects of a gradient in chemical composition through the films cannot be overlooked. Fick's laws as applied to the time-lag method of measuring gas transmission parameters are probably not applicable if a significant gradient in chemical composition exists through a polymer film.

2. Glassy Polymers

Vieth and Sladek have recently developed a new technique for estimating diffusion constants in glassy polymers from transient sorption data [67]. The technique applies to solutes that are sorbed according to an isotherm that consists of two components: the linear component corresponding to Henry's law and the nonlinear part corresponding to the Langmuir equation (cf. p. 40). As discussed above, the model involves ordinary dissolution in the polymer and hole-filling sorption.

The mathematics for this sorption model was extended to the case of diffusion and developed under the assumption that gas trapped in microvoids is immobilized and that the driving force for diffusion is the concentration gradient of dissolved molecules. A nonlinear partial differential equation resulted, and it was necessary to resort to a numerical technique to obtain solutions to this equation with the aid of a computer. These solutions of the equation were obtained in terms of three independent parameters. Since these solutions could not be further simplified, several empirical correlations were tried in an attempt to find a general correlation and one proved highly successful. It involves plotting the parameters Γ versus $\sqrt{\theta'/D}$ and calculating the diffusivity by comparing the results against a standard Γ versus $\sqrt{\theta'}$ curve. These dimensionless parameters are defined as follows:

$$\Gamma = \frac{(p_0 - p)}{(p_0 - p_f)} \tag{23}$$

and

$$\theta' = \frac{Dt}{L^2}\left[1 + \frac{C'_H b/S_D}{(1 + bp)^2}\right]^{-1} \tag{24}$$

where p_0 = initial pressure (atm), p_f = final pressure (atm), D = diffusivity, (cm^2/sec), t = time (sec), L = half-thickness of polymer sheet

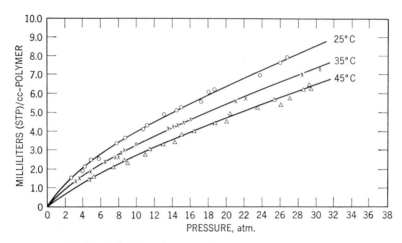

Fig. 23   Solubility of methane in oriented polystyrene [64].

(cm), and other units are the same as in (17).   This correlation has been very successful for carbon dioxide diffusion in polyethylene tereph-thalate [67].

Similar work on the system, methane-polystyrene [68], at tempera-tures from 25 to 45°C and pressures from 3 to 30.5 atm confirmed this model of diffusion in glassy polymers.

Sorption isotherms are shown in Figure 23 and the calculated values of the sorption parameters S_D, b, C_H' are given in Table 9.   A typical Langmuir plot, used to obtain values of b and C_H', is shown in Figure 24.

TABLE 9

Calculated Values of the Sorption Parameters
for Methane in Glassy Oriented Polystyrene

T (°C)	S_D [ml (STP)/cc-atm]	C_H' [me (STP)/cc polymer]	b (atm^{-1})
25	0.193	3.25	0.164
35	0.175	2.53	0.146
45	0.165	1.98	0.112

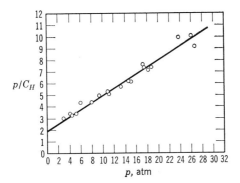

Fig. 24   Langmuir plot for methane sorption by hole filling in oriented poly-styrene at 25°C [64]. Slope = 0.308, intercept = 1.875, $C_H' = 3.247$, $b = 0.164$.

A plot of the solution to the diffusion equation appears in Figure 25. This plot was used in conjunction with the data presented in Figure 26 and similar others to obtain values of the diffusion constants of methane in polystyrene at several temperatures.

The first step in analyzing the result requires the determination of the constants for sorption in (17) for this system.

The slope of the linear portion of an isotherm in the high-pressure region in Figure 23 is set equal to S_D. By subtracting $S_D p$ from the total solubility C, the hole-filling contribution C_H is determined for each pressure.

$$C_H = \frac{C_H' bp}{1 + bp} \qquad (25)$$

or

$$\frac{p}{C_H} = \frac{1}{C_H' b} + \frac{p}{C_H'}. \qquad (26)$$

Then plots of p/C_H as a function of p yield C_H' and b (Figure 24). This procedure is continued until the best fit of the data on the Langmuir plot is achieved; the values of S_D, C_H', and b so determined for several temperatures are presented in Table 9.

An analysis of the accuracy of the direct sorption method indicates that the initial pressures are the main source of error.   At the 95 % level of confidence error limits on solubility C of $\pm 6.5\%$ are obtained for methane.

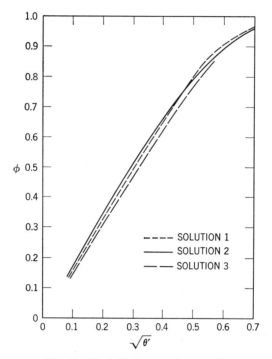

Fig. 25   Vieth-Sladek correlation [64].

To calculate diffusion constants the measured values of p were converted to the dimensionless pressure decay ϕ and then plotted against

$$\left(\frac{\theta'}{D}\right)^{1/2} = \left\{\frac{t}{L^2}\left[1 + \frac{C_H' b/K_D}{(1 + bp)^2}\right]^{-1}\right\}^{1/2} \tag{27}$$

by using the values of C_H', b, and S_D from Table 9. The scaling factors giving the best fit of these data to the general correlation (Figure 25) are listed for the corresponding theoretical curve fits of the data presented in Figure 26, along with the values of the diffusion constants, calculated as $D = (1/\text{scaling factor})^2$.

As a check on the calculated values of the diffusion constants, an attempt was made to apply Crank's solution [25] of the diffusion equation by using the data from sequential runs at the limit of the pressure range studied. It was possible to estimate diffusion constants at all

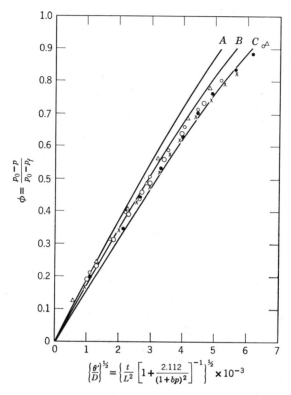

Fig. 26   Kinetics of methane sorption in oriented polystyrene at 35°C [64]. Scaling factors: A, 8750 $D = 1.31 \times 10^{-8}$; B, 9400 $D = 1.13 \times 10^{-8}$; C, 10250 $D = 0.95 \times 10^{-8}$ cm²/sec.   Experimental point sets: $\triangle$ 6.73 → 4.76, $\bigcirc$ 8.80 → 6.31, $\bigcirc$ 11.54 → 8.48, ● 13.52 → 10.01, $\times$ 19.71 → 15.12 atm.

three temperatures, and they agreed within a factor of 2 with those calculated from the correlation in Figure 25.   Thus the method used here would appear to provide a satisfactory estimation of diffusion constants over a wide range of nonideal sorption behavior and supplements the other currently available methods [25] that apply to very low or very high pressure experimentation for the methane-polystyrene system.

Previous work on gas diffusion in glassy polymers [46, 51, 69, 70] showed that gas diffusion constants are abnormally low and activation energies are anomalously high when compared with their values in typically rubbery polymers, such as polyethylene, over the same temperature range.

To explain these results, Michaels, Vieth, and Barrie [46] proposed that the low-pressure sorption and time-lag measurements they had performed actually measured D_{eff} rather than D. Thus the diffusion constant for methane in glassy polyethylene terephthalate at 45°C was found to be of the order of 10^{-9} cm^2/sec, whereas the corresponding value in polyethylene would be of the order of 10^{-7} cm^2/sec. It was felt that if the true value of D could be obtained for methane in glassy polymer, it might be closer to the value for polyethylene than had originally been supposed.

The diffusion constant for methane in glassy polystyrene was found to be 1.45×10^{-8} cm^2/sec, whereas the effective diffusivity was calculated to be 6.2×10^{-9} cm^2/sec, in good agreement with the measured value for methane in polyethylene terephthalate.

Thus it would appear that the anomalously low diffusion constants previously observed for gases in glassy polymers can be partially attributed to the existence of the dual modes of sorption and their influence on the diffusion process, and thus it may not be necessary to invoke wholly new concepts of the nature of the diffusion process in glassy polymers in order to explain the data. In particular, the Brandt model [64] of cooperative chain-and-diffusing molecule micromotion, applied successfully to rubbery polymers, may also be valid for glassy polymers but with the proviso that the motions be severely damped, enough so to permit the existence of microvoids as a stable phase in the glassy polymer.

A consistent picture emerges of gas diffusion in a glassy polymer which features dual modes of sorption in compact amorphous regions and microvoids with weak dispersion forces binding the solutes to the polymer matrix. The intensity of this binding is greater in the microvoids, perhaps because of the focusing of force fields of polymer chain segments that compose the hole periphery. The diffusion process is controlled by the activated jumping of penetrant molecules down the concentration gradient in the compact amorphous regions, with the microvoids acting as sinks to trap diffusing molecules and impede the penetration of a diffusion front. Polymer micromotions probably do contribute to the diffusion process but may be of a lower frequency or amplitude in glassy systems than in normally rubbery systems.

V. Transport of Liquids through Polymer Films

Although a considerable amount of study has been concentrated on the transport of gases, little has been directed toward understanding the physical factors controlling the sorption and diffusion of liquids, particularly in polycrystalline polymers which represent a significant portion of those in use. In earlier work on this problem the effect of thermal and solvent treatment of polyethylene on the permeability of liquid xylene isomers was studied [71]. These results indicated that treatment of polymer films before permeation by annealing at an elevated temperature in the presence of a swelling solvent led to increased permeation rates and sorption of the xylenes in the polymer. In addition to increasing liquid permeation rates as much as tenfold, it was found that under certain conditions the selectivity of the polymer for p-xylene over the other isomers could be enhanced. The mechanism responsible for these changes in film properties has been clarified, in addition to determining the structural aspects of a crystalline polymer that affect liquid sorption and diffusion generally [72], for both p- and o-xylene in polyethylene and for a second liquid isomeric pair, *cis*- and *trans*-1,2-dichloroethylene.

Steady-state permeation of a solvating liquid through a polymer film can be described by Fick's law:

$$J = -D\frac{dc}{dx};\qquad(2)$$

J is the steady-state mass flux per unit area, and D and dc/dx are the local diffusion coefficient and the concentration gradient, respectively, at any point in the film. When the permeant concentration is maintained at zero at the downstream face, integration gives

$$Jl = \int_0^{c_1} D \, dc,\qquad(28)$$

wherein l is the liquid-swollen film thickness, and c_1 is the concentration of permeant in the polymer at equilibrium with the pure liquid in contact with the upstream face. Defining an integral diffusivity, $\bar{D}$, as

$$\bar{D} = \frac{1}{c_1}\int_0^{c_1} D \, dc\qquad(29)$$

and calling Q the permeation flux per unit film thickness, equal to Jl, substitution into (28) yields

$$Q = \bar{D}_{c_1}. \tag{30}$$

In calculating Q from experimental data the dry-film thickness was used. Values of Q calculated in this manner will be lower than the true values, but the maximum error has been estimated at 5%.

Equation (30) shows that the liquid-permeation flux consists of a diffusion and a sorption contribution. By combining equilibrium swelling data, giving c_1, with steady-state permeation measurements, giving Q, information may be obtained about both the sorption and diffusion aspects of liquid transport in polymers.

The apparatus used to obtain liquid permeation data is illustrated in Figure 27. The cell in which the film was mounted (Figure 28) served as the liquid reservoir. The downstream section of the system was initially evacuated to a pressure less than 0.1 mm Hg, following which the run was begun by slowly charging liquid to the cell. Permeation flux determinations were made by timing the rate of fall of the upstream liquid level in a uniform-bore glass capillary, replications being made by refilling the capillary. A series of such tubes, ranging from 30 to 85

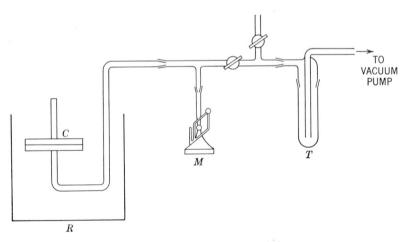

Fig. 27 Diagram of apparatus for liquid permeation studies: C is the cell (Fig. 28), M is a McLeod gage, T is a trap, and R is a thermostatted oil bath ($\pm 0.2°C$ for 20 hr).

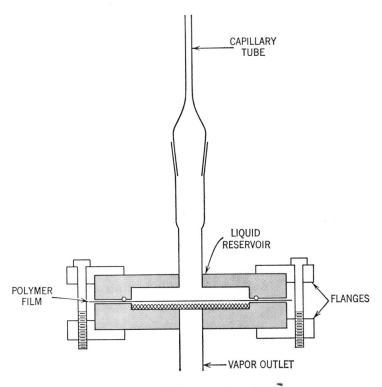

Fig. 28   Liquid permeation cell *C*.

mils in diameter, could be interchanged so that under given conditions of permeation the rate could be determined over a sufficiently small time period (usually 100 to 300 sec).   The thickness of the dry polymer film was determined to 0.01 mil by means of a Pratt and Whitney super-micrometer.   The precision of the values of Q obtained by this technique was estimated to be 5.3 % at the 95 % confidence level.   A material balance on the system, obtained by collecting the permeated liquid in the downstream trap, checked well [72].

To measure liquid sorption gravimetrically a technique was developed which limited the exposure of the swollen film to the atmosphere before weighing, hence minimized evaporation losses of liquid from the polymer.   Such desorption losses can be considerable if suitable precautions are not taken.   After equilibrium with liquid at the desired temperature an immersed film strip was wound directly from the liquid bath onto a spool with a sealed cylindrical container.   By weighing the assembly and knowing the tare weight of the container, the dry film weight, and the polymer density, the liquid sorption in grams of liquid per cubic centimeter of polymer could be computed.   From replicate measurements the precision of the technique at the 95 % confidence level was estimated to be 8.6 %.   The integral diffusivities calculated from (30) were precise to 15 %.

A. Sorption

Sorption data for the xylene and dichloroethylene isomers are summarized in Table 10.   Important physical properties of the sorbing liquids are given in Table 11.   The volume of liquid sorbed per unit volume of amorphous polymer C^* has been calculated from the relationship

$$C^* = \frac{c}{\rho_1 \alpha},\qquad(31)$$

where c is the experimentally determined sorption in grams per cubic centimeter of total dry polymer, ρ_1 is the liquid density in grams per cubic centimeter, and α is the amorphous volume fraction in the dry polymer.   The validity of basing sorption on amorphous content of the polymer rests on three important assumptions: (a) polyethylene consists of two distinguishable ordered phases, the crystalline and amorphous, each possessing unique properties at a given temperature; (b) liquid sorption occurs in the amorphous phase only, with additive liquid and amorphous-polymer volumes; and (c) the change in the relative proportions of crystalline and amorphous material at low temperatures caused by the introduction of a swelling liquid is small.

In several studies involving solutions of liquids or solvating vapors in polyethylene the assumptions outlined above have been applied successfully for several different purposes.   Chiang and Flory, investigating the melting behavior of polyethylene-liquid systems, found good agreement with melting-point depression theory when liquid concentrations were based on amorphous phase content obtained by dilatometry [73]. Rogers, Stannett, and Szwarc have used an amorphous-phase basis in interpreting their vapor-sorption data in polyethylene [37].   Finally,

TABLE 10

Equilibrium Sorption of Liquids in Polyethylene ($T = 30°C$) [72]

| Solvent | Treatment | | C^* (Volume sorbed liquid/volume amorphous polyethylene) | | | | α_{25} [b] |
	Concentration (g/g[a])	Temperature (°C)	p-Xylene	o-Xylene	trans-1,2-Dichloroethylene	cis-1,2-Dichloroethylene	
Untreated polymer			0.297	0.253	0.208	0.183	0.396
None	0	80	0.333	0.326	0.232	0.199	0.348
None	0	97	0.370	0.355	0.226	0.191	0.305
None	0	115	0.359	0.389	0.260	0.186	0.242
p-Xylene	0.10	80	0.397	0.377	0.269	0.250	0.346
p-Xylene	0.10[c]	97	0.455	0.445	0.302	0.305	0.297
p-Xylene	0.10	115	0.490	0.563	0.346	0.311	0.226
p-Xylene	0.04	97	0.384	0.377			0.301
p-Xylene	0.20	97	0.570	0.544	0.436	0.409	0.287
p-Xylene	0.30	97	0.615	0.620	0.477	0.481	0.282
n-Decane	0.085[c]	97	0.447	0.479			0.294
Ethylene bromide	0.26[c]	97	0.488	0.481			0.281

[a] Treatment concentration expressed as grams of solvent per gram of total dry polymer, at stated temperature for 20 hr.

[b] Amorphous volume fraction of dry polymer at 25°C.

[c] Represents a volume concentration of 0.10 ml of solvent per cc of polymer plus solvent.

65

TABLE 11

Physical Properties of Permeants

Permeant	p_1 (30°C) (g/cc)	V_m (25°C) (cc/g-mole)	δ^a (cal/cc)$^{1/2}$	A_D (Å)^b
o-Xylene	0.872	121	9.0	31
p-Xylene	0.852	124	8.6	28
cis-1,2-dichloroethylene	1.267	76	9.1	25
trans-1,2-dichloroethylene	1.240	78	9.0	21

ᵃ Hildebrand solubility parameter [75].

ᵇ Empirical characteristic diffusion cross-sectional area, V_m'/l_{max}, where l_{max} is maximum length of extended molecule and V_m' is the molecular volume of the liquid at 25°C [75].

Barrer and Fergusson [74] found it necessary to consider vapor sorption as occurring only in the amorphous phase of polyethylene in order to obtain thermodynamic consistency in their data.

Comparison among liquids in the untreated polymer in Table 10 shows sorption decreasing in the order of p-xylene > o-xylene > trans-1,-2-dichloroethylene > cis-1,2-dichloroethylene. The trend is qualitatively consistent with the trend in the differences between the solubility parameters of liquid and polymer ($\delta_p \simeq 8.0$); a small difference in solubility parameters is indicative of low energy of interaction in mixing, hence good compatibility and high solubility.

In all cases thermal and solvent treatment cause an increase in C^*. Sorption increases with increasing treatment temperature, and the rate of increase is greater in solvent-treated film than in dry-annealed film. Sorption increases strongly with increasing treatment concentration, whereas the choice of specific treatment solvent appears to have little effect.

Before considering a morphological argument for the increases in C^* values with thermal and solvent treatment, it is worthwhile to summarize briefly the thermodynamics of limited swelling of a polycrystalline polymer by a compatible solvent. A comparison can be made between a polycrystalline and a chemically crosslinked polymer. In the latter, swelling in the presence of a compatible solvent is ultimately limited by the restriction placed on network expansion by crosslinks. The polymer will expand until the osmotic pressure generated by the mixing of liquid with polymer is balanced by the elastic retractive force of the network. Similarly, when a polycrystalline polymer is diluted with a compatible solvent, a balancing force must be exerted or swelling would continue without limit and would result in ultimate dissolution of the polymer. In terms of a structural model of polyethylene, expansion of the amorphous phase is limited because the ends of amorphous chain segments are fixed in adjacent crystalline lamellae. In this way the interlamellar segments are physically crosslinked.

From these considerations Flory's theory of equilibrium swelling may be applied to the sorption of liquids in a polycrystalline polymer. By equating the change in chemical potential of the liquid due to mixing to the change caused by the elastic reaction of the polymer network, and assuming isotropic swelling, we obtain

$$- [\ln(1 - v_2) + v_2 + \chi_1 v_2^2] = \frac{V_1 \rho_a}{M_c}\left(1 - \frac{2M_c}{M}\right)\left(v_2^{1/3} - \frac{2}{f} v_2\right), \quad (32)$$

where v_2 = volume fraction of polymer in the amorphous phase solution = $1/(1 + C^*)$, ρ_a = amorphous polymer density, V_1 = liquid molar volume, χ_1 = Flory-Huggins interaction parameter = $(V_1/RT)(\delta_1 - \delta_2)^2$, δ_1 = solubility parameter of solvent, δ_2 = solubility parameter of amorphous polyethylene $\simeq 8.0$ (cal/cc)$^{1/2}$, M_c = average molecular weight of the polymer chain between crosslinks, M = primary molecular weight of the polymer, and f = functionality of crosslink [76].

Equation 32 can be solved for M_c, which gives a measure of the average amorphous chain segment length in the swollen polymer. Assuming $M_c \ll M$, and taking the crosslink functionality f as large, yields

$$M_c = -V_1 \rho_a v_2^{1/3} [\ln(1 - v_2) + v_2 + \chi_1 v_2^2]^{-1} \quad (33)$$

or, in terms of the C^* defined by (31),

$$M_c = V_1 \rho_a \left(\frac{1}{1 + C^*}\right)^{1/3} \left[\ln\left(\frac{C^*}{C^* + 1}\right) + \left(\frac{1}{1 + C^*}\right) + \chi_1 \left(\frac{1}{1 + C^*}\right)^2\right]^{-1}.$$

$$(34)$$

This model has been applied by Rogers, Stannett, and Szwarc to vapor sorption in polyethylene [37]. Its applicability rests on the assumption that the swelling theory is valid in the case of the relatively short chain segments present in a highly crystalline polymer. This limitation develops from the fact that it is necessary to assume a given distribution of root-mean-square end-to-end distances of chain segments in order to arrive at an expression for the elastic free energy, which appears in the right-hand side of (32). For sufficiently long polymer chains this distribution may be approximated by a Gaussian function. As the segment length decreases, however, the approximation becomes less accurate, and the validity of (32) becomes questionable. Flory has estimated that 10 chain units (methylene groups in this case) constitute

TABLE 12

Calculated Average Number of Methylene Units per
Amorphous Chain Segment [72]

Treatment			From p-xylene sorption		From o-xylene sorption	
Solvent	Concentration $(g/g)^a$	Temperature $(°C)$	$M_c{}^b$	Methylene units	M_c	Methylene units
Untreated polymer			154	11	142	10
None	0	80	169	12	174	12
None	0	97	185	13	190	14
None	0	115	181	13	205	15
p-Xylene	0.10	80	197	14	199	14
p-Xylene	0.10^c	97	224	16	234	17
p-Xylene	0.10	115	240	17	294	21
p-Xylene	0.04	97	192	14	199	14
p-Xylene	0.20	97	276	20	284	20
p-Xylene	0.30	97	303	22	334	24
n-Decane	0.085^c	97	218	16	252	18
Ethylene bromide	0.26^c	97	238	17	253	18

[a] Treatment concentration expressed as grams of solvent per gram of total dry polymer at stated temperature for 20 hr.

[b] Polymer molecular weight between lamellar crystallites.

[c] Represents a volume concentration of 0.10 ml of solvent per cc of polymer plus solvent.

the lower limit for which the Gaussian distribution holds if the polymer chain is regarded as freely jointed. Restriction of bond angles, which is significant, and hindered rotation about the bond, which can be neglected for polyethylene, effectively raise the lower limit for which (32) is applicable.

Despite these uncertainties the theory should serve to indicate relative values in treated and untreated polymers in the swollen state. Those calculated from (34) are given in Table 12. The data indicate the magnitude of the increase in the amorphous-chain-segment length in treated films that is necessary to account for the observed increases in sorption. These results show that substantial changes in polymer morphology must occur during film treatment to allow segment lengths to increase to this degree in the swollen polymer.

The segment lengths calculated from p- and o-xylene sorption in a given polymer are for all practical purposes the same. The value of M_c equivalent to 11 methylene units for the untreated polymer agrees with that obtained by Rogers, Stannett, and Szwarc from p-xylene vapor sorption in a polyethylene of the same degree of crystallinity [37].

1. Effect of Treatments on Polymer Morphology and Sorption

When polyethylene is crystallized from the melt by quenching or rapid cooling, the initial nucleation rate is high because of a high degree of supercooling. The crystalline growth rate is rapid as long as the polymer temperature is in the range in which this is permissible, but growth ceases rather abruptly because of the quench cooling. Consequently the majority of the crystallinity develops under conditions of rapid growth. In addition, the degree of crystallinity is low because of the short time period during which nucleation and growth can occur. Therefore the size of crystallites formed under these conditions, or, more correctly, the extent of regions of perfect order, is limited.

The body of evidence for a folded chain type of crystallite in bulk polymer is growing (Volume I, Chapter 7), although the resulting lamellae are probably much less perfect than in solution-grown crystallites. A spectrum of lamellar step heights (corresponding to the c or chain axis of the crystallites) would be expected. According to Hoffman and Lauritzen, a unique lamellar step height is formed at each crystallization temperature, the step height decreasing with decreasing temperature [77]. In a quenched polymer the amount of crystallization that can occur at a given temperature is small before the temperature drops and lamellae characterized by a lower step height begin to form. These characteristics of the crystallization kinetics probably profoundly affect the physical properties of the uncrystallized amorphous polymer. The continual change in step height should cause a large number of defects within lamellae, and these regions of imperfection must be considered part of the amorphous phase of the resulting polymer.

The high degree of nucleation and rapid growth would be expected to establish strong competition for uncrystallized polymer among the number of growing lamellae. Thus the probability that a given polymer chain will be incorporated into more than one lamella is very high, and the matrix should be characterized by a high degree of interlamellar linking by amorphous chain segments. In addition, there is probably considerable entanglement of polymer chains in these interlamellar links which can exert considerable interchain friction when the polymer is subjected to stress.

In contrast to quenched material, polyethylene crystallized from the melt by slow cooling will exhibit a distinctly different morphology. The degree of crystallinity of the resulting polymer is higher, since the polymer is exposed for a longer time to the temperature range in which crystallization is rapid. The crystals formed are fewer in number and greater in extent. Since, at high temperatures, a greater amount of crystallinity can form at each temperature, the distribution of lamellar step heights about the mean should be narrow. The lower crystallization rates at high temperatures will reduce the competition among growing lamellae for amorphous polymer, and the degree of interlamellar linking by amorphous chains should be low relative to that of quenched polymer. Entanglement of interlamellar chains is likewise considerably reduced, thus eliminating the interchain friction referred to above. Finally, because conditions of slow cooling will lead to more nearly perfect lamellae, the amount of intralamellar amorphous polymer in the form of crystal imperfections will be less.

From these considerations the structural aspects of the polymer matrix that would be expected to affect the properties of amorphous material are the degree of interlamellar linking and entanglement and the amount of intralamellar structural imperfection. The role played by the latter in the process of liquid sorption may be small, since its ability to expand to accommodate a diluent is restricted by the rigidity of the lamellae in which it is imbedded. On the other hand, it appears that interlamellar linking and entanglement would be of major importance in sorption. With reference to the correspondence between a crystalline and a crosslinked polymer, it seems likely that quenched polyethylene, with a high degree of interlamellar linking and entanglement, is analogous to a tightly crosslinked network. The latter, of course, would limit swelling by a liquid to a greater degree than would a weakly crosslinked material.

The effect of annealing on interlamellar linking can be clarified by considering the melting behavior of polyethylene. It has been found experimentally that the disappearance of crystallinity begins at temperatures considerably below the final melting point. Hoffman and others have shown that this can be caused by the melting of small crystals, since their melting points are depressed by high surface free-energy per unit volume [77]. When the polymer is heated to a suitable temperature, a certain fraction of the crystalline regions initially present is melted. If it is annealed at that temperature, recrystallization will occur under conditions of relatively slow growth. Thus annealing not only increases the total crystallinity (as indicated by the decreased α values shown in Table 10) but it effects an exchange of a portion of the initial structure containing a high degree of interlamellar linking for a more nearly perfect crystalline morphology characterized by a lower degree of interlamellar linking. As the annealing temperature is increased, the degree of initial melt-out and recrystallization increases; hence the fraction of the polymer structure exhibiting a low degree of interlamellar linking increases.

The change in polyethylene morphology induced by annealing in the swollen state, or on solvent treating, becomes clear when it is realized that the polymer melting point is depressed by the presence of solvent. Since the activity of the swollen polymer is lowered by dilution, the melting point of crystals of any size will be reduced and the entire melting range is shifted to lower temperatures. Consequently, at a given annealing temperature, as the solvent concentration increases, the amount of melt-out of the original crystalline structure, and its replacement by recrystallized lamellae, increases, and the degree of interlamellar linking in the structure decreases.

Thus, qualitatively, the effects of heat and solvent treatment on interlamellar linking account for the observed increase in equilibrium sorption per unit volume of amorphous polymer with an increase in annealing temperature or solvent concentration. These effects can also be considered in terms of the average amorphous chain length. The osmotic pressures generated by sorption, in the range of 50 to 100 atm, are probably sufficient to cause some fragmentation of the internal crystalline structure of the polymer, which will lengthen amorphous chain segments until the network exerts a retractive force to balance the osmotic stress. In treated films, in which the degree of interlamellar linking is reduced, the osmotic force per amorphous chain is greater, and thus the greater disruption of crystallinity allows the mean chain-segment length to increase. This is responsible for the changes in M_c with treatment shown in Table 12. That structural changes induced in the polymer by treatment are manifested primarily when the film is reswollen and that these changes lead to disorientation within the structure are both supported by light-scattering data.

B. DIFFUSION

A summary of steady-state integral diffusivities at 30°C for p- and o-xylene and $trans$- and cis-1,2-dichloroethylene is given in Table 13. Comparison of diffusivities between liquids within each isomeric pair illustrates the important influence of molecular shape on diffusion in polymers; p-xylene exhibits a diffusion coefficient about twice that of the $ortho$ isomer. Between the dichlorides the ratio of $\bar{D}$ for $trans$ to that for cis is approximately 2.5. These observations are qualitatively consistent with the characteristic diffusion cross section A_D for those molecules presented in Table 11. In developing this characteristic dimension, an effort has been made to allow for partial preferred orientation of anisometric molecules diffusing in a polymer matrix.

Thermal and solvent treatment of the polymer have a marked effect on the integral diffusivities. These effects can be characterized by the effect of treatment on C^*. It has been found [72] that the integral diffusivities can be correlated by the following empirical relationship:

$$\bar{D} = D_{c=0} \exp\{\gamma C^*\}. \tag{35}$$

The values of D_c and γ found are shown in Table 14.

It should be pointed out that (35) is inconsistent with a similar relationship used by other investigators to correlate concentration-dependent local diffusivities; that is,

$$D = D_{c=0} \exp\{\gamma' c'\}, \tag{36}$$

TABLE 13

Integral Diffusivities and Permeation Rates ($T = 30°C$) [72]

Treatment			p-Xylene		o-Xylene		trans-1,2-Dichloroethylene		cis-1,2-Dichloroethylene	
Solvent	Concentration (g/g[a])	Temperature (°C)	$\bar{D} \times 10^7$ cm²/sec	$Q \times 10^7$ g-cm/cm²-sec	$\bar{D} \times 10^7$ cm²/sec	$Q \times 10^7$ g-cm/cm²-sec	$\bar{D} \times 10^7$ cm²/sec	$Q \times 10^7$ g-cm/cm²-sec	$\bar{D} \times 10^7$ cm²/sec	$Q \times 10^7$ g-cm/cm²-sec
Untreated polymer			1.1	0.11	0.59	0.051	4.5	0.48	1.8	1.16
None	0	80	1.3	0.13	0.59	0.058				
None	0	97	1.5	0.14	0.72	0.068	5.4	0.46	2.3	0.17
None	0	115	1.5	0.11	0.57	0.047				
p-Xylene	0.10	80	1.5	0.18	0.78	0.09	5.5	0.65	2.0	0.22
p-Xylene	0.10[b]	97	1.9	0.22	0.91	0.11	5.5	0.62	2.0	0
p-Xylene	0.10	115	1.9	0.23	0.83	0.09	6.7	0.66	2.7	0.24
p-Xylene	0.04	97	1.5	0.15	0.69	0.068				
p-Xylene	0.20	97	2.6	0.36	1.4	0.19	6.3	1.0	2.5	0.36
p-Xylene	0.30	97	2.8	0.41	1.5	0.23	7.0	1.2	2.8	0.47
n-Decane	0.085[b]	97	2.0	0.22	0.89	0.11				
Ethylene bromide	0.26[b]	97	2.0	0.23	1.1	0.13				

[a] Treatment concentration expressed as grams of solvent per gram of total dry polymer, at stated temperature for 20 hr.
[b] Represents a volume concentration of 0.10 ml of solvent/cc of polymer plus solvent.

TABLE 14

Sorption-Diffusion Correlation Parameters

	$D_{c=0} \times 10^7$ cm²/sec	γ (cc amorphous polymer/ml sorbed liquid)
p-Xylene	0.51	2.8
o-Xylene	0.30	2.5
trans-1,2-Dichloroethylene	3.8	1.3
cis-1,2-Dichloroethylene	1.7	1.1

since the substitution of (36) into (29) and integration yields

$$\bar{D} = \frac{D_{c=0}}{\gamma'c'} (e^{\gamma'c'} - 1). \tag{37}$$

On testing the data of McCall and Slichter [78, 79] in which local diffusivities were determined by application of (36) it has been found that (35) is a reasonably satisfactory approximation of (37) for aromatics in linear and branched polyethylenes. Since both (35) and (36) are empirical, this inconsistency is of no consequence except when comparing data from different sources.

The constants for (35) derived from several investigations are presented in Table 15. The striking point to note is that γ in Table 15 [72] is appreciably lower than the value obtained in similar systems, whereas the extrapolated volume of $D_{c=0}$ is appreciably higher. Since p-xylene is a slightly better solvent for polyethylene than benzene (solubility constant = 9.2), it would be expected that if γ were to differ it would be higher for the former if this constant were related to a plasticizing effect on the polymer. The slightly higher temperature and lower density polymer used in this work can account for the slightly higher value of $\bar{D}_{sat}$ for p-xylene in comparison with the value obtained for benzene in a linear polyethylene at 25°C. The markedly higher extrapolated value of $D_{c=0}$ for p-xylene cannot be accounted for by these small differences, however. Although the increased diffusivities caused by thermal and solvent treatment are apparently correlated by the increased permeant solubility per unit volume of amorphous polymer, these treatments must also cause structural modifications to the polymer, that

TABLE 15

Diffusion Parameters for Aromatic Hydrocarbons in Polyethylene

Estimated polymer density (g/cc)	Permeant	Temper-ature (°C)	$\bar{D}_{sat} \times 10^{7a}$ cm²/sec	$D_{c=0} \times 10^7$ cm²/sec	γ (cc amorphous polymer/ml sorbed liquid)	Reference
0.92	Benzene	0	0.6	0.021	12.2	[80]
0.92	Benzene	25	2.1	0.16	8.5	[78, 79]
0.92	Benzene	50		2.4	7.1	[74]
0.97	Benzene	25	1.0	0.033	11.7	[78, 79]
0.95	p-Xylene	30	1.1	0.51	2.8	[72]

[a] Integral diffusivity with saturated liquid or vapor in contact with film.

do not occur or do not influence the transport process at lower permeant activities.

In attempting to develop a model to account for this behavior, it has been found that Q is linear in C^*, at least for the xylene isomers studied. For both o- and p-xylene a twofold increase in C^* results in approximately a fourfold increase in Q at 30°C. It is conceivable that the increase in C^* with thermal and solvent treatment reflects the inbibing of xylene into regions of the film in which the amorphous polymer chain concentration is vanishingly small. This amounts to saying microvoids are formed in swollen treated films that contain essentially pure sorbed liquid. The remaining liquid would be sorbed into amorphous regions in which the polymer-liquid solution is the same as that existing in the untreated film. The liquid-filled microvoids would offer a minor barrier to diffusion. If the microvoids were randomly distributed, the microvoid area per unit area of amorphous polymer would be equal to the microvoid volume per unit volume of amorphous polymer. Under these conditions

$$Q_{treated} = Q_{untreated} \frac{C^*_{treated}}{C^*_{untreated}},$$

since a constant permeation area and the dry-film thickness have been used in computing Q. If this model were correct, a twofold increase in C^* should result in a twofold rather than the observed fourfold increase in Q.

In all likelihood the increase in C^* accompanying film treatment results in a heterogeneous distribution of liquid in the amorphous matrix but not to the extent outlined above. Rather than a simple two-region type of sorption proposed above to account for the increased C^*, a gradation in the amorphous polymer concentration within the polymer-liquid solution is encountered. In some regions the polymer concentration is similar to that in the untreated film. In other regions progressively lower polymer concentrations are encountered, and finally some regions exist which for all practical purposes are microvoids. Since the local increase in C^* would be greatest in the regions of lowest polymer concentration, the major barrier to diffusion could still be offered by the regions of high polymer concentration. Accordingly, the low concentration-dependence of integral diffusivities could be explained. Although it has been proposed that thermal and solvent treatment result in a net decrease in the degree of interlamellar cross-linking, steric factors can prevent this decrease from being homogeneous on a scale comparable to diffusion path lengths or zone sizes.

According to this latter model, the degree to which a permeant is capable of swelling an untreated film should reflect the ability of a permeant to reduce the diffusion barrier in treated films. It will be noted in Tables 10 and 14 that there is a direct correlation between C^* values in the untreated films and the values of γ obtained from integral diffusivities in treated films.

C. Permeability

Having discussed the factors affecting diffusion and sorption in treated polyethylene films, we may now consider how these factors contribute to the liquid permeation flux in the polymer. Combining (30) with the concentration dependence of diffusivity gives

$$Q = D_{c=0} \, \rho_1 \alpha C^* \exp(\gamma C^*). \tag{38}$$

Figure 29, which gives the xylene permeation flux as a function of treatment temperature, shows that Q passes through a maximum at about 95°C. The initial rise is due to increased sorption (and also diffusivity) with treatment temperature; at high temperatures, however, the decrease in α caused by treatment overtakes the sorption increase and a reduction in Q results. In spite of the decrease in interlamellar linking through the temperature range, Q begins to fall off because less material is

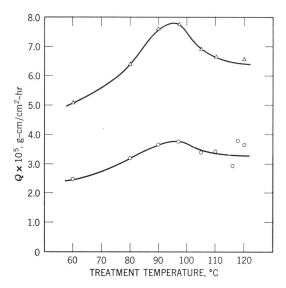

Fig. 29   Permeation flux as a function of temperature of treatment for *o*- (*lower*) and *p*-xylene (*upper curve*).

available for sorption and diffusion.   This effect has not been observed for the *cis*- and *trans*-1,2-dichloroethylenes, in which the increase in sorption with treatment temperature is not so marked [72].   Rather, the permeation flux remains almost constant, since the decrease in α and the sorption increase with treatment temperature remain in balance.

1. Transient Permeation

A series of experiments was done to determine the behavior of the permeation flux during the approach to steady state.   In order to study the effects of temperature and film history, permeation measurements were made in this fashion under a variety of conditions: temperature, film untreated, treated with various amounts of permeant, preswollen, and preswollen and dried [72].

The effects of permeation temperature on the flux-time curves for both *o*- and *p*-xylene showed that *first*, within the temperature range studied, the permeation flux rises to a maximum within about 1 hr, followed by a decline to the steady-state value in 2 to 5 hr; *second*, as the permeation temperature increases from 50 to 90°C, the time over which the permeation flux is elevated above the final steady-state value decreases until at 90°C the maximum no longer appears; and, *third*, as the permeation temperature increases, the time required to attain steady state decreases.

Certain additional data are of interest also. The density of the polymer after permeation (measured after drying and cooling) is increased over that of the initial untreated film, and the density increase is greater at higher permeation temperatures (Table 16). The density

TABLE 16

Effect of Permeation on Polymer Amorphous Volume Fraction

Polymer treatment	$\alpha_{25°C}$
Untreated, before permeation	0.396
Permeated to steady state at 40°C	0.388
Permeated to steady state at 50°C	0.381
Permeated to steady state at 80°C	0.344
Permeated to steady state at 90°C	0.315
Permeated to maximum rate only, 50°C	0.377
Preswollen at 50°C before permeation	0.380
Preswollen at 50°C after permeation	0.376
Preswollen at 50°C permeated without drying	0.376

of the polymer permeated to the maximum flux only also shows the increase. An estimate of the time required for 95% of saturation of the film with liquid, using a diffusivity of 10^{-8} cm^2/sec (which is an order of magnitude lower than that prevailing at steady state at 50°C) gives a value of less than 1 min. Therefore the changes with time observed are not caused by diffusion transients alone.

2. Origins of Structural Changes Occurring in the Polymer During Permeation

It is likely that the causes of the transient behavior are related to slow relaxation processes occurring in the polymer as the structure adjusts in configuration to the swollen state. Undoubtedly similar factors operate to give rise to both the non-Fickian behavior in sorption-desorption kinetics and various time-dependent effects before steady state that have been cited many years ago [74, 79, 81, 82]. Indeed, typical sigmoid curves of sorption and nonlinear desorption as a function of $\theta^{1/2}$ have been obtained for liquid p-xylene in polyethylene [72]. Two other studies describe a type of behavior in which the permeation flux passes through a maximum before the attainment of steady state [81, 82].

Park noted this effect for methylene chloride vapor in polystyrene below the glass transition [82]. The maxima were attributed to other types of flow during the transient period because of an internal capillary network frozen into the glassy structure. This effect can be ruled out for polyethylene which is considerably above its second-order transition at room temperature.

Meares has obtained a very sharp peak at times that were short relative to the time required to reach steady state for allyl chloride vapor in polyvinyl acetate at 40°C, about 10°C above the glass transition [81]. He has proposed that the stress imposed on the underlying dry polymer when the upstream face of the film expands causes the formation of microheterogeneities through which permeation is rapid. As the stressed polymer relaxes through the elastic reaction of the network to the applied force, the permeation rate falls. This mechanism might be applicable to polyethylene, except that Meares has noted that the maximum is achieved within seconds, and after the flux rate falls it gradually rises again to the steady-state value over a period of hours or, under some conditions, even days. With polyethylene, permeation was carried out for 24 hr after steady state was attained, with no change in flux. From this comparison it appears that the effects differ significantly from those outlined by Meares.

Since the transient behavior cannot be explained completely by mechanisms hitherto proposed, it may be instructive to consider the possible factors that can account for the effects described.

a. Permeant Osmotic Pressure. As mentioned above, the driving force for the expansion of the polymer network in sorption is osmotic pressure. These pressures may possibly become high enough to cause disruption and fragmentation of the lamellar crystalline structures in the polymer. An osmotic stress, like any applied stress, is transmitted in the polymer by the slow rearrangement of amorphous chain segments into extended configurations. The extended chains exert a force on the crystalline structures to which they are connected, possibly causing fracture and further lengthening of the chain segment by unraveling from the crystal lattice. Additional amorphous chain segments form, connecting the crystalline fragments. This process should result in a disorientation of lamellar crystallites from the normal twisted-ribbon structure envisioned in spherulitic untreated polymer (this disorientation has been qualitatively supported by light-scattering measurements). The process continues until sufficient intercrystalline links are formed such that the polymer network can exert a retractive force equal to the osmotic pressure, in this way limiting swelling. If diffusion is rapid, the rate at which swelling occurs should be controlled by the rate at which chain segments can assume extended configurations; in this respect swelling is analogous to mechanical creep. The permeation rate through the polymer should increase during this state because of three factors: (a) increasing sorption, (b) increasing diffusivity because of the concentration effect on D, and (c) the creation of more direct flow paths resulting from fragmentation of crystallites.

   b. Osmotic-Pressure Gradient Across the Film. Since a concentration gradient exists during permeation, a gradient in osmotic pressure must be established similarly through the film. The diffusion coefficient increases with concentration, and consequently the concentration and osmotic pressure gradients must be steepest near the downstream face of the film. Thus an expanding force is imposed on the unswollen region of the film by the swollen polymer adjacent to it, which may tend to increase the permeation rate in the unswollen region. If the tensile stresses imposed on this region can gradually be relieved by creep, its permeability may slowly decrease, with consequent reduction in net flux through the film.

   c. Crystalline Melt-Out. In addition to the mechanical disruption of the crystalline phase brought about by osmotic pressure (which may cause the loss of only a relatively small fraction of the total crystallinity), some melting or dissolution of lamellae will occur on the introduction of liquid into the polymer. This may be possible at a temperature as low as 30°C, and undoubtedly occurs at higher temperatures (above 70–80°C). Crystalline melt-out will increase the permeation flux through increased sorption because of the formation of additional amorphous polymer and the removal of crosslinks. If the latter effect allows increased sorption per unit amorphous volume, the effective integral diffusivity will increase because of its concentration dependence.

d. Recrystallization. Recrystallization, or at least further crystallization, of the polymer occurs during permeation, as can be seen from the data in Table 16. It should be noted, however, that the reduction in α presented in this table need not all have resulted during the permeation process, since some crystallization could also occur when the film was cooled and dried before density measurement. Consequently the density data cannot serve as a true indication of the recrystallization history during permeation. Crystallization may begin to take place as soon as concentration of permeant is present in sufficient concentration to increase amorphous chain mobility significantly. At low permeant concentration chain rearrangement may be restricted simply to extension brought on by the osmotic stress, thus hindering crystallization. The rate of crystallization should increase with time, since the permeant concentration increases with time. This should continue until the amount of crystallized material imposes restrictions on both the mobility and the number of possible configurations of the remaining amorphous chain segments. Crystallization should cause a reduction in permeation flux by removal of permeable amorphous polymer and limitation of mobility of the remaining amorphous segments. At low temperatures the amount of additional crystallization may be quite small, and the most important effect may be a reordering of the structure to yield crystal lamellae that are larger in the b and c crystallographic directions and more anisotropic. This could be brought about by elimination of a fraction of the intralamellar imperfections, which, in turn, would reduce the permeation rate by elongating flow paths through a film. Through a combination of these effects recrystallization probably is contributory to the decline in permeation flux from the maximum to the steady-state value.

3. Effect of Permeation Temperature and Solvation History

The considerations outlined in the preceding paragraphs suggest explanations for the effect of permeation temperature on transient behavior. As the permeation temperature increases, polymer segmental mobility becomes higher. Therefore the rate of expansion of the polyethylene network increases and leads to a decrease in the time required to achieve the maximum permeation flux. First, this is supported by the fact that the rate of attainment of swelling equilibrium in the polymer increases with temperature. Second, the rate of crystallization increases with temperature, and this is reflected in a decrease in the time required to reach steady-state permeation.   At 90°C crystallization is rapid enough so that the decline in permeation flux is not observed and the attainment of a steady state is controlled only by the rate of expansion of the network.

The transient behavior of solvent-treated polyethylene films is shown in Figure 30 [72].   The major differences between these and the untreated films is a reduction in the decrement between the maximum and the steady-state permeation rates.   As the treatment concentration increases, going from untreated, to 10% solvent, to excess solvent, it has

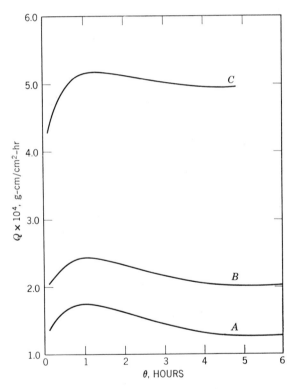

Fig. 30   Approach to steady-state permeation in polyethylene; *p*-xylene, 50°C: *A*, untreated film; *B*, film treated with 10% *p*-xylene at 97°C; *C*, film treated with excess *p*-xylene at 97°C.

been postulated that the degree of interlamellar linking in the dry polymer decreases. Consequently, during swelling, the osmotic stress per
amorphous chain segment would be higher in treated films, with an
attendant reduction in the ability of the polymer to recrystallize. Solvent treatment itself results in a recrystallization of the polymer which
should be more stable under conditions of permeation. If recrystallization during permeation causes the reduction in permeation rate after
the maximum, a film that has been recrystallized during treatment
should exhibit less tendency to recrystallize during permeation and the
trends in Figure 30 are to be expected. It is somewhat surprising that
the time to attain the maximum permeation rate is not decreased by
treatment, although this may result from the fact that the increment in
permeation rates between the initial and maximum values increases
with increasing solvent concentration.

For films in which preswelling was followed by permeation without
drying no maximum in permeation rate was observed. Since the polymer was allowed to equilibrate with liquid for 16 hr, the structural
changes outlined above would have been completed when permeation
began. There is, however, a gradual decline in permeation rate with
time. This phenomenon could be repeatedly observed in the same film
by discontinuing pervaporation, allowing the entire film to reswell, and
then again commencing pervaporation. If the flux continues to decrease
toward the steady-state permeability of the preswollen and dried film,
the gradual decrease in flux could be ascribed to the slow concentration
of the downstream low-concentration region of the film due to reduction
in osmotic stress. The high gradient in osmotic stress established by
rapidly reducing the concentration in the downstream face to zero
could be responsible for the fact that films permeated in this manner
consistently ruptured after several hours of permeation. This steep
gradient does not form when the dry film is permeated, since the rate of
swelling is low.

An alternative explanation must be considered, however, if the flux-
time curves for preswollen films are interpreted as leveling off at a
steady-state flux higher than that attained in the preswollen and dried
films. In each case the solvation history of the downstream face was
identical; each had been preswollen and dried. In the former, however,
the osmotic pressure gradient had been established by desorption and in
the latter, by sorption. Conceivably, because of the loss of solvent
from the downstream region of the film while regions further upstream

remain highly swollen, the downstream region (which normally provides the major resistance to transport) is prevented from shrinking and densifying and thereby offers much less resistance to diffusion than normal polymer of equal solvent content.

Thus, in summary, the effect of solvent and thermal treatment of polyethylene films, consisting of annealing in the presence or absence of a solvent, is to increase the transport rate of liquids through the polymer. The sorption of liquid per unit amorphous phase is increased by structural changes in the polymer and an attendant increase in integral diffusivity results. The heat of solution of xylene in polyethylene, obtained from the temperature-dependence of sorption, is consistent with values reported for other aromatic liquids.

The model proposed relates the loss in effective crosslinking of the amorphous polymer by crystallites brought about by heat or solvent treatment to liquid sorption in the resulting polymer. There is evidence that fragmentation and disorientation of the polyethylene crystalline structure is brought about by the osmotic stresses imposed during swelling.

Before attaining steady-state permeation the flux of the xylene and 1,2-dichloroethylene isomers passes through a maximum. The initial increase with time appears to be attributable to the effects on polymer structure of the buildup of osmotic pressure and some dissolution or fragmentation of crystallites. The decline in flux to the steady-state value may be the result of further crystallization and stress relaxation after the rearrangement of chain segments in the swollen state.

APPENDIX TABLE 1

Transport Parameters for Gases in Polymers[a]

1. POLYCRYSTALLINE POLYMERS

(listed in order of decreasing oxygen permeability)

Polymer	Crystallinity class[b]	Gas	T (°C)	$\bar{P}$ (barrer)[c]	$E_{\bar{P}}$ (kcal/mole)	D (cm²/sec × 10⁶)	E_D (kcal/mole)	S [ml(STP)/cc-cm Hg × 10⁴]	ΔH (kcal/mole)
Polydimethyl siloxane "silicone rubber"[d] (10 wt% silica aerogel filler)	III	He	25	400	3.5	83	1.9	4.8	1.6
		H_2	25	770	3.3	78	2.2	10	1.1
		Ne	25	310	3.3	26	2.0	12	1.3
		O_2	25	660	2.1	21	2.2	32	-0.1
		A	25	790	2.3	22	2.5	35	-0.2
		N_2	25	340	2.6	16	2.7	22	-0.1
		Kr	25	1300	1.5	16	2.9	82	-1.4
		Xe	25	2850	0.8	10.0	3.3	290	-2.5
		n–C_4H_{10}	25	12,500	-2.6	4.0	3.4	3100	-6.0
Polydimethyl siloxane[e] (33 wt% silica aerogel filler)	III	He	25	340	3.7	76	3.1	4.5	0.6
		H_2	25	650	3.0	49	3.0	13	0.0
		Ne	25	260	3.7	21	3.1	12	0.6
		O_2	25	600	2.7	19	3.1	32	-0.4
		A	25	680	2.6	19	3.5	36	-0.9
		N_2	25	280	3.2	14.4	3.3	19	-0.1
		Kr	25	1080	1.7	14.3	4.1	76	-2.4
		CO_2	25	3230	-0.3	11.0		293	
		CH_4	25	940		12.7		74	
		Xe	25	2250	1.1	9.0	4.4	250	-3.3
		n–C_4H_{10}	25	ca. 10,000		ca. 5.0		2000	

APPENDIX TABLE 1 (*continued*)

Polymer	Crystal-linity class[b]	Gas	T (°C)	$\bar{P}$ (barrer)[c]	$E_{\bar{P}}$ (kcal/mole)	D (cm²/sec × 10⁶)	E_D (kcal/mole)	S [ml(STP)/cc-cm Hg × 10⁴]	ΔH (kcal/mole)
Poly-*cis*-isoprene "natural rubber"	III	He	25	31	6.1	22	4.3	1.4	−1.8
		H_2	25	49	6.8	10	6.0	4.9	0.8
		O_2	25	24	7.5	1.6	8.3	15	−0.8
		A	25	23	7.8	1.4	7.9	16	−0.1
		CO	25	16	7.4	1.4	7.4	11	0.0
		N_2	25	8.1	8.8	1.1	8.7	7.4	0.1
		CO_2	25	131	6.1	1.1	8.9	119	−2.8
		CH_4	25	30	7.4	0.90	8.7	33	−1.3
		C_2H_2	25	100	7.3	0.6	9.5	166	−2.2
		C_3H_6	25	252	6.9	0.18	10.2	1400	−3.3
		C_3H_8	25	168	5.5	0.21	11.1	800	−5.6
		SF_6	25	3.6	8.5	0.12	12.0	30	−3.5
Polybutadiene "*Cis*-4"	III	He	25	33	4.8	16	4.1	2.1	0.7
		Ne	25	19	5.2	6.6	4.2	2.9	1.0
		A	25	41	4.6	4.1	5.1	10	−0.5
		N_2	25	19	5.1	3.0	6.0	6.3	−0.9
Polybutadiene (emulsion)	III	H_2	25	43	6.6	10.1	5.1	4.3	1.5
		O_2	25	20	7.1	1.7	6.8	12	0.3
		N_2	25	7.3	8.2	1.3	7.2	5.6	1.0
		CO_2	25	145	5.2	1.3	7.3	112	−2.1

Polyethylene/ propylene (47/53) "EPR 3806"	III	He	27	21	7.1	9.4	6.5	2.2	0.6
		Ne	27	9.2	8.1	3.5	7.6	2.6	0.5
		A	27	12.2	8.9	1.06	9.5	11.5	−0.6
		N_2	27	4.9	9.8	0.79	10.5	6.2	−0.7
Polyethylene/ propylene (37/63) "EPR 3480"	III	He	24	29	6.7	18	5.9	1.6	0.8
		Ne	24	11.3	7.8	5.3	7.6	2.1	0.2
		A	24	14.5	9.0	1.3	10.0	11.1	−1.0
		N_2	24	5.6	10.1	0.94	10.2	5.95	−0.1
Polyethylene/ propylene (29/71) "EPR 3418"	III	He	25	32	6.6	17	3.6	1.9	3.0
		Ne	25	12.0	7.6	4.5	6.8	2.7	0.8
		A	25	13.1	9.5	1.06	8.9	12.4	0.6
		N_2	25	4.9	10.5	0.67	10.9	7.3	−0.4
Polystyrene/ butadiene (23/77) "Ameripol 1507"	III	He	25	23	6.4	16	5.7	1.4	0.7
		Ne	25	9.7	7.0	5.5	7.0	1.8	0.0
		A	25	13	8.1	1.4	8.0	9.1	0.1
		N_2	25	5.1	8.7	1.1	8.6	4.6	0.1
Polystyrene/ butadiene (23/77) "Buna–S"	III	H_2	25	40	6.8	9.8	5.1	4.1	1.7
		O_2	25	17	7.5	1.39	7.1	12.2	0.4
		A	25	16		0.57		28	
		N_2	25	6.3	8.6	1.01	7.8	6.2	0.8
		CO_2	25	123	5.6	1.02	8.1	121	−2.5
Polystyrene/ butadiene (50/50) "Hycar 2001"	III	He	25	13.4	7.0	16	4.3	0.84	2.7
		Ne	25	5.0	7.1	4.2	6.1	1.19	1.0
		A	25	4.5	10.1	0.58	9.9	7.8	0.2
		N_2	25	1.7	9.9	0.43	10.1	4.0	−0.2

APPENDIX TABLE 1 (*continued*)

Polymer	Crystal-linity class[b]	Gas	T (°C)	$\bar{P}$ (barrer)[c]	$E_{\bar{P}}$ (kcal/mole)	D (cm²/sec × 10⁶)	E_D (kcal/mole)	S [ml(STP)/cc-cm Hg × 10⁴]	ΔH (kcal/mole)
Ethyl cellulose, plasticized, "Ethocel 610"	III	N_2	25	3.9		0.30		13	
		H_2S	25	150	1.8	0.09	6.9	1700	−5.1
Rubber hydro-chloride, unplasticized, "Pliofilm FM"	II	N_2	25	0.13		0.03		4.3	
Rubber hydro-chloride, unplasticized, "Pliofilm NO"	II	N_2	25	0.0081		0.004		2.0	
Polyvinyl butyral (unplasticized)	III	N_2	25	2.8	8.6	0.56	8.6	5.0	0.0
		H_2S	30	7.0	4.3	0.005	8.5	1400	−4.2
Polybutadiene/ acrylonitrile (80/20) "Perbunan 18"	III	He	25	17	6.8	15.5	4.2	1.1	2.6
		H_2	25	25	7.2	6.4	6.2	3.9	1.0
		O_2	25	8.2	8.6	0.79	8.1	10.4	0.5
		N_2	25	2.5	9.9	0.51	8.5	4.9	1.4
		CO_2	25	63	7.0	0.43	9.2	147	−2.2

Polymer		Gas							
Polybutadiene/ acrylonitrile (73/27) "Perbunan (Ger)"	III	He	25	12.3	7.0	11.7	5.2	1.1	1.8
		H_2	25	16	7.9	4.5	6.9	3.6	1.0
		O_2	25	3.9	9.7	0.43	9.2	9.1	0.5
		N_2	25	1.1	11.4	0.25	10.4	4.4	1.0
		CO_2	25	31	8.1	0.19	10.7	163	−2.6
		C_2H_2	25	25	9.8	0.076	12.3	329	−2.5
Polybutadiene/ acrylonitrile (68/32) "Hycar OR–25"	III	He	25	9.9	7.4	11.2	5.2	0.88	2.2
		H_2	25	11.7	8.2	3.9	7.0	3.0	1.2
		O_2	25	2.3	10.5	0.28	10.3	8.2	0.2
		N_2	25	0.62	12.3	0.15	11.7	4.1	0.6
		CO_2	25	19	9.0	0.11	12.0	173	−3.0
Polychloroprene, "Neoprene G"	III	H_2	25	13.6	8.1	4.0	6.6	3.4	1.5
		O_2	25	4.0	9.9	0.40	9.4	10.0	0.5
		N_2	25	1.17	10.6	0.25	10.3	4.7	0.3
		CO_2	25	26	8.5	0.24	10.8	11.0	−2.3
Polyisoprene/ methacrylonitrile (74/26, emulsion)	III	H_2	25	14	8.1	3.6	6.9	3.9	1.2
		O_2	25	2.4	11.0	0.24	9.6	10	1.4
		N_2	25	0.6	12.8	0.123	11.6	4.9	1.2
		CO_2	25	14	10.1	0.09	12.2	160	−2.1
Polydimethyl- butadiene, "Methyl rubber"	III	H_2	25	17	8.0	4.0	7.5	4.3	0.5
		O_2	25	2.1	11.3	0.14	11.1	15	0.2
		N_2	25	0.47	13.3	0.079	12.4	5.9	0.9
		CO_2	25	7.5	11.2	0.063	12.8	119	−1.6

APPENDIX TABLE 1 (continued)

Polymer	Crystallinity class[b]	Gas	T (°C)	$\bar{P}$ (barrer)[c]	$E_{\bar{P}}$ (kcal/ mole)	D (cm²/sec × 10⁶)	E_D (kcal/ mole)	S [ml(STP)/ cc-cm Hg × 10⁴]	ΔH (kcal/ mole)
Polyvinylidene fluoride/hexa-fluoropropylene "Viton A"	III	O_2	25	1.45		0.082		18	
		N_2	25	0.44		0.034		11.3	
		CO_2	25	7.8		0.033		236	
Polyisoprene/ acrylonitrile (74/26, emulsion)	III	He	25	7.8	7.6	8.0	4.9	0.98	2.7
		H_2	25	7.5	9.1	2.5	7.4	3.0	1.7
		O_2	25	0.85	12.7	0.09	12.1	9.4	0.6
		N_2	25	0.18	15.0	0.045	14.5	4.0	0.5
		CO_2	25	4.3	12.5	0.031	14.4	140	1.1
Polyisobutylene/ isoprene (98/2) "Butyl rubber"	III	He	25	8.4	7.6	5.9	5.8	1.42	1.8
		H_2	25	7.3	8.7	1.5	8.1	4.9	0.6
		O_2	25	1.3	10.7	0.081	11.9	16	-1.2
		N_2	25	0.33	12.5	0.045	12.1	7.3	0.4
		CO_2	25	5.2	9.9	0.058	12.0	90	-2.1
		C_2H_2	25	1.7	11.6	0.020	12.8	85	-1.2
		C_3H_8	25	1.9	10.7	0.005	13.0	380	-2.3
Polyvinyl acetate, "Gelva 145"[c]	III	He	30	13	7.5	9.6	5.4	1.4	2.1
		H_2	30	9.1	10.0	2.6	7.5	3.5	2.5
		Ne	30	2.8	9.5	1.7	8.5	1.6	1.0
		O_2	30	0.5	13.4	0.058	14.5	8.6	-1.1
		A	30	0.2	14.6	0.016	16.5	13	-1.9
		Kr	30	0.07	18.2	0.0025	19.4	28	-1.2
		CH_4	30	0.05	19.8	0.0029	19.3	17	0.5

90

Polymer		Gas	Temp (°C)						
Polyvinyl chloride, vinyl acetate, "VY HH"		He	25			0.57	7.6	3.2	1.4
		H₂	25			2.27	6.84	1.6	2.8
		Ne	25			0.241	10.2		
		O₂	25			0.013	10.6		
		NH₃	25			0.007	10.5		
		CO	25			0.0015	17.1		
		CO₂	25			0.0005	20.6		
Polyvinyl chloride, "Corvic D 65/6" (40% tricresyl phosphate)	III	H₂	25	0.48	6.4	0.15	5.0		
		Ne	25	1.4	9.8	0.87	7.0		
Polytetrafluoro-ethylene/hexa-fluoropropylene (86/14) "FEP Teflon" ($\rho = 2.15,\ \alpha = 0.67$)	II	CO₂	25	1.7	9.1	0.061	11.0	213	−2.9
		CH₃Cl	25	0.018	13.2	0.008	12.0	106	−2.6
		SF₆	25			0.0017	15.8		
Polybutene-1,^{g} extracted, slow-cooled	II	He	25	56		3.7		15	
		A	25	14		0.11		127	
Polybutene-1,^{g} extracted, slow-cooled and annealed	II	He	25	16		8.8		1.8	
		A	25	1.8		0.13		14	
Polybutene-1,^{g} extracted, quenched and annealed	II	A	25	5.0		0.26		19	

APPENDIX TABLE 1 (*continued*)

Polymer	Crystal-linity class[b]	Gas	T (°C)	$\bar{P}$ (barrer)[c]	$E_{\bar{P}}$ (kcal/mole)	D (cm²/sec × 10⁶)	E_D (kcal/mole)	S [ml(STP)/cc-cm Hg × 10⁴]	ΔH (kcal/mole)
Polypropylene, quenched ($\rho = 0.893$, $\alpha = 0.5$)	II	CO_2	25	3.0	11.1	0.085	12.6	35	−1.5
Polypropylene, quenched ($\rho = 0.893$, $\alpha = 0.5$) "Profax 651A"	II	He	30	10.1	6.4	13.0	4.9	0.78	1.5
		A	30	1.09	11.4	0.15	12.2	7.3	−0.8
		CF_4	30	0.010	18.3	0.0048	15.7	2.08	2.6
Same sample, annealed at 150°C ($\rho = 0.913$, $\alpha = 0.29$)	I	He	30	6.7	6.4	11	4.8	0.61	1.6
		A	30	0.94	10.8	0.15	11.3	6.3	−0.5
		CF_4	30	0.0093	18.3	0.0053	15.6	1.75	2.7
Polypropylene, slow-cooled ($\rho = 0.906$, $\alpha = 0.39$) "Profax 651a"	I	He	30	8.0	6.3	10.3	4.8	0.78	1.5
		A	30	1.05	10.9	0.14	11.4	7.5	−0.5
		CF_4	30	0.0093	18.3	0.0047	16.3	2.0	2.0
Same sample, annealed at 150°C ($\rho = 0.913$, $\alpha = 0.29$)	I	He	30	6.4	6.4	11.0	4.5	0.58	1.9
		A	30	0.91	10.8	0.14	11.8	6.5	−1.0
		CF_4	30	0.0127	17.2	0.0068	15.2	1.9	2.0
Polyethylene, branched, "Alathon 14" ($\rho = 0.914$, $\alpha = 0.57$)	II	He	25	4.9	8.3	6.8	5.9	0.72	2.4
		O_2	25	2.9	10.2	0.46	9.6	6.3	0.6
		A	25	2.7	10.8	0.36	10.1	7.5	0.8
		CO	25	1.5	11.1	0.33	9.5	4.5	1.7

Polymer	Gas	Temp (°C)						
Polyethylene, branched (ρ = 0.921, α = 0.51) (ρ = 0.918, α = 0.54) (ρ = 0.910, α = 0.61)	N₂	25	0.97	11.8	0.32	9.9	3.0	1.9
	CO₂	25	12.6	9.3	0.37	9.2	34	0.1
	CH₄	25	2.9	11.3	0.19	10.9	15	0.4
	C₂H₆	25	6.8	11.3	0.068	12.8	100	−1.5
	C₃H₄	25	42	9.3	0.105	11.9	400	−2.6
	C₃H₆	25	14.5	10.4	0.058	12.5	250	−2.1
	C₃H₈	25	9.5	11.2	0.032	13.3	300	−2.1
	SF₆	25	0.17	14.3	0.0135	14.8	12.5	−0.5
II	C₂H₆	25	2.8	12.3	0.041	14.4	68	−2.1
	C₂H₆	25	3.5	13.7	0.056	18.4	63	−4.7
	C₂H₆	25	5.0	12.0	0.067		75	
Polyethylene, branched, "Alathon 35," molded (ρ = 0.945, α = 0.46) II	He	30	3.2	7.1	5.0		0.64	
	O₂	30	1.5	10.0	0.35	9.3	4.3	0.7
	N₂	30	0.48	11.1	0.22	10.2	2.2	0.9
Polyethylene, branched, "Alathon 3 NC–10" (ρ = 0.917, α = 0.55) II	He	25	5.7	7.4	7.7	5.6	0.74	1.8
	N₂	25	1.03	10.6	0.29	9.7	3.6	0.9
	CH₄	25	3.7	10.9	0.18	11.3	21	−0.4
	C₃H₈	25	9.1	11.7	0.26	15.1	350	−3.4
"Alathon 3, NC–10," irradiated 10⁸ r (ρ = 0.928, α = 0.55) II	He	25	5.7	7.7	5.4	4.7	1.06	3.0
	N₂	25	0.92	11.5	0.19	10.5	4.8	1.0
	CH₄	25	2.5	11.5	0.10	11.5	25	0.0
	C₃H₈	25	5.1	12.6	0.012	15.3	430	−2.7

APPENDIX TABLE 1 (*continued*)

Polymer	Crystallinity class[b]	Gas	T (°C)	$\bar{P}$ (barrer)[c]	$E_{\bar{P}}$ (kcal/mole)	D (cm²/sec × 10⁶)	E_D (kcal/mole)	S [ml(STP)/cc-cm Hg × 10⁴]	ΔH (kcal/mole)
Polyethylene, linear "Grex" ($\rho = 0.964$, $\alpha = 0.23$)	I	He	25	1.14	7.1	3.1	5.6	0.37	1.5
		O_2	25	0.41	8.4	0.170	8.8	2.4	−0.4
		A	25	0.38	9.0	0.116	9.3	3.3	−0.3
		CO	25	0.19	9.4	0.096	8.8	2.0	0.6
		N_2	25	0.14	9.5	0.093	9.0	1.5	0.5
		CO_2	25	1.7	7.2	0.124	8.5	13.7	−1.3
		CH_4	25	0.39	9.7	0.057	10.4	6.8	−0.7
		C_2H_6	25	0.59	10.2	0.015	12.5	39	−2.3
		C_3H_4	25	4.0	7.9	0.025	11.3	160	−3.4
		C_3H_6	25	1.15	9.3	0.0106	12.5	108	−3.2
		C_3H_8	25	0.54	10.7	0.0049	13.6	110	−2.9
		SF_6	25	0.0034	13.2	0.0016	15.0	5.3	−1.8
Polyethylene linear, ($\rho = 0.951$, $\alpha = 0.33$)	I	CH_4	25	0.62	12.1	0.086	14.2	7.2	−2.1
		C_2H_6	25	0.87		0.020		44	
		C_3H_8	25	0.81		0.0068		120	
		C_4H_{10}	25	2.6		0.0036		720	
Polyethylene linear, "Grex" molded ($\rho = 0.967$, $\alpha = 0.21$)	I	O_2	30	0.47	8.3	0.19	8.6	2.5	−0.3
		N_2	30	0.15	9.9	0.18	9.3	0.83	0.6
		CO_2	30	1.9	6.9	0.13	8.2	14.6	−1.3

Sample		Gas							
"Grex" molded (ρ = 0.955, α = 0.25)	I	He	30	2.2	7.0	4.5	8.8	0.49	−0.8
		O₂	30	0.84	8.0	0.29	9.3	2.9	0.1
		N₂	30	0.29	9.4	0.15	8.9	1.9	−1.8
		CO₂	30	3.5	7.0	0.21		17	
"Super Dylan," molded (ρ = 0.953, α = 0.31)	I	O₂	30	0.43		0.14		3.1	
"Super Dylan," molded (ρ = 0.945, α = 0.37)	I	O₂	30	0.65		0.17		3.8	
Polyethylene linear, "Grex" molded, quenched (ρ = 0.952, α = 0.33)	I	He	25	1.8	5.7	5.8	5.6	0.31	1.1
		A	25	0.67	8.7	0.17	9.6	3.9	−0.9
		C₂H₆	25	1.10	10.5	0.022	13.0	50	−2.5
"Gex" annealed at 130°C (ρ = 0.978, α = 0.14)	I	He	25	1.3	6.4	8.1	3.8	0.16	2.6
		A	25	0.50	9.2	0.18	9.2	2.8	0.0
		C₂H₆	25	0.76	12.3	0.027	13.4	28	−1.1
"Grex" molded, slow-cooled (ρ = 0.969, α = 0.21)		He	25	0.79	7.1	3.6	4.3	0.22	2.8
		A	25	0.29	6.6	0.110	6.2	2.6	0.4
		C₂H₆	25	0.43	10.3	0.0123	12.3	35	−2.0
"Gex" annealed at 130°C (ρ = 0.976, α = 0.17)		He	25	0.68	6.9	4.2	4.4	0.16	2.7
		A	25	0.22	9.5	0.11	9.4	2.0	0.1
		C₂H₆	25	0.40	10.4	0.014	12.3	29	−1.9

APPENDIX TABLE 1 (continued)

Polymer	Crystallinity class[b]	Gas	T (°C)	P̄ (barrer)[c]	$E_{\bar{P}}$ (kcal/mole)	D (cm²/sec × 10⁶)	E_D (kcal/mole)	S [ml(STP)/cc-cm Hg × 10⁴]	ΔH (kcal/mole)
Ethyl cellulose, plasticized, Ethocel 610	III	N_2	25	3.9		0.30		13	
		H_2S	25	150	1.8	0.09	6.9	1700	−5.1
Polyethylene terephthalate, "Mylar," 3–10 mils ($\rho = 1.38$, $\alpha = 0.58$)	II	He	100	9.2	6.8	12	6.6	0.77	0.2
		O_2	100	0.55	10.6	0.19	10.6	2.9	0.0
		A	100	0.41	12.6	0.090	12.6	4.6	0.0
		N_2	100	0.145	14.0	0.085	14.0	1.7	0.0
		CO_2	100	2.0	12.2	0.046	16.9	43	−4.7
		CH_4	100	0.21	14.7	0.028	17.8	7.5	−3.1
Rubber hydrochloride, plasticized, "Pliofilm FM"	II	N_2	25	0.13		0.03		4.3	
Rubber hydrochloride, unplasticized, "Pliofilm NO"	II	N_2	25	0.0081		0.004		2	
Polyvinylidene chloride/vinyl chloride "Saran"	II	N_2	25	0.0012		0.000092		13	
2. GLASSY POLYMERS[h]									
Ethyl cellulose (49.5 ethoxyl)	III	He	25	53		2.2		24	
		O_2	25	15		0.61		25	

		Gas							
		A	25	10		0.39		26	
		N_2	25	4.4		0.21		21	
		CO_2	25	113		0.55		210	
		NH_3	25	700		0.12		5800	
		SO_2	25	264		0.06		4400	
		C_2H_6	25	9.2		0.015		610	
		C_3H_8	25	3.7		0.003		1250	
Cellulose acetate									
2.75 acetyl		CO_2	30	6.2		0.030		210	
2.45 acetyl		CO_2	30	5.1		0.025		200	
2.29 acetyl		CO_2	30	4.5		0.018		250	
Cellulose acetate/butyrate (1.41 acetyl/1.44 butyryl)		CO_2	30	7.1		0.030		240	
Cellulose butyrate (2.8 butyryl)		CO_2	30	14.8		0.060		250	
Polystyrene, molded, "Polystyrol"	III	He	24	17		10		1.7	
		H_2	24	15		4.6		3.3	
		N_2	24	1.0		0.1		10	
		A	24	0.8		0.05		16	
Polystyrene, molded, "Styroflex"		Kr	20	0.50	5.9	0.0069	5.8	73	0.1
Polystyrene, biaxially oriented, "Trycite"	III	CH_4	25	0.84	1.3	0.0078	6.4	110	−5.1

APPENDIX TABLE 1 (*continued*)

Polymer	Crystallinity class[b]	Gas	T (°C)	$\bar{P}$ (barrer)[c]	$E_{\bar{P}}$ (kcal/mole)	D (cm²/sec × 10⁶)	E_D (kcal/mole)	S [ml(STP)/cc-cm Hg × 10⁴]	ΔH (kcal/mole)
Polydimethylmethylene-di-*p*-phenylene carbonate, "Lexan Polycarbonate" (5-mil cast film)	III	H_2	25	12.0	5.4	0.64	5.0	18.5	0.4
		O_2	25	1.4	4.6	0.021	7.7	67	−3.1
		A	25	0.8	5.4	0.015	6.0	53	−0.6
		CO_2	25	8.0	3.8	0.0048	9.0	1660	−5.2
		SF_6	25	0.0000065		0.0000001	ca. 20	65	
		Kr	20	0.24	5.8	0.0017	9.3	140	−3.5
Polytetrafluoroethylene, "TFE Teflon" ($\rho = 2.144$, $\alpha = 0.48$)	I	CO_2	25	20.4	4.7	0.17	8.4	120	−3.7
		CH_3Cl	25	10	5.0	0.03	9.6	334	−4.6
		CF_3H	25	21	3.9	0.03	9.5	700	−5.6
		SF_6	25	0.14	9.4	0.00034	14.6	410	−5.2
Polytetrafluoroethylene, "TFE Teflon" ($\rho = 2.236$, $\alpha = 0.11$)	I	CH_3Cl	25	0.17	7.7	0.0018	11.4	94	−3.7
Polychlorotrifluoroethylene, "Kel F" ($\rho = 2.1$, $\alpha = 0.45$)	I	N_2	25	0.091		0.0036		25	
Polyvinyl butyral	III	CO_2	15	5.5		0.018		300	
Polyethyl methacrylate[l]	III	He	25	15	6.4	11.0	3.7	1.4	2.7
		Ne	25	3.1	6.9	1.6	5.7	1.9	1.2
		O_2	25	1.4	8.7	0.11	7.6	13	1.1

Polymer	Class	Gas	Temp (°C)						
		A	25	0.68	9.1	0.028	10.3	24	−1.2
		N₂	25	0.25	9.7	0.020	10.2	13	−0.5
		CO₂	25	5.0	6.9	0.036	7.9	138	−1.0
		Kr	25	0.38	10.3	0.080	11.0	48	−0.7
		SF₆	25	0.017	10.7	0.00027	15.4	63	−4.7
		H₂S	25	4.02	9.4	0.0038	11.4	1060	−2.0
Cellulose diacetate		Kr	20	0.10		0.0020		50	
Poly-1,2-bis-(4-hydroxyphenyl)-methane epichlorohydrin	III	He	23	7.8					
		H₂	23	8.3					
		CO₂	23	2.5	4.7	0.0075	9.1	330	−4.4
		O₂	23	0.48		0.0220		22	
Poly-1,8-bis (4-hydroxyphenyl)-methane epichlorohydrin	III	He	23	3.3					
		H₂	23	2.2					
		CO₂	23	0.57	6.2	0.0015	9.5	380	−3.2
		O₂	23	0.12		0.0090		13	
Poly-2,8-bis (4-hydroxyphenyl)-methane epichlorohydrin	III	He	23	6.1					
		H₂	23	5.0					
		CO₂	23	1.1	4.1	0.0030	8.7	370	−4.6
		O₂	23	0.27		0.0140		20	
Polyvinyl acetate,[j] "Gelva 145"	III	He	15	7.1	3.2	7.7	4.2	0.92	−1.0
		H₂	15	4.7	2.8	1.7	5.2	2.8	−2.4
		Ne	15	1.2	2.7	1.1	7.4	1.1	−4.7
		O₂	15	0.2	4.8	0.024	11.1	8.3	−6.3
		A	15	0.089	7.7	0.0063	11.4	14	−3.7
		Kr	15	0.025	6.4	0.00076	14.5	33	−8.1

Polymer	Crystallinity class[b]	Gas	T (°C)	$\bar{P}$ (barrer)[c]	$E_{\bar{P}}$ (kcal/mole)	D (cm²/sec × 10⁶)	E_D (kcal/mole)	S [ml(STP)/cc-cm Hg × 10⁴]	ΔH (kcal/mole)
Polyvinyl chloride, unplasticized, "Corvic D65/6"	III	CO_2	20	0.035		0.0021		170	
		H_2	25	1.5	6.9	0.33	7.7	4.5	−0.8
		Ne	25	0.29	6.7	0.13	7.8	2.2	−1.1
Chlorinated polyvinyl chloride (62% Cl)	III	Kr	20	0.029		0.0005		58	
Nitrocellulose	III	O_2	25	1.0		0.103		10	
		A	25	0.11		0.0069		16	
		N_2	25	0.12		0.0129		9.3	
		CO_2	25	2.12		0.0162		130	
		NH_3	25	57		0.0046		12400	
		SO_2	25	1.8		0.0008		2300	
		C_2H_6	25	0.063		0.0001		630	
		C_3H_8	25	0.0084		0.00002		420	
Epoxy, amine-cured, "EGK 19"		Kr	20	0.0067		0.00023		29	
Polyethylene terephthalate, "Mylar" (3–10 mils) ($\rho = 1.383$, $\alpha = 0.58$)	II	He	25	1.45	4.7	2.0	4.5	0.73	0.2
		O_2	25	0.035	7.7	0.0036	10.8	9.7	−3.1
		N_2	25	0.0067	7.8	0.0013	12.2	5.2	−4.4
		CO_2	25	0.16	4.4	0.00051	11.9	310	−7.5
		CH_4	25	0.0036	8.8	0.00015	14.1	24	−5.3
		Kr	20	0.024	10.3	0.00015	8.1	160	−2.2

	[b]		T (°C)	[c]		S[h]			
"Mylar 100A" (1.0 mil)	II	O_2	23	0.020	7.3	0.0024	9.9	8.4	−2.6
		CO_2	23	0.114		0.00038		300	
"Mylar 65 HS" (0.65 mil)	II	O_2	23	0.042		0.0035		12	
		CO_2	23	0.197		0.0010		197	
Polyhexamethylene adipamide,[k] "Nylon 6, 6"	II	N_2	25	0.0081	2.0	0.0027	6.7	3.0	−4.7
		CO_2	25	0.07		0.00083		85	
		Kr	2.0	0.0033		0.00028		11.8	
Poly-ε-caprolactam, "Nylon 6" ($\rho = 1.165$, $\alpha = 0.63$)		N_2	25	0.0025	9.7	0.00025	5.3	10	−4.4
		CO_2	30	0.10		0.0018		55	
Polyvinylidene chloride/ vinyl chloride									
(85/15)		CO_2	20	0.003		0.0003		100	
(70/30)		CO_2	20	0.007		0.0006		120	
(60/40)		CO_2	20	0.015		0.0012		150	

[a] Most data obtained on molded sheets > 10 mils thick.
[b] I—highly crystalline (>50 vol %); II—low to moderate crystallinity (<50%); III—completely amorphous (<20% crystallinity).
[c] Barrer: [ml(STP)-cm/cm²-sec-cm Hg] × 10¹⁰.
[d] Values apply between −40 and 40°C. Polymer begins to crystallize at −50°C.
[e] Upper transition temperature is close to 26°C. Values apply above this temperature.
[f] After crystal transformation is complete.
[g] Values apply only to rubbery polymer above about 90°C.
[h] Henry's law does not apply to solubilities; values of S should not be extrapolated above 1 atm.
[i] T_g is approximately 65°C, but no changes are observed in E_P and E_D between 25 and 80°C.
[j] Lower transition temperature is approximately 17°C. Values given apply below this temperature.
[k] At 0% relative humidity; transport parameters are sensitive to relative humidity.

APPENDIX TABLE 2

Permeabilities of Polymers
(listed approximately in order of decreasing oxygen permeability)

Polymer	Trade name	Crystallinity class[a]	Gas	T (°C)	P[b] (barrer)	E_P (kcal/mole)
Polydimethyl siloxane	RTV	III	O_2	23	440	
			CO_2	23	2300	
					4	
Polydimethyl siloxane (electron-irradiated, 10^8 r)	Silicone rubber	III	O_2	25	60	
			N_2	25	220	
			CO_2	25	2600	
Polydimethyl siloxane/phenyl siloxane (95/5)		III	O_2	25	450	
			N_2	25	200	
(80/20)			H_2	25	190	
			O_2	25	130	
			N_2	25	50	
			CO_2	25	720	
(67/33)			H_2	25	77	
			O_2	25	45	
			N_2	25	14	
			CO_2	25	300	

			Gas	Temp	Value
(50/50)			H_2	25	44
			O_2	25	14
			N_2	25	4.9
(33/67)			H_2	25	12.3
			O_2	25	3.6
			N_2	25	0.81
			CO_2	25	19
(20/80)			H_2	25	12.3
			O_2	25	2.1
			N_2	25	0.44
			CO_2	25	10.2
Nitrile-silicone rubber		III	He	25	86
			H_2	25	123
			O_2	25	85
			N_2	25	33
			CO_2	25	670
Fluoro-silicone rubber		III	O_2	25	114
			N_2	25	49
Poly-4-methylpentene-1	TPX	II	O_2	25	24
			CO_2	25	96
Polypropylene/ethylene(55/45)		III	O_2	25	14
Polybutadiene/styrene (86/14)		III	O_2	25	13
Polyisobutylene		III	O_2	25	9.0

APPENDIX TABLE 2 (*continued*)

Polymer	Trade name	Crystallinity class[a]	Gas	T (°C)	$\bar{P}^{b}$ (barrer)	$E_{\bar{P}}$ (kcal/mole)
Polysulfide	Thiokol	III	O_2	23	7.7	
Ethyl cellulose, unplasticized		II	He	30	30	
			O_2	25	7.2	
			N_2	25	2.1	
			CO_2	28	46	
			CH_4	30	6.4	
Ethyl cellulose, plasticized	Ethocel 610	II	H_2	25	87	
			O_2	25	8.0	4.0
			N_2	25	2.5	4.2
			CO_2	25	48	4.4
			NH_3	25	705	
			SO_2	25	264	
			C_2H_4O	0	4.2	(p = 33 cm Hg)
Polytetrafluoroethylene/ hexafluoropropylene (86/14)	FEP Teflon	II	He	25	50	4.7
$\rho = 2.15$			H_2	25	14.1	6.1
$\alpha = 0.67$			O_2	25	5.9	5.7
			N_2	25	2.2	7.0
			CH_4	25	0.11	7.7
			C_2H_6	33	0.65	
Polytetrafluoroethylene	Teflon		He	30	59	
$\rho = 2.14$			N_2	30	5.0	
$\alpha = 0.5$			CO_2	30	25	
			HCl	30	0.11	

Material		Gas	Temp	Value	
$\rho = 2.18$ $\alpha = 0.35$		He	30	25	
		N_2	30	1.9	
		CO_2	30	9.0	
		HCl	30	0.038	
$\rho = 2.22$ $\alpha = 0.25$		He	30	11	
		H_2	33	12	
		N_2	30	0.79	
		CO_2	30	2.6	
		HCl	30	0.014	
		C_2H_6	33	0.65	
Polystyrene/butadiene (73/27) Lustrex	III	O_2	25	6.0	
Cellulose acetate-propionate	III	O_2	25	6.0	
Cellulose acetate-butyrate Kodapak F-298	III	He	25	14	4.6
		H_2	25	21	4.1
		O_2	25	5.6	
		N_2	30	1.0	
		CO_2	25	31	4.2
Cellulose butyrate		He	20	17.0	
		CO_2	20	14.8	
Polyethylene/vinyl acetate (85/15) $\rho = 0.930$ Elvax	II	O_2	25	5.0	
		N_2	25	1.4	
		CO_2	25	18	

APPENDIX TABLE 2 (*continued*)

Polymer	Trade name	Crystallinity class[a]	Gas	T (°C)	$\bar{P}^{b}$ (barrer)	$E_{\bar{P}}$ (kcal/mole)
(70/30) $\rho = 0.940$			He	30	21	
			O_2	25	7.2	
			N_2	25	2.4	
			CO_2	25	30	
			CH_4	30	11	
Polybutadiene/acrylonitrile (80/20)		III	O_2	25	4.8	
Polychloroprene	Neoprene W	III	He	23	13	
			H_2	23	20	
			O_2	23	4.0	
			N_2	23	1.1	
			CO_2	23	22	
			CH_4	23	2.6	
			C_3H_8	23	16	
			C_4H_{10}	23	47	
			NH_3	23	200	
			SO_2	23	320	
			Freon 12	23	9.0	
			Freon 22	23	18	
Polystyrene/butadiene (82/18)		III	O_2	25	3.6	
(90/10)		III	O_2	25	3.0	

Material	Trade name	Class	Gas	Temp (°C)	Value	
Polyethylene/sodium acrylate (96/4)	Surlyn A	II	He	25	7.5	
			O₂	25	3.5	
			N₂	25	1.0	
			CH₄	25	3.0	
			CO₂	25	11	
Polyvinyl butyral 30% OH		II	He	20	7.6	
			CO₂	20	4.0	
39% OH			He	20	6.7	
			O₂	30	2.49	
			N₂	27	0.25	
			CO₂	27	2.60	
Polystyrene, molded	Lustrex	III	O₂	25	2.5	
			N₂	25	0.83	
			CO₂	25	9.7	
Cast, annealed		III	A	25	0.95	4.0
Biaxially oriented	Polyflex Trycite	III	H₂	25	21	
			O₂	25	1.8	
			A	25	1.6	
			N₂	25	0.4	
			CO₂	25	8.3	
			CH₄	25	2.0	
Molded	Styron 666	III	O₂	25	2.4	4.3
			N₂	25	0.3	7.3

APPENDIX TABLE 2 (continued)

Polymer	Trade name	Crystallinity class[a]	Gas	T (°C)	$\bar{P}^{b}$ (barrer)	E_P^{b} (kcal/mole)
Molded	Styron 475	III	O_2	25	1.8	5.1
			N_2	25	0.3	7.5
Isotactic		III	O_2	25	0.54	
Foamed (20 mil)	Styrofoam	III	O_2	25	1.5	
Chlorosulfonated polyethylene	Hypalon	II	He	23	7.2	
			H_2	23	10.8	
			O_2	23	2.1	
			N_2	23	0.92	
			CO_2	23	15.8	
			CH_4	23	1.7	
			C_3H_8	23	7.5	
			C_4H_{10}	23	47	
			NH_3	23	97	
			CO_2	23	128	
			Freon 12	23	4.6	
			Freon 23	23	8.2	
Polybutene-1	Bu-Tuf	II	O_2	25	2.0	
Polyvinylidene fluoride/ hexafluoropropylene	Viton	III	He	25	11.2	7.2
			O_2	30	1.45	
			N_2	25	0.44	
			CO_2	30	7.8	

Material	Trade name	Type	Gas	Temp (°C)		
Poly-4,4'-isopropylidene-di-p-phenylene carbonate, cast film (ca. 5 mils)	Lexan Merlon	II	O_2	25	1.4	
			N_2	25	0.28	
			CO_2	25	5.6	
Cast film (<1.0 mil)	Lexan	II	He	25	15	
			H_2	25	14	
			O_2	25	1.7	
			N_2	25	0.37	
			CO_2	25	12	
			CH_4	25	0.45	
			C_2H_4	25	0.8	
			C_2H_6	25	0.4	
			C_3H_8	25	0.2	
Polypropylene, molded		I	He	25	48 (?)	9.2
			H_2	25	54 (?)	7.7
			O_2	25	1.7	11.4
			N_2	25	0.3	13.3
			CO_2	25	7.2	9.1
Extruded, quenched film (2.0 mil)	Moplefan (Avisun)	II	H_2	25	12.0	11.5
			O_2	25	2.0	15
			N_2	25	0.45	3.3
			CO_2	25	7.4	
			NH_3	25	9.2	
			C_2H_4O	23	9.0	
			C_2H_6	33	1.6	
Extruded, quenched, and annealed	Pro-Fax	I	H_2	33		
			O_2	25	1.4	
			CO_2	25	4.5	

APPENDIX TABLE 2 (*continued*)

Polymer	Trade name	Crystallinity class[a]	Gas	T (°C)	$\bar{P}$[b] (barrer)	$E_{\bar{P}}$ (kcal/mole)
Biaxially oriented film	Pro-Fax	I	H_2	33	5.7	
			O_2	25	1.0 (0.8)	
			N_2	25	0.2 (0.11)	
			CO_2	25	2.8	
			C_2H_6	33	0.014	
Biaxially oriented film	Cryovac Y	I	O_2	23	0.54	
			N_2	23	0.060	
			CO_2	23	1.85	
Chlorinated polyethylene	Tyrin					
25% Cl		II	O_2	23	1.4	15
33% Cl		II	O_2	25	1.75	
42% Cl			CO_2	25	6.2	
		III	CO_2	25	0.29	
			O_2	23	0.36 (0.1)	9.0
Polyxylene oxide (10 mil)	PPO	II	O_2	25	1.2	
Polyxylene oxide (0.15 mil)		II	He	25	90	
			H_2	25	130	
			O_2	25	18	
			N_2	25	3.6	
			CO	25	6.0	
			A	25	7.0	
			CH_4	25	4.3	
			CO_2	25	85	

Polymer	Trade name	Class	Gas	T (°C)		
Polybisphenol-A/bisphenol sulfone	Polysulfone	III	H_2	25	10.8	
			O_2	25	1.3	
			N_2	25	0.25	
			CO_2	25	5.7	
Polyvinyl butyral/vinyl alcohol-acetate	Butvar	III	O_2	25	1.17	
Polystyrene/acrylonitrile/butadiene (60/25/15)	Lustran	III	O_2	25	0.75	
			N_2	25	0.15	
			CO_2	25	3.0	
(50/35/15)	Tyril 767	III	O_2	25	0.50	6.1
			N_2	25	0.045	5.5
Polystyrene/acrylonitrile (74/26)	Lustran SAN	III	O_2	25	0.42	
			N_2	25	0.088	
			CO_2	25	1.7	
Cellulose acetate (unplasticized)	Lumarith P 912 Kodacel A–30	III	He	20	13.6	
			H_2	25	8.4	5.2
			O_2	25	0.68	5.0
			N_2	25	0.23	6.5
			CO_2	25	5.0	4.5
			H_2S	25	3.0	5.1
			C_2H_4O	0	14	($p = 33$ cm Hg)
Cellulose acetate (15% dibutyl phthalate)		III	He	25	13.6	5.3
			H_2	25	12.3	
			N_2	25	0.5	
			CO_2	25	2.5	4.3

111

APPENDIX TABLE 2 (continued)

Polymer	Trade name	Crystallinity class[a]	Gas	T (°C)	P[b] (barrer)	E_P (kcal/mole)
Cellulose triacetate (43% acetyl)	Kodapak IV F404	III	O_2	25	1.0	
			N_2	25	0.17	
			CO_2	25	5.7	
Polyethylene/graft styrene (branched polyethylene) 4.8% styrene			N_2	30	1.5	11.5
			CO_2	30	23	8.6
13.4			N_2	30	1.35	11.3
			CO_2	30	18	8.4
20.9			N_2	30	0.92	10.4
			CO_2	30	15	8.4
34.4			N_2	30	0.97	9.6
			CO_2	30	16	6.7
41.3			N_2	30	1.05	9.4
			CO_2	30	20	6.0
Polyethylene/graft acrylonitrile (branched polyethylene) 1.8% acrylonitrile			N_2	30	1.73	11.4
			CO_2	30	25.4	8.6
4.1			N_2	30	1.57	11.7
			CO_2	30	20	8.8

			Gas	Temp (°C)		
		4.3	N_2	30	1.26	11.5
			CO_2	30	17.1	8.7
		20.8	N_2	30	1.06	12.0
			CO_2	30	13.6	9.0
		31.3	N_2	30	0.71	11.9
			CO_2	30	9.7	8.9
Polyethylene, branched, molded, $\rho = 0.922$	II		H_2S	30	43	10.0
			NH_3	20	33	8.0
			SO_2	30	39	10.6
			C_2H_4O	0	0.49	($p = 33$ cm Hg)
			Freon 12	20	1.1	
Irradiated, $\rho = 0.92$ Cryovac L200	II		O_2	23	3.5	
			N_2	23	0.6	
			CO_2	23	15.5	
Biaxially oriented Cryovac D 925	II		O_2	23	1.0	
			N_2	23	0.29	
			CO_2	23	4.6	
Polyester polyurethane Estane 5701	III		He	23	4.9	
			O_2	23	1.24	
			N_2	23	0.44	
			CO_2	23	9.6	
Estane 5702	III		He	23	6.7	10.9
			O_2	23	2.0	12.2
			N_2	23	0.71	9.4
			CO_2	23	21	

APPENDIX TABLE 2 (continued)

Polymer	Trade name	Crystallinity class[a]	Gas	T (°C)	$\bar{P}$[b] (barrer)	$E_{\bar{P}}$ (kcal/mole)
	Estane 5740 X070	III	He	23	2.2	
			O$_2$	23	0.66	
			N$_2$	23	0.24	
			CO$_2$	23	2.8	
	Adiprene L-100		O$_2$	29	4.8	
			CO$_2$	29	40	
			Freon 12	25	60	
			Freon 22	25	1900	
Polyethylene, linear, blown film ($\rho = 0.953$, $\alpha = 0.31$)	Grex	II	He	30	1.9	6.9
			O$_2$	30	0.78	8.0
			N$_2$	30	0.25	9.9
			CO$_2$	30	3.3	6.9
			Freon 12	20	0.46	
Machine cast ($\rho = 0.943$, $\alpha = 0.38$)	Grex	II	He	30	3.4	6.9
			O$_2$	30	1.5	7.9
			N$_2$	30	0.47	9.7
			CO$_2$	30	7.0	6.9
Polyvinyl chloride/ acrylonitrile (60/40)		III	O$_2$	23	0.66	
			CO$_2$	23	2.0	
Polybutadiene/acrylonitrile (60/40)	Nitrile rubber	III	O$_2$	23	0.6	

Material	Trade name		Gas	Temp (°C)		
Cellulose nitrate		III	He	25	6.9	5.2
			H_2	25	2.0	5.7
			O_2	25	0.6	
			N_2	25	0.12	
			CO_2	20	1.5	
Methyl cellulose	Methocel	II	O_2	25	0.50	
			N_2	25	0.17	
			CO_2	25	2.6	
Polyethylene oxide	Carbowax Polyox	I	O_2	25	0.46	
Polyepichlorohydrin	Hydrin 100	III	O_2	25	0.42	
Polyvinyl acetate	Gelva V100	III	O_2	25	0.36	
			N_2	25	0.06	
Polyvinylchloride/vinyl acetate	VYGH and VYHH	III				
(25/75)			O_2	25	0.78	
(50/50)			O_2	25	0.36	
(60/40)			He	25	0.85	7.2
			H_2	25	0.92	7.4
			O_2	25	0.24	9.7
			N_2	25	0.06	12.3
			CO_2	25	1.6	8.2
(68/32)			O_2	25	0.15	

APPENDIX TABLE 2 (continued)

Polymer	Trade name	Crystallinity class[a]	Gas	T (°C)	$\bar{P}$[b] (barrer)	$E_{\bar{P}}$ (kcal/mole)
(80/20)			O_2	25	0.12	
Poly-p-xylylene	Parylene N	I	O_2	23	0.18	
			N_2	23	0.05	
			CO_2	23	1.35	
Polychloro-p-xylylene	Parlene C	III	O_2	23	0.03	
			N_2	23	0.004	
			CO_2	23	0.07	
Polymethyl methacrylate/styrene	Zerlon 150 "XT"	III	O_2	25	0.18	7.4
			N_2	25	0.020	7.8
Polymethyl methacrylate	Lucite	III	O_2	25	0.102	
			N_2	25	0.023	
Polyvinyl acetal (20% OH)		II	He	20	2.5	
			CO_2	20	0.52	
Polyoxymethylene (polyacetal)	Delrin	I	O_2	25	0.09 (0.05)	
			N_2	25	0.024	
			CO_2	25	0.45	
Polychlomethyloxetane	Penton	II	O_2	25	0.072	

Polymer	Trade name		Gas	Temp (°C)	Permeability	
Poly-4,4'-diphenylether pyromellitimide	Kapton (formerly "H Film")	III	H₂	25	1.5	
			O₂	25	0.15 (0.07)	
			N₂	25	0.036	
			CO₂	25	0.27	
Polyvinyl chloride (rigid)	Geon 101 EP	III	O₂	30	0.05 (0.10)	
			N₂	30	0.02 (0.04)	
			A	18	0.0033	
			CO₂	30	0.24 (1.0)	
			H₂S	20	0.2	
			NH₃	20	4.9	
Polyvinyl chloride, plasticized, (31% DOP)	Geon 202	III	He	30	1.4	
			H₂	30	2.0	14.0
			N₂	30	0.17	
			O₂	30	0.60	
			CO₂	30	3.1	13.0
			CH₄	30	0.2	
Polyvinyl chloride (plasticized)	Prime Wrap	III	O₂	25	5.8	
			He	20	1.3	
Polyvinyl formal (7.4% OH)		III	CO₂	20	0.16	
Polyvinyl formal/vinyl alcohol-acetate	Formvar	III	O₂	25	0.028	
Rubber hydrochloride	Pliofilm NO	II	N₂	25	0.0081	
			O₂	30	0.025	
			CO₂	25	0.13	8.6
			H₂S	30	0.10	17.6

APPENDIX TABLE 2 (*continued*)

Polymer	Trade name	Crystallinity class[a]	Gas	T (°C)	$\bar{P}$[b] (barrer)	$E_{\bar{P}}$ (kcal/mole)
Rubber hydrochloride (plasticized)	Pliofilm FM	II	H_2	25	1.6	5.6
			O_2	25	0.43	8.4
			N_2	25	0.106	10.0
			CO_2	25	1.02	8.6
Polyvinylidene fluoride	Kynar	I	H_2	25	0.41	
			O_2	25	0.04	
			CO_2	25	0.36	
Polychlorotrifluoroethylene extruded film (2–5 mils)	Kel F–300	I	O_2	25	0.02	10.6
			N_2	25	0.0006	13.7
			CO_2	25	0.072	11.2
molded, quenched ($\alpha \simeq 0.7$)		II	O_2	25	0.04	11.2
			N_2	25	0.01	14.3
			CO_2	25	0.078	11.8
molded, annealed ($\alpha \simeq 0.2$)		I	O_2	25	0.01	10.9
			N_2	25	0.003	11.9
			CO_2	25	0.018	11.1
Polychlorotrifluoroethylene/ vinylidene fluoride (97/3)	Kel F X500	I	O_2	25	0.02	13.8
			N_2	25	0.01	13.6
			CO_2	25	0.08	13.1

			Gas	Temp.		
(75/25)	Kel F X800	II	N_2	25	0.03	15.5
			CO_2	25	0.18	15.1
(68/32)	Kel F X3700	II	O_2	25	0.42	12.8
			N_2	25	0.16	13.5
			CO_2	25	2.7	11.9
(50/50)	Kel F X5500	II	O_2	25	0.93	13.1
			CO_2	25	4.0	13.9
(20/80)	—	II	O_2	25	0.57	
	Aclar 33 C	I	O_2	25	0.042	
			CO_2	25	0.096	
	Aclar 22 C	II	H_2	33	4.7	
			O_2	25	0.072	
			N_2	25	0.015	
			CO_2	25	0.18	
	Aclar 22 A	II	O_2	25	0.089	
			N_2	25	0.015	
			CO_2	25	0.24	
	Trothene B	II	He	25	0.34	
			N_2	25	0.012	
			CH_4	25	0.0084	

119

APPENDIX TABLE 2 (*continued*)

Polymer	Trade name	Crystallinity class[a]	Gas	T (°C)	$\bar{P}$[b] (barrer)	$\bar{E}_{\bar{P}}$ (kcal/mole)
Polyethylene terephthalate	Mylar	I	He	25	1.1	4.6
			H_2	25	0.6	5.5
			O_2	25	0.030	6.4
			N_2	25	0.005	7.5
			CO_2	25	0.10	6.2
			H_2S	25	0.058	7.4
			Freon 12	20	0.04	
Poly-1,4-cyclohexylene-dimethylene terephthalate		II	O_2	25	0.116	
Phenoxies		III				
Polybisphenol-L/epichlorohydrin			O_2	25	0.45	
Polybisphenol-V/epichlorohydrin			O_2	25	0.09	
Polybisphenol-ACP/epichlorohydrin			O_2	25	0.072	
Polybisphenol-A/epichlorohydrin	Phenoxy		O_2	25	0.030	
			N_2	25	0.020	
			CO_2	25	0.070	
Polytetrachlorobisphenol-A/epichlorohydrin			O_2	25	0.024	
Polybisphenol-F/epichlorohydrin			O_2	25	0.024	

Polydichlorobisphenol-A/epichlorohydrin		O_2	25	0.018	
Polybisphenol sulfone/epichlorohydrin		O_2	25	0.012	
Polyhydroquinone/epichlorohydrin		O_2	25	0.003	
Polyamides					
Molded sheets (annealed)					
Polyaminododecanoic acid	Nylon 12	O_2	25	0.36	
Polyaminoundecanoic acid	Nylon 11 (Rilsan)	O_2	25	0.12	
		N_2	25	0.03	
		CO_2	25	0.06	
Polymethoxymethylhexamethylene adipamide	Nylon 8	O_2	25	0.058	
Polyhexamethylene azelamide	Nylon 6, 9	O_2	25	0.035	
Polyhexamethylene adipamide	Nylon 6, 6 Chemstrand Blue C	H_2	25	1.0	8.1
		O_2	25	0.034	10.4
		N_2	25	0.008	11.2
		CO_2	25	0.17	9.7
Polyhexamethylene sebacamide	Nylon 6, 10 Chemstrand Blue C 10V	O_2	25	0.02	
		CO_2	25	0.074	
Polyhexamethylene adipamide/hexamethylene sebacamide	Zytel 2345	O_2	25	0.015	
Poly-ε-caprolactam	Nylon 6	H_2	25	0.54	
		O_2	25	0.0060	
		N_2	25	0.0015	
		CO_2	25	0.021	

APPENDIX TABLE 2 *(continued)*

Polymer	Trade name	Crystallinity class[a]	Gas	T (°C)	$\bar{P}$[b] (barrer)	$E_{\bar{P}}$ (kcal/mole)
Films						
Poly-ε-caprolactam, quenched, oriented	Capran 77C	II	O_2	25	0.04	
			N_2	25	0.02	
			CO_2	25	0.18	
quenched		II	H_2	30	0.94	
			O_2	30	0.083	
			N_2	30	0.11	
			CO_2	30	0.42	
Epoxy (amine-cured)	LA 16		O_2	25	0.054	
Epoxy (amine-cured)	Epon 1001		O_2	29	0.049	
			CO_2	29	0.086	
Polyvinyl fluoride, highly oriented film	Tedlar	II	He	25	0.97 (1.8)	
			H_2	25	0.36	
			O_2	25	0.02	
			N_2	25	0.004 (0.02)	
			CO_2	25	0.09	
			CH_4	25	0.0065	
molded			O_2	25	0.2	
			N_2	25	0.04	
Polyvinylidene chloride		I	O_2	25	0.0020	
			N_2	25	0.0010	
			CO_2	25	0.012	

Polyvinylidene chloride/vinyl chloride					
(90/10)	Saran	I	He	25	0.066	
			H$_2$	25	0.076	1.59
			O$_2$	25	0.0042	16.8
			N$_2$	25	0.001	12.3
			CO$_2$	25	0.029	
			CH$_4$	25	0.00025	
			H$_2$S	30	0.03	17.8
(85/15) (unplasticized)		II	He	34	0.3	
			O$_2$	25	0.012	
			CO$_2$	34	0.06	
(85/15) (plasticized)		II	CO$_2$	20	0.16	
			Freon 12	20	0.018	
(70/30)		II	O$_2$	25	0.0108	
(50/50)		III	O$_2$	25	0.030	
	Saran 517	I	O$_2$	30	0.0051	15.9
			N$_2$	30	0.00094	16.8
			CO$_2$	30	0.029	12.3
			H$_2$S	30	0.03	17.8
			C$_2$H$_4$O	30	8.0	
	Saran 7	II	He	25	0.24	
			O$_2$	25	0.007	
			N$_2$	25	0.0009	
			CO$_2$	25	0.022	
	Saran 26	III	He	25	0.55	
			O$_2$	25	0.024	
			N$_2$	25	0.007	
			CO$_2$	25	0.11	

APPENDIX TABLE 2 (*continued*)

Polymer	Trade name	Crystallinity class[a]	Gas	T (°C)	$\bar{P}$[b] (barrer)	E_P (kcal/mole)
Polyvinylidene chloride/ acrylonitrile (80/20)	Saran Wrap	II	Freon 12	25	0.018	
			O_2	25	0.0078	
	Saran	II	O_2	25	0.0042	
			N_2	25	0.00050	
			CO_2	25	0.0106	
(60/40)		III	O_2	25	0.021	
			N_2	25	0.0025	
			CO_2	25	0.047	
Polyacrylonitrile		II	O_2	25	0.00022	
Regenerated Cellulose 15% glycerol (dry cellulose basis)	Cellophane		O_2	25	0.0003–0.006	
			N_2	25	0.0002	
			CO_2	25	0.003	
Polyvinyl alcohol (bone dry)		I	O_2	25	0.000001	
			CO_2	25	0.00001	

[a] I: highly crystalline ($>$ 50 volume %). II: low to moderate crystallinity ($<$50%). III: completely amorphous ($<$20% crystallinity).
[b] Barrer: [ml (STP)-cm/cm²-sec-cm Hg] $\times 10^{10}$.

APPENDIX TABLE 3

Permeabilities of Moisture-Sensitive Films[a]

Polymer	Trade name	Gas	T (°C)	% R.H.[b]	$\bar{P}$ (barrer)[c]
Poly-ε-caprolactam	Nylon 6	CO_2	30	0	0.1
				44	0.17
				95	0.29
Polyvinyl alcohol	Vinol 125	O_2	25	0	ca. 0.000001
		O_2	23	50	0.0018
		CO_2	23	75	0.042
		CO_2	23	84	0.052
		O_2	23	90	0.11
		CO_2	23	90	1.3
		CO_2	23	94	11.9
Regenerated cellulose (unsoftened)		O_2	24	48	0.0058
		O_2	24	87	0.066
		O_2	24	100	7.0
7% glycerol (dry cellulose basis)	Cellophane	O_2	24	48	0.0043
		O_2	24	87	0.13
12% glycerol (dry cellulose basis)	Cellophane	O_2	24	48	0.0039
		O_2	24	87	0.23

APPENDIX TABLE 3 (continued)

Polymer	Trade name	Gas	T (°C)	% R.H.[b]	$\bar{P}$ (barrer)[c]
15% glycerol (dry cellulose basis)	Cellophane	O_2	20	0	0.00014
		N_2	20	0	0.00014
		CO_2	20	0	0.0010
		O_2	20	32	0.00077
		N_2	20	32	0.00027
		CO_2	20	32	0.014
		O_2	20	54	0.0045
		N_2	20	54	0.0011
		CO_2	20	54	0.093
		O_2	20	76	0.055
		N_2	20	76	0.011
		CO_2	20	76	1.7
		O_2	20	92	0.58
		N_2	20	92	0.134
		CO_2	20	92	15.0
21% glycerol (dry cellulose basis)	Cellophane	CO_2	24	0	0.001
		O_2	24	31	0.0005
		CO_2	24	31	0.017

O_2	24	46	0.0035
CO_2	24	46	0.15
O_2	24	83	0.4
CO_2	24	83	14
O_2	24	93	2.0
CO_2	24	93	59
O_2	24	100	6.0

[a] Accuracy of low-humidity data is questionable. Permeabilities may be time-dependent over a period of many days.
[b] Percent relative humidity. Data selected where relative humidity is approximately constant on both sides of the film.
[c] $\bar{P}$ calculated by using dry film thicknesses which in all cases were 0.9 to 1.0 mil.

References

1. H. A. Daynes, *Proc. Roy. Soc. (London)*, **A97**, 286 (1920) et seq.
2. R. M. Barrer, *Diffusion in and Through Solids*, Macmillan, Cambridge University Press, 1941.
3. G. J. van Amerongen, *Rubber Chem. Technol*, **37**, 1065 (1964). A review with 312 references.
4. O. Kedem and A. Katchalsky, *Biochim. Biophys. Acta*, **27**, 229 (1958).
5. A. Katchalsky and P. F. Curran, *Nonequilibrium Thermodynamics in Biophysics*, Oxford University Press, London, 1966.
6. N. Lakshminarayanaiah, *Chem. Rev.*, **65**, 491 (1965). A review with 609 references.
7. V. Stannett, M. Szwarc, R. L. Bhargava, J. A. Meyer, A. W. Myers, and C. E. Rogers, *Permeability of Plastic Films and Coated Papers to Gases and Vapors*, Tappi Monograph Series No. 23, New York, 1962.
8. C. E. Rogers, "Permeability and Chemical Resistance," in E. Baer, *Engineering Design for Plastics*, Reinhold, New York, 1964, Chapter 9.
9. H. Yasuda, "Permeability Constants," in J. Brandrup and E. H. Immergut, *Polymer Handbook*, Wiley-Interscience, New York, 1966, Sect. V–13.
10. J. Crank and G. S. Park, *Diffusion in Polymers*, Academic, New York, 1968.
11. T. Graham, *Trans. Roy. Soc. (London)*, **156**, 399 (1866).
12. A. Fick, *Ann. Phys. Leipzig*, **170**, 59 (1855).
13. S. von Wroblewski, *Ann. Phys. Chem.*, **8**, 29 (1879).
14. W. Jost, *Diffusion, Methoden der Messung u. Auswertung*, Fortschritte der physikalischen Chemie, **Vol. 1**, Steinkopf, Darmstadt, 1957.
15. W. R. R. Park, *Anal. Chem.*, **29**, 1897 (1957).
16. 1968 *ASTM Standards*, Am. Soc. for Testing and Materials, Philadelphia, Part 27, Method D 1434.
17. S. A. Stern, P. J. Garies, T. F. Sinclair, and P. H. Mohr, *J. Appl. Polymer Sci.*, **7**, 2035 (1963).
18. N. N. Li and E. J. Henley, *A. I. Ch. E. J.*, **10**, 666 (1964).
19. H. Yasuda, V. Stannett, H. L. Frisch, and A. Peterlin, *Makromol. Chem.*, **73**, 188 (1964).
20. A. E. Sickinger, *TARA*, **172**, (1963).
21. T. L. Caskey, *Mod. Plastics*, **45**, No. 4, 148 (Dec. 1967).
22. A. S. Michaels and R. B. Parker, Jr., *J. Phys. Chem.*, **62**, 1604 (1958).
23. R. M. Barrer, *Trans. Faraday Soc.*, **35**, 628 (1939).
24. R. M. Barrer, *Trans. Faraday Soc.*, **35**, 644 (1939).
25. J. Crank, *The Mathematics of Diffusion*, Clarendon, Oxford, 1956 (1964).
26. W. Jost, *Diffusion in Solids, Liquids, Gases*, Academic, New York, 1960.
27. A. S. Michaels and H. J. Bixler, *J. Polymer Sci.*, **50**, 393 (1961).
28. A. S. Michaels and H. J. Bixler, *J. Polymer Sci.*, **50**, 413 (1961).
29. H. J. Bixler, A. S. Michaels, and R. B. Parker, Jr., *Rev. Sci. Instr.*, **31**, 1155 (1960).
30. D. R. Paul and A. T. Di Benedetto, *J. Polymer Sci.*, **C**, No. 10, 17 (1965).
31. V. Stannett and H. Yasuda, "The Measurement of Gas and Vapor Permeation and Diffusion in Polymers," in J. V. Schmitz and W. E. Brown, Eds., *Testing of Polymers*, Wiley-Interscience, New York, 1967, **Vol. II**, Chapter 13, p. 393.

32. J. Weiner, S. Wilkinson, and C. L. Brown, *Permeability of Organic Materials to Gases and Vapors*, Bibliography Ser. No. 169 (June 1948), Suppl. I (April 1956), Suppl. II (1964), Institute of Paper Chemistry, Appleton, Wis.
33. A. A. Taylor, M. Karel, and B. E. Proctor, *Mod. Packaging*, **33**, No. 10, 131 (June 1960).
34. J. H. Hildebrand and R. L. Scott, *Solubility of Nonelectrolytes*, Reinhold, New York, 1950.
35. J. E. Jolley and J. H. Hildebrand, *J. Am. Chem. Soc.*, **80**, 1050 (1958).
36. A. S. Michaels and H. J. Bixler, *J. Polymer Sci.*, **50**, 393 (1961).
37. C. E. Rogers, V. Stannett, and M. Szwarc, *J. Phys. Chem.*, **63**, 1406 (1959).
38. J. Horiuti, *Sci. Papers, Inst. Phys. Chem. Research, Tokio*, No. 341, **17**, 125 (1931).
39. G. J. van Amerongen, *J. Polymer Sci.*, **5**, 207 (1950).
40. T. K. Kwei and W. M. Arnheim, *J. Polymer Sci.*, **A2**, 1873 (1964).
41. S. A. Stern, J. T. Mullhaupt, and P. J. Gareis, *A. I. Ch. E. J.*, **15**, 64 (1969).
42. H. Burrell and B. Immergut, "Solubility Parameter Values," in J. Brandrup and E. H. Immergut, Eds., *Polymer Handbook*, Wiley-Interscience, New York, 1966, Sect. IV-341.
43. A. S. Michaels and R. B. Parker, *J. Polymer Sci.*, **41**, 53 (1959).
44. H. A. Bent, *J. Polymer Sci.*, **24**, 387 (1957).
45. C. H. Klute, *J. Appl. Polymer Sci.*, **1**, 340 (1959).
46. A. S. Michaels, W. R. Vieth, and J. A. Barrie, *J. Appl. Phys.*, **34**, 1 (1963).
47. W. F. Wuerth, Sc. D. Thesis, Department of Chemical Engineering, Massachusetts Institute of Technology, Cambridge, 1967.
48. D. Jeschke and H. A. Stuart, *Z. Naturforsch.*, **16a**, 37 (1961).
49. A. S. Michaels, H. J. Bixler, and H. L. Fein, *J. Appl. Phys.*, **35**, 3165 (1964).
50. C. L. Beatty, M. S. Thesis, Engineering Division, Case Western Reserve University, Cleveland, Ohio, 1968.
51. P. Meares, *J. Am. Chem. Soc.*, **76**, 3415 (1954).
52. R. J. Kokes, F. A. Long, and J. L. Hoard, *J. Chem. Phys.*, **20**, 1711 (1951).
53. F. A. Long and R. J. Kokes, *J. Am. Chem. Soc.*, **75**, 2232 (1953).
54. F. J. Norton, *J. Appl. Polymer Sci.*, **7**, 1649 (1963).
55. W. R. Vieth, H. H. Alcalay, and A. J. Frabetti, *J. Appl. Polymer Sci.*, **8**, 2125 (1964).
56. W. R. Vieth, P. M. Tam, and A. S. Michaels, *J. Colloid Interface Sci.*, **22**, 360 (1966).
57. C. A. Kumins and J. Roteman, *J. Polymer Sci.*, **55**, 683 (1961).
58. R. M. Barrer, R. Mallinder, and P. S-L. Wong, *Polymer*, **8**, 321 (1967).
59. B. Takeda and B. Yamaguchi, *Kogyo Kagaku Zasshi*, **62**, 321, 1897 (1959).
60. G. J. van Amerongen, *Rubber Chem. Technol.*, **8**, 821 (1955).
61. P. Y. Hsieh, *J. Appl. Polymer Sci.*, **7**, 1743 (1963).
62. H. J. Bixler, A. S. Michaels, and M. Salame, *J. Polymer Sci.*, **A-1**, 895 (1963).
63. R. M. Barrer and G. Skirrow, *J. Polymer Sci.*, **3**, 549 (1948).
64. W. W. Brandt, *J. Phys. Chem.*, **63**, 1080 (1959).
65. S. Glasstone, K. J. Laidler, and H. Eyring, *The Theory of Rate Processes*, McGraw-Hill, New York, 1941.
66. K. Chmutov and E. Finkel, *Zhur. Fiz. Khim.*, **33**, 93 (1959).
67. W. R. Vieth and K. J. Sladek, *J. Colloid Sci.*, **20**, 9 (1965).

68. W. R. Vieth, C. S. Frangoulis, and J. A. Rionda, Jr., *J. Colloid Interface Sci.*, **22**, 454 (1966).
69. P. Meares, *Trans. Faraday Soc.*, **53**, 101 (1957).
70. P. Meares, *Trans. Faraday Soc.*, **54**, 40 (1958).
71. A. S. Michaels, R. F. Baddour, H. J. Bixler, and C. Y. Choo, *IEC Proc. Design Dev.*, **1**, 14 (1962).
72. R. F. Baddour, A. S. Michaels, H. J. Bixler, R. P. De Filippi, and J. A. Barrie, *J. Appl. Polymer Sci.*, **8**, 897 (1964).
73. R. Chiang and P. J. Flory, *J. Am. Chem. Soc.*, **83**, 2857 (1961).
74. R. M. Barrer and R. R. Fergusson, *Trans. Faraday Soc.*, **54**, 989 (1958).
75. J. H. Hildebrand and R. L. Scott, *Regular Solutions*, Prentice-Hall, Englewood Cliffs, N. J., 1962.
76. P. J. Flory, *Principles of Polymer Chemistry*, Cornell University Press, Ithaca, N.Y., 1953.
77. J. Hoffman and J. Lauritzen, *J. Res. Natl. Bur. Stds.*, **65A**, 297 (1961).
78. D. W. McCall, *J. Polymer Sci.*, **26**, 151 (1957).
79. D. W. McCall and W. P. Slichter, *J. Am. Chem. Soc.*, **80**, 1861 (1958).
80. C. E. Rogers, V. Stannett, and M. Szwarc, *J. Polymer Sci.*, **45**, 61 (1960).
81. P. Meares, *J. Polymer Sci.*, **27**, 405 (1958).
82. G. S. Park, *Trans. Faraday Soc.*, **48**, 11 (1952).

CHAPTER 2

POLYETHYLENE

ORVILLE J. SWEETING*

Yale University, New Haven, Connecticut

* Formerly Associate Director of Research, Film Division, Olin Corporation, New Haven, Connecticut.   Present address: Quinnipiac College, New Haven 06518.

I. Introduction

Long considered unpolymerizable, ethylene was first converted to solid polymers in March 1933 by Fawcett and Gibson, who were working in the laboratories of Imperial Chemical Industries, Ltd., at Winnington [1]. As a part of an extensive study of the effect of high pressures on chemical reactions, benzaldehyde dissolved in ethylene had been subjected to pressures up to 1400 atm and temperatures up to 170°C without apparent reaction. When the autoclave was opened, however, a white waxy solid was found on the inner walls. The benzaldehyde was recovered unchanged and the solid product appeared to be a polymer of ethylene. Repetition of the experiment with ethylene gave only trifling amounts of solid product, yet violent explosions occurred when the pressure was increased, and the experiments were abandoned.

After an interval during which other research efforts were pressed, experimentation with ethylene at high pressure was resumed, and in December 1935 polyethylene was rediscovered [2–4]. Further experiments showed that explosions could be prevented by close control of minute amounts of oxygen in the ethylene. Formidable problems of design of high-pressure equipment were solved and by the end of 1938 a ton of polymer had been made. The potential industrial value of the polymer was recognized, and I.C.I. registered the trade mark *Alketh* (later *Alkathene*) for their material. The electrical and mechanical properties of polyethylene made it a preferred molding material for applications in electronics, especially radar, and with the outbreak of World War II the supply was pre-empted for military use.

An extensive review of the properties of early high-pressure polyethylenes was published in 1957 [5] at about the same time as Raff and

Allison's treatise appeared [6].   Since then several other excellent books and shorter reviews have appeared [6a–15] which may be consulted for more detailed discussion of certain aspects of the subject than is possible in this chapter.

Since the introduction of polyethylene as a packaging film in about 1946, the use of this inexpensive clear film has grown to approximately one billion pounds in 1970, valued at about $600 million [16].   This single material (though perhaps it would be more accurately viewed as many *polyethylenes*) accounts for about 60% of the packaging film market at the present time, or more than twice as much as the cellophane poundage used.   Seven others, polypropylene, polystyrene, pliofilm, vinyl polymers, polyvinylidene chloride, polyesters, and cellulose acetate, share the remaining market; only one of them (vinyl) has as much as 10% of the market.

For convenience and logically, but with a measure of arbitrariness, polyethylene film is classified as having low, medium, or high density (Table 1).

TABLE 1

Properties of Polyethylenes

	Low density	Medium density	High density
Specific gravity	0.910–0.925	0.926–0.940	0.941–0.965
Clarity	Transparent to	Transparent to	Translucent to
	translucent	translucent	opaque
Heat-Seal, °F.	250–350	260–310	275–310

Low-density film, made by the high-pressure processes described later, is relatively inexpensive, easily heat-sealed, and not optically clear; it has a balance of strength and optical properties and finds its greatest use in the manufacture of bags for bulk packaging.   Thousands of products ranging from peanuts to hardware are packaged in low-density polyethylene.

Medium-density film is used chiefly as overwrapping film for packages filled on high-speed machines (up to 500/min.).   All sorts of baked goods, confectionery, toilet articles, and dry goods are wrapped on these machines, and the film must be stiff enough to be handled at high speed, yet seal well, cut well, and have a nice balance of stiffness and clarity.

High-density film is used for special purposes. The clarity is poor, the heat-seal temperature range is narrower than with low- or medium-density films, rigidity is high, and tear strength is low, all of which add up to special applications. Grease resistance and heat resistance are exploited, for instance, in the use of this film for boilable packages (e.g., vegetables) and the added strength is advantageous for use as strong shipping sacks and drum liners.

II. Production of Ethylene

Except for small amounts obtained as a by-product in refinery cracking operation, ethylene is produced by pyrolysis of selected hydrocarbon fractions in petroleum refining. Alumina catalysts at 340 to 400°C crack and dehydrogenate feed materials to produce ethylene of very high purity in 90 to 95% yields. Though in principle any hydrocarbon may be used, ethane, available in vast amounts from natural gas and refinery waste gas, is the most economical feed. Catalysts last longer than with other raw materials of higher molecular complexity, and ethylene is the principal product, thus making purification a simpler operation. A great deal of propane and butane is also cracked, for in recent years propylene has become an important raw material in the manufacture of C_3 chemicals and for increasing the output of polypropylene (Chapter 3).

Ethylene for polymer manufacture must be of high purity, at least 99% and commonly 99.9%. The I.C.I. high-pressure process requires a feed of the latter purity, though it is said that the low-pressure processes can function satisfactorily on ethylene of slightly lower purity, provided that chain-transfer and chain-stopping agents are absent. The maximum amount of methane plus ethane permissible is 0.1%, of propylene, 30 ppm, and of oxygen, 5 ppm. Despite these stringent specifications, the manufacturing cost of ethylene falls in the range of 2 to 3.5 cents a pound over-all. The low cost of the monomer, coupled with the desirable properties of the multitude of polyethylenes possible by variations in polymerization techniques, has persuaded more than 70 companies to manufacture polyethylene.

III. Polymerization of Ethylene

A. Early Work

Polymethylene, $(-CH_2-)_n$, was probably the first synthetic hydrocarbon high polymer made; it was produced from diazomethane in 1898 to 1900 and attracted little interest [18, 19]. The first hydrocarbon polymer that resembled present-day high-pressure polyethylenes was very likely the solid obtained by Lind and Glockler [20], when they subjected ethane to a semicorona discharge. These investigators did not characterize the product as polyethylene, however, and the experiment was forgotten.

Most of the early work on ethylene polymerization was directed toward the production of gasoline or lubricants by high pressures in the presence of various free-radical generators, with or without ultraviolet light or photosensitizing agents. As early as 1927 ethylene and its homologs had been converted to oils by boron fluoride catalysts [21, 22]. Liquid mixtures of open-chain, cyclic, and aromatic hydrocarbons were formed by polymerizing ethylene at high pressures in the presence of phosphoric acid or anhydride at temperatures up to 800°C (23–28); other catalysts included air [29], oxygen [30], aluminum chloride [31], nickel [32], and titanic acid [33]. At room temperature olefins of low molecular weight were formed and stabilized by hydrogenation [23]. The explosive decomposition and polymerization of ethylene to oils at 50 atm and 380°C was reported [34], and it appears to be one of the unlucky chances of chemistry that polyethylenes of high molecular weight were not discovered in the 1920s.

McDonald and Norrish [35] found in 1936 that when ethylene was irradiated at pressures below 1 mm with light from a hydrogen discharge tube a solid polymer was deposited. This discovery led to much photochemical work of great academic interest but little practical value. The formation of aluminum alkyls during polymerization was noted [36], but the discoverers failed to realize the significance of this finding, which might be considered the forerunner of the later commercially successful work of Ziegler and Natta. Catalytic polymerization at prevailing atmospheric pressures with cobalt and iron catalysts containing promoters was investigated extensively by Russian and Japanese chemists [37–40], but the results were poor and the products were of low molecular weight. Experiments at pressures up to 1800 atm gave no better results, and the work was abandoned [41, 42].

B. HIGH-PRESSURE POLYMERIZATION

Until about 1956 all useful polymerization processes for preparing solid polymers of ethylene were based on the fundamental discovery of chemists at Imperial Chemical Industries that ethylene could be converted to solid polymers in the presence of small concentrations of oxygen, organic peroxides, or other suitable catalysts by the application of high pressures [1–4].

Conventional high-pressure polyethylene has been made in the United States since World War II, principally by the Bakelite Division of Union Carbide and Carbon Chemicals Corporation and by E. I. du Pont de Nemours and Company under license from Imperial Chemical Industries. Imperial Chemical Industries controlled the basic patents in the United States, which covered in a comprehensive way polymers consisting essentially of $-CH_2-$ groups, melting in the range 100 to 120°C, characterized by a crystalline structure as revealed in x-ray analysis, and having a molecular weight about 6000. These patents have now expired, and therefore several other companies have entered the field with their own variations of high-pressure processes.

Solid polymers are obtained by mixing ethylene with approximately 0.01 to 5 % oxygen, compressing the mixture to at least 500 atm pressure, and heating in a well-stirred autoclave at 200°C or above [2, 4]. Upon release of the pressure a solid polymer separates, and the unpolymerized ethylene may be recycled or sent to an adjoining process to be made into ethylene oxide and its derivatives. It is of the highest importance that the incoming ethylene stream be free from impurities, particularly acetylene (normally 99.9 % ethylene is required). Usually the amount of oxygen is controlled by reducing the oxygen content of the incoming ethylene to 0.001 % and adding the required amount.

It was found that increasing the pressure increased the molecular weight of the product and accelerated polymerization and that increasing the temperature accelerated polymerization but gave a product of lower molecular weight [4]. Increasing the oxygen content also reduced the molecular weight, presumably by increasing the number of free-radical centers at which polymerization could start. The optimum conditions were difficult to define but seemed to be 1500 atm pressure, 0.03 to 0.10% oxygen, and a temperature of 190 to 210°C. Conversion varied from 6% at 0.01% oxygen to 25% at 0.13% oxygen.

C. Low-pressure Polymerization

In the early 1950s the polymerization of ethylene was accomplished at 1 atm and somewhat higher pressures by Karl Ziegler and his associates at the Max Planck Institute for Coal Research at Mülheim am Ruhr. One of these processes utilized triethylaluminum and titanium(IV) chloride in an elegantly simple laboratory procedure [43–45].

Holzkamp, Breil, and Martin were graduate students who worked with Ziegler over a period of years, beginning about 1952, in experiments with aluminum alkyls and mixed catalyst systems. Ziegler remarks owlishly in one of the papers [45]:

> As we were led from one discovery to the next one, we did not think it was right to collect all of the work in one single master's thesis. Thus we split up the work so that Mr. Breil did mostly the scientific part of the work, and Dr. Martin . . . took care of the technical aspects. He found out within a few days that certain highly active types of catalysts, especially this one with titanium, made possible a polymerization at rather low pressures. When we succeeded at 10–20 atmospheres, we were exceedingly triumphant, and we did an experiment then at 1 atmosphere, simply because it had to appear somewhere in the theses. We never doubted that this experiment was just a waste of time. When we got polymerization, we thought we had done something wrong! But it was the right thing all right, as we found again and again, when we watched the faces of visitors to whom we showed our polymerization at ordinary pressures.

A remarkable discovery was the effect of traces of other metals. Polymerization-active catalysts were not obtained in general if the reaction between the aluminum alkyl and heavy metal ion present in traces resulted in the free colloidal metal. Addition of nickel chloride, for example, yields free nickel, and this combination with triethylaluminum leads to dimers of olefins only [45, 46]; a similar result had usually occurred in previous years in exhaustive studies by other chemists with metal, metal-oxide, and organometallic catalysts [5]. Some metals showed no activity whatsoever.

The most active catalysts were found to be triethylaluminum with a few hundredths of a per cent of added titanium(IV) chloride. The effect of such an amount of titanium is marked, but zirconium and almost any of the elements of the fourth, fifth, and sixth groups of the periodic system, including thorium and uranium, are effective. Diethyl-aluminum chloride does not polymerize ethylene even at 100°C and 100 atm pressure, but if a trace of titanium(IV) chloride is added poly-merization proceeds exothermically at normal pressure [45, 47].

The chemistry of the formation of the catalyst follows from the relative electronegativities of aluminum, 1.6, and titanium(IV), 1.5, in a metathetical reaction

$$R_3Al + TiCl_4 \longrightarrow R_2AlCl + RTiCl_3.$$

Further substitution does not occur, for alkyltitanium halides are unstable (as are the alkyl derivatives of the transition metals generally). The next reaction to occur is decomposition

$$RTiCl_3 \longrightarrow R + TiCl_3.$$

Titanium(III) chloride is a stable solid that precipitates. The free alkyl radicals disappear rapidly and they play no part in the polymerization reaction that follows.

Titanium(III) chloride is a complex substance, and there is much confusion about the structures reported. Four modifications have been described [48]. The β form is low in stereospecificity, but the α, γ, and δ forms all are capable of producing linear olefin polymers of high molecular weight.

The violet forms of $TiCl_3$, prepared in a variety of ways, are used today in Ziegler polymerizations with carefully controlled added amounts of aluminum alkyl halides. Both the rate of polymerization and the stereospecificity of the catalyst are influenced decisively by catalyst composition and ratio of Al to Ti, of Ti(III) to Ti(IV), and of Al to Cl. Pure titanium(III) chloride, in the absence of organometallic halide, has no catalytic activity. Natta has published a review of his work on the nature of the catalyst and the reaction kinetics [49].

The mechanism of polymerization is still a matter for discussion. All of the theories talk about defects in the surface of the crystals of $TiCl_3$, either unfilled coordination points at the edges [50–53] or active sites in "perturbed regions" of the crystal [54, 55].

Hundreds of publications have appeared which speculate on the action of Ziegler-Natta catalysts. This mass of data has been discussed and correlated [12–15, 49–55] with kinetic, x-ray, and other data.

In the polymerization described by Ziegler [45, 47] about 1% of the catalyst by weight, based on ethylene, was suspended in a suitable oil and ethylene was bubbled through rapidly at room temperature. Ethylene was completely absorbed and the yield of polymer was quantitative; no recycle was required.

In a large 5-liter glass container for home canning we stirred 2 liters of the solution or suspension of our catalyst in a suitable medium, which was often Aliphatin, a diesel oil from the Fischer-Tropsch synthesis. We introduced ethylene at room temperature. The temperature rose right away, and flakes of polyethylene were noticeable after only a few minutes. After the temperature reached 70°C, we cooled by aid of an air blast. One can absorb easily 200 liters of ethylene per hour per liter of solution (in our case 400 liters per hour). All the gas is absorbed; there is nothing leaving the solution. The mixture becomes thicker and thicker. After 1–1.5 hr, one can no longer stir it. Now one has a mushy suspension which is gray to brown, the color depending upon the catalyst. Upon admission of air it becomes white. One treats it preferably with dry alcohol, to which it loses all inorganic components, presumably by the formation of aluminum alkoxides and titanic acid esters, which are both easily soluble. The filtered and dried polymer will always show ash analyses of 0.01 % or even lower.

The experiment just described will furnish approximately 400 g. of polyethylene and can be run in the same way on a large scale. The first technical experiment with 1000 times these amounts was done by Ruhrchemie and gave the same results as the laboratory experiments had given. [45]

Other patents describe the use of Ziegler's catalysts and illustrate the variety of combinations of organometallic compounds and salts that can be used: simple alkylaluminums with titanium(IV) chloride; iron with RMX in which M is a metal of the third periodic group; a variety of substituted aluminum hydrides (e.g., $RAlH_2$, R_2AlH, R_2AlX, R_2AlOR) with R_2AlB [in which B may be $-N(CH_3)_2$, $-N(CH_3)-C_6H_5$, $-NR(COR)$, $-SR$, $-OCOR$, $-OSO_2R$] and a compound (usually a halide) of a metal of Group IV, V, or VI; dimethylmagnesium or propylmagnesium chloride, diethylzinc, phenylmagnesium bromide with titanium(IV) or zirconium(IV) chloride, or tetrabutyl orthotitanate [56–59].

In contrast to the high-pressure polymers the low-pressure, high-density polyethylenes have few short-chain branches. This can be shown by the absence of an absorption peak at 7.25 μ in the infrared spectrum [45, 60]. The molecular weight can be varied from 10,000 to several million.

The work of Natta at Milan and Ziegler at Mülheim with 1-olefins [57–60] has resulted in crystalline, high-melting polymers with characteristics very different from those shown by the commercial polymers hitherto produced (Table 1). The Ziegler catalyst system can be employed to produce polyethylenes with a wide range of properties. In 1963 Ziegler and Natta received the Nobel prize in chemistry for this work.

Two other developments, one by Phillips Petroleum and the other by Standard Oil of Indiana, have produced polyethylenes similar to the Ziegler polyethylenes by processes that also use a heterogeneous catalyst system. The polymers appear to be polymethylenes (i.e., with almost no branching) with significantly higher densities than those previously produced by the I.C.I. high-pressure process.

The basic Phillips patents disclose a calcined chromium(III) oxide catalyst supported on silica-alumina; it can be used in a liquid- or vapor-phase polymerization [61, 62]. Conversion of ethylene to polymer, using a fixed-bed catalyst at 88°C and a pressure of 600 psi, is quantitative. The catalyst can also be used in suspension. Numerous promoters (nickel, thorium, iron, manganese, uranium, vanadium, molybdenum, tungsten, zirconium, etc.) may be used. The crystalline melting point of the resulting polymers such as Marlex 50 is 113 to 127°C, the density is about 0.96, and the degree of crystallinity as measured by x-ray analysis is 50% greater than the crystallinity of the conventional high-pressure polyethylenes, or more than 90% [63]. These polymers are discussed in detail in Section VII.

A series of patents issued to Standard Oil of Indiana claims the use of a wide variety of metals and metallic oxides as catalysts for the polymerization of ethylene at moderate temperatures and pressures in a suitable liquid medium. Included are combinations of an alkali metal and an oxide of periodic group VIA [64], an alkali metal hydride and an oxide of Group VIA [65], a nickel-cobalt alloy supported on activated carbon in a hydrocarbon medium [66], partially reduced nickel and cobalt oxides [67], lithium (or sodium) aluminum hydride and an oxide of Group VA [68], γ-alumina, titania, and zirconia with partially reduced molybdena [69, 70], lithium or sodium borohydride and an oxide of Group VA [71], borohydrides with an oxide of Group VIA [72, 73], nickel oxide with an alkali metal hydride [74], nickel oxide with an alkali metal borohydride [75], an alkaline earth carbide with an oxide of Group VIA [76], and the hydrides of Group II together with a metallic oxide of Group VIA [77].

Laboratory polymers similar to the Ziegler linear polyethylenes have been made at extremely high pressures (7000–75,000 atm), using as catalysts α,α′-azobisisobutyronitrile, α,α′-azobis(α,γ-dimethylvaleronitrile), and 1,1'-azobiscyclohexanecarbonitrile or benzoyl peroxide in various diluents such as benzene, isooctane, and methanol [78]. These polymers had crystallinity values 30% greater than those of the conventional high-pressure polyethylenes. Patent claims [79] covered linear polymers with less than one side chain to 200 carbon atoms, mp > 127°C, and densities of 0.95 to 0.97.

IV. Crystal and Molecular Structure of Polyethylene

The properties of polyethylenes prepared under various conditions of polymerization vary widely. This variation in properties is inherently determined by the differences in the molecular structure that result under different conditions of polymerization. Consideration of the properties of the bulk polymer must include not only the detailed molecular structure but also the arrangement of the polymer chains into ordered or crystalline regions and the arrangement of these crystalline regions.

In considering this section, reference to Chapters 2, 3, and 7 of Volume I may be helpful.

The following structural features are of importance in considering the properties of polyethylenes:

1. *Crystal structure of the unit cell.* The precise geometric arrangement of the methylene groups with respect to each other in the crystalline regions.

2. *Percentage of crystallinity.* The weight per cent of the total material that is in a state of high order.

3. *Size and shape of crystallites or aggregates of crystallites.* The crystallites formed on cooling molten polyethylene tend to form aggregates. Under certain conditions the crystallites are massed in spherical clusters of radiating needles, called spherulites. The conditions of heating and cooling and the molecular parameters that determine the form of the spherulites in polyethylene are of considerable interest because the latter have great influence on the physical properties of the finished film, both visual and mechanical.

4. *Branching.* Because of the various kinetic possibilities of chain transfer by a growing polymer chain the polyethylene molecules may not be linear.

5. *Presence of unsaturation and oxygen-containing groups.* Either because of the kinetics of termination or because of other attendant

reactions with initiator or other impurities the polymer molecules may contain oxygen or unsaturated groups.

Ever since the discovery of polyethylene considerable effort has been devoted to the study of these structural details of the polymer by infrared absorption, x-ray diffraction, microscopy, solution properties of the polymer, and other means.

A. INFRARED SPECTRUM

Infrared absorption has been found to be a valuable tool in the study of the fine structure of polyethylene. In Table 2 are listed the absorption bands in the 2-to-15 μ region which have been observed in polyethylene [60, 80–94] and their assignment to the various molecular vibrations.

Of particular interest are the bands from 1300 to 1500 cm^{-1} and the doublet at 720 to 730 cm^{-1}. The doublet is characteristic of solid polyethylene and crystalline paraffin hydrocarbons. When the sample has been cooled to 4°K, the band at 730 cm^{-1} becomes very sharp, whereas its companion at 720 cm^{-1} remains unchanged [83]. When polyethylene melts, the doublet with maxima at 720 and 730 cm^{-1} is replaced by a broad band centered at 720 cm^{-1} [81, 87, 88]. This characteristic change in the infrared spectrum of polyethylene on melting has been shown to result from the crystalline nature of polyethylene, that is, segments of polymer chains exhibit order like that of crystalline solids. The absorption band at 720 cm^{-1}, observed in molten polyethylene and in solution, has been assigned to the rocking of the hydrogen atoms in the methylene groups [95]. This vibration was found to be perpendicular to the polymer chain axis from studies by polarized infrared radiation [96, 97]. Splitting of the single absorption band at 720 cm^{-1} into a doublet absorbing at 720 and 730 cm^{-1} occurs on crystallization because of the interaction of the methylene groups in the crystalline regions [98, 99]. This conclusion is supported by the studies of C_{36} hydrocarbon crystals with polarized infrared radiation [93].

With C_{36} hydrocarbon crystals it was shown [93] that the 730-cm^{-1} band has maximum absorption when the infrared radiation is polarized parallel to the a-axis of the crystals and the 720-cm^{-1} band has maximum absorption with radiation polarized parallel to the b-axis of the crystals. Theory and experimental evidence from crystalline hydrocarbons suggest that in polyethylene the absorption band at 730 cm^{-1}

TABLE 2

Infrared Absorption Bands of Polyethylene

Wavelength (μ)	Wave number (cm^{-1})	Assignment	Dichroic properties
3.38	2958	Unsymmetrical stretching in methyl groups	No dichroism
3.42	2920	Unsymmetrical stretching in methylene groups	Weak perpendicular dichroism
3.48	2880	Symmetrical stretching in methyl groups	Weak perpendicular dichroism
3.51	2858	Symmetrical stretching in methylene groups	Weak perpendicular dichroism
3.66	2735	Methyl groups	
3.74	2678	Methylene groups	
6.79	1470[a] }	Deformation of methylene groups perpendicular to chain axis	Perpendicular dichroism in all three bands
6.82	1465 }		
6.84	1463		
6.87	1458	Deformation of methyl groups	No dichroism
7.25	1375	Symmetrical deformation of methyl groups	No dichroism
7.30	1372 }		
7.39	1355 }		
7.45	1340 }	Deformation of methylene groups	Parallel dichroism
7.68	1300 }		
11.22	890	Rocking of methyl groups	
13.70	730[a] }	Rocking of methylene groups perpendicular to chain direction	Perpendicular dichroism
13.88	720 }		

[a] Purely crystalline bands.

arises from the crystalline regions only and the 720-cm^{-1} band arises from both the crystalline and the noncrystalline (disordered) regions. Further, for an oriented sample of polyethylene in which the polymer chains are aligned in a particular direction absorption for both the 730- and the 720-cm^{-1} bands would show dichroic properties; that is, absorption would depend on the direction of polarization. The absorption would be a maximum for radiation polarized perpendicular to the direction of the polymer chains, that is, the absorption would show perpendicular dichroism [86, 93, 100].

Other absorption bands in the spectrum of polyethylene also show dichroic properties [87, 96, 99, 101, 102]. In particular, the 1463-to-1470-cm^{-1} bands show behavior similar to that of the 720-to-730-cm^{-1} doublet. In Table 2 the types of dichroism shown by the various absorption bands are indicated. (The words parallel dichroism signify that the absorption for radiation polarized parallel to the direction of the polymer chains is a maximum.) The study of the dichroism of various absorption bands, using polarized infrared radiation, is of much importance in the study of orientation of polymer chains in various samples of polyethylene and in the proper assignment of the molecular vibrations responsible for the different absorption bands [97, 100–104].

In the 1300-to-1372-cm^{-1} region there are four absorption bands centered approximately at 1300, 1340, 1355, and 1372 cm^{-1}. The 1340-cm^{-1} absorption appears only as a weak shoulder on the 1355-cm^{-1} band. These bands develop from the deformation frequencies of the methylene groups [99, 102]. The band at 1372 cm^{-1} has been interpreted as arising from a superposition of absorption from both the crystalline and the noncrystalline regions. The other bands at 1300, 1340, and 1355 cm^{-1} come from disordered regions only; they are weak in highly crystalline polyethylene and are absent in crystalline hydrocarbons. All four bands show parallel dichroism [87, 102] and therefore should be important in the study of the noncrystalline content and orientation of the polymer segments in the unordered regions.

The absorption band at 1375 cm^{-1} arises from the symmetrical deformation of the methyl groups [82, 87, 105]. Its presence is an indication of branched structure, and quantitative measurements at a wavelength corresponding to it have been used, as discussed later, for the determination of the degree of branching in polyethylene.

In addition to the absorption bands discussed above and listed in Table 2, several others corresponding to unsaturated and oxygen-containing groups have been identified in polyethylene, especially in those samples of polyethylene that had been subjected to thermal or photochemical degradation.   Table 3 lists the absorptions correspond-

TABLE 3

Absorption Bands in Polyethylene Corresponding to Olefinic Unsaturation and Oxygen-Containing Groups

Wavelength (μ)	Wave number (cm^{-1})	Assignment
2.81	3559	Hydroperoxide, R_3COOH
5.692	1757	Anhydride, $(RCO)_2O$
5.738	1715	Ester carbonyl, RCOOR
5.771	1733	Aldehyde carbonyl, RCHO
5.798	1725	Ketonic carbonyl, R_2CO, close to chain end
5.812	1720	Ketonic carbonyl, R_2CO, internal
5.838	1713	Carboxylic carbonyl, RCOOH
6.08	1644	Olefinic unsaturation, total
10.07	990	Vinyl unsaturation, $RCH{=}CH_2$
10.35	964	Trans internal unsaturation, $RCH{=}CHR$
11.00	908	Vinyl unsaturation, $RCH{=}CH_2$
11.25	888	Vinylidene unsaturation, $R_2C{=}CH_2$

ing to the unsaturated and oxygen-containing groups that have been observed [81, 84, 90, 92, 100, 106, 107].   A typical infrared absorption spectrum of polyethylene is shown in Figure 1.   Besides the bands marked, about 17 other very weak bands appear in the polyethylene spectrum.   These are probably carbon-carbon and methylene frequencies and combination bands [94].

B. CRYSTAL STRUCTURE AND CRYSTALLINITY

In the discussion of the infrared spectrum of polyethylene the disappearance of certain bands on melting has been mentioned.   This was attributed to the presence of regions in polyethylene in which the segments of the polymer chains are able to arrange themselves in the regular three-dimensional order characteristic of crystalline solids.   Thus, like many other polymers, polyethylene is a polycrystalline material (Volume I, Chapter 7).

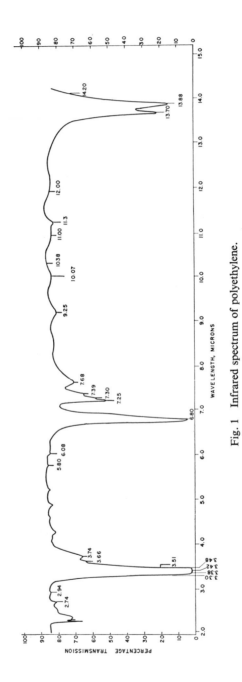

Fig. 1 Infrared spectrum of polyethylene.

146

The x-ray diffraction pattern of polyethylene consists of sharp diffraction lines from the crystalline regions and a halo from the non-crystalline regions [108, 109]. The crystal structure of polyethylene has been determined from x-ray studies. The unit cell is orthorhombic with $a_0 = 7.40$ Å, $b_0 = 4.93$ Å, and $c_0 = 2.534$ Å, four CH_2 groups in a unit cell (Figure 2). The symmetry elements possessed by the unit cell are found to correspond to the space group Pnam. The important elements of symmetry present in the unit cell are three sets of twofold axes parallel to each unit cell axis, a diagonal glide plane perpendicular to the a-axis, a glide plane perpendicular to the b-axis with glide along the a-axis, and a mirror plane perpendicular to the c-axis. The calculated

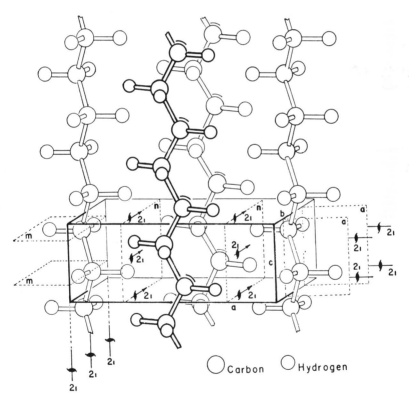

Fig. 2   Crystal structure of polyethylene.   Screw axes of symmetry and glide planes are indicated [110].

density is 1.00 g/cc. (For definitions of space group symbols and various symmetry elements [110] and [111] may be consulted.) The polymer chains in the crystal are parallel to the c-axis of the unit cell. The observed intensities of a number of lines in the diffraction pattern of polyethylene are not in agreement with those calculated from the structure of the unit cell. This failure has been attributed to the non-spherical character of the electron cloud drawn out in the plane of the three nuclei of the methylene groups [108]. Essentially the same crystal structure has been found in the Ziegler-Natta highly crystalline polyethylenes [112], though minor differences exist.

The proportion of crystalline and noncrystalline domains in polyethylene (percentage of crystallinity) is not constant but depends on factors in polymerization as well as on conditions after extrusion (e.g., tensilizing and annealing). Values for the older types of polyethylene manufactured at moderate pressures (5–20,000 psi) were commonly about 60% crystalline (the value varying somewhat with the method of measurement; see below), but the extent of order can be rather closely controlled by catalyst and conditions in the manufacture of Ziegler-Natta high-density linear polyethylenes [12].

Many important properties of polyethylenes are significantly dependent on the percentage of crystallinity of the sample, and several methods have been developed to determine this quantity. Disagreement among the values determined on similar samples by different methods arises not only from the experimental difficulties but also from the theoretical point of view inherent in the various methods. Methods that have been used for the determination of polyethylene crystallinity include x-ray diffraction [113–124], density [125], determinations of heat content [126, 127], infrared absorption [128–130], and nuclear magnetic resonance [131].

The x-ray diffraction method is particularly well suited to determinations of crystallinity in polyethylene because the peak corresponding to scattering from the noncrystalline regions is well separated from the peaks corresponding to the reflections from the crystalline regions. The x-ray methods used for the determination of crystallinity in polyethylenes are based either on a comparison of the intensity of radiation scattered from noncrystalline areas to the intensity scattered by molten polyethylene [115, 116, 120–122] or on a comparison of intensities of x-rays scattered by the noncrystalline areas with the intensity of the crystalline reflections [113, 117–119].

Many of the experimental difficulties in the latter method, when the intensities are recorded on a photographic film, have been overcome by the use of the x-ray diffractometer [124]. In the x-ray diffraction methods the main uncertainty is in the background correction to be applied. The scattering from noncrystalline regions may extend to a lower angular range than has been considered in a number of determinations. The values for percentage of crystallinity as determined by x-ray methods tend to be higher than those determined by other independent methods. Also, in these methods the sample has to be free from orientation of the crystalline and noncrystalline regions or appropriate corrections must be applied.

The determination of crystallinity from density is based on a knowledge of the densities of the crystalline and noncrystalline regions of polyethylene. The density of crystalline regions may be unequivocally computed from the dimensions of the unit cell [108], but the estimation of density of polycrystalline polyethylene at a temperature appreciably lower than the melting point is subject to considerable uncertainty. The usual method for calculating the density has been to construct a plot of the density (or specific volume) as a function of temperature of molten polyethylene and extrapolate to lower temperatures [121, 125, 126]. Other methods of obtaining the density of the noncrystalline regions for use in calculations of crystallinity from density of the sample have been used [132, 133]. The values of the density of noncrystalline polyethylene at 25°C from extrapolation of densities of polyethylene melts are appreciably higher than those estimated by other methods. Also, in applying the density method for determining crystallinity, the possibility of the inclusion of microvoids in the samples and incomplete wetting of the samples by the flotation liquid should not be ignored.

The crystallinity from measurements of heat content is calculated by dividing the difference between the heat content of noncrystalline polyethylene and the measured heat content of the sample by the heat of fusion of entirely crystalline polyethylene [126, 127]. The heat content of the amorphous polyethylene is obtained by extrapolation of the results obtained at temperatures above the melting point. The uncertainty in this method is in the exact value of the heat of fusion of entirely crystalline polyethylene, but fairly accurate estimates of this quantity from the study of crystalline long-chain hydrocarbons are available [134].

In principle all absorption bands in the infrared spectrum of polyethylene that arise solely from crystalline or amorphous regions may be used for the determination of crystallinity. The absorption bands at 730 and 1300 cm^{-1} are most suitable [128–130]. In these methods absorptions at 1300 cm^{-1} by a polyethylene sample before and after complete melting are compared. If proper precautions are taken, this method should give accurate crystallinity values, for it seems to be free from the inherent uncertainties that are generally present in other methods.

The variation of crystallinity with temperature of polyethylenes with branching has been compared with that of the linear polyethylenes [123]. Linear polyethylenes keep their high crystallinity up to about 110°C, but a gradual change in crystallinity with temperature is observed in polyethylenes which have short-chain branching.

The structural feature that primarily determines the crystallinity of polyethylene is the number of short branches present. Molecular weight and molecular-weight distribution have only a minor effect on the crystallinity of polyethylene, if they exert any effect at all [135–137].

On cooling low-density polyethylene, a transition at -38°C has been observed from proton magnetic resonance [138]. This probably corresponds to the second-order transition temperature of polyethylene, though the value appears to be a bit too low. Studies of the coefficient of expansion of samples with densities of 0.92 to 0.97 g/cc and melting points from 100 to 138°C have given values of -21 to -25°C for the glass transition temperature [125, 139]. The glass transition temperature of highly crystalline Ziegler-Natta polyethylenes appears to be approximately the same.*

C. Branched Structure

Soon after the first successful polymerization of ethylene the branched structure of polyethylene was inferred from the estimation of the concentration of methyl groups by infrared absorption [80]. Infrared absorption at 3.38 μ by polyethylene samples indicated that there were more methyl groups per molecule than can be accounted for by the terminal groups. Another absorption band characteristic of the methyl groups in the infrared spectrum of polyethylene is at 7.25 μ.

* Much lower values have been reported, but they are probably in error (cf. Volume I, p.249).

Later studies of infrared absorption at 7.25 and 3.38 μ have established that molecules of most commercial polyethylenes have an appreciably branched structure [81, 84, 140]. In contrast polyethylene prepared from diazomethane [141] and the polymers prepared at pressures from 1 to about 50 atm appear to be free from short branches [45, 62, 63, 142].

One mechanism for chain branching in vinyl polymers postulates that branching arises because of the kinetic possibility that an active growing chain transfers its activity to another dead polymer chain by abstracting a hydrogen atom [143].

$$\sim CH_2CH_2CH_2\cdot + CH_3(CH_2)_mCH_2(CH_2)_nCH_3 \longrightarrow$$

$$\uparrow$$

(growing
polymer chain)   (site of chain transfer)

$$\sim CH_2CH_2CH_3 + CH_3(CH_2)_m \underset{\bullet}{CH}(CH_2)_nCH_3$$

$$\uparrow$$

New site for chain growth

It was realized, however, that this mechanism could not be entirely responsible for all chain branching present in polyethylenes. If it were, it can be shown from kinetic analysis that the average number of branches per molecule, as inferred from infrared measurements on many polyethylenes, would be sufficient to give a polymer system of infinite molecular weight [144]. Also, under certain conditions the observed extent of branching, occurring according to the above mechanism, would lead to a system crosslinked into an infinite network [145].

It was to be expected that an increase in branching frequency would give polymers of high melt viscosity, low tensile strength for a given molecular weight, and low density of the polymer, that is, a lower percentage of crystallinity because of the increase in methyl group content. It was observed, however [136], that under certain conditions of polymerization polymers differing appreciably in density, hence crystallinity, could be prepared to show approximately the same melt viscosity and tensile strength. By carrying polymerization to different extents it was also possible to prepare samples with the same density but appreciably different melt viscosity and tensile strengths. It was concluded that to reconcile these facts it is necessary to postulate a second mechanism by which branching in polyethylenes may occur; namely, by intramolecular chain transfer through formation of transient cycloparaffin structures of four or five carbon atoms. The former mechanism leads to long-chain branching, whereas the latter gives rise to short-chain branching only. During the polymerization of ethylene both mechanisms of chain branching may operate simultaneously to produce both long and short branches in a polyethylene sample. The mechanism leading to short-chain branching may be represented as follows:

$$\sim CH_2CH_2CH_2CH_2CH_2CH_2 \cdot \longrightarrow \sim CH_2CH_2CH_2$$

$$\cdot H_2C \quad CH_2$$

$$CH_2$$

$$\sim CH_2CH \cdot$$

$$(CH_2)_3$$

$$CH_3$$

V. Optical Properties of Polyethylene

A. SPHERULITIC FORM OF CRYSTALLINE AGGREGATES

An estimate of the size of polyethylene crystallites from a broadening of the x-ray reflections has shown that the largest dimensions of the crystallites are not greater than 300 Å and may even be less than 100 Å [109]. The presence of crystalline regions of such small dimensions would not be visible in the optical microscope because of the limit on its resolving power, and a sample of the polymer would be transparent, since the crystallites are smaller than the wavelengths of visible light.

In contrast to these expectations low-density polyethylenes at room temperature are at least slightly translucent, and the optical microscope reveals in thin crystallized films organized structures considerably bigger than 300 Å. The crystallites appear to have become associated into clusters of characteristic form. These aggregates have two principal features: the assemblage of crystallites radiates in all directions from a point and in the polarizing microscope with crossed Nicol prisms exhibits a black Maltese cross. By analogy with aggregates of similar characteristic features, observed in many organic materials of low molecular weight and in many minerals [14, 146–149], these aggregates of crystallites in polyethylene and other polycrystalline polymers are called spherulites [cf. Volume I, p. 334 ff].

Though a spherulitic structure is easily observed only in thin films, it should not be inferred that in massive specimens and molded shapes spherulites are not formed. In fact, in the case of some polymers the presence of spherulites in thick specimens has been confirmed from a surface study of blocks or microtomed sections [150, 151].

Ideally, the spherulites would be spherical, but either interference from adjacent growing spherulites or interruption in the growth process may give rise to open sheaflike or other intermediate structures. The sizes of spherulites in a sample of polyethylene, or of any other crystal-line polymer, may vary over a considerable range from submicroscopic to a few tenths of a millimeter. Microscopy studies can give an estimate of the approximate average size of the spherulites in a given sample, and it has been possible in this way to establish some general conclusions regarding the factors that affect spherulite size and the effect of spherulite size on the properties of the crystallized polymer.

At least six important factors affect the size of the spherulites: molecular weight, percentage of crystallinity, the degree of branching of the polyethylene sample, the thermal history, the mechanical working of the melt, and the rate of cooling [109, 140, 152–154]. Polyethylenes with the least branching form the largest spherulites. In general, spherulite size increases with molecular weight and crystallinity of linear polyethylenes. Slow cooling of the melt through the melting range produces large spherulites, whereas sudden quenching to low temperatures produce small spherulites. Mechanical working of the melt, followed by shock cooling, gives materials with very small spherulite size. Clarity of the films improves, for the smaller the size of the spherulites, the clearer the sample.

X-ray diffraction studies with a special microbeam camera have shown that in the spherulites of polyethylene the polymer chains are perpendicular to the radii of the spherulites [155]. In a few special cases careful observations of the crystallized polymer melts showed closely banded fibrillar structures [156]. On the basis of the results from these studies Keller proposed that the spherulites of crystallizing polymers consist of fibrillar units arranged along a helical path. This helical arrangement of the fibrillar units was believed to arise from a regular branching of the fibrils with a constant branching period and angle in the two-dimensional case and by a constant branching period, two constant angles, and a constant direction of rotation in the three-dimensional case. The fibrils themselves were believed to consist of a closely coiled ropelike arrangement of the chains. According to this mechanism, in the formation of the spherulites the polymer chains first coil themselves tightly to form the fibrils ("small-scale helices") which get arranged further along a helical path ("large-scale helices"), with the axes of the large-scale helices parallel to the radii of the sperulites.

In contrast to Keller's complex mechanism of spherulite growth, a comparatively simple mechanism involving statistically outward radial growth from a nucleus has been proposed [157]. According to this mechanism, a spherulite originates from a single nucleus. From this nucleus a crystallite grows by lateral accretion of molecular segments of polymer chains. Fine strands of crystalline order proceed outward from the fringes of this crystallite, seeding further crystallizable domains and initiating the growth of additional crystallites. These crystallites in turn send out nucleating streamers, and growth takes place in a statistically radial fashion. Growth stops either because of the high viscosity of the system or because of interference from the neighboring growing spherulites. The optical properties and shapes of the spherulites are determined by the shapes of the crystallites.

An important property related to the structure of the spherulites of polyethylenes is the sign of optical birefringence. The spherulites show negative birefringence, that is, the refractive index along the tangent to the spherulite is higher than that along the radius [109, 140, 155]. Since the highest refractive index in both crystalline and noncrystalline regions of polyethylene is along the chain axis [158, 159], the negative birefringence of the spherulites implies that the polymer chains are arranged perpendicularly to the radii of the spherulites. This is confirmed by x-ray diffraction studies of the spherulites by use of the microbeam camera [155].

B. Light Scattered by Polyethylene Films

The light scattered by polyethylene film determines its clarity. Some of this light is scattered because of the roughness of the surface, an aspect that can be protected during extrusion and handling (cf. Volume I, Chapters 6 and 8). In addition to the surface scattering, polyethylene, in common with other crystalline polymers, scatters light because of the inherent inhomogeneities in the structure resulting from the different refractive indexes of the crystalline and noncrystalline regions. Thus the crystalline polymers show more turbidity than the corresponding melts [125, 160]. The scattering ability of polyethylene films depends appreciably on the quenching conditions and thermal history [161, 162]. Quantitative studies have shown that the light scattered by a thin polyethylene sample quenched to 0°C from 125°C is much lower than that scattered by an annealed sample [163, 164].

These observations have been used to guide practices in film extrusion that lead to clear polyethylene films from a wide variety of commercial polymers that differ to a considerable extent in density, viscosity, and molecular weight, consequent on the method of manufacture.

VI. Orientation in Polyethylene Films

The primary and most extensive studies of orientation and effects of orientation in polymer films have been performed on polyolefins. This work is treated extensively in Volume I, Chapter 7. Here we mention only a few specific considerations as they apply to polyethylene.

A. a-Axis Orientation in Polyethylene Films

Polyethylene film prepared by extrusion, followed by quenching in air or water, shows an unexpected preferential orientation of the crystalline regions. We would expect that as the melt issues from a die the polymer chains would become oriented in the direction of flow (commonly referred to as the machine direction), and on crystallization the crystalline regions would be preferentially oriented with the c-axes parallel to the machine direction, since the c-axis of the crystals is parallel to the polymer chains (see Figure 2 for the relation between the crystallographic axes and the direction of the polymer chains). This is not true for extruded polyethylene films or for samples prepared under similar conditions of flow and crystallization.

Two important techniques used in establishing the orientation of the crystalline regions in polyethylene films are x-ray diffraction and the dichroism of infrared absorption. To appreciate the interpretation of the results from these methods in terms of the orientation it is helpful to consider the two limiting cases of orientation in polyethylene films:

1. a-axis orientation, according to which the a-axes of the crystalline regions are preferentially oriented parallel to the machine direction and the b- and c-axes are in a plane perpendicular to the machine direction.

2. c-axis orientation, according to which the c axes of the crystallites are preferentially oriented parallel to the machine direction (or to the stretching direction in the study of stretched films).

For the study of orientation in polyethylene films from the dichroism of polarized infrared radiation the doublet with absorption peaks at 13.70 and 13.88 μ is particularly suitable.   As mentioned above (Section IV–A), the absorption at 13.70 μ is from the crystalline regions, whereas both the crystalline and the noncrystalline regions contribute to the absorption at 13.88 μ.   It has been established that the absorptions at 13.70 and 13.88 μ from the crystalline regions of polyethylene correspond to the dipole change vectors parallel to the a- and c-axes of the crystal unit cell, respectively.   The absorption of infrared radiation is proportional to the scalar product of the vectors $\mathbf{E}$ and $\mathbf{D}$, where $\mathbf{E}$ is the electric vibration vector of the polarized infrared radiation and $\mathbf{D}$ is the dipole change vector corresponding to the molecular motion that gives rise to the absorption.   Therefore, when the orientation is such that the crystalline regions are preferentially oriented with their a-axes parallel to the machine direction, the 17.70-μ band will exhibit maximum absorption for the radiation that has its vibration vector parallel to the machine direction, compared with the absorption for the radiation that has its vibration vector perpendicular to this direction.   Thus for a-axis orientation of the crystalline regions in the film the 13.70-μ band will exhibit parallel dichroism.   From similar considerations it can be shown that the 13.88-μ band will show perpendicular dichroism for a-axis orientation of the crystalline regions.   Since both the crystalline and the noncrystalline regions contribute to the absorption at 13.88 μ, any orientation of the noncrystalline regions would also affect the dichroic properties of the 13.88-μ band.   The presence of orientation of the noncrystalline regions, however, can be inferred from the dichroic properties of other bands from the noncrystalline regions alone, namely, at 7.68, 7.45, and 7.39 μ [99, 101].   For the c-axis orientation of the crystalline regions the a- and b-axes will be perpendicular to the machine (or stretching) direction and both the 13.70- and 13.88-μ bands will exhibit perpendicular dichroism; that is, maximum absorption will occur for radiation with the electric vibration vector perpendicular to the machine direction.   Table 4 summarizes the dichroic properties of the 13.70- and 13.88-μ bands for the a-axis and c-axis orientation of the crystalline regions in polyethylene films.

TABLE 4

Dichroism of the Absorption Bands at 13.70 and 13.88 Microns for a-Axis and c-Axis Orientations of Polyethylene Crystalline Regions

Orientation in crystalline regions	Electric vibration vector **E** with respect to machine or stretching direction	Absorption at 13.70 μ	Absorption at 13.88 μ	Type of dichroism
a-Axis parallel, b- and c-axes perpendicular, to machine direction	**E** parallel to machine direction	Maximum	Minimum	Parallel for 13.70-μ band and perpendicular for 13.88-μ band
	E perpendicular to machine direction	Minimum	Maximum	
c-Axis parallel to machine direction; a- and b-axes distributed about c-axis	**E** parallel to machine direction	Minimum	Minimum	Perpendicular dichroism for both 13.70-μ and 13.88-μ bands
	E perpendicular to machine direction	Maximum	Maximum	

From x-ray diffraction and dichroism of the 13.70- and the 13.88-μ infrared absorption bands it was deduced that in extruded polyethylene films the crystalline regions are preferentially oriented with their a-axes parallel to the machine direction [159, 165, 166]. After the film has been stretched approximately 200%, the orientation changes to c-axis orientation; that is, the crystalline regions become preferentially oriented with their c-axes parallel to the stretching direction (see Section B).

The birefringence of extruded polyethylene films does not seem to fit the above inferences concerning a-axis orientation in extruded polyethylene films. Calculations of the principal refractive indexes of the polyethylene crystal and molecule showed that for the crystal the highest refractive index (γ_c) is parallel to the chain axis, the lowest refractive index (α_c) lies along the a-axis of the crystal, and the intermediate refractive index (β_c) is along the b-axis [158, 159]. For polyethylene molecules not in the crystalline regions the highest refractive index (γ_a) is also along the chain axis, the lowest refractive index (α_a) is perpendicular to both the chain axis and the plane containing the carbon atoms, and the intermediate refractive index (β_a) is parallel to the plane of carbon atoms but perpendicular to the chain axis. In polyethylene films in which the a-axes of the crystalline regions are preferentially oriented parallel to the machine direction and the b- and c-axes are perpendicular to this direction (a-axis orientation) it would therefore be expected that a lower refractive index would be exhibited parallel to the machine direction than in the transverse direction when the segments of molecules in the noncrystalline regions are completely disordered; that is to say, polyethylene films should be expected to show a negative birefringence. Extruded polyethylene films generally show a slight positive birefringence instead.

The small positive birefringence shown by polyethylene films should be interpreted with considerable caution, however, before concluding that this evidence is contrary to that shown by x-ray diffraction and polarized infrared absorption, since a multiplicity of effects may contribute to the birefringence of films. Some of the factors, other than orientation, that may affect birefringence of films are (a) form birefringence [167, 168], which arises purely from the anisotropic shape of regions of one refractive index embedded in a matrix of another refractive index, (b) the introduction of stresses, microvoids, or microcracks during crystallization of the films, and (c) the branched structure of polyethylene. Branch points are sites at which discontinuity in the ordering of chains to form crystallites occurs. When the chains in the crystallites pack preferentially perpendicular to the machine direction, the short branches containing the methyl groups may be aligned parallel to the machine direction. The polarizability of the methyl group is large and even a small degree of alignment of the methyl groups parallel to the machine direction will tend to make the birefringence of polyethylene films positive.

In the early studies on the orientation of polyethylene films it was suggested that the chains in the noncrystalline regions are oriented parallel to the machine direction, that is, perpendicular to the orientation of the chains in the crystalline regions of the film [159]. The perpendicular dichroism of the 13.88-μ band was considered evidence for this cross-orientation of the chains in the film. It now seems that this conclusion was based on erroneous interpretation of the absorption at 13.88 μ as arising solely from the noncrystalline regions. Both noncrystalline and crystalline regions contribute to the 13.88-μ absorption band, and the perpendicular dichroism of this band can arise from a-axis orientation of crystalline regions alone. No dichroism of the absorption bands that arise solely from the disordered chains has been observed in extruded polyethylene films [165, 166].

The a-axis orientation in extruded polyethylene films may occur by the following mechanism, based on the preferred direction of crystal growth [165]. In the molten polyethylene the polymer molecules are entangled and intertwined. Because of the shearing stresses present during extrusion, some alignment of chain segments occurs. As the molten polymer cools, the crystals probably do not grow in the direction of extrusion, since it is unlikely that the exact spatial arrangement between chains of neighboring molecules, necessary for crystallization, would be present along sufficient lengths of the chains. Crystal growth in the cooling sheet along the direction of extrusion will therefore be inhibited because of the small chance that molecules over extended chain lengths will be in a favorable spatial arrangement. Yet an arrangement conducive to crystallization is likely over short lengths of adjacent chains of carbon atoms. The crystallites in the cooling polymer will therefore have a tendency to grow preferentially in a direction perpendicular to the length of the polymer chains; that is, the crystallites will have a tendency to grow in the form of needles with their long dimension perpendicular to the direction of flow. This is schematically shown in

Figure 3*a*.   Since these needlelike crystallites are formed in a flowing matrix, they tend to become oriented further in a direction governed now by their shape.   Thus the crystallites that preferred to grow at right angles to the direction of flow may now slue round in the flowing matrix and line up with their long dimensions parallel to the direction

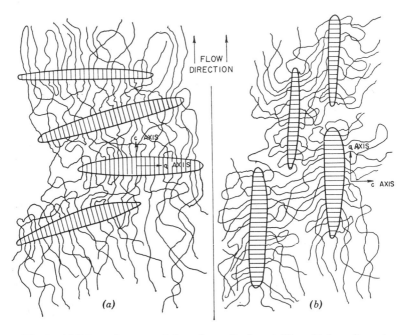

Fig. 3   (a) Schematic representation of growth of crystallites with long dimension perpendicular to flow direction of the melt; (b) alignment of crystallites with long dimension parallel to direction of flow, resulting in chains in the crystalline regions perpendicular to the machine direction of extruded polyethylene film.

of flow.   Schematically, the crystallites shown in Figure 3*a* will attain a position in the final film, as shown in Figure 3*b*.   This results in a preferential orientation with the polymer chains in the crystalline regions perpendicular to the direction of extrusion, that is, with the *c*-axis of the crystallites perpendicular to the extrusion (machine) direction.

The diffuse character of the reflections from planes containing an *l* index (0*kl*, *hkl*, etc.) in the x-ray diffraction patterns of polyethylene films suggests and supports the inference that the lengths of the crystallites along the direction parallel to the chain direction are smaller than in the direction perpendicular to the chains.

It has also been shown that crosslinked molten polyethylene film, allowed to crystallize in a stretched condition, results in an orientation of the crystalline regions similar to that discussed above: namely, with the *a*-axis of the crystalline regions preferentially oriented parallel to, and with the *c*-axis perpendicular to, the direction of stretching of the melt [169]. The results have been interpreted to mean that this type of orientation is a thermodynamic effect and represents the equilibrium orientation of crystals in a stretched structure.

B. STRETCHING AND RELAXATION OF POLYETHYLENE

The changes that occur in the orientation of the crystalline regions of polyethylene films on stretching and relaxation have been studied rather carefully by x-ray diffraction, infrared dichroism, and birefringence [166, 170–173].

As discussed in the preceding section, in extruded polyethylene films the preferred orientation of the crystalline regions may be interpreted as *a*-axis orientation. After the films have been stretched two- to three-fold, however, the orientation changes to *c*-axis orientation, that is, with the *c*-axis of the crystallites parallel to the stretching direction.

Two important questions arise from these observations:

1. How is this change of orientation on stretching brought about and how are the crystalline regions oriented at intermediate stages of elongation?
2. What is the effect produced on orientation when the stretched samples are allowed to relax?

Answers to these questions were first sought by x-ray diffraction studies of stretched and relaxed filaments of polyethylene [174, 175] and later by similar experiments [166] with films, using a Norelco micro-beam x-ray diffraction camera with a collimator size of 100 μ. (The use of a microscope allowed any selected point in the sample to be positioned within 5 μ in front of the collimator.) Thus is was possible to follow the orientation changes closely in the necked-down region of a stretched sample. The stretching of polyethylene proceeds by alignment of the a-axes of the crystallites perpendicular to the stretching direction with the planes containing b- and c-axes perpendicular to the a-axis but randomly oriented, followed by a gradual change to a state in which the c-axes become more nearly aligned in the stretching direction —although perfect alignment is not achieved. On relaxation either the orientation of the c-axis parallel to the original stretching direction becomes less perfect or folding of the oriented crystallites about the b-axis takes place. In the latter case the preferred crystallographic directions parallel to the stretching direction have indices such as $[\mu 0 \omega]$.

The changes in orientation of the crystallites during stretching and relaxation of polyethylene films suggest that in polyethylene the crystallites are stacks of a microstructure containing b- and c-axes. First the stacks become oriented parallel to the stretching direction. followed by orientation to bring the c-axis parallel to the stretching direction.

VII. Other Properties of Polyethylene

A. DENSITY

As indicated in Table 1, polyethylenes are more or less arbitrarily classified on the basis of density into "low-density," "medium-density," and "high-density" polymers. The designations grew historically as new polymerization processes were developed, beginning with the original I.C.I. work that gave densities around 0.91 g/cc and culminating in Ziegler-Natta polymers with densities of 0.97+, close to the theoretical value, 1.00 g/cc, calculated from the lattice dimensions of the unit cell.

The final density of a film depends primarily on the structure of the polymer but secondarily on the manufacturing process and concomitant or subsequent treatment, which may be mechanical, or thermal, or both.

Density of bulk polymer has been found to be determined primarily by the amount of short-chain branching and the molecular weight [176]. Short-chain branching in the free-radical polymerization depends solely on the temperature and pressure; branching increases as pressure is reduced and temperature is increased. At least up to $\overline{M}_n = 50,000$, density increases as molecular weight decreases. At constant temperature and pressure the type and amount of free-radical initiator and chain-transfer agent have no effect on short-chain branching, though the latter does, of course, affect molecular weight (and therefore density). Different initiators (at the same temperature and pressure) have no effect on density, which is contrary to many claims in the patent literature that specific initiators increase the density of polyethylene. The following equation has been found to relate the pressure, temperature, and molecular-weight data to density, for 20,000 psi $< p <$ 30,000 psi, 130°C $< T <$ 200°C, and 25,000 $< \overline{M}_n <$ 50,000.

$$d = [0.963 + 2.4 \times 10^{-7}\, p - 1.18 \times 10^{-4}\, T - 4.0 \times 10^{-7}\, \overline{M}_n] \pm 0.001.$$

B. SOLUBILITY

Polyethylenes generally are not soluble in any solvent below 50 to 60°C, but at temperatures above 70°C the high-pressure (low-density) types are soluble in many solvents. Carbon tetrachloride and highly chlorinated ethylenes are the most effective solvents [177]. In toluene and xylenes these polyethylenes show appreciable solubility above 60°C, the solubility decreasing rapidly with increasing chain length. Polyethylene samples with branched structure and heterogeneity in molecular weight have a higher solubility than shown by linear polyethylenes of the same average molecular weight and of narrower distribution.

The effects of various chemical substances on polyethylenes is summarized in Table 5 [178]. Water and aqueous solutions of salts have no effect on polyethylenes in general. Most polar organic liquids (alcohols, aldehydes, esters, and ketones) do not attack polyethylene, but many of them are environmental stress-cracking agents. Strong oxidizing agents may have no visible effect on polyethylene (e.g., 30% hydrogen peroxide), but they attack the polymer chemically and embrittle it.

The solubility curves of polyethylenes in good solvents rise monotonically as a function of temperature, indicating an equilibrium between liquid and crystalline solid. With poor solvents such as amyl acetate and nitrobenzene, however, the solubility curve passes through a maximum at low concentration. This is indicative of liquid-liquid equilibrium between concentrated and dilute solution phases. The phase diagram of polyethylenes with amyl acetate or nitrobenzene exhibits two concentration regions, one over which liquid-liquid phase separation occurs and the other over which equilibrium between liquid and crystalline solid phases exists [179, 180]. It is unfortunate that this important

TABLE 5

Effects of Chemical Substances on Low-, Medium-, and High-Density
Polyethylenes[a]

	Increase in weight (%)			
Solvent	$d = 0.918^b$	$d = 0.935^c$	$d = 0.960^d$	Appearance
*Acetone	1.2	1.2	1.0	No change
Acetic acid, 5%	0.8	0.8	0.9	No change
Aqueous ammonia, 10%	0.0	0.0	0.0	No change
*Butyl acetate	4.1	3.4	3.4	No change
Calcium chloride, 2.5%	0.0	0.0	0.0	No change
*Carbon disulfide	36.8	21.4	12.9	Swollen
*Carbon tetrachloride	37.9	22.8	16.3	Swollen
*Cellosolve	0.3	0.5	0.5	No change
*Chloroform	25.1	16.2	12.0	Swollen and warped
*Citric acid, 10%	0.0	0.0	0.0	No change
*Ethanol, 50%	0.1	0.1	0.1	No change
*Ethanol, 95%	0.0	0.2	0.2	No change
*Ethyl acetate	2.8	2.5	2.5	No change
*Ethylene glycol	0.0	0.0	0.0	No change
*Ethylene dichloride	6.9	5.4	5.0	Very slight yellowing, slightly swollen
*Formaldehyde, 35%	0.1	0.1	0.1	No change
*Gasoline (regular)	13.5	8.8	6.7	Swollen and yellowed
*Gasoline (aviation)	15.1	10.0	5.7	Swollen and yellowed
*Glycerol	0.0	0.0	0.0	No change
*Green soap solution	0.0	0.1	0.1	No change
*Heptane	10.0	6.9	0.7	Swollen
Hydrochloric acid, 10%	−0.2	0.0	0.0	No change
Hydrogen peroxide, 30%	0.0	0.1	0.0	No change
Methanol, 5%	0.0	0.0	0.1	No change
*Methanol, 100%	0.0	0.1	0.1	No change
*Motor oil	5.0	2.2	1.1	No change
Nitric acid (conc.)	4.8	1.9	1.4	Yellowed
*Oleic acid	2.4	1.7	1.4	No change
*Olive oil	0.3	0.2	0.2	No change
*Phenol, 5%	0.2	0.1	0.2	No change
Sodium chloride, 10%	0.0	0.0	0.0	No change
Sodium carbonate, 2%	0.0	0.0	0.0	No change
Sodium hydroxide, 10%	0.1	0.0	0.1	No change
Sodium hydroxide, 1%	0.0	0.0	0.1	No change
Sodium hypochlorite, 2%	0.0	0.0	0.0	No change
Sulfuric acid, 30%	0.0	0.0	0.0	No change

TABLE 5 (*continued*)

| Solvent | Increase in weight (%) | | | Appearance |
	$d = 0.918^b$	$d = 0.935^c$	$d = 0.960^d$	
Sulfuric acid, 3%	0.0	0.0	0.1	No change
*Toluene	15.1	9.8	7.5	Swollen
*Turpentine	14.5	9.1	7.2	Swollen
Water	0.0	0.0	0.0	No change
*Wesson oil	0.2	0.2	0.1	No change
*Xylene	15.4	10.3	7.9	Swollen

* Stress-cracking agent.
[a] Results on injection-molded disks, 2 in. in diameter, 1/8 in. thick, immersed and stored at 73°F for 1 yr.
[b] Low-density.
[c] Medium-density.
[d] High-density polymer.

difference in phase equilibria between polyethylene and different solvents has not been appreciated in selecting solvents for fractionation studies [181].

C. FRACTIONATION†

For the characterization of polymers subdivision into fractions of a narrow molecular-weight range is usually required.

In the fractionation of polyethylenes several experimental difficulties develop. Most fractionations must be carried out at elevated temperatures because of the limited solubility at room temperature. Furthermore, a main difficulty in obtaining consistent fractions is that from most solvents polyethylene separates as a semicrystalline phase instead of as the concentrated solution, or gel phase, that usually occurs with most amorphous polymers such as polystyrene. The separation of the crystalline phase is usually a poorly reversible process, attended by large supercooling effects. In a case like this the crystallization rate plays a role comparable in importance to the equilibrium solubility. Whereas equilibrium factors favor separation of species of the higher molecular weights, the inherent slow rate of crystallization may prevent them from

† See Volume I, Chapter 15.

separating. Unless proper care is taken in the selection of a solvent for fractionation, the intermediate fractions may have molecular weights higher than those removed in the early stages of the fractionation and the fractionation is consequently misleading. Unfortunately, in some of the reported fractionation studies of polyethylenes, this has not been taken into consideration and the solvents used are not those that give liquid-liquid separation.

Another procedure based on the addition of a nonsolvent (e.g., propanol) to a solution of polyethylene in toluene has also been described [182, 183]. In these studies the values of the intrinsic viscosity of the intermediate fractions were not found to be higher than for the fractions that separated first, but in a number of fractionations of DYNH polyethylene by the procedure described [182] it was found that some of the intermediate fractions did have a higher intrinsic viscosity than that of the fractions which had separated earlier [5]. It appears that solvents that do not give liquid-liquid separation of phases should be avoided, and neither of these two procedures is suitable for efficient fractionation.

Other work describes successful fractionation of several commercial polyethylenes, xylene being used as solvent and triethylene glycol as nonsolvent. The fractionation was done at 130°C by application of both extraction and precipitation techniques [184]. Another procedure that may be used successfully involves cooling slowly a solution of polyethylene in amyl acetate [5]. Starting at about 130°C, most of the fractions of high molecular weight separate in gel form down to about 105°C. Below 105°C the phase that separates is crystalline, but because of the prior removal of most of the material of higher molecular weight the fractions separating below 105°C in crystalline form are satisfactory.

Marlex 50 (a Hercules linear polyethylene) has been successfully fractionated in p-xylene at 85 and 90°C [185].

D. MOLECULAR WEIGHT AND MOLECULAR-WEIGHT DISTRIBUTION

The determinations of the molecular weights of polyethylenes from measurements of such properties as osmotic pressure and light scattering are beset with considerable experimental difficulty because of the necessity of making measurements at temperatures above 70°C. Techniques have been developed, however, to make osmotic pressure measurements at elevated temperatures to determine the number-average molecular weight of both unfractionated and fractionated samples [182, 186–188].

Osmometers of the Zimm-Myerson type [189] modified for use at high temperatures [183, 184] have been found quite suitable for osmotic pressure measurements of polyethylene solutions. For low-density polyethylenes of low molecular weight (number-average < 6000) methods based on elevation of the boiling point [186–190] and depression of the freezing point [191] have been used.

Because of the convenience of measuring intrinsic viscosity, in contrast to measurements of molecular weights by methods such as osmotic pressure and light scattering, attempts have been made to relate intrinsic viscosity to molecular weights determined by other methods. From osmotic pressure and ebulliometric measurements on a number of unfractionated polyethylenes the following relation between intrinsic viscosity $[\eta]$, measured in xylene at 75°C, and number-average molecular weight $\overline{M}_n$ was proposed early [186].

$$[\eta](\text{g/liter})^{-1} = 1.35 \times 10^{-4}\overline{M}_n^{0.63}.$$

The viscosity-average molecular weight calculated from intrinsic viscosity is closer to a weight-average than to a number-average molecular weight (cf. Volume I, Chapters 2 and 15). It is doubtful that the relation between number-average molecular weight and intrinsic viscosity can be of general validity for unfractionated polyethylene samples. At best, relations of this type can be valid only for polyethylenes that have similar molecular-weight distribution curves [192].

A more generally valid relation between molecular weight and intrinsic viscosity has been obtained from the measurements of osmotic pressure of polyethylene fractions obtained by cooling amyl acetate solutions [124]. This relation is for $[\eta]$ measured in toluene at 80°C. Since this relation

$$[\eta](\text{g/100 cc})^{-1} = 7.25 \times 10^{-5}\overline{M}_n^{0.85}$$

was obtained on consistent fractions of polyethylene, application of this relation to unfractionated polyethylene can be expected to give molecular weights closer to the viscosity average. It should also be pointed out here that if the polyethylene has long-chain branching its intrinsic viscosity will be lower than that of a linear polymer of the same weight-average molecular weight [193–195]. The molecular weight of polyethylene with long-chain branching calculated from an intrinsic viscosity-molecular weight relation will therefore be lower than the true viscosity-average molecular weight.

Other relationships of molecular weight and intrinsic viscosity have been published. For linear polyethylene in the range of molecular weights of 50,000 to 6 million, measured in 1-chloronaphthalene at 125°C, the following relationship was valid [196]:

$$[\eta](g/100 \text{ cc})^{-1} = 4.3 \times 10^{-4} \overline{M}_w^{0.67}.$$

Another relationship

$$[\eta] = 2.36 \times 10^{-4} \overline{M}_w^{0.78}$$

has been reported for linear polyethylenes of molecular weights varying from 50,000 to 1.5 million; viscosities were measured in tetralin at 120°C [197].

Gloor gives the following relationship for Ziegler polyethylenes, measured in decahydronaphthalene at 135°C [198]:

$$[\eta] = 4.6 \times 10^{-4} \overline{M}_v^{0.73},$$

where $\overline{M}_v$ is the viscosity-average molecular weight of whole polymer polyethylenes. The relation between $\overline{M}_v$ and $[\eta]$ is by definition independent of the molecular-weight distribution.

For high-density (linear) polyethylenes the following relation has been used [199]. Measurements were made in tetrahydronaphthalene at 130°C.

$$[\eta] = 3.78 \times 10^{-4} \overline{M}_w^{0.72}.$$

Light-scattering measurements have been made on solutions of low-density polyethylene in 1-chloronaphthalene at temperatures of about 125°C to determine weight-average molecular weights [200–202]. The polyethylene solutions are difficult to clarify by filtration and light-scattering measurements may be in error because of the unavoidable presence of dust or other large scattering particles. In spite of this uncertainty, the noteworthy fact is that the ratios of the weight-average to number-average molecular weights for most of the polyethylenes are notably higher than for most of the common polymers. Ratios of weight-average to number-average molecular weights as high as 40 to 70 have been reported for some samples of polyethylene [191, 200, 201]. For polymers that show exponential distribution of molecular weights $[W(m) = ame^{-am}$, where $W(m)$ is the weight fraction of a species of molecular weight m] the ratio between weight- and number-average molecular weights is 2. The high ratios between weight-average and number-average molecular weights is indicative of broad molecular-weight distribution curves. The molecular-weight distribution curves in the high-molecular-weight range for most polyethylenes are of the form

$$W(m) = bm^{-n}e^{-am},$$

where n has values between 0 and 1.0. This broad distribution of molecular weights when $W(m)$ is plotted against m is a consequence of chain transfer by a growing radical to a polymer molecule already formed, thus leading to the development of long-chain branching.

VIII. Rheological Properties of Molten Polyethylene

Methods of extruding molten polymers have been described in detail in Volume I, Chapter 8, and the basic rheology of such molten polymers forms the content of Chapter 6. Although Chapter 8 is not concerned exclusively with polyethylene extrusion, to be sure, most of the discussion and many of the data apply directly to polyethylenes.

Most polyethylene film is manufactured today by the blown tubing process (Volume I, p. 410 ff), and of considerable importance is the design of extruders of high capacity which are suited to polyethylenes of different flow properties. Considerable attention has been devoted to the development of principles for the design of extruders and extrusion dies which take into account the flow properties of the molten polymer.

The importance of the melt viscosity of polyethylene has been appreciated in an empirical way ever since the commercial manufacture of this polymer began. In the industry the measurements of *melt index*,* which is an empirical measure of the flow properties of the polymer, have long been utilized in the characterization and grading of polyethylenes [203]. Melt index, though useful in grading polyethylenes of widely different flow characteristics, is not an adequate measure of the rheological properties or of viscosity of the molten polymer. It is not uncommon to find polyethylenes of the same melt index with markedly different rheological properties.

* The melt index is the weight in grams of polyethylene extruded in 10 min at 190°C through an orifice of specified diameter when a given weight is placed on the driving piston. Melt flow is used for linear polyethylenes; it has the same meaning as melt index, but at 202°C.

Since melt index is used so commonly in the characterization of polyethylenes, it is pertinent to mention the limitations of this criterion. A study of the melt viscosities of a variety of polyethylene resins by capillary viscometers and the results compared with the values of the melt index of the same samples showed [5] a lack of correspondence between the melt-viscosity and melt-index values that could be explained by the following considerations:

1. Molten polyethylene shows a notably non-Newtonian behavior; that is, at shearing stresses of the magnitude of those under which the melt-index apparatus operates the rate of shear is not linear with shearing stress. (In the measurements by capillary viscosimeters the shearing stress used is so small that the flow behavior is nearly Newtonian.) The relations between shearing stress and rates of shear (a measure of the non-Newtonian character) for the melts from different polyethylenes may vary appreciably.

2. The deformation of the polymer melt under a given shearing stress is dependent on time, and in the measurements of the melt index account is not taken of the entrance and exit corrections in the flow of the melt through the orifice used in the apparatus. These corrections would be expected to vary for samples of different flow characteristics.

Other sources of error also exist in the use of the melt-index apparatus which limit its ability to characterize the flow properties of polyethylene samples adequately, but convenience in manufacturing operations makes it a useful guide.

Considering the viscosity of the melt measured at low rates of shear in which the contribution of elastic shear strain is negligible and the flow is Newtonian, we should expect that a relation between melt viscosity and weight-average molecular weight exists. For polymers in general it has been proposed that melt viscosity varies as the 3.4 power of the weight-average molecular weight [204], provided that the molecular weight is above a critical value. This is in approximate agreement with the dependence of isothermal melt viscosity on molecular weight expected from theoretical considerations [205]. It has been shown that the melt viscosity of polyethylene at a given temperature also varies as the 3.4 power of the weight-average molecular weight, provided that appropriate correction for the number of short chains per hundred methylene groups is made [206]. Long-chain branching of polyethylene was shown to have no appreciable effect on melt viscosity. The relation between melt viscosity and weight-average molecular weights was found to be

$$\eta_0 = 3.01 \times 10^{-12} \, \overline{M}_w^{3.4} \, e^{-2.35 N_c},$$

where η_0 represents the isothermal viscosity of the melt under Newtonian conditions of flow, $\overline{M}_w$ is the weight-average molecular weight, and N_c is the number of methyl groups (equal to the number of short-chain branches) per hundred methylene groups [206].

IX. Degradation and Environmental Protection of Polyethylene

The degradation of polyethylene has been studied over a wide temperature range in the presence and absence of air or oxygen, with and without the influence of light [207–210].

In the absence of oxygen polyethylene is stable up to about 300°C; at higher temperatures degradation occurs progressively, but unlike polystyrene and the acrylates, which are degraded to monomer, polyethylene yields products that are like the original polymer but of a lower degree of polymerization. Further degradation yields liquids but only above 370°C are appreciable amounts of gaseous products formed. About 30 such gaseous products consisting of straight-chain alkenes, alkanes, and dienes result.

It is likely that weak links in the polymer chain are first ruptured and that pyrolysis then proceeds by a chain reaction, particularly at higher temperatures. Such weak links may be at branches or in positions adjacent to carbonyl or other oxygenated structures in the chain. Unbranched polymethylene prepared by the decomposition of diazomethane is more stable than ordinary commercial polyethylenes, polypropylene, or polyisobutylene and shows some differences in degradation products, notably in an increase in the amount of unsaturated products [211–216].

The oxidative aging of polyethylene appears to be an autocatalytic free-radical reaction, which can be inhibited by carbon black (known to be effective in deactivating free radicals) and by standard antioxidants [207]. Even more effective are combinations of 0.5 to 5% carbon black of particle size < 1000 Å with dodecylmercaptan or RSSR' compounds (e.g., diphenyl sulfide and dibenzothiazolyl sulfide); either carbon or RSSR' alone is much less effective [217]. A variety of other antioxidants, complex phenols or phosphorus ester S-amides or S-sulfenamides, for example, 4,4'-benzylidenebis(2-*tert*-butylphenol) [218] and $(C_2H_5O)_2P$-$(=S)SSN(CH_3)_2$ [219] has been recommended.

The oxidation of polyethylene has been investigated in oxygen and air at temperatures of 140 to 225°C, in oxygen-ozone mixtures at 25 to 109°C, and in 90% fuming nitric acid at 25 to 80°C [207–210, 220]. Structural changes were followed by infrared spectroscopy [84, 221]. In oxygen the O—H band at 2.9 μ developed rapidly at 150 to 210°C and soon reached a maximum, indicating the formation of hydroperoxide links. The 5.9-μ band ($\overset{\backslash}{C}O$ absorption) continued to increase throughout aging, but C—O—C bands at 8–8.5 μ were the last to appear, signifying crosslinking by secondary reactions. Among the evolved products water, carbon dioxide, formaldehyde, and aliphatic acids with ketone and aldehyde groups present were identified. Mixtures of ozone and oxygen degraded polyethylene faster than oxygen, but the chemical processes involved appear to be the same. Fuming nitric acid at low temperatures produced mainly chain scission, since the C=C band at 6.08 μ became pronounced; at higher temperatures nitration became the principal reaction.

The rate of oxidation of polyethylene at room temperature in the absence of light is negligible, but photooxidation is rapid. The degradation can be inhibited by special antioxidants that permit the manufacture of transparent film.

X. Surface Treatment of Polyethylene for Ink and Coating Adhesion

A. INTRODUCTION

The film surfaces of polyolefins are notably indifferent to the adhesion of inks or coatings, and since the first introduction of polyethylene to the packaging film market this has been a handicap. Later the problems were found to be more difficult to solve with isotactic polypropylene.

In the early days nonadhesion was believed to be the result of an almost total lack of polar groups in the chemical surface of the film and most research was directed toward the introduction of such groups through oxidation of the surface by various means, though mechanical devices such as prestretching and printing on heated rolls were also used.

A bewildering variety of chemical agents was tried, usually in the presence of air and frequently in ultraviolet light. These agents included use of chlorine [222, 223], ozone—alone or with halogens or halogen acids, nitrous oxide with ultraviolet radiation [224, 225], fluorine [226], acidic dichromate solutions [227], chlorinated hydrocarbons in ultraviolet radiation [228], chlorine dioxide, thionyl chloride, sulfuryl chloride, nitrosyl chloride, and nitrogen dioxide [229].

One searches the literature in vain for the underlying causes of effectiveness of treatment. Most patents make obeisance to oxidation as the primary effect, without stating convincing proof, and this easy explanation becomes less convincing when considered with respect to the various heating and electric treatments that have become standard practice in the industry. The single brave attempt to ascribe the whole effect to ultraviolet light [230] includes this sentence: "Bright sunlight, at sea-level elevation and in the latitude of London, necessitates a very long exposure of three months or longer, and for this reason is not recommended as a source of ultraviolet light for the purposes of the present invention." That may well qualify for an award as the patent understatement of midcentury.

B. COMMERCIAL TREATING METHODS

The earliest successful process in commercial use for treating polyethylene was that employed in Cambridge, Massachusetts, in 1951 by the Harwid Co., acquired by Olin Industries in 1953, and operated for many years at the Pisgah Forest plant of the Olin Corporation. This process, as first practiced, used the original Traver design in which the film was passed at less than 150 fpm beneath a bank of electric discharge tubes. The film was carried by an electrically conducting material so that the combination formed a condenser in which an electric discharge field was established about the material to be treated [231, 232]. The fundamental cause of the effectiveness of treatment was vague (e.g., the treating area reeked of ozone), but the process was effective in taking the film directly from a quench tank in line with the extruder and treating and winding it immediately. Some trouble with blocked finished rolls was encountered, but by adjustment of the time and intensity of treatment this was minimized. It was a great improvement over the separate operation of unwinding extruded rolls of film, treating, and rewinding and could be done with rolls up to about 56 in. wide.

It soon became apparent from laboratory experiments that the expensive, fragile glow-discharge tubes were superfluous. Equally good results were achieved by use of a brass electrode with its knife edge mounted above the moving film; the glow discharge set up at 15 to 20,000 V and 2 to 5 kc could be more easily controlled than by use of discharge tubes.   The cost of treatment was thereby reduced from about 27 to 3 cents a pound of film [233] at speeds up to about 250 fpm.   The corona discharge method of treating polyethylene film (utilizing various types of electrode) rapidly became the preferred one [234–238] and later was applied to isotactic polypropylene [239].

In addition to the claims made for infrared radiation [240], it was shown that effective treatment was achieved by impingement of an oxidizing gas flame [241–244].   These methods of treatment stressed the need for maintaining a temperature differential on the two sides of the film, one face in contact with a chilled roll, but any theoretical scientific basis for this belief is lacking.   The method was also much used for treatment of irregularly shaped objects, such as bottles and tubes, which could not readily be treated in any other way.   A nice control was required to avoid melting a film while a flame at 3500°F was impinging on one surface, and flame treatment of film never found wide acceptance.

Several investigators, but especially Kreidl, maintained that the effect of treatment was oxidation of the film surface, but the evidence presented [245] for polar groups such as $\diagdown\!C\!O$, $-CO_2H$, and $-OH$ or unsaturation such as $-CH\!=\!CH-$ or $-CH\!=\!CH_2$ was not convincing. It was stated that a total of 12 to 20 surfaces of treated film had to be superimposed to give a measurable effect, and even then the reduction in transmission was only about 2% in the region of 6 μ (there is a $\diagdown\!C\!O$ absorption at 5.85 μ and a $-CH\!=\!CH-$ absorption at 6.10 μ).

We were unable to detect by the most sensitive methods known to us a measurable increase in oxygen after treatment of polyethylene *by the commercial electric discharge treatment* or chromic acid treatment, though we well knew that *heavily overtreated* film subjected to an electric discharge for 15 min or more exhibits a substantial increase in carbonyl oxygen.   Washing the overtreated film in cold methyl alcohol removed about 85% of the carbonyl compounds, yet the adhesion of ink and coatings was not decreased by the washing process.   This indicated that factors other than oxidation of the surface are important for the adhesion of ink and coatings.

It has also been reported [246] that by exposing polyethylene films at 400°C for short contact times in a flow system comparable to commercial equipment it was found that both high- and low-density polyethylenes readily undergo random scission of the hydrocarbon chain to create vinyl unsaturation sites. Only small amounts of volatile materials were produced, but significant amounts of vinyl, vinylidene, and internal unsaturation were all produced during degradation of branched resins and only vinyl unsaturation with unbranched resins.

In addition to electric discharge, flame treatment, and treatment with various oxidizing agents, it is also possible to achieve some measure of ink adhesion by dipping the article in hot solvents (e.g., tetrachloroethylene, trichloroethylene, or toluene) at 85°C for 30 sec and printing or coating within 24 hr. This method is effective in swelling or softening the olefin surface slightly but is of no practical value for film treatment.

C. EVIDENCES OF FILM TREATMENT

Empirically it has been found that treatment either by flame or by electric discharge causes changes in several properties of a film that must be accounted for in any theory of treatment.

1. Printing inks and coatings are more adherent than to untreated film. This has been the basis of extent-of-treatment testing. In the "Scotch-tape" test an ink is rolled on with a hand-proofing roll, allowed to dry for 5 to 8 min, a strip of tape is pressed over the ink, and stripped off smartly. The percentage of ink remaining on the film can be judged in several ways as a measure of treatment sufficiency.

2. Treated film is more readily stained by methylene blue, Ziehl-Neelsen stain, or several other solutions used in bacteriological work. In the test 1 to 5 drops of solution are applied to the film, blotted off after 30 sec, and the intensity of stain left is measured.

3. The heat-seal strength of overtreated film is decreased to near-zero, and the efficacy of treatment for ink retention can be judged by a comparison of the strength of heat seals of treated and untreated film over time.

4. Water forms discrete droplets on untreated film but tends to form a continuous layer on treated film. Discrete drops form in lightly treated samples but the contact angle is much greater on treated film than on untreated.

5. Carbonyl compounds can be eluted from the surface of heavily treated samples by solvents such as methanol and ethyl ether. The ability of the base sheet to hold ink is not reduced after elution, but heat-seal strengths are greatly increased (i.e., stronger seals exist between layers).

6. Effectiveness of a given type of treatment is fairly consistent in lot to lot of polymer from one manufacturing process but varies greatly with the type of polymer from different suppliers.

D. DEGRADATION AND AGGREGATION MECHANISMS

A full explanation of these observations is complex, but an attempt at understanding must consider the ease with which polyethylene undergoes various degradative and aggregative changes on exposure to high temperature and to ultraviolet light in the presence or absence of oxygen. The degradative changes proceed by chain scission and result in the breakdown of molecules into small segments, whereas the aggregative changes occur by progressive interlinking of chains to build up large, highly branched molecules. In polyethylene these two opposite changes probably proceed simultaneously when the film is exposed to high temperature (which can be produced locally either by a flame or by electric discharge bombardment with electrons) or ultraviolet light. The presence of oxygen, as seen from the following discussion, plays an important part. Also, these processes are autocatalytic and depend markedly on the nature of the polymer constituting the polyethylene film.

When polyethylene is exposed to high temperatures, the following changes occur:

1. Low-molecular-weight fragments, of an average molecular weight of 700, form. These fragments can be separated by molecular distillation [208, 247].

2. There is an increase in melt viscosity of the polymer on heating, and material forms which swells considerably in xylene at 60°C, whereas almost all the material before heating in air is soluble under these conditions [229].

3. An increase in the carbonyl content of the polymer occurs, as measured by infrared absorption [84, 92].

The mechanism by which the degradation of polyethylene and other olefin polymers proceeds is now fairly well understood. In thermal degradation free-radicals are formed first. The exact mechanism of the initiation of free-radicals is not well understood but it is not difficult to visualize. In polyethylene, for example, a small concentration of $\diagdown \!\!\!\! C\!=\!C \diagup$ bonds and therefore some hydroperoxide groups

—OOH will probably always be present.   Initiation of free-radicals can proceed at any of these sites.   After the initiation of free-radicals further breakdown of the polymer proceeds by an autocatalytic process in which oxygen plays an important part.   The following steps illustrate the degradation in the case of polyethylene.

1.   The free-radical R· may stabilize itself by transferring its activity to a carbon atom of a polyethylene chain which has a tertiary hydrogen or is adjacent to a double bond.   In low- and medium-density polyethylene there are many branch points at which this can happen; and the greater difficulty of treatment of linear polyethylene is contributory evidence to the correctness of the mechanism described, since it may be expected to possess fewer such sites.   The greater susceptibility of polypropylene to oxidation can also be explained by the frequency of tertiary hydrogen atoms along the carbon chain.

$$R\cdot + \sim CH_2-CH_2-CH-CH_2-CH_2\sim CH_3 \longrightarrow$$
$$\qquad\qquad\qquad\quad\; | $$
$$\qquad\qquad\qquad\quad CH_2$$
$$\qquad\qquad\qquad\quad\; \wr$$
$$\qquad\qquad\qquad\quad CH_3$$

$$RH + \sim CH_2-CH_2-\overset{\cdot}{C}-CH_2-CH_2\sim CH_3$$
$$\qquad\qquad\qquad\qquad | $$
$$\qquad\qquad\qquad\quad\; CH_2$$
$$\qquad\qquad\qquad\qquad \wr$$
$$\qquad\qquad\qquad\quad\; CH_3$$
$$\qquad\qquad\qquad\qquad \textbf{A}$$

2.   The new free-radical **A** acquires an oxygen molecule, which may then extract a tertiary hydrogen atom from another molecule at a branch site by intermolecular transfer, forming a hydroperoxide and a new radical **D**.

$$\qquad\qquad\qquad\qquad\qquad O_2$$
$$\qquad\qquad\qquad\qquad\qquad | $$
$$A + O_2 \longrightarrow \sim CH_2-CH_2-C-CH_2-CH_2\sim CH_3$$
$$\qquad\qquad\qquad\qquad\qquad | $$
$$\qquad\qquad\qquad\qquad\quad CH_2$$
$$\qquad\qquad\qquad\qquad\qquad \wr$$
$$\qquad\qquad\qquad\qquad\quad CH_3$$
$$\qquad\qquad\qquad\qquad\qquad \textbf{B}$$

$$\text{B} + \sim\!CH_2\!-\!CH_2\!-\!\underset{\underset{\displaystyle CH_3}{\overset{\displaystyle |}{\overset{\displaystyle CH_2}{|}}}}{CH}\!-\!CH_2\!-\!CH_2\!\sim\!CH_3 \longrightarrow$$

$$\sim\!CH_2\!-\!CH_2\!-\!\underset{\underset{\displaystyle CH_3}{\overset{\displaystyle |}{\overset{\displaystyle CH_2}{|}}}}{\overset{\overset{\displaystyle O\!-\!O\!-\!H}{\displaystyle |}}{C}}\!-\!CH_2\!-\!CH_2\!\sim\!CH_3 + \text{D}$$

$$\text{C}$$

Alternatively, peroxide free-radicals such as **B** can be stabilized by intramolecular reaction to form hydroperoxide free-radicals **E**.

$$\text{B} \longrightarrow \sim\!CH_2\!-\!CH_2\!-\!\underset{\underset{\displaystyle CH_3}{\overset{\displaystyle |}{\overset{\displaystyle CH_2}{|}}}}{\overset{\overset{\displaystyle O\!-\!O\!-\!H}{\displaystyle |}}{C}}\!-\!CH_2\!-\!\overset{.}{C}H\!\sim\!CH_3$$

$$\text{E}$$

3. The hydroperoxide in structures such as **C** and **E** is quite unstable and undergoes decomposition by breaking at the peroxide bond; this is followed by scission of chains, which produces other free-radicals and stable products of lower molecular weight.

$$\text{C} \longrightarrow \cdot OH + \sim\!CH_2\!-\!CH_2\!-\!\underset{\underset{\displaystyle CH_3}{\overset{\displaystyle |}{\overset{\displaystyle CH_2}{|}}}}{\overset{\overset{\displaystyle O}{\displaystyle |}}{C}}\!\big\}\!CH_2\!-\!CH_2\!\sim\!CH_3$$

$$\downarrow$$

$$\overset{\displaystyle O}{\underset{\displaystyle \parallel}{}}$$

$$\text{E} \longrightarrow \sim CH_2-CH_2-\overset{\overset{\displaystyle O}{|}}{\underset{\underset{\underset{\displaystyle CH_3}{\overset{\displaystyle |}{CH_2}}}{|}}{C}}\!\!\!\}\ CH_2-\overset{\overset{\displaystyle OH}{|}}{CH}\sim CH_3$$

$$\downarrow$$

$$\sim CH_2-CH_2-\overset{\overset{\displaystyle O}{\|}}{\underset{\underset{\underset{\displaystyle CH_3}{\overset{\displaystyle |}{CH_2}}}{|}}{C}} \quad + \quad \cdot CH_2-\overset{\overset{\displaystyle OH}{|}}{CH}\sim CH_3$$

This process may lead to degradations such as the following sequence.

$$\cdot CH_2-\overset{\overset{\displaystyle OH}{|}}{CH}\sim CH_3 \longrightarrow CH_3-\overset{\overset{\displaystyle O}{|}}{CH}\sim CH_3 \longrightarrow$$
$$CH_3\cdot + O{=}CH\sim CH_3 \text{ etc.}$$

Another mechanism by which degradation may occur, and which does not require oxygen to proceed, is as follows.

$$\sim CH_2-CH_2-\overset{\overset{\displaystyle \cdot}{\underset{\underset{\displaystyle CH_2}{|}}{C}}}{}-CH_2-CH_2\sim \longrightarrow$$
$$\sim CH_2-CH_2-\underset{\underset{\displaystyle CH_2}{|}}{C}{=}CH_2 + \cdot CH_2\sim$$

Which of these two mechanisms operates in the degradation of a polyolefin depends on the conditions and the structure of the polymer. Also, the size of the fragments produced as a result of such processes depends on the conditions under which degradation occurs and the extent of branching in the polymer. It has been shown by careful studies [84, 92] that heat degradation in the presence of oxygen proceeds by the first mechanism and results in products containing carbonyl groups, whereas degradation by ultraviolet light proceeds through the second mechanism and results in an increase of the double-bond concentration.

Degradation of the polymer is not the only change that is brought about by heat. It has been generally observed that heating of polyethylene, especially in the presence of air or oxygen, results in an appreciable increase in the melt viscosity. This suggests that there is an accompanying aggregative buildup of molecules by crosslinking to give molecules of higher molecular weight. The buildup may be a result of the intermolecular combination of two radicals or the addition of a free-radical to a residual double bond in another molecule. Thus

$$2 \sim CH_2-CH_2-\overset{\displaystyle .}{C}-CH_2\sim \longrightarrow$$
$$\underset{\displaystyle \overset{|}{CH_3}}{\overset{|}{CH_2}}$$

$$\sim CH_2-CH_2-\underset{\displaystyle \overset{|}{CH_2}}{\overset{\overset{\displaystyle CH_2}{|}}{C}} - \underset{\displaystyle \overset{|}{CH_2}}{\overset{\overset{\displaystyle CH_2}{|}}{C}} -CH_2-CH_2\sim$$

with CH_3 groups at the ends.

or

$$\sim CH_2-CH_2-\overset{\displaystyle .}{\underset{\displaystyle \overset{|}{CH_3}}{\overset{|}{C}}}-CH_2\sim \; + \; \sim CH=CH-CH_2-\underset{\displaystyle \overset{|}{CH_2}}{\overset{|}{CH}}\sim \longrightarrow$$

$$\sim CH_2-CH_2-\overset{\displaystyle CH_2}{\underset{\displaystyle \overset{|}{CH_3}}{\overset{|}{\underset{\displaystyle \overset{|}{CH_2}}{C}}}}-CH-\overset{\displaystyle .}{CH}-CH_2-\underset{\displaystyle \overset{|}{CH_2}}{\overset{|}{CH}}\sim$$

It appears that treatment of polyethylene by exposure to a transient gas flame or to an electric discharge would in all probability result in degradation and crosslinking at the surface. If the degradation proceeds as a degradative oxidation, it will produce some carbonyl and other groups such as hydroxyl or hydroperoxide.

Electron micrography provides evidence that the electric-discharge treatment produces mechanical roughening or pitting of the surface of polyethylene.

E. Experimental Evidence for a Treatment Mechanism

A tremendous amount of research has been done to elucidate the mechanism of flame and corona discharge treatment, but most of the work was done on heavily overtreated samples and serves only to confirm the possible explanations given above. The classic studies of these polymers [207–210] have been confirmed repeatedly (e.g., see [248, 249]), and probably no one doubts the essential correctness of the likelihood of surface oxidation of polyolefins in the corona discharge over a period of time; but it is manifestly doubtful if this effect is the primary cause for ink and coating adhesion in the *commercial* treatment of polyolefin films when the time of exposure is of the order of 10^{-7} sec.

1. Contact Angles of Liquids

As noted above, the contact angle of water on untreated film is noticeably less than on treated film; quantitative data are given in Table 6.

TABLE 6

Contact Angles of Water on Polyethylene Film Surfaces

Sample	Contact angles with water
1. Untreated polyethylene film	104°
2. Polyethylene film treated by corona discharge[a]	122°
3. Polyethylene film treated by Traver process[b]	112°
4. Untreated side of sample 3	102°
5. Marlex-50 film, untreated	98°

[a] Berthold-Pace process [233].
[b] [232].

To measure such angles a low-power microscope, fitted with an eyepiece (with crosshair and graduated circle) and a suitable stage, was employed. The crosshair was first aligned with the surface of the polyethylene sample at the point of contact of a drop.   The eyepiece was then rotated until the crosshair was tangent to the side of the water drop at the point of contact with the film.   The angle of rotation of the eyepiece was taken as the contact angle.

Contact angles measured for several other liquids gave qualitatively similar results, but the differences in angles for water were larger.

The contact angles for a variety of base sheets averaged about 104° for untreated film and about 122° for film treated by the electric-discharge method.   (The lower efficiency of the Traver process is reflected in the intermediate value of 112°.)

2. Solvent Extraction

Samples of commercially treated films were washed with several solvents and the refractive index of the solvents was compared with the analogous wash liquid from untreated film.   Table 7 presents some of the data.

TABLE 7

Refractive Index Differences between Solvents and
Extracts of Treated and Untreated Polyethylene Films

Solvent	Extracts	Refractive index difference between solvent and extract (± 0.00001)
Carbon tetrachloride	Treated	0.00002
	Untreated	0.00001
Toluene	Treated	0.00014
	Untreated	0.00012
0.1% aqueous sodium lauryl sulfate	Treated	0.00006
	Untreated	0.00006
1-Chloronaphthalene (25°C)	Treated	0.00001
	Untreated	0.00001
1-Chloronaphthalene (65°C)	Treated	0.00000
	Untreated	0.00000

The differences in the refractive indexes between the solvents and the extracts were determined with the Brice-Phoenix Differential Refractometer [250]. From the sensitivity of the refractometer it was estimated that about 0.1% polyethylene in the extracts could be detected by this method. The samples of films were shaken with 5 ml of the solvent in small weighing bottles, the solution was transferred to the cell of the differential refractometer with a hypodermic syringe, and the difference between the refractive indexes of the extract and the solvent was measured.

These and a variety of similar results with other solvents showed refractive indexes within experimental error. Thus it appears that the same amount of extractable material is present on the surface of the treated and untreated films, and the electric-discharge treatment of the film does not result in the formation of low molecular weight material in appreciable amount. These results, however, do not preclude the possibility of formation during a discharge process of products of molecular weight which are not extracted by solvents under the experimental conditions used.

3. Infrared Studies

Extensive studies were made by infrared spectroscopy in the 2- to 15-μ region in an attempt to find chemical differences between untreated and commercially treated polyethylene films. Both reflection and transmission spectra were made.

Reflection spectra of a set of two films showed that the treated film gives the same spectrum as the untreated film.

Transmission spectra were obtained from samples ranging from 15 to 37 mils in thickness. All spectra were carefully examined in the 5.8- to 6.1-μ region to ascertain any differences in the content of $\searrow CO \nearrow$ and $\searrow C{=}C \nearrow$ unsaturation in the treated and untreated films. Spectra of the treated and untreated films were also obtained in the 2- to 3-μ region to determine the relative amounts of other oxygen-containing groups such as hydroperoxide $-OOH$. Within experimental error no differences were found between the treated and the untreated polyethylene films in the amounts of carbonyl and $\searrow C{=}C \nearrow$ double bonds. Neither the treated nor the untreated polyethylene showed any absorption at 2.81 μ, where hydroperoxide absorption should appear.

When treated and untreated film samples were held at 150°C for 30 to 40 min, a small peak at 2.81 μ developed in untreated polyethylene film, as reported previously [92], but no peak developed in the treated film. Thus there is some indication that the thermal degradative reactions that give rise to the hydroperoxide groups in polyethylene are suppressed in the treated film.

4. Polymerization by Active Centers on the Film Surface

To investigate the possibility of the formation of hydroperoxide groups and an increase in $\overset{\backslash}{\underset{/}{C}}{=}\overset{/}{\underset{\backslash}{C}}$ concentration on the film surface by the electric-discharge treatment, studies were made of the polymerization rate of styrene and the molecular weights of polystyrene formed by polymerizing styrene in the presence and absence of treated or untreated polyethylene films.

It was expected that if hydroperoxide groups were present in the treated films and absent in the untreated films the polymerization rate in the presence of treated films would be greater than that in the presence of untreated films because of the catalytic effect of the hydroperoxide groups.

Further, if the treated polyethylene films had a greater number of unsaturated groups than the original untreated film, this might be detectable from the relative efficiencies of these films as chain-transfer agents, for when a monomer like styrene is polymerized in the presence of a polymer (polyethylene) the latter may act as a chain-transfer agent; that is, the active chains for the polymerizing monomer may abstract a hydrogen from the polymer, thereby terminating the growing chain but creating a new site from which polymerization may proceed. This would result in lowering the molecular weight of the polymer formed from the polymerizing monomer. The greater the chances of chain transfer, the lower the molecular weight of the polymer formed. The probability of chain transfer in the presence of polyethylene would increase with increasing concentration of the unsaturated groups.

Reasoning thus, we performed the following experiments.

1. The rate of polymerization of styrene without added peroxide catalyst was studied in the presence of (a) untreated polyethylene film, (b) treated polyethylene film, (c) treated film degassed under ultrahigh vacuum to remove both adsorbed and absorbed oxygen, (d) untreated

film similarly degassed, and (e) treated film which was then thoroughly washed with toluene and degassed under ultrahigh vacuum. For control purposes a run with no polyethylene film present was also made.

The polymerization rates were followed dilatometrically; 3.5 to 4 g of styrene and 0.2 g of polyethylene film cut into small pieces were introduced into the dilatometer bulbs. Mercury was used as the confining liquid. On polymerization, contraction in the volume of the system occurs, and the rate of polymerization is proportional to the rate of decrease in volume.

2. After polymerization the dilatometer bulbs were opened, and the contents were extracted with toluene at room temperature. This separated the polyethylene film (some of which may now carry grafted polystyrene because of chain-transfer possibilities) from polystyrene and unreacted monomer. (Extraction with toluene at room temperature removes polystyrene and unreacted monomer from the polyethylene film, in which the latter is not soluble at room temperature.) The toluene extract was stirred into an excess of methanol to precipitate the polystyrene and to remove the unreacted monomer. The precipitated polystyrene was washed with methanol and dried in a vacuum oven to constant weight. The intrinsic viscosities of the polystyrene samples obtained in this manner were determined in toluene at 30°C. The polyethylene films recovered from these experiments were examined by infrared absorption to detect the presence of any polystyrene grafts on them.

3. Since the polymerization rates of styrene in the absence of added catalyst were small, a series of polymerization runs with styrene containing 0.5% by weight of benzoyl peroxide was also made. Polymerization rates and molecular weights of the polystyrenes formed in the presence of treated and untreated polyethylene fims before and after degassing and in the presence of treated polyethylene film which had been washed and then degassed were studied.

Polymerizations without added catalyst were followed for 20 to 22 hr. All five conditions produced polymerizations with the same kinetics—rates shown by identical straight lines when change in volume was plotted as a function of time. Identical results (but faster rates) were found for the polymerizations with added catalyst for (a) styrene, (b) styrene plus treated film, and (c) styrene plus untreated film, when the rates were followed for periods up to 2 hr.

Table 8 presents the viscosity-average molecular weight data for a few typical polymerizations with and without added catalysts. The small differences between the values are within the experimental error in the determination of intrinsic viscosities, and the uncertainties in the unavoidable loss during separation of polystyrene by precipitation. In any case there are no systematic differences between the molecular weights of the polystyrene isolated from the systems in which the treated or the untreated polyethylene film was present. This indicates that as far as the concentration of groups such as hydroperoxides is concerned, there is no difference between the treated and untreated polyethylene films.

The infrared absorption spectra of the polyethylene films recovered from the polymerization systems showed a strong absorption at 6.26 μ which was not present in the original films. This absorption is characteristic of the phenyl group, and since this absorption persisted unchanged on repeated extraction with methyl ethyl ketone (a solvent in which polystyrene is soluble at room temperature) it can only be concluded that polystyrene units have become grafted to polyethylene through chain transfer of growing polystyrene chains. Calculations from the infrared absorption at 6.26 μ showed that in the grafted films approximately 14% polystyrene is present. No significant difference between the amounts of polystyrene grafted on treated and untreated polyethylene films was found, however. The melting point of the films with grafts was 112°C (as compared with 115°C for the original samples).

The lack of any major difference in the molecular weights of polystyrenes from the polymerization runs in which polyethylene films were present and in which there was no polyethylene is rather surprising when we consider that the isolated polyethylene films contained about 14% grafted polystyrene. Since grafting could have taken place only through chain transfer, the presence of as much as 14% polystyrene in the grafted polyethylene films would lead us to expect appreciable chain transfer and as a consequence an appreciable lowering of molecular weight when polymerization occurred in the presence of films. A likely explanation for the results seems to be that chain transfer occurs at only few sites and for some unapparent reason the polystyrene chains that grow at these sites are considerably longer than the unattached polystyrene chains.

Thus even the indirect techniques described do not reveal any differences in the concentration of hydroperoxide and $\diagdown C = C \diagup$ groups in either treated or untreated films.

TABLE 8

Intrinsic Viscosity and Molecular Weight of Polystyrenes

Polymerization system containing	$[\eta]$, dl/g[c]	Huggins's constant, k'[d]	$\bar{M}_v$[e]
Styrene[a]	1.97	0.27	603,000
Styrene + untreated polyethylene film[a]	1.89	0.34	576,000
Styrene + treated film[a]	1.80	0.26	537,000
Styrene + untreated film after degassing[a]	1.88	0.26	575,000
Styrene + treated film after degassing[a]	1.92	0.31	603,000
Styrene + treated film after washing with methanol and degassing[a]	1.90	0.36	501,000
Styrene[b]	0.66	0.36	132,000
Styrene + untreated film[b]	0.66	0.37	132,000
Styrene + treated film after degassing[b]	0.69	0.38	147,000
Styrene + treated film after washing with methanol and degassing[b]	0.63	0.32	129,000

[a] Polymerized at 84°C without added catalyst.

[b] Polymerized at 30°C with 0.5% benzoyl peroxide.

[c] Measured in toluene at 30°C: $\eta_{sp}/c = [\eta] + k'[\eta]^2 c$, where η_{sp} is the specific viscosity at concentration c in g/100 ml and k' is Huggins's constant [251].

[d] $[\eta]$ is obtained in the usual way from the ordinate intercept of a plot relating η_{sp}/c and c; $k'[\eta]^2$, the slope, is found, from which k' is readily calculated.

[e] Viscosity-average molecular weight calculated from $[\eta] = k'M^a = 1.03 \times 10^{-4} M^{0.74}$ [181, p. 312].

5. Optical Microscopy

The surfaces of treated and untreated film samples were carefully examined with microscopes with magnifications up to 1500 X. No characteristic differences were observed either with the phase contrast microscope or when the polarizing microscope was used with reflected light. With transmitted light it was found that the polyethylene film surfaces produced such profuse scattering that details could not be easily observed. When the film was immersed in a liquid with a refractive index of about 1.60, however, we could focus on either the upper of lower surface without interference from scattered light. Again, no characteristic differences were observed between treated and untreated films in either unpolarized or polarized light or with oblique lighting. Apparently any differences in surface irregularities are beyond the resolution of microscopes operating with visible light (see Section 6).

6. Electron Microscopy of the Surface of Treated and Untreated Polyethylene Films

Of all the techniques used in the search for experimental evidence to support a theory of treatment mechanism electron microscopy was found to be most successful in revealing differences between the untreated polyethylene films and those treated for ink adhesion. Only a few electron micrographs, which are typical of more than a hundred obtained for this study, can be reproduced here, but no inconsistencies appear among them.

For making the electron micrographs of polyethylene films Formvar replicas were made of the surfaces. The films were lightly brushed with a fine camel's-hair tip before making the replicas to remove surface contamination. The replicas were shadowed with chromium at a 30° angle to introduce contrast. All the electron micrographs were made with an RCA electron microscope (EMU Universal) at a direct magnification of 6000 diameters and were then photographically enlarged to 24,000 diameters.

Figures 4 to 9 represent the electron photomicrographs of the following samples: (4) untreated Olin film extruded from Du Pont Alathon 16A resin; (5) Olin film treated by corona discharge in-line on the extruder, same resin, same roll; (6) untreated commercial Visking film; (7) Visking film treated by corona discharge; (8) similarly treated commercial Durethene film; (9) similarly treated commercial Celanese film.

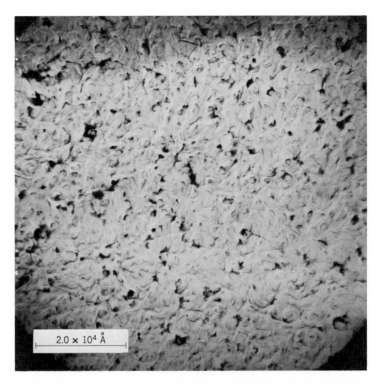

Fig. 4   This plate shows the uniformly rough surface of untreated Olin polyethylene
film.   Original print 24,000 X.

All samples of untreated film from several manufacturers and various lots of polymer showed only minor variations in the uniform rough surface, which can be considered normal for current extrusion practice.

All of the films treated by the corona discharge method [233] show an etched surface.   The uniform texture of the surface has been modified and islands are present, similar in texture to the untreated film, interspersed with smooth-bottomed craters 20 to 40,000 Å in diameter.   The depth of etching in the treated films estimated from these electron micrographs is 800 to 1500 Å: (Fig. 5, 1100–1150 Å; Fig. 7, 800–1000 Å; Fig. 8, 1250–1500 Å; Fig. 9, 1250–1500 Å).   The minor variations in the size, depth, and distribution of the etched areas might be the result of

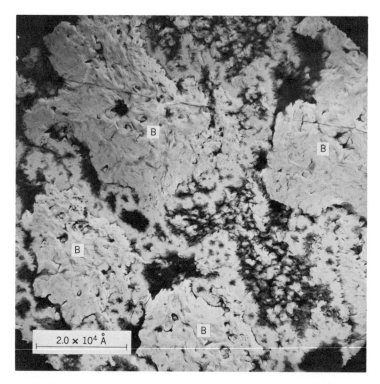

Fig. 5   Olin-treated film from the same roll as the sample in Fig. 4.   This photomicrograph shows clearly the islands whose structure is similar to the surface of the untreated film surrounded by areas (B) with relatively smooth surfaces.   In the original film these areas are smooth-bottomed craters 20 to 40,000 Å in diameter and 1100 to 1150 Å deep.

several factors involving resin composition, conditions of extrusion, or conditions of treating but are probably the results of small variations in the corona discharge treatment on different days under varying humidities.

Thus it appears that under conditions of commercial electric-discharge treatment the principal change that effects ink adhesion results from a micropitting of the surface.   The background texture of the film remains, but indented into the surface at random are comparatively smooth-bottomed craters about 1000 Å deep and 25,000 Å across.

Fig. 6   Untreated Visking film.   Compare uniformly rough surface with that shown in Fig. 4.

It is well known that uniformity of treatment requires careful regulation of time and intensity of the corona impinging on the film surface, and the mechanical change observable is sufficient to explain an enhanced ink adhesion over areas that range from a few square microns to the entire film area (as in background printing).   Also explained is the fact that ink adhesion of a wide variety of inks is enhanced by treatment, as is the adhesion of all sorts of coatings which may be rich or poor in polar groups.

Another way of viewing the result is that although inks and coatings bridge the microroughness (Figs. 4 and 6) of untreated film and do not strongly adhere, the same inks and coatings penetrate the larger craters

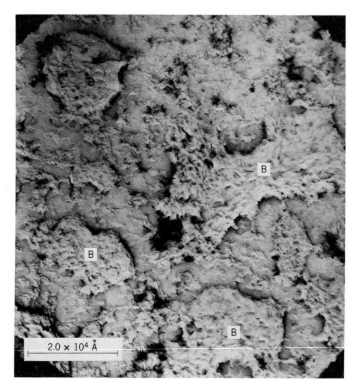

Fig. 7   Treated Visking film.   The smooth craters in this sample (B) are 800 to
1000 Å deep.

of the treated surface up to about 1000 Å and when dried are not easily
dislodged.

Excessive treatment, on the other hand, interferes with the adhesion
of certain types of coating, perhaps by the formation of degradation
products of low molecular weight.   If the latter are not soluble in the
coating vehicle, they may interfere with adhesion of coating to substrate.

Excessive treatment also reduces heat-seal strength, which can be
explained in that the etched craters cover essentially the entire film surface and thus smooth out the over-all microroughness on a larger scale.

The larger contact angles of liquids on treated film can be explained
by the liquids' spreading readily out into the craters.

Fig. 8   Treated Durethene film.   The island structure (B) is not so pronounced as in Figs. 5 and 7 but is clearly discernible, nevertheless.   These craters are 1250 to 1500 Å deep in the film sample.

It would be presumptuous to say that chemical changes do not occur on the film surface during commercial corona discharge treatments of olefin films (they probably do), but it appears that formation of oxygenated polar groups or double bonds cannot be invoked to explain the major effects of corona treatment in the improved adhesion of inks and coatings.   In one of the very early patents the mere presence of polar groups was found insufficient to secure ink adhesion; incorporation of small amounts of chlorinated polyethylene, copolymerized vinyl acetate/vinyl chloride, etc., in the polyethylene resin before extrusion was found to have no measurable effect whatsoever [223].

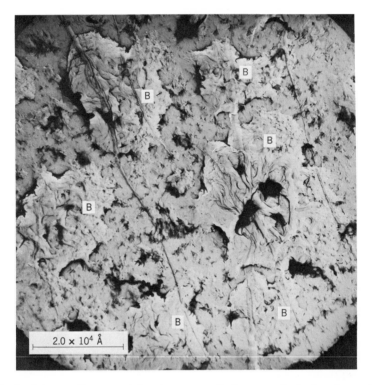

Fig. 9   Celanese film treated by the corona discharge method also shows the characteristic craters, here 1250 to 1500 Å deep.

XI. Film Properties

As we have seen, polyethylene is a complex molecular substance of widely varying molecular weight and molecular-weight distribution, with more or less branched structure and more or less incorporated catalyst, depending on the method of manufacture. We have shown that the properties of the polymers have a great effect on their extrudability into film, which results in a bewildering variety of conditions for extrusion and a corresponding variety of film products available in the market.

The most important applications in the film field depend on density, transparency, resistance to penetration by chemical substances in liquid or gaseous form, heat-sealability, favorable low-temperature properties, and tensile and thermal properties, all of which have already been discussed.

It is difficult to state general properties for a polymeric film as varied as the large number of polyethylenes. We shall attempt here to tabulate some general properties for low-, medium-, and high-density polyethylene films (Tables 9 to 11), followed by properties for two specialty films (Tables 12, 13).

The permeability to organic vapors (such as odors and flavors) varies widely and seems to follow no simple relationship to structure or size of the small molecule, though in general nonpolar, compact, planar molecules (e.g., benzene) diffuse more rapidly than nonplanar, less compact ones of similar molecular weight (e.g., cyclohexane and n-hexane), and much more rapidly than those with appreciable dipole moments (e.g., water, acetone, propanol, and 1,2-dichloroethane) [252–256].

Variation in the properties of polyethylene film may be related to (1) molecular weight and distribution of species of different molecular weight in the polymer, (2) proportion of crystalline material in the polymer, (3) texture of the film, that is, size distribution, form, and arrangement of crystalline and amorphous regions, (4) short- and long-chain branching of the molecules, and (5) presence of oxygen-containing groups and unsaturation.

It is improbable that any functional or mechanical property of polyethylene is determined solely by any of these parameters. The properties are affected simultaneously by all of them to some extent. It is expected, however, that certain physical properties will show a primary dependence on one of the molecular and textural characteristics.

1. The molecular weight and distribution of species of various molecular weights may be expected primarily to affect properties such as tensile strength, low-temperature brittle point, and tear resistance [257, 258], for these properties involve either extensive movements of the polymer segments or rupture of the sample. The percentage of crystalline material in the polymer has a less important effect on these properties. Melt viscosity is determined primarily by the weight-average molecular weight, hence depends on the distribution of polymer species of different molecular weights. The low-temperature brittle point (a measure of flexibility at low temperatures) is also dependent to an important extent on the degree of crystallinity and the texture of the sample.

TABLE 9

Some Properties of Polyethylene Films[a,b]

Property	Low density (0.910–0.925)	Medium density (0.926–0.940)	High density (0.941–0.965)	ASTM method[d]
Clarity	Transparent to translucent	Transparent to translucent	Translucent to opaque	
Coverage, in.²/lb	30,000	29,500	29,000	
Tensile strength, psi	1000–3500	2000–5000	3000–10,000	D 882–64T
Extensibility, %	225–600	225–500	5–400	D 882–64T
Impact strength, kg–cm	7–11	4–6	1–3	[260]
Tear strength (Elmendorf), g/mil	100–400	50–300	15–300	D 1922–61T
Stiffness (Handle-O-Meter), g, MD	2.5–4.5	5–10	8–16	
TD	3–7	6–14	10–20	
Heat-seal range, °F	250–350	260–310	275–310	
Permeability, H_2O, g/24 hr-m² (100°F – 90% rh)	18[c]	8–15[c]	5–10[c]	E 96–63T Method EI
O_2, ml/mil-m²-24 hr-1 atm (73°F – 0% rh)	3900–13,000	2600–5200	520–3900	D 1434
CO_2, ml/mil-m²-24 hr-1 atm (73°F – 0% rh)	7700–77,000	7700–13,000	3900–10,000	D 1434
Maximum use, °F	150	180–220	230	
Minimum use, °F	–60	–60	–60	
Coefficient of friction	0.30–0.50	0.30–0.50	0.30–0.50	

[a] All values for 1-mil film.
[b] [16], p. 164–165, and other sources.
[c] To convert to g/100 in.² divide by 15.5.
[d] Details in Volume I, Chapter 13; cf. also [17].

TABLE 10

Gas Transmission Rates of Low-Density Polyethylene Film[a, b]

	Thickness (in.)						
	0.001	0.0015	0.002	0.003	0.004	0.005	0.006
Water	1.4	0.93	0.7	0.46	0.35	0.28	0.23
Nitrogen	180	120	90	60	45	36	30
Oxygen	550	370	275	180	135	110	90
Helium	1,225	820	610	410	305	240	205
Hydrogen	1,960	1,300	980	655	490	390	325
Carbon dioxide	2,900	1,940	1,450	970	725	580	485
Sulfur dioxide	6,200	4,130	3,100	2,070	1,550	1,240	1,035
Ethylene oxide	29,300	19,530	14,650	9,800	7,375	5,860	4,900
Methyl bromide	79,100	52,730	39,550	26,370	19,775	15,850	13,185

[a] Water-vapor transmission rate is reported in g/100 in.^{2}-24 hr and measured at 90% relative humidity and at 100°F (ASTM Method E 96–63T Procedure E). Other gas transmission rates are reported in ml/100 in.^{2}-24 hr and measured at 77°F with a pressure differential at 725 mm of mercury across the film. Data are reduced to standard temperature and pressure.
   [b] [261].

TABLE 11

Transmission of Essential Oils Through
2-Mil Low-Density Polyethylene Film[a]

Oil	g/100 in.^{2}-24 hr
Vanillin	0.018
Methyl salicylate	1.150
Coriander oil	0.879
Petit grain	0.606
Nutmeg oil	7.470
Lemon oil	14.241
Orange oil	14.033
Grapefruit oil	16.756
Cassia oil	2.262
Geraniol	0.168
Geranium oil	0.423

   [a] [261].

TABLE 12

Properties of Special Clysar[a] EH Heat-Shrinkable Polyethylene Films

Property	Film Designation[b]		
	400 EH–30	300 EH–30	200 EH–30
Approximate thickness, mils	0.75	1.00	1.50
Yield, in.^{2}/lb	40,000	30,000	20,000
Shrinkage at 212°F, %	25–35	25–35	25–35
Haze, %	2.5	3.0	3.0
Gloss, 20°	105	100	90
Water vapor transmission rate, g/100 in.^{2}-24 hr	1.1–1.3	0.8–1.0	0.5–0.7
Tensile strength, psi	8000–10,000	8000–10,000	8000–10,000
Elongation, %	120–150	120–150	120–150
Tear (Elmendorf), g/mil	4–5	6–8	10–12
Hot wire seal, g/in.	2000	2000	2000
Shrink stress, psi at 212°F	350–400	350–400	350–400

[a] Du Pont shrinkable film, [262].

[b] Code designations: First series of digits: Yield (400 EH–30 has a yield of 40,000 in.^{2}/lb); first letter E—polyethylene; second letter H—heat-shrinkable; second set of digits 30—not treated for printing.

2. The proportion of crystalline material in the polymer (i.e., percentage crystallinity) affects primarily those properties that are concerned with slight movements of the polymer segments relative to one another. Such properties as the softening temperature, Young's modulus in tension, bending modulus (related to stiffness), yield point (related to the tension to cause cold drawing), surface hardness, permeability to water vapor, and the sorption of gases and liquids fall in this category. The density of the polymer is determined primarily by the extent of crystallinity, although the thermal history of the sample also has an important effect. The dynamic mechanical properties are significantly affected by the percentage crystallinity of polyethylene [259]. The more crystalline materials have the higher dynamic shear moduli and higher melting points.

3. The size and arrangement of the crystalline regions and the properties of the regions between the crystalline aggregates have important effects on the properties of the products made from polyethylene resins. These textural properties of polyethylene in the solid state are determined not only by the molecular properties of polyethy-

TABLE 13

Comparative Properties of Low-Density Polyethylene and Coated Polyethylene

	Coated polyethylene	Polyethylene
Gauge, mils	1.7[a]	1.5
Yield, in.^{2}/lb	17,500	20,000
Clarity	Excellent transparency	Some haze; dull surface
Water-vapor permeability, g/m^2-24 hr	8.0	11.6
Gas permeability, ml/m^2-24 hr-atm		
Air	10	1,400
O$_2$	150	6,000
Odor resistance, hours to detection		
Clove	2	0.03
Vanilla	6	0.07
Onion	>8	0.03
Vinegar	>8	0.07
Oil permeability, days to penetration		
Cottonseed	>90	25
Mineral	>90	4
Heat-seal strength, g/1.5 in., coated-coated, 300°F, 30 psi, 0.5 sec, 75°F/35% rh	2310	Weld (> 2400)
Anchorage of coating, hours to fail, 75°C	>4	
Pendulum impact strength, kg-cm[b]	4.0	9.0
Falling-ball impact strength, kg–cm	18.2	20.5
Coefficient of friction	0.20–0.25	0.25–0.30
Tensile strength, to break, psi MD	3700	3100
TD	1900	2200
Tear strength, g. MD	450	420
TD	510	400
Elongation, % MD	520	420
TD	860	1320

[a] 0.2 mil of coating on one side.
[b] [260].

lene but also by the conditions under which crystallization of the melt occurs. Polyethylene films which have small crystalline aggregates are more transparent than films with large aggregates. Films with the aggregates aligned preferentially in a given direction tear easily in that direction. The low-temperature flexibility and the stress cracking of films may depend very much on the polycrystalline nature of the films.

4. The important molecular property that affects the percentage crystallinity is the degree of short-chain branching. Long-chain branching, on the other hand, has no significant effect on crystallinity (Section IV).

5. The presence of unsaturation and oxygen-containing groups has but a minor effect on the mechanical properties of films but an important bearing on the crosslinking and degradation at elevated temperatures or on the effect of exposure to ultraviolet and high-energy particle radiation. The presence of oxygen-containing groups or catalyst residues has a marked effect on electrical properties such as power factor and dielectric loss, but these are not of great importance in film applications.

Coated Polyethylene. After suitable treatment (cf. Section X) polyethylene can be coated with a variety of functional coatings, the most successful of which has been a copolymer of polyvinylidene chloride, applied either from solution or emulsion [263].

Two such highly effective coatings are the following:

1. An emulsion polymer consisting of 87 % vinylidene chloride, 11 % butyl acrylate, and 2 % methacrylic acid to give a good balance among properties of blocking, adhesion, and permeability.

2. A similar formulation in which acrylonitrile and itaconic acid are substituted for the two latter ingredients.

Table 13 [263] presents some comparative data for 1.5-mil polyethylenes, one coated with formulation (1).

One of the desirable features of polyethylene is its excellent resistance to the transmission of water vapor, which makes it a good protective material for both dry and moist foods. This low permeability of polyethylene is complemented by the addition of a protective coating. On the other hand, its chief disadvantage which precludes its use as an effective gas barrier is its extremely high gas permeability. The application of a coating lowers the permeability of polyethylene to air, oxygen, and other common gases to approximately 5 % of the value found with uncoated film.

Certain foods require a packaging film that resists attack by fats, greases, and oils. The poor oil resistance of ordinary polyethylene film has been overcome by the polymer coating.

Protection of packaged products from contamination by foreign odors penetrating the package, as well as protection of adjacent materials from contamination by odors from materials within the package, is a shortcoming of commercial polyethylene films. Tests performed with such flavors as cloves, vanilla, and onions, as well as with vinegar, have shown that the coated polyethylene film is far superior to uncoated polyethylene.

In addition, coating improves the film's appearance, properties referred to often as surface gloss, clarity, and liveliness of printing.

XII. Markets and Applications

Since the introduction of polyethylene as a packaging film in about 1946, as we have already mentioned, the demand has grown to approximately one billion pounds in 1970, valued at about $600 million. This material accounts for about 60% of the packaging film market at the present time, more than twice as much as the cellophane poundage used (Table 14).

TABLE 14

Consumption of Films in Packaging, 1960–1968
(millions of pounds)

	1960	1961	1962	1963	1964	1965	1966	1967	1968
Polyethylenes	275	370	400	455	500	615	730	735	795
Cellophanes	439	423	404	405	410	405	395	385	360
Vinyls	18	20	22	24	30	40	40	70	90
Polypropylenes		3	15	25	28	40	45	50	65
Sarans	17	19	21	23	26	30	30	30	32
Polyesters	8	8	10	10	10	20	20	20	20
Cellulose acetates	10	10	11	12	13	15	15	15	15
Polystyrenes	3	5	7	11	15	15	15	15	15
Pliofilm	12	13	14	15	15	15	15	10	7
Miscellaneous	2	2	3	3	4	4	5	8	10
Total packaging uses	784	873	907	983	1051	1199	1310	1338	1409

Low-density film, made by a high-pressure process is inexpensive, easily heat-sealed, has good optical clarity and a balance of strength and optical properties, and finds its greatest use in the manufacture of bags for bulk packaging. Medium-density film is employed chiefly as overwrapping film for packages filled on high-speed machines, up to 500 a minute. High-density film is used for special purposes as a result of the relatively poor clarity, narrow heat-seal temperature range, high rigidity, and low tear strength. Grease and heat resistance, however, are higher than in the low- and medium-density varieties, and this advantage is exploited, for example, in the use of this film for boilable packages.

The low-to-medium-density film is used to package produce, textiles, frozen foods, baked goods, candy, macaroni, toys, tools, hardware, paper products, and many other materials.

In Table 15 the principal uses of polyethylene film are summarized

TABLE 15

Consumption of Polyethylene Film in
Packaging in 1965, 1968, and 1970
(millions of pounds)

	1965	1968	1970[a]
Food Packaging			
Candy	16	20	24
Bread, cake	80	120	140
Crackers, biscuits	5	5	9
Meat, poultry	16	20	25
Fresh produce	145	140	165
Snacks	2	5	8
Noodles, macaroni	6	8	10
Cereals	3	7	9
Dried vegetables	8	10	12
Frozen foods	14	28	40
Dairy products	8	10	13
Other foods	7	22	35
Total food uses	310	395	490
Nonfood Packaging			
Shipping bags, liners	80	90	100
Rack and counter	40	60	85
Textiles	60	75	85
Paper	25	50	65
Laundry, dry cleaning	60	80	90
Miscellaneous	40	45	60
Total nonfood uses	305	400	485
Total packaging uses	615	795	975

[a] Estimate

TABLE 16

Comparative Costs: Films, Foils, Laminates[a]

Material	Cost ($/lb)	Coverage (in.^{2}/lb)	Cost (cents/10^3 in.^{2})
Cellophane			
MS, 195	0.64	19,500	3.3
MS, 220	0.64	22,000	2.9
MSA or MSB, 195	0.71	19,500	3.6
Saran-coated, 140	0.81	14,000	5.8
Saran-coated, 195	0.81	19,500	4.2
Saran-coated, 250	0.75	25,000	3.0
Polyethylene-coated, 182	0.77	18,250	4.2
Vinyl-coated, 210	0.71	21,000	3.4
Cellulose acetate			
Cast, 1 mil	0.96	22,000	4.4
Extruded, 1 mil	0.70	22,000	3.2
Cryovac S, 0.6 mil	1.09	27,300	4.0
Fluorohalocarbon, 1 mil	8.26	13,000	64.0
Nylon, 1 mil	2.15	24,500	8.8
Pliofilm, 0.75 mil	0.95	33,000	2.9
Polycarbonate, 1 mil	2.07	23,100	8.9
Polypropylene			
Cast, 1 mil	0.60	31,000	1.9
Balanced, uncoated, 1 mil	0.90	31,000	2.9
Balanced, Saran-coated, 1 mil	1.20	28,000	4.3
Polystyrene, oriented, 1 mil	0.63	26,300	2.4
Polyester			
Nonheat-sealing, 0.92 mil	1.80	21,500	8.4
Nonheat-sealing, 0.5 mil	2.09	40,000	5.2
Heat-sealing, 0.5 mil base	2.00	27,600	7.3
Polyethylene			
Low-density, 1 mil	0.27	30,000	0.9
Medium-density, 1 mil	0.28	30,000	0.9
High-density, 1 mil	0.32	29,000	1.1
Polyethylene-cellophane			
1 mil/195 MS	1.05	11,800	8.9
Polyethylene-polyester			
1.5 mil/0.5 mil	1.59	13,200	12.0
Saran, 1 mil	1.08	16,300	6.6
Vinyl			
Cast, 1 mil	0.83	21,600	3.8
Extruded, 1 mil	0.55	21,500	2.6
Water-soluble film, 1.5 mil	1.38	15,000	9.2
Aluminum foil			
0.00035 in.	0.62	29,300	2.1
0.001 in.	0.53	10,250	5.3
Foil-acetate, 1 mil/1 mil	1.49	6,750	22.1

[a] Compare with Volume I, p. 13.

and the years 1965 and 1968 are compared to show market trends. Of the total approximately half finds application in food packaging.

Table 16 indicates clearly on the basis of cost of coverage, why, unless special properties of other films demonstrate a clear advantage in a given application, the usage of polyethylene is so great.

The newer products based on polyethylene include a variety of laminates for special uses, a high-density film oriented in one direction only and coated with a copolymer of ethylene and polyvinyl acetate, in addition to the coated film described in the preceding section. The former coated product is intended to combine the stiffness of the high-density base sheet with wide-temperature sealing characteristics of the coating. Both products were offered on the market at the time when great attention was being given to polypropylene and did not find wide acceptance.

Acknowledgment

We wish to express thanks to former colleagues in the Film Research and Development Department of Olin at New Haven for many of the data and results included in Section X–C of the text, but especially to the late George H. Berthold and to Sundar L. Aggarwal, Matthew J. Carrano, Leon Marker, Roman Mykalojewycz, Anderson Pace, Jr., Howard A. Scopp, Marvin C. Tobin, and George P. Tilley, who from time to time over several years illuminated the murky problems of film treatment by a series of carefully planned crucial experiments. The full experimental details are given in Olin Research Reports, mostly unpublished. We are also indebted to Dr. Joseph Fleischer, then Olin's senior patent attorney in New Haven, for many helpful discussions of the problems and for locating obscure related patents. The electron micrographs were made by Mrs. Althea Revere of Washington.

References

1. M. W. Perrin, *Research (London)*, **6,** 111 (1953); R. O. Gibson, *Roy. Inst. Chem. (London) Lecture Series*, **1964,** No. 1.
2. E. W. Fawcett, R. O. Gibson, and M. W. Perrin: U.S. Patent 2,153,553 (April 11, 1939).
3. E. W. Fawcett, R. O. Gibson, M. W. Perrin, J. G. Paton, and E. G. Williams: British Patent 471,590 (Sept. 6, 1937).
4. M. W. Perrin, J. G. Paton, and E. G. Williams: U.S. Patent 2,188,465 (Jan. 30, 1940).

5. S. L. Aggarwal and O. J. Sweeting, *Chem. Rev.*, **57**, 665 (1957), 550 references. Translated and reprinted in *Uspek. Khim.*, **27**, 1115 (1958), with additional Russian references.

6. R. A. V. Raff and J. B. Allison, *Polyethylene*, Interscience, New York, 1956.

6a. T. O. J. Kresser, *Polyethylene*, Reinhold, New York, 1957.

7. George E. Ham, Ed., *Copolymerization*, Wiley-Interscience, New York, 1964, Chapters III, IV.

8. J. K. Stille, *Chem. Rev.*, **58**, 541 (1958), 205 references.

9. A. Renfrew and P. Morgan, *Polythene*, 2nd ed., Iliffe, London, 1960.

10. H. Hagen and H. Domininghaus, *Polyäthylen*, Brunke Garrels, Hamburg, 1961.

11. J. M. McKelvey, *Polymer Processing*, Wiley, New York, 1962.

12. N. G. Gaylord and H. F. Mark, *Linear and Stereoregular Addition: Polymerization with Controlled Propagation*, Interscience, New York, Wiley, 1959.

13. R. A. V. Raff and K. W. Doak, *Crystalline Olefin Polymers*, Wiley-Interscience, New York, 1965.

14. P. H. Geil, *Polymer Single Crystals*, Wiley-Interscience, New York, 1963.

15. W. Franke, H. Logemann, G. Pieper, and H. Webber, "Polymerization of Ethylene," in Houben-Weyl *Methoden der Organischen Chemie*, Vol. 14 (1), Thieme, Stuttgart, 1962, pp. 562–621.

16. *Modern Packaging Encyclopedia, 1970, Modern Packaging*, **43**, No. 7A (July 1970).

17. *1966 Standards*, American Society for Testing and Materials, Philadelphia, Vols. 26 and 27.

18. H. von Pechmann, *Ber.*, **31**, 2643 (1898).

19. E. Bamberger and F. Tschirner, *Ber*, **33**, 959 (1900).

20. S. C. Lind and G. Glockler, *J. Am. Chem. Soc.*, **51**, 2811 (1929).

21. M. Otto, *Brennstoff-Chem.*, **8**, 321 (1927).

22. F. Hofmann, *Chem. Ztg.*, **57**, 5 (1933).

23. E. Desparmet, *Bull. soc. chim.* [5], **3**, 2047 (1936).

24. A. D. Dunstan, *Trans. Faraday Soc.*, **32**, 227 (1936).

25. V. N. Ipatieff and B. B. Corson, *Ind. Eng. Chem.*, **28**, 860 (1936).

26. V. N. Ipatieff and H. Pines, *Ind. Eng. Chem.*, **27**, 1364 (1935).

27. V. N. Ipatieff and H. Pines, *J. Gen. Chem.* (U.S.S.R.), **6**, 321 (1936).

28. B. W. Malisher, *Petroleum Z.*, **32**, No. 19, 1 (1936).

29. P. J. Wiezevich and J. M. Whiteley, U.S. Patent 1,981,819 (Nov. 20, 1934).

30. S. Lenher, U.S. Patent 2,000,964 (May 14, 1935).

31. H. I. Waterman and A. J. Tulleners, *Chim. Ind.* (*Paris*), Special No., June, **1933**, 496.

32. V. I. Komarewski and N. Balai, *Ind. Eng. Chem.*, **30**, 1051 (1938).

33. K. Peters and K. Winzer, *Brennstoff-Chem.*, **17**, 366 (1936).

34. G. Egloff and R. E. Schaad, *J. Inst. Petroleum Technol.*, **19**, 800 (1933).

35. R. D. McDonald and R. G. W. Norrish, *Proc. Roy. Soc.* (*London*), A**157**, 480 (1936).

36. F. C. Hall and A. W. Nash, *J. Inst. Petroleum Technol*, **24**, 471 (1938).

37. M. Koidzumi, *J. Chem. Soc. Japan*, **64**, 257 (1943).

38. Y. Konaka, *J. Soc. Chem. Ind., Japan*, **43**, Suppl. binding 363 (1940).

39. S. Y. Pshezhetskiĭ, *J. Phys. Chem.* (*U.S.S.R.*), **14**, 1151 (1940).

40. T. Shiba and A. Ozaki, *J. Chem. Soc. Japan*, Pure Chem. Sect., **74**, 295 (1953).
41. S. Kodama, and I. Taniguchi, *J. Chem. Soc. Japan*, Ind. Chem. Sect., **53**, 385 (1950).
42. H. Tani and C. Sato, *Chem. High Polymers (Japan)*, **5**, 57 (1948).
43. K. Ziegler, *Brennstoff-Chem.*, **35**, 321 (1954).
44. K. Ziegler, E. Holzkamp, H. Breil, and H. Martin, *Angew Chem.*, **67**, 426 (1955); U.S. Patent 3, 113, 115 (Dec. 3, 1963).
45. K. Ziegler, E. Holzkamp, H. Breil, and H. Martin, *Angew. Chem.*, **67**, 541 (1955); *Chim. Ind. (Milan)*, **37**, 881 (1955).
46. K. Ziegler and H.-G. Gellert, U.S. Patent 2,695,327 (Nov. 23, 1954); German Patents 878,560 and 889,299.
47. K. Ziegler and H.-G. Gellert, U.S. Patent 2,699,457 (Jan. 11, 1955).
48. (a) R. W. G. Wyckoff, *Crystal Structures*, Vol. 2, Wiley-Interscience, New York, 1964, pp. 50, 55, 126; (b) G. Natta, P. Corradini, and G. Allegra, *J. Polymer Sci.*, **51**, 399 (1961).
49. G. Natta and I. Pasquon, "The Kinetics of the Stereospecific Polymerization of α-Olefins", in *Advances in Catalysis*, Academic, New York, Vol. XI, 1959, p. 1.
50. P. Cossee, *Trans. Faraday Soc.*, **58**, 1226 (1962).
51. P. Cossee, *J. Catalysis*, **3**, 80 (1964).
52. E. J. Arlman, *J. Catalysis*, **3**, 89 (1964).
53. E. J. Arlman and P. Cossee, *J. Catalysis*, **3**, 99 (1964).
54. H. M. Van Looy, L. A. M. Rodriguez, and J. A. Gabant, *J. Polymer Sci.*, **A1, 4**, 1927 (1966).
55. L. A. M. Rodriguez and H. M. Van Looy, *J. Polymer Sci.*, **A1, 4**, 1951 (1966); *ibid.*, 1971.
56. K. Ziegler, Belgian Patent 533,362 (May 16, 1955).
57. K. Ziegler, Belgian Patent 534,792 (Jan. 31, 1955).
58. K. Ziegler, Belgian Patent 534,888 (Jan. 31, 1955).
59. K. Ziegler, Belgian Patent 542,658 (May 8, 1956).
60. D. C. Smith, *Ind. Eng. Chem.*, **48**, 1161 (1956).
61. A. Clark, J. P. Hogan, R. L. Banks, and W. C. Lanning, *Ind. Eng. Chem.*, **48**, 1152 (1956).
62. Phillips Petroleum Co., Belgian Patent 530,617 (Jan. 24, 1955), Australian Pat. application 6365/55.
63. R. V. Jones, and P. J. Boeke, *Ind. Eng. Chem.*, **48**, 1155 (1956).
64. R. A. Mosher, U.S. Patent 2,725,374 (Nov. 29, 1955).
65. E. Field and M. Feller, U.S. Patent 2,726,231 (Dec. 6, 1955).
66. Standard Oil Company (Indiana), British Patent 721,046.
67. B. L. Evering, A. K. Roebuck, and A. Zletz, U.S. Patent 2,727,023, (Dec. 13, 1955).
68. E. Field and M. Feller, U.S. Patent 2,727,024 (Dec. 13, 1955).
69. B. L. Evering and E. F. Peters, U.S. Patent 2,728,754 (Dec. 27, 1955).
70. E. F. Peters, U.S. Patent 2,700,663 (Jan. 25, 1955).
71. E. Field and M. Feller, U.S. Patent 2,728,757 (Dec. 27, 1955).
72. E. Field and M. Feller, U.S. Patent 2,728,758 (Dec. 27, 1955).
73. E. Field and M. Feller, U.S. Patent 2,731,453 (Jan. 17, 1956).

74. M. Feller and E. Field, U.S. Patent 2,717,888 (Sept. 13, 1955).
75. M. Feller and E. Field, U.S. Patent 2,717,889 (Sept. 13, 1955).
76. H. S. Seelig, U.S. Patent 2,710,854 (June 14, 1955).
77. E. Field and M. Feller, U.S. Patent 2,731,452 (Jan. 17, 1956).
78. A. W. Larchar, D. C. Pease, R. A. Hines, and W. M. D. Bryant, *Ind. Eng. Chem.*, **49**, 1071 (1957).
79. A. W. Larchar and D. C. Pease, U.S. Patent 2,816,883 (Dec. 17, 1957).
80. J. J. Fox and A. E. Martin, *Proc. Roy. Soc. (London)*), **A175**, 208 (1940).
81. H. W. Thompson and P. Torkington, *Proc. Roy. Soc. (London)*, **A184**, 21 (1945).
82. H. W. Thompson and P. Torkington, *Trans. Faraday Soc.*, **41**, 246 (1945).
83. G. W. King, R. M. Hainer, and H. O. McMahon, *J. Applied Phys.*, **20**, 559 (1949).
84. L. H. Cross, R. B. Richards, and H. A. Willis, *Discussions Faraday Soc.*, **9**, 235 (1950).
85. L. Kellner, *Proc. Phys. Soc. (London)*, **64A**, 521 (1951).
86. F. M. Rugg, J. J. Smith, and J. C. Atkinson, *J. Polymer Sci.*, **9**, 579 (1952).
87. F. M. Rugg, J. J. Smith, and L. H. Wartman, *J. Polymer Sci.*, **11**, 1 (1953).
88. R. S. Stein and G. B. B. M. Sutherland, *J. Chem. Phys.*, **21**, 370 (1953).
89. N. Acquista and E. K. Plyler, *J. Optical Soc. Am.*, **43**, 977 (1953).
90. W. M. D. Bryant and R. C. Voter, *J. Am. Chem. Soc.*, **75**, 6113 (1953).
91. E. Borello, E. and C. Mussa, *J. Polymer Sci.*, **13**, 402 (1954).
92. F. M. Rugg, J. J. Smith, and R. C. Bacon, *J. Polymer Sci.*, **13**, 535 (1954).
93. S. Krimm, *J. Chem. Phys.*, **22**, 567 (1954).
94. K. Rossman, *J. Chem. Phys.*, **23**, 1355 (1955).
95. N. Sheppard and G. B. B. M. Sutherland, *Nature*, **159**, 739 (1947).
96. A. Elliott, E. J. Ambrose, and R. B. Temple, *J. Chem. Phys.*, **16**, 877 (1948).
97. G. B. B. M. Sutherland and A. V. Jones, *Nature*, **160**, 567 (1947).
98. R. S. Stein, *J. Chem. Phys.*, **22**, 734 (1955).
99. M. C. Tobin, *J. Chem. Phys.*, **23**, 819 (1955).
100. A. G. Nasini and E. Borello, *Simposio Internazionale di Chimica Macromolecolare, Ricerca sci.*, **25**, Suppl. A, 686 (1955).
101. E. J. Ambrose, A. Elliot, and R. B. Temple, *Proc. Roy. Soc. (London)*, **A199**, 183 (1949).
102. M. C. Tobin and M. J. Carrano, *J. Chem. Phys.*, **25**, 1044 (1956).
103. A. Keller and I. Sandeman, *J. Polymer Sci.*, **15**, 133 (1955).
104. V. N. Nikitin, M. V. Vol'kenshteïn, and B. Z. Volchek, *Zhur. Tekh. Fiz.*, **25**, 2486 (1955).
105. H. W. Thompson and P. Torkington, *Proc. Roy. Soc. (London)*, **A184**, 3 (1945).
106. R. S. Silas, Abstracts of papers, Symposium on Molecular Structure and Spectroscopy, Ohio State University, Columbus, June, **1956**, p. 49.
107. N. Formigoni, *Simposio Internazionale di Chimica Macromoleculare, Ricerca Sci.*, **25**, Suppl. A, 854 (1955).
108. C. W. Bunn, *Trans Faraday Soc.*, **35**, 482 (1939).
109. C. W. Bunn and T. C. Alcock, *Trans. Faraday Soc.*, **41**, 317 (1945).
110. C. W. Bunn, *Chemical Crystallography*, Oxford University Press, London, 1946.
111. N. F. M. Henry and K. Lonsdale, *International Tables for X-Ray Crystallography*, Vol. I, Kynoch, Birmingham, England, 1952.

112. W. P. Slichter, *J. Polymer Sci.*, **21,** 141 (1956).
113. J. L. Matthews, H. S. Peiser, and R. B. Richards, *Acta Cryst.*, **2,** 85 (1949).
114. P. H. Hermans, *Kolloid-Z.*, **120,** 3 (1951).
115. P. H. Hermans and A. Weidinger, *J. Polymer Sci.*, **4,** 709 (1949).
116. P. H. Hermans and A. Weidinger, *J. Polymer Sci.*, **5,** 269 (1950).
117. J. J. Trillat, S. Barbezat, and A. Delalande, *J. chim. phys.*, **47,** 877 (1950).
118. J. J. Trillat, S. Barbezat, and A. Delalande, *Compt. rend.*, **231,** 853 (1950).
119. W. M. D. Bryant, J. P. Tordella, and R. H. H. Pierce, Jr., *J. Am. Chem. Soc.*, **74,** 282 (1952).
120. S. Krimm and A. V. Tobolsky, *Textile Research J.*, **21,** 814 (1951).
121. S. Krimm and A. V. Tobolsky, *J. Polymer Sci.*, **7,** 57 (1951).
122. S. Krimm, *J. Phys. Chem.*, **57,** 22 (1953).
123. J. Natta and P. Corradini, *Simposio Internazionale di Chimica Macromolecolare*, *Ricerca sci.*, **25,** Suppl. A, 695 (1955).
124. S. L. Aggarwal and G. P. Tilley, *J. Polymer Sci.*, **18,** 17 (1955).
125. E. Hunter and W. G. Oakes, *Trans. Faraday Soc.*, **41,** 49 (1945).
126. H. C. Raine, R. B. Richards, and H. Ryder, *Trans. Faraday Soc.*, **41,** 56 (1945).
127. M. Dole, W. P. Hettinger, Jr., N. R. Larson, and J. A. Wethington, Jr., *J. Chem. Phys.*, **20,** 781 (1952).
128. R. G. J. Miller and H. A. Willis, *J. Polymer Sci.*, **19,** 485 (1956).
129. V. N. Nikitin and E. L. Pokrovskiĭ, *Doklady Akad. Nauk S.S.S.R.*, **95,** No. 1, 109 (1954).
130. M. C. Tobin and M. J. Carrano, *J. Polymer Sci.*, **24,** 93 (1957).
131. C. W. Wilson, III, and G. E. Pake, *J. Polymer Sci.*, **10,** 503 (1953).
132. R. Roberts and F. W. Billmeyer, Jr., *J. Am. Chem. Soc.*, **76,** 4238 (1954).
133. J. B. Nichols, *J. Appl. Phys.*, **25,** 840 (1954).
134. G. S. Parks and J. R. Mosley, *J. Chem. Phys.*, **17,** 691 (1949).
135. R. B. Richards, *J. Applied Chem. (London)*, **1,** 370 (1951).
136. M. J. Roedel, *J. Am. Chem. Soc.*, **75,** 6110 (1953).
137. C. A. Sperati, W. A. Franta, and H. W. Starkweather, Jr., *J. Am. Chem. Soc.*, **75,** 6127 (1953).
138. R. Newman, *J. Chem. Phys.*, **18,** 1303 (1950).
139. F. Danusso, G. Moraglio, and G. Talami, *J. Polymer Sci.*, **21,** 139 (1956).
140. W. M. D. Bryant, *J. Polymer Sci.*, **2,** 547 (1947).
141. C. E. H. Bawn and T. B. Rhodes, *Trans. Faraday Soc.*, **50,** 934 (1954).
142. E. Grams and E. Gaube, *Angew. Chem.*, **67,** 548 (1955).
143. P. J. Flory, *J. Am. Chem. Soc.*, **59,** 241 (1937).
144. J. K. Beasley, *J. Am. Chem. Soc.*, **75,** 6123 (1953).
145. T. G. Fox and S. Gratch, *Ann. N. Y. Acad. Sci.*, **57,** 367 (1953).
146. H. W. Morse, C. H. Warren, and J. D. H. Donnary, *Am. J. Sci.*, **23,** 421 (1932).
147. H. W. Morse and J. D. H. Donnary, *Am. Mineral.*, **21,** 391 (1936).
148. W. H. Bryan, *Proc. Roy. Soc. Queensland*, **52,** No. 6, 41 (1941).
149. H. E. Buckley, *Crystal Growth*, Wiley, New York, 1951, p. 35.
150. C. M. Langkammerer and W. E. Catlin, *J. Polymer Sci.*, **3,** 305 (1948).
151. G. L. Clark, M. H. Mueller, and L. L. Stott, *Ind. Eng. Chem.*, **42,** 831 (1950).
152. W. Banks, J. N. Hay, A. Sharples, and G. Thomson, *Nature*, **194,** 542 (1962).
153. H. Kojima and A. Abe, *Kobunshi Kagaku*, **20 (217),** 289 (1963).

154. W. Banks, J. N. Hay, A. Sharples, and G. Thomson, *Polymer*, **5**, 163 (1964).
155. A. Keller, *J. Polymer Sci.*, **17**, 351 (1955).
156. A. Keller and J. R. S. Waring, *J. Polymer Sci.*, **17**, 447 (1955).
157. W. M. D. Bryant, R. H. H. Pierce, Jr., C. R. Lindegren, and R. Roberts, *J. Polymer Sci.*, **16**, 131 (1955).
158. C. W. Bunn, *Fibre Science*, **1949**, 144.
159. D. R. Holmes, R. G. Miller, R. P. Palmer, and C. W. Bunn, *Nature*, **171**, 1104 (1953).
160. A. Van Rossem and J. Lotichius, *Kautschuk*, **5**, 2 (1929).
161. P. Debye and F. Bueche, *Phys. Rev.*, **81**, 303 (1951).
162. F. P. Price, *J. Phys. Chem.*, **59**, 191 (1955).
163. J. J. Keane, F. H. Norris, and R. S. Stein, *J. Polymer Sci.*, **20**, 209 (1956).
164. J. J. Keane and R. S. Stein, *J. Polymer Sci.*, **20**, 327 (1956).
165. S. L. Aggarwal, G. P. Tilley, and O. J. Sweeting, *J. Appl. Polymer Sci.*, **1**, 91 (1959).
166. S. L. Aggarwal, G. P. Tilley, and O. J. Sweeting, *J. Polymer Sci.*, **51**, 551 (1961).
167. W. J. Schmidt, *Kolloid-Z.*, **96**, 135 (1941).
168. P. H. Hermans, *Physics and Chemistry of Cellulose Fibres*, Elsevier, Houston, 1949, p. 220.
169. T. T. Li, R. J. Volungis, and R. S. Stein, *J. Polymer Sci.*, **20**, 199 (1956).
170. R. S. Stein and F. H. Norris, *J. Polymer Sci.*, **21**, 381 (1956).
171. R. S. Stein, *J. Polymer Sci.*, **31**, 327 (1958).
172. R. S. Stein, *J. Polymer Sci.*, **31**, 335 (1958).
173. R. S. Stein, *J. Polymer Sci.*, **34**, 709 (1959).
174. A. Brown, *J. Applied Phys.*, **20**, 552 (1949).
175. A. Brown, *J. Applied Phys.*, **23**, 287 (1952).
176. G. A. Mortimer and W. F. Hamner, *J. Polymer Sci.*, **A2**, 1301 (1964).
177. C. S. Myers, *J. Polymer Sci.*, **13**, 549 (1954).
178. *Technical Report TR–17A*, Eastman Chemical Products, Kingsport, Tenn., 1968.
179. H. C. Raine, R. B. Richards, and H. Ryder, *Trans. Faraday Soc.*, **41**, 56 (1945).
180. R. B. Richards, *Trans Faraday Soc.*, **42**, 10 (1946).
181. P. J. Flory, *Principles of Polymer Chemistry*, Cornell University Press, Ithaca, N.Y., 1953, p. 574.
182. K. Ueberreiter, H. J. Orthmann, and G. Sorge, *Makromol. Chem.*, **8**, 21 (1952).
183. M. Socci, G. Lanzavecchia, *Simposio Internazionale di Chimica Macromolecolare, Ricerca sci.*, **25**, Suppl. A, 497 (1955).
184. L. H. Tung, *J. Polymer Sci.*, **20**, 495 (1956).
185. R. Koningsveld and A. J. Pennings, *Rec. Trav. Chim.*, **83**, 552 (1964).
186. I. Harris, *J. Polymer Sci.*, **8**, 353 (1952).
187. S. H. Pinner and J. V. Stabin, *J. Polymer Sci.*, **9**, 575 (1952).
188. J. V. Stabin and E. H. Immergut, *J. Polymer Sci.*, **14**, 209 (1954).
189. B. H. Zimm and I. Myerson, *J. Am. Chem. Soc.*, **68**, 911 (1946).
190. H. Morawetz, *J. Polymer Sci.*, **6**, 117 (1951).
191. C. E. Ashby, J. S. Reitenour, and C. F. Hammer, *J. Am. Chem. Soc.*, **79**, 5086 (1957).
192. V. Kokle, F. W. Billmeyer, Jr., L. T. Muus, and E. J. Newitt, *J. Polymer Sci.*, **62**, 251 (1962).

193. B. H. Zimm and W. H. Stockmayer, *J. Chem. Phys.*, **17**, 1301 (1949).
194. C. D. Thurmond and B. H. Zimm, *J. Polymer Sci.*, **8**, 477 (1952).
195. W. H. Stockmayer and M. Fixman, *Ann. N.Y. Acad. Sci.*, **57**, 334 (1953).
196. J. T. Atkins, L. T. Muus, C. W. Smith, and E. T. Pieski, *J. Am. Chem. Soc.*, **79**, 5089 (1957).
197. E. Duch and L. Z. Küchler, *Elektrochem.*, **60**, 220 (1956).
198. W. E. Gloor, "Polyethylenes Made by the Ziegler Process," in Kirk-Othmer *Encyclopedia of Chemical Technology*, Vol. 14, p. 259 (1967).
199. C. J. Stacey and R. L. Arnett, *J. Polymer Sci.*, **A2 (2)**, 167 (1964).
200. F. W. Billmeyer, Jr., *J. Am. Chem. Soc.*, **75**, 6118 (1953).
201. L. D. Moore, Jr., *J. Polymer Sci.*, **20**, 137 (1956).
202. L. T. Muus and F. W. Billmeyer, Jr., *J. Am. Chem. Soc.*, **79**, 5079 (1957).
203. *1966 Standards*, American Society for Testing and Materials, Philadelphia, Method D 1238–65T.
204. T. G. Fox and S. J. Loshaek, *J. Applied Phys.*, **26**, 1080 (1955).
205. F. Bueche, *J. Chem. Phys.*, **20**, 1959 (1952).
206. W. L. Peticolas and J. M. Watkins, *J. Am. Chem. Soc.*, **79**, 5083 (1957).
207. *Polymer Degradation Mechanisms*, National Bureau of Standards (U.S.) Circular 525, Washington, 1953.
208. S. L. Madorski, *Thermal Degradation of Organic Polymers*, Wiley-Interscience, New York, 1964.
209. H. H. G. Jellinek, *Degradation of Vinyl Polymers*, Academic, New York, 1955.
210. N. Grassie, *Chemistry of High Polymer Degradation Processes*, Butterworths-Interscience, New York, 1956.
211. S. L. Madorsky, S. Straus, D. Thompson, and L. Williamson, *J. Polymer Sci.*, **4**, 639 (1949); *J. Research Natl. Bur. Standards*, **42**, 499 (1949).
212. W. G. Oakes and R. B. Richards, *J. Chem. Soc.*, **1949**, 2929.
213. R. Simha and L. A. Wall, *J. Polymer Sci.*, **6**, 39 (1951).
214. R. Simha, L. A. Wall, and P. J. Blatz, *J. Polymer Sci.*, **5**, 615 (1950).
215. S. L. Madorsky, *J. Polymer Sci.*, **9**, 133 (1952).
216. L. A. Wall, S. L. Madorsky, D. W. Brown, S. Straus, and R. Simha, *J. Am. Chem. Soc.*, **76**, 3430 (1954).
217. W. L. Hawkins, V. L. Lanza, and F. H. Winslow, German Patents 1,118,451 (Nov. 29, 1956), 1,131,007 and 1,131,008 (June 7, 1962).
218. A. D. Dietzler and C. L. Stacey, U.S. Patent 3,082,188 (March 19, 1963).
219. H. Malz, F. Lober, O. Bayer, and H. Schuerlen, U.S. Patent 3,044,981 (July 17, 1962).
220. H. C. Beachell and S. P. Nemphos, *J. Polymer Sci.*, **21**, 113 (1956).
221. F. M. Rugg, J. J. Smith, and R. C. Bacon, *J. Polymer Sci.*, **13**, 535 (1954).
222. J. R. Myles and D. Whittaker, British Patent 581,717 (Oct. 22, 1946).
223. W. F. Henderson, U.S. Patent 2,502,841 (April 4, 1950).
224. L. E. Wolinski, U.S. Patents 2,715,075, –076, –077 (Aug. 9, 1955); 2,801,446, –447 (Aug. 6, 1957); 2,805,960 (Sept. 10, 1957).
225. W. E. F. Gates, British Patent 800,714 (Sept. 3, 1958).
226. S. P. Joffre, U.S. Patent 2,811,468 (Oct. 29, 1957).
227. P. V. Horton, U.S. Patent 2,688,134 (Feb. 2, 1954) and Reissue 24,062 (Sept. 20, 1955).

228. Farbenfabriken Bayer A. G. Leverkusen, Belgian Patent 545,444 (Feb. 22, 1956).
229. Unpublished results from the author's files.
230. W. Berry, R. A. Rose, and C. R. Bruce, British Patent 723,631 (Feb. 9, 1953).
231. G. W. Traver, U.S. Patent 2,910,723 (Nov. 3, 1959); French Patent 1,065,670 (May 28, 1954).
232. G. W. Traver, U.S. Patent 3,018,189 (Jan. 23, 1962).
233. G. H. Berthold and A. Pace, Jr., U.S. Patent 2,935,418 (May 3, 1960); filed June 3, 1953.
234. Visking Corp., British Patent 722,875 (Feb. 2, 1955).
235. F. N. Rothaker, U.S. Patent 2,802,085 (Aug. 6, 1957).
236. R. F. Pierce and V. G. Potter, U.S. Patent 2,810,933 (Oct. 29, 1957).
237. S. Plonsky, U.S. Patent 2,923,964 (Feb. 9, 1960).
238. E. Sauter and H. C. Grossmann, German Patents 1,019,017, 1,065,952 (March 24, 1960).
239. Montecatini and K. Ziegler and F. Ranalli, Australian Patent 231,061 (Oct. 24, 1960); this appears to be the first patent issued for polypropylene treatment by electric discharge.
240. K. S. Hoover, U.S. Patent 2,715,363 (Aug. 16, 1955).
241. W. H. Kreidl, U.S. Patents 2,632,921 (March 31, 1953), 2,704,382 (March 22, 1955).
242. W. H. Kreidl and F. Hartmann, *Plastics. Technol.*, **1**, 31 (1955).
243. M. F. Kritchever, U.S. Patents 2,648,097 (Aug. 11, 1953), 2,683,894 (July 20, 1954).
244. J. C. Von der Heide and H. L. Wilson, *Modern Plastics*, **38**, No. 9, 199 (1961).
245. K. Rossman, *J. Polymer Sci.*, **19**, 141 (1956).
246. W. J. Tabar, *Proc. Battelle Symp. Thermal Stability of Polymers*, Columbus, Ohio, 1963, G 1.
247. Ref. 207, especially pp. 221–235.
248. G. D. Cooper and M. Prober, *J. Polymer Sci.*, **44**, 397 (1960).
249. L. R. Hougen, *Nature*, **188**, 577 (1960).
250. B. A. Brice and M. Halwer, *J. Opt. Soc. Am.*, **41**, 1033 (1951).
251. M. C. Huggins, *J. Am. Chem. Soc.*, **64**, 2616 (1942).
252. P. M. Doty, W. H. Aiken, and H. Mark, *Ind. Eng. Chem., Anal. Ed.*, **16**, 686 (1944).
253. W. H. Aiken, P. M. Doty, and H. Mark, *Mod. Packaging*, **18**, No. 12, 137 (1945).
254. B. Katchman and A. D. McLaren, *J. Am. Chem. Soc.*, **73**, 2124 (1951).
255. J. A. Cutler, E. Kaplan, A. D. McLaren, and H. Mark, *Tappi*, **34**, 404 (1951).
256. M. C. Slone and F. W. Reinhart, *Mod. Plastics*, **31**, No. 10, 203 (1954).
257. R. B. Richards, *J. Applied Chem. (London)*, **1**, 370 (1951).
258. C. A. Sperati, W. A. Franta, and H. W. Starkweather, Jr., *J. Am. Chem. Soc.*, **75**, 6127 (1953).
259. L. E. Nielsen, *J. Applied Phys.*, **25**, 1209 (1954).
260. K. W. Ninnemann, *Mod. Packaging*, **30**, No. 3, 163 (1956).
261. *Visqueen Film Technical Data*, Ethyl Corporation, 1968.
262. "Du Pont Clysar EH Polyolefin Film", Technical Bulletin, E. I. du Pont de Nemours and Co., Film Department, Wilmington, Del., 1968.
263. O. J. Sweeting and J. J. Levitzky, *Mod. Packaging*, **34**, No. 7, 128 (1961).

CHAPTER 3

POLYPROPYLENE

ORVILLE J. SWEETING*

Yale University, New Haven, Connecticut

and

HARRY A. KAHN†

Monroe, Connecticut

* Formerly Associate Director of Research, Film Division, Olin Corporation, New Haven, Connecticut.   Present address: Quinnipiac College, New Haven 06518.

† Formerly Director of Research and Applications, USI Film Division, National Distillers and Chemical Corp, Stratford, Connecticut.

I. Manufacture of Polypropylene

Early studies on the polymerization of propylene, before 1950, led only to polymers of very low molecular weight which were oils and low-melting solids and which found little practical use.

In the early 1950's Natta at the Polytechnic Institute of Industrial Chemistry (Milan) joined with Ziegler of the Max Planck Institute (Mühlheim) (cf. Chapter 2) in a series of brilliant researches that led to an entirely new class of polymers from the 1-olefins. In startling contrast to the previous polymers of low molecular weight these Ziegler-Natta polymers could be made in a controlled way. Molecular weights could be varied up to several million monomer units, but even more remarkable was the configuration of the polymers. They were nearly unbranched and could be made with regular structure and high crystallinity, with regular structure and a lower crystallinity, or with no extensive regular structure and very low order.

The first disclosure of solid special catalysts had been made by Ziegler in 1954 [1], and this was rapidly followed by a series of patents and other publications [2–9] reflecting an enormous amount of research on polymerization of several 1-olefins, the most important commercially being ethylene (Chapter 2) and propylene.

An immediate collaboration of Ziegler and Natta of the Milan Polytechnic began, and in his earliest papers Natta refers to the heterogeneous catalyst obtained when titanium(IV) chloride is added to diethylaluminum chloride (to form $TiCl_3$ *in situ*) in a carefully controlled predetermined ratio as the Ziegler catalyst [10–25]. This work has been reviewed elsewhere [26–29].

Crystalline polypropylenes have also been made by oxide catalysts developed by Phillips Petroleum [30, 31] and Standard Oil of Indiana [32–45]. The stereospecificity of these processes is less than that for the Ziegler catalysts, however, and they have not been extensively exploited commercially. Two composition-of-matter patents were issued to Montecatini in 1963 [46, 47] and the basic Ziegler catalyst patent was granted in 1966 [48].

The feed material in the Ziegler-Natta process consists of propylene of a purity greater than 99.6%. Small amounts of saturated hydrocarbons can be tolerated (though they reduce efficiency), but electron donor molecules such as allene, butadiene, methyl acetylene, oxygen, carbon monoxide, carbon dioxide, and various sulfur compounds such as carbonyl sulfide, carbon disulfide, dimethyl sulfide, mercaptans, and hydrogen sulfide must be excluded, for they have a most deleterious effect on the catalyst [49–51]. Traces of water also destroy the catalyst system, as might be expected.

The process is carried out in its essentials just as the laboratory experiments were first done [8] by passing the monomer at 1 atm or slightly above into a stirred slurry of the catalyst, violet titanium(III) chloride with diethylaluminum chloride or triethylaluminum suspended in a suitable diluent at 40 to 80°C. Diluents may be any of the paraffins and naphthenes of suitable boiling point; usually cyclohexane or heptane is used. With a properly prepared catalyst system (much depends on the state of subdivision and the purity of the violet form of $TiCl_3$ used),

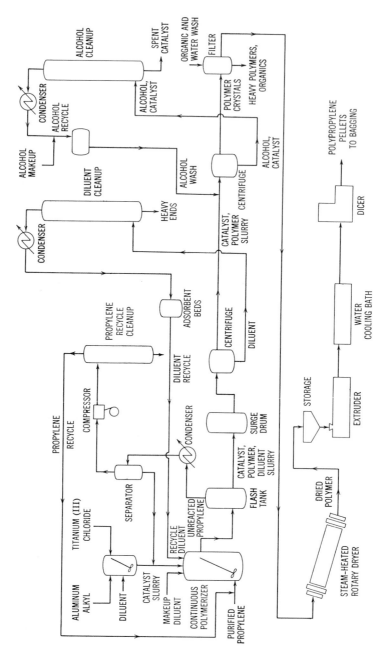

Fig. 1   Flow sheet for polypropylene manufacture.

216

crystalline polypropylene forms continuously and precipitates as a solid on the catalyst surface until the slurry becomes very thick (up to 40% solids). The slurry is then stripped of unreacted monomer, the catalyst is dissolved by adding a reactive material such as an alcohol or other reactive hydrogen compound [e.g., $(RCO)_2CH_2$] [52], and the polymer is separated by filtration and dried.

Molecular weight can be controlled by adding hydrogen [53], by use of diethylzinc as cocatalyst [54], or by raising the polymerization temperature [55].

The isotactic purity of the product depends on the specificity of the catalyst and the particular catalyst system used. Normally the material produced contains more atactic and syndiotactic material than is desirable and the product must be purified. The less crystalline polymers are more soluble than the desired isotactic material, and therefore differential solubility in the diluent at higher temperatures provides effective purification. Several such extractions of the more soluble components may be required.

Finally the product is combined with stabilizers, extruded in coarse strands, and chopped into pellets for shipment. Figure 1 presents a typical flow sheet [56].

Worldwide commercialization of the Ziegler-Natta processes for

TABLE 1

U.S. Polypropylene Plant Capacity in 1968

Location	Capacity (million lb)	Trade name
Alamo Industries, Houston, Texas[a]	70	Marlex
Avisun Corp., New Castle, Del.	200	Olemer, Oleform
The Dow Chemical Co., Torrance, Calif.	30	
Enjay Chemical Co., Baytown, Texas	150	Escon
Hercules, Inc., Lake Charles, La.	200	Pro-fax
Novamont Corp., Neal, W. Va.[b]	60	Moplen
Rexall Chemical Co., Odessa, Texas	55	El Rexene
Shell Chemical Co., Woodbury, N.J.	110	
Texas Eastman Co., Longview, Texas	90	Tenite
	965	

[a] Affiliate of Phillips Petroleum Co. and National Distillers (now owned by Diamond Shamrock).
[b] Affiliate of Montecatini SG.

manufacture of isotactic polypropylene has been carried on by Monte-catini who hold the basic composition-of-matter patents [46, 47]. At the end of 1968 plant capacity in the United States was near a billion pounds annually (Table 1); consumption for all uses was approximately 800 million pounds a year (Table 2) [57, 58], of which about 10% was extruded into packaging film.

TABLE 2

Uses of Polypropylenes

Market	Consumption (million lb)			
	1965	1966	1967	1968
Injection molding	155	210	310	360
Filament and fiber	90	115	160	175
Film and sheeting	55	60	70	90
Blow molding	5	8	10	12
Wire and cable	2	4	8	10
Pipe and profiles	10	10	12	15
Extrusion coating		3	5	8
Export	30	50	75	80
Miscellaneous	18	30	40	50
Totals	365	490	690	800

II. Properties of the Polymer

A. Structure

The Ziegler-Natta polypropylenes constituted the first stereoregular polymers ever made from an unsymmetrical monomer by a nonliving process, and Natta recognized that the properties (such as melting point

$$n \text{ CH}_2{=}\text{CHCH}_3 \xrightarrow[\text{1 atm, 20°C.}]{\text{Al(C}_2\text{H}_5)_3\text{-TiCl}_4} (-\text{CH}_2\text{-}\overset{\overset{\text{CH}_3}{|}}{\underset{\underset{\text{H}}{|}}{\text{C}}}-)_n$$

head-to-tail addition

and density) could be accounted for by an alignment of the molecules $\text{CH}_2{=}\text{CHCH}_3$ in a helical pattern with all of the methyl groups (a) in analogous positions (b) in alternating positions, or (c) in random positions with respect to the ($-CH_2-CH-$) backbone (see Volume I, Chapter 2, p. 38 ff).   To these three configurations Natta assigned the terms *isotactic, syndiotactic,* and *atactic,* respectively.   Figure 2 illustrates the situation with the continuous chain of carbon atoms considered to be projected into a plane.   In (*a*) the methyl groups all lie above the plane, in (*b*) they are alternately but regularly above and below the plane, and in (*c*) they are above or below in random sequence.   (Figures 4 and 5 in Chapter 2, Volume I, p. 39, show the spatial relationships somewhat better.)

An enormous amount of structure work has been done at Milan by Natta and his collaborators.   In isotactic polypropylene the repeat

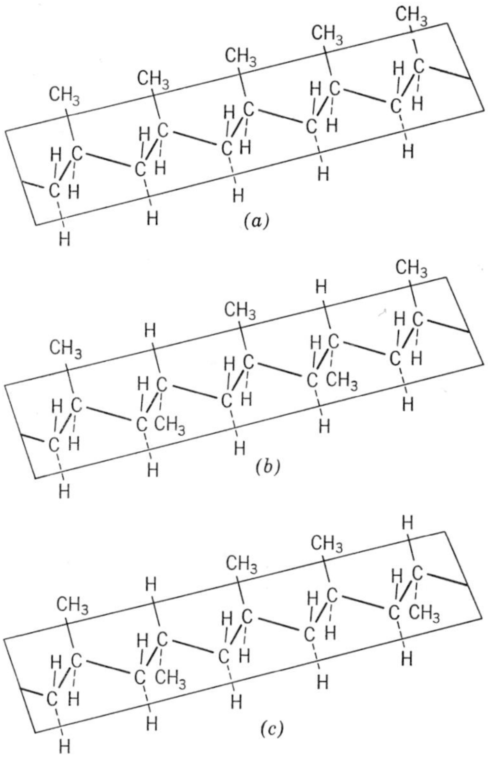

Fig. 2   This diagram illustrates the spatial disposition of methyl groups in (*a*) isotactic, (*b*) syndiotactic, and (*c*) atactic polypropylene chain segments.

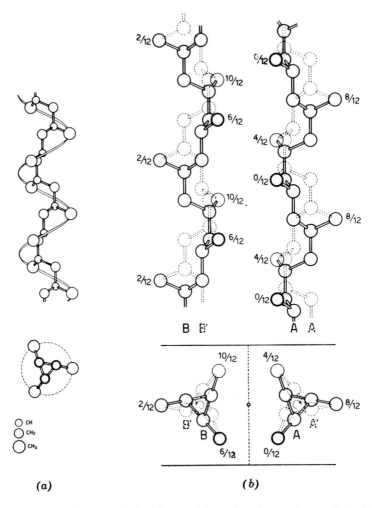

Fig. 3 (a) One of the four helical forms of isotactic polypropylene. (b) Packing of chains in isotactic polypropylene [18].

220

distance is 6.50 ± 0.05 Å, which corresponds to three polypropylene units of a helix (not a zigzag chain as in polyethylene). Four such helical arrangements are possible: two right-hand and two left-hand helixes (Figure 3) each of the right and left pairs differing in the regular aspect of methyl groups [11, 15, 18].

Syndiotactic polypropylene has a repeat distance of 7.4 Å, which corresponds to four monomer units. The syndiotactic material is usually made by homogeneous catalysis at very low temperatures [23, 24, 59] and differs markedly in properties from the isotactic material. The latter is highly crystalline, as measured by x-rays, high-melting, and of low solubility.

The unit cell of isotactic polypropylene has the cell constants, $a = 6.65$ Å, $b = 20.96$ Å, $c = 6.50$ Å, and $\beta = 99°20'$ [14, 60]. The unit cell is monoclinic, contains 12 monomer units, and four chains pass through the unit cell, each with the helical configuration [61–63]. The density calculated is 0.936 g/cm.^{3}

For their work Ziegler and Natta received the Nobel prize for chemistry in 1963.

B. MORPHOLOGY

The morphology of polypropylene is complex [64–71]. Spherulites (Volume I, Chapter 7, p. 334 ff) form with great ease as the molten polymer is cooled and under suitable conditions may grow to 400 μ in diameter [64] (Figure 4). Other studies have distinguished at least four different types of spherulite [65, 66], the growth of which must be controlled in the manufacture of film or physical properties may be seriously impaired [67, 68].

The types of spherulite are distinguished on the basis of superficial appearance, sign of birefringence, and extinction pattern in a cross-polarized field. In addition, mixed forms (i.e., not classifiable into the four groups) appear in profusion; they show no appreciable areas of pure positive or negative birefringence but, instead, a coarse fibrous mixture of the two.

In all cases, varying with temperature and spherulite type, a fibrous texture is found. It is radial, visible in both polarized and unpolarized light, and often profusely branched. The fibrous structures appear to fan out first by dendritic growth and the interstices become filled in later.

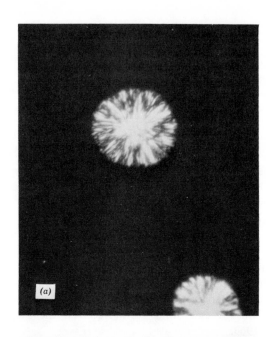

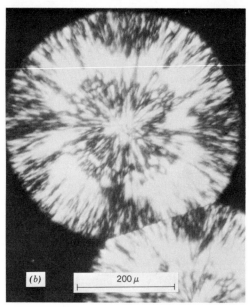

Fig. 4 Photomicrographs of polypropylene spherulites grown at 145°C: (a) after 5.3 hr; (b) after 25 hr [64].

222

Also, as a second characteristic, distinct cracks appear between fibrils and at spherulite boundaries.

Experiments with bulk polypropylene crystallized both at high pressures (60,000 psi), to form a triclinic structure, and from dilute solution, to form a monoclinic structure, show that crystallization takes place in the form of lathlike lamellar crystals. The similarities of morphology of samples crystallized under such widely differing conditions as those stated suggest that the chain molecules are folded. X-ray analysis of the two crystal structures confirms that they are closely related and that conversion is easy to explain as a partial slip along fold planes [69].

The rate of crystallization of polypropylene is phenomenally rapid. It is almost impossible to quench the molten polymer to less than 50% crystallinity with any practical commercial method. Use of liquid nitrogen as a quenching medium has produced crystallinities in the 20 to 30% range, as estimated from density measurements. With slow quenching, as in air-cooled film or molded articles, crystallinity approaches 75 to 80%. Even rapidly quenched materials, such as chill-roll or water-bath cast film, continue to crystallize at a steady rate (constant half-life) until the equilibrium value of 75 to 80% crystallinity is reached in about a year at normal temperatures.

The consistently high crystallinity of polypropylene, and the inability to reduce it by process control, is responsible for many of the weaknesses of polypropylene as well as its desirable properties. Crystal formation during working, flow, and shaping of polypropylene articles produces physically distorted and oriented crystal milieus quite different in character from a situation in which crystals are formed from amorphous material that is molecularly oriented, such as polyethylene. High-density polyethylene shares this property of high crystallization rate and is similarly limited in use by physical weaknesses related to crystallization during forming.

There is reason to believe that a partially ordered state—a two-dimensional or so-called smectic crystal—exists in some phases of polypropylene.

Polypropylene shows non-Newtonian plastic flow to a very high degree (decrease of viscosity with increasing shear rate) which can be surprisingly variable in different polymers and lots of material. In order to develop more reliable quality control, many vendors and users measure and report melt-flow measurements at several piston weights. Such measurements are usually indicated by a subscript giving the piston weight in kilograms. The usual procedure is to run flow tests with three additional weights, 2.5, 5.0 and 10.0 kg to give $MI_{2.5}$, MI_5, and MI_{10}. Shear rates in actual processing are usually quite different from those obtained in the conventional melt-flow apparatus, and matching several points rather than one gives a better indication of applicability.*

Polypropylene can suffer significant thermal degradation during a melt-flow test, enough to change the measurement if the timing of the various steps is not meticulously reproduced.

C. INFRARED SPECTROSCOPY

Polypropylene shows, as expected, many of the usual infrared absorption bands shown by polyethylene and other aliphatic hydrocarbon polymers (Table 3) [72–77].

The intense parallel bands at 1166, 998, and 841 cm^{-1} are characteristic of crystallized isotactic polypropylene and are ascribed to the isotactic helical structure [75, 76] since they decrease in intensity or disappear completely on melting. The two bands at 1152 and 975 cm^{-1} appear also in the molten isotactic polymer and may be ascribed to the chemical structure of head-to-tail sequence of monomer units; they are ascribed to the isotactic configuration, independent of the helical arrangement [75–77]. The bands at 1166 and 998 cm^{-1} can be used to estimate the degree of orientation by the use of polarized infrared from the monochromator.

TABLE 3

Infrared Absorption Bands of Crystalline Polypropylene [73, 75]

Frequency (cm^{-1})	Relative intensity	Interpretation
807	Weak	
841	Medium strong	
899	Weak	
940	Very weak	
975	Medium strong	CH$_2$ rocking, C—C stretching, and CH$_3$
998	Medium strong	wagging and rocking
1043	Very weak	
1100	Weak	
1152	Shoulder	
1166	Strong	

* A discussion of melt index and melt flow appears in Chapter 2, Section VIII.

TABLE 3 (*continued*)

Frequency (cm^{-1})	Relative intensity	Interpretation
1218	Very weak	CH$_2$ twisting and bending
1256	Weak	
1295	Shoulder	
1304	Weak	CH$_2$ wagging, assym. C—C stretching, and CH bending
1328	Very weak	
1358	Medium	
1378	Strong	Sym. CH$_3$ bending
1437	Shoulder	CH$_2$ bending and antisym. bending
1456	Strong	
1465	Strong	
2840	Strong	Sym. CH$_2$ stretching
2867	Strong	Sym. CH$_3$ stretching, antisym. CH$_2$ stretching and CH stretching
2877	Strong	
2907	Very strong	
2918	Very strong	
2926	Shoulder	
2950	Very strong	Antisym. CH$_3$ stretching
2962	Very strong	

D. Molecular Weight

Intrinsic viscosity of polypropylenes may conveniently be determined in decahydronaphthalene at 135°C and the viscosity-average molecular weight, $\overline{M}_v$ can be calculated from the intrinsic viscosity [η] (Volume I, Chapter 2) by the following relationship [78].*

$$[\eta] = 1.62 \times 10^{-4} \, \overline{M}_v^{0.77}$$

The weight-average molecular weight $\overline{M}_w$ of commercial polypropylenes is about 30% higher.

* Natta gives [η] = $1.75 \times 10^{-4} \, \overline{M}_v^{0.74}$ for determinations in tetralin at 135°C [21].

Commercially, melt-flow rate [79] is used to characterize a polymer for extrusion (melt-flow rate is the same as melt index, q.v., except that the measurement is made at 230°C instead of at 190°C. It is subject to all of the caveats that apply to the melt index (Chapter 2) but is quickly determinable and is useful. The ranges that may be expected are illustrated in Table 4, for a series of Hercules Pro-fax samples [78].

TABLE 4

Melt-Flow Rate, Intrinsic Viscosity, and Average Molecular
Weights of a Series of Pro-Fax Polypropylenes

Sample	Melt-flow rate[a] (dg/min)	Intrinsic viscosity[b] (dl/g)	$\overline{M}_v \times 10^{-3}$	$\overline{M}_w \times 10^{-3}$
61XX	60	1.3	120	156
62XX	30	1.5	140	180
63XX	15	1.7	170	220
64XX	6.5	2.05	210	270
65XX	4	2.25	250	325
66XX	2	2.6	290	380
67XX	0.8	3.1	360	470
68XX	0.4	3.5	420	550

[a] ASTM D 1238–65T, 230° [79].
[b] Viscosity in decahydronaphthalene at 135°C.

E. COPOLYMERS

Though the properties of crystalline polypropylene are advantageous for certain uses, the narrow melting range makes processing difficult, particularly in film extrusion. Of even greater importance in processing, however, is the high rate of crystallization mentioned above. Often the most desirable combinations of stiffness, tensile strength, elongation, low-temperature impact strength, and ability to absorb punishment have been achieved by copolymerization and blending, which are intended to reduce the crystallization rate and broaden the melting range. These two procedures overcome some of the deficiencies of crystalline homopolymers, but they degrade other desirable characteristics of the polymers. The copolymerization of propylene with even small amounts of ethylene or other 1-olefin by standard polymerization processes causes a pronounced loss of crystallinity and amorphous, rubbery polymers unsuitable for film extrusion generally result.

Blending has similar effects and frequently adds haze and color to the plastic.

A special polymerization process developed by Eastman Chemical Products produces copolymers of 1-olefins that exhibit the degree of crystallinity normally associated only with homopolymers [80]. The term *polyallomer* was coined to identify the polymers manufactured by this process and to distinguish them from polymer blends and previously known copolymers. The present commercial polyallomer formulations are based on block copolymers of propylene and ethylene (Volume I, Chapter 2).

The various formulas of propylene-ethylene polyallomer are unique materials that are quite different in physical properties and crystallinity from blends of polypropylene and polyethylene and also distinctly different from copolymers produced from propylene and ethylene by other polymerization processes (Table 5). Insolubility in hexane and heptane establishes that the polyallomer plastics are highly crystalline materials. Infrared spectra indicate that polyallomer chains contain polymerized segments of each of the monomers employed. These segments exhibit crystallinities normally associated only with homopolymers of these monomers.

The propylene-ethylene polyallomer formulations have many of the best properties of both high-density polyethylene and crystalline polypropylene and provide a number of important advantages over each of these polymers. With an annealed density of 0.91 g/cc, polyallomer is one of the least dense solid polyolefins.

Polyallomer overcomes the most serious property deficiencies of crystalline polypropylene, offering brittleness temperatures as low as $-40°$, impact strengths as high as 12 ft-lb/in. (65 cm-kg/cm) of notch, and less notch sensitivity. It is superior in resistance to fatigue from flexing, particularly at low temperatures.

Polyallomer is superior to linear polyethylene in flow characteristics and moldability, softening point, hardness, stress-crack resistance, and mold shrinkage. Stress-cracking tests (ASTM D 1693 [79]) of annealed specimens show no failures after 1200 hr [80].

Polyallomer exhibits a substantial loss of modulus as compared with polypropylene, but is still an improvement over polyethylene.

All of the propylene homopolymers have minor differences among them, as might be expected; Eastman's Tenite is noticeably different in processing behavior from all of the others as a group. The difference appears to be caused by a somewhat slower crystallization rate. In

TABLE 5

Physical Properties of Tenite Polyallomer 502[a]

Property[b]	Value	ASTM method [79]
Flow rate, g/10 min	2.0	D 1238–65T (Cond. L)
Density, g/cc	0.899	D 1505–63T
Specific heat	0.5	
Refractive index, n_D^{23}	1.492	
Coefficient of linear expansion, $\times 10^5$ in./in.-°C		
$<10°C$	15–18	
$>10°C$	9–10	
Softening point, Vicat, °C	131	D 1525–65T
Deflection temperature, 264 psi load, °C	51	D 648–56 (1961)
Tensile strength at yield, psi	4100	D 638–64T
Stiffness in flexure		
$\times 10^{-5}$ psi	1.1	D 747–63
$\times 10^{-3}$ kg/cm²	7.7	D 747–63
Rockwell hardness, R scale	76	D 785–65
Izod impact strength		
Notched, at 23°C (73°F)		
ft-lb/in. of notch	1.7C	D 256–68
cm-kg/cm of notch	9.4C	D 256–68
Notched, at 23°C (73°F)		
ft-lb/in. of width	No break	D 256–68
cm-kg/cm of width	No break	D 256–68
Unnotched at $-18°C$ (0°F)		
ft-lb/in. of width	6C	D 758–48 (1961)
cm-kg/cm of width	33C	D 758–48 (1961)
Water absorption, %	<0.01	
Burning rate	Slow	
Mold shrinkage, %	1–2	

[a] Available also with heat stabilizers and ultraviolet stabilizers added. Some of these formulations are acceptable for packaging food materials.

[b] Measurements made on samples injection-molded according to ASTM Method D 2146–65T [79].

most laboratory tests of molecular weight, ultimate crystallinity, etc., there is no detectable difference, but the performance on production machinery quite plainly stands apart. Any production crew can identify a Tenite resin from a group of resins by performance. The difference is sometimes, but not always, beneficial; therefore it cannot be said that Tenite is better or worse than others—only different. This claim of uniqueness is also made by the manufacturer but has not been supported by any details. It may be that Tenite homopolymer is actually a block copolymer consisting of blocks of right-hand and left-hand polypropylene, which could be brought about by the same process used for the polyallomers, namely intermittent feed of monomer to the reactors. In the case of polyallomers the monomers are alternated. For the homopolymer starting and stopping the feed permits occasional reversal of the "hand" during each period of chain growth.

The properties inherent in propylene-ethylene polyallomer formulations are not found in blends of polyethylene and polypropylene, but some of them are approached by blends of polypropylene and rubber. Polyallomer is superior to rubber-modified polypropylene in color, clarity, moldability, electrical properties, and resistance to blushing when bent or stretched.

The initiation of crystallites and the growth of crystalline regions seem to be slower in polyallomer than in highly crystalline homopolymers. Significant increases in the degree of crystallinity of polyallomer occur for a considerable period after solidification of the material from the melt. Basically, this means that the crystalline characteristics are determined more by the material than by the rapidity with which the molten plastic is cooled. From a practical standpoint it means that the processing conditions for producing tough products are less critical for polyallomer than for most formulations of polypropylene and high-density polyethylene.

The slower rate of crystallization allows the congealing polymer to cease flow movement before crystals begin to form. The crystals are thus protected from flow orientation, which occurs with homopolymer, and the consequent establishment of stress-failure planes. When the crystals do finally form, they are more stable and without built-in strains which contribute to failures. Slower crystallization also results in larger numbers of nuclei and generally smaller crystal aggregates. Some of these properties can be induced in homopolymer by the addition of nucleating agents that promote small spherulites.

The heat-sealing range of polyallomer is wider than that of either polyethylene or polypropylene.

F. Physical Properties of Commercial Polymers

Table 6 exhibits the properties of unmodified polypropylenes ranging in melt flow from an index of 0.2 to 20.0 g/10 min. Listed are the results that may be expected from quality-control tests commonly performed on commercial polymer [82].

TABLE 6

Typical Properties of Unmodified Polypropylenes

Property	Grade designation[a] : all food grades (F, FT, FN)								ASTM Method[b]
	002	007	015	025	040	080	120	200	
Nominal melt flow, g/10 min	0.2	0.7	1.5	2.5	4.0	8.0	12.0	20.0	D 1238–65T (Cond. L)
Density, 23°C, g/cc	0.905	0.905	0.905	0.905	0.905	0.905	0.905	0.905	D 792–60T
Mold shrinkage, in./in.	0.015–0.025	0.015–0.025	0.015–0.025	0.015–0.025	0.010–0.020	0.010–0.020	0.010–0.020	0.010–0.020	D 955–51
Water absorption, %	<0.02	<0.02	<0.02	<0.02	<0.02	<0.02	<0.02	<0.02	D 570–63
Environmental stress cracking	None	None	None	None	None	None	None	None	D 1693–60T
Tensile yield strength, psi	4900	4900	5000	5000	5000	5100	5200	5300	D 638–61T
Elongation at yield, %	15	12	12	10	10	9	9	9	D 638–61T
Ultimate elongation, %	>200	>200	>200	>200	>200	>200	>100	>20	D 638–61T
Izod impact strength, 23°C, ft-lb/in.	3	1.9	1.3	1.0	0.9	0.8	0.7	0.6	D 256–56
Stiffness in flexure, $\times 10^{-5}$ psi	1.60	1.70	1.85	1.95	2.10	2.20	2.25	2.25	D 747–63
Rockwell hardness, R	92	92	93	95	97	97	98	100	D 785–62

Property									ASTM method[b]
Heat deflection temperature, 66 psi, °C	100	100	105	107	110	110	110	110	D 648-56
Heat deflection temperature, 264 psi, °C	57	57	59	61	65	65	65	65	D 648-56
Vicat softening point, 1 kg load, °C	150	150	150	152	152	155	155	155	D 1525-65T
Coefficient of linear expansion, $\times 10^{-5}$ in./in.-°F	5	5	5	5	5	5	5	5	D 696-44
Thermal conductivity, Btu/hr-ft^2-°F-in.	1.15	1.15	1.15	1.15	1.15	1.15	1.15	1.15	
Specific heat, cal/g-°C	0.46	0.46	0.46	0.46	0.46	0.46	0.46	0.46	
Flammability	Slow	Slow	Slow	Slow	Slow	Slow	Slow	Slow	D 635-63
Melting point, °C	168	168	168	168	168	168	168	168	D 2117-64
Dielectric constant at 10^6 cps	2.25	2.25	2.25	2.25	2.25	2.25	2.25	2.25	D 150-65T
Dissipation factor at 10^6 cps	<0.0001	<0.0001	<0.0001	<0.0001	<0.0001	<0.0001	<0.0001	<0.0001	D 150-65T
Volume resistivity $\times 10^{-16}$ Ω/cm	6.5	6.5	6.5	6.5	6.5	6.5	6.5	6.5	D 257-61
Dielectric strength, short time 1/8-in. thickness, V/mil	660	660	660	660	660	660	660	660	D 149-59

[a] F, general purpose, food grade; FT, GP, food grade, antistatic; FN, GP, food grade, nucleated.

[b] American Society for Testing and Materials, 1966 *Standards* [79].

TABLE 7
Typical Properties of Modified Polypropylenes[a]

Property	FM-GP			FH-GP		FX-GP		CM		CH		CX
	002	040	080	040	080	040	002	040	080	040	080	040
Nominal melt flow, g/10 min	0.2	4.0	8.0	4.0	8.0	4.0	0.2	4.0	8.0	4.0	8.0	4.0
Density, 23°C, g/cc	0.905	0.905	0.905	0.905	0.905	0.905	0.905	0.905	0.905	0.905	0.905	0.905
Mold shrinkage, in./in.	0.015–0.025	0.010–0.020	0.010–0.020	0.010–0.020	0.010–0.020	0.010–0.020	0.015–0.025	0.010–0.020	0.010–0.020	0.010–0.020	0.010–0.020	0.010–0.020
Tensile yield strength, psi	4000	4300	4500	3900	4000	3500	4000	4300	4500	3900	4000	3500
Elongation at yield, %	15	10	9	10	9	9	15	10	9	10	9	9
Ultimate elongation, %	>200	>100	>100	>100	>100	>100	>200	>100	>100	>100	>100	>100
Izod impact strength, 23°C ft-lb/in.	12.0	1.6	1.4	2.5	2.4	3.5	12.0	1.6	1.4	2.5	2.4	3.5
Impact strength, unnotched: 23°C ft-lb/in.	>30	22	20	25	22	>30	>30	22	20	25	22	>30
0° ft-lb/in	30	15	14	20	15	25	30	15	14	20	15	25
−17 ft-lb/in	10	8	6	12	8	15	10	8	6	12	8	15
−29 ft-lb/in.	8	6	4	10	6	12	8	6	4	10	6	12
−40 ft-lb/in.	6	4	4	6	4	8	6	4	4	6	4	8
Stiffness in flexure $\times 10^{-5}$ psi	1.25	1.60	1.70	1.50	1.60	1.30	1.25	1.60	1.70	1.50	1.60	1.30
Rockwell hardness, R	80	93	95	85	90	75	80	93	95	85	90	75

Property												
Heat deflection temperature, 66 psi, °C	93	105	107	102	105	100	93	105	107	102	105	100
Vicat softening point, 1 kg load, °C	143	150	150	146	150	144	143	148	150	146	150	144
Coefficient of linear expansion $\times 10^{-5}$ in./in.-°F	5.5	5.5	5.5	5.5	5.5	5.5	5.5	5.5	5.5	5.5	5.5	5.5
Flammability	Slow	Slow	Slow	Slow	Slow	Slow	Slow	Slow	Slow	Slow	Slow	Slow
Melting point, °F	168	168	168	168	168	168	168	168	168	168	168	168
Oven aging, time to failure of 0.050-in. specimens, forced air draft, stationary exposure, 302°F, hr							>1500	>1500	>1500	>1500	>1500	>1500

a For test method references see final column of Table 6.

b FM, GP, food grade, medium impact; FH, GP, food grade, high impact; FX, GP, food grade, extra impact; CM, extended heat stability, medium impact; CX, extended heat stability, high impact; CH, extended heat stability, medium impact; CH, extended heat stability, high impact; CX, extended heat stability, extra impact.

Similarly, typical data for a series of modified polypropylenes with a 40-fold range of melt-flow rates [82] are presented in Table 7.

G. RESISTANCE TO CHEMICALS

Polypropylene has exceptionally good resistance to attack by a wide variety of chemical substances, including acids, alkalis, salt solutions, and organic solvents, even at elevated temperatures (80°C). Environmental stress cracking does not occur with solvents and detergents. Some chlorinated compounds and aromatic hydrocarbons and the high boiling aliphatic hydrocarbons produce swelling and attack the surface slightly. Table 8 [81, 82] presents some of the available data on chemical resistance.

TABLE 8

Effect of Chemicals on Polypropylene[a]

	Percent Change in		
Chemical	Weight	Thickness	Appearance
Acetone	2.2	1.0	No change
Acetic acid			
5%	0.1	0.1	Slightly bleached
50% (80°C, 30 days)	0.5		
Ammonium hydroxide 10%	0.0	−0.1	Slightly bleached
Benzene (22°C, 90 days)	12.0		Swelled
Butyl acetate	6.3	1.8	Slightly bleached
Butyl phthalate			
(22°C, 90 days)	0.30		No change
Calcium chloride, 2.5%	0.1	−0.4	Slightly yellow
Carbon disulfide	18.3	3.9	Swollen and slightly bleached
Carbon tetrachloride	43.0	7.3	Swollen and slightly bleached
Cellosolve solvent	0.4	0.2	Slightly yellow
Chloroform	26.7	4.8	Severely swollen and slightly bleached
Citric acid, 10%	0.0	0.3	No change
Copper sulfate, 10%	0.0	0.2	No change
Ethanol			
50%	0.2	0.1	No change
95%	0.2	0.5	No change
Ethyl acetate	5.0	1.6	Slightly swollen
Ethyl ether (22°C, 30 days)	8.5		
Ethylene glycol	0.1	0.3	No change

TABLE 8 (*continued*)

Chemical	Percent Change in		Appearance
	Weight	Thickness	
Ethylene dichloride	9.2	2.0	Swollen and slightly bleached
Ferric chloride, 10%	0.1	−0.1	No change
Formaldehyde, 35%	0.2	−0.1	Slightly yellow
Gasoline (regular)	13.7	4.5	Swollen and very slightly bleached
Gasoline (aviation)	12.2	4.5	Swollen and bleached
Glycerol	0.0	0.2	No change
Green soap solution	0.1	0.2	No change
Heptane	11.1	4.4	Swollen and bleached
Hydrochloric acid, 10%	0.0	0.3	No change
Hydrogen peroxide, 30%	0.0	0.2	Yellow
Methanol			
5%	0.0	0.3	No change
100%	0.4	0.5	Slightly bleached
Motor oil	0.2	0.0	No change
Nitric acid (conc.)	1.0	0.3	Yellow
Oleic acid	0.2	0.3	No change
Olive oil	0.0	0.5	No change
Phenol, 5%	0.1	0.1	No change
Phosphoric acid (conc.)	0.0	−0.2	No change
Potassium dichromate, 10%	0.0	0.3	No change
Sodium chloride, 10%	0.0	−0.2	No change
Sodium carbonate 2%	0.0	−0.2	Very yellow
Sodium hydroxide			
10%	0.0	0.1	No change
1%	0.0	0.6	No change
Sodium hypochlorite	0.0	−0.3	No change
Sulfuric acid			
30%	0.0	−0.2	No change
3%	0.0	0.0	No change
(conc.)	0.0	0.2	Stained
Toluene	12.8	3.6	Swollen and slightly bleached
Triethyl phosphate	0.1	0.2	No change
Turpentine	14.2	5.0	Swollen and bleached
Water	0.0	0.2	No change
Wesson oil	0.0	−0.2	No change
Xylene	12.7	3.7	Swollen and slightly bleached

[a] Results reported are on samples of a general-purpose polypropylene stored at 23°C for one year (unless another condition is given in parentheses).

III. Polypropylene Film

A. INTRODUCTION

Investigation of polypropylene as a film-forming material began immediately after the material had been widely publicized by Natta and Montecatini. The analogy of the material to polyethylene was obvious and apparent advantages existed in the higher melting point and more highly crystalline structure. The high crystallinity not only gave rise to a stiffer and stronger film but also made possible the prospect of very substantial improvements in these properties through orientation.

Polypropylene has been commercialized both as cast and oriented film. The cast film has made the major penetration in the marketplace, and the oriented product has attracted most of the attention. The high degree of crystallinity and associated spherulitic formation inherent in polypropylene has generally kept this material out of the blown tubing area because of the very poor physical properties that are developed through the slow cooling inherent in the manufacture of air-cooled tubing.

Polypropylene markets and products are still in a state of rather frenetic change as fundamental modification of the polymer structure, through copolymers and blending, shift the product properties and displace older formulations. The main efforts at modification have been directed at reducing the tendency of polypropylene to be brittle at temperatures below freezing and at reducing the brittleness induced by slow cooling as encountered in blown tubing and in the molding of heavy-walled articles. So far every gain in the desired properties has extracted a severe toll in the existing satisfactory properties of the homo-polymers, particularly in a serious reduction of stiffness.

B. CAST FILMS

Cast polypropylene film was the most obvious target for initial work with the new polymer since a great deal of film-casting equipment and technology was already in existence based on polyethylene. The immediate incentives for cast polypropylene were both esthetic and economic. Polypropylene proved to have inherently better optical properties than polyethylene, plus enough additional stiffness and tensile strength that appreciable reductions in required film thickness might be contemplated. The comparative properties of cast polypropylene are discussed below.

1. Clarity and Gloss

Polypropylene in the proper molecular weight ranges proved to be capable of exceptionally high gloss and clarity, in general much clearer and more transparent than polyethylene. The optical properties are nearly the same as those of cellophane, polyvinyl chloride, and polystyrene. Gardner haze levels below 1 % and gloss values in the 90s are not difficult to obtain with polypropylene films cast on a chill-roll or into a water bath. The clarity and gloss caused much initial excitement about cast polypropylene and have, indeed, carried the product far into several markets, such as breadwrapping. Other properties, however, affect the optical impact of the film, and cast polypropylene has been hampered in the high-grade printing area because of a tendency encountered in all manufacturers' resins to form a scattering of pimples or specks in high-speed casting operations. These defects are similar in appearance to gels in polyethylene but should be referred to by another name, such as blotches. Gels in polyethylene have a specific structure related to the formation of oxidatively crosslinked polymer, either during polymer processing or during extrusion of film. In polypropylene blotches in most instances appear to be centered in discrete particles of an impurity.

2. Tensile Strength

The tensile strength of cast polypropylene is roughly twice that of polyethylene. It will usually run from 5000 to 7000 psi, depending on the direction of measurement and the conditions of manufacture. Polypropylene shows a much stronger tendency than polyethylene to become oriented during manufacturing stages. This tendency is aggravated at higher operating speeds. Hence most cast polypropylene has distinctly unbalanced properties which reflect the preferential orientation in the machine direction.

3. Tear Strength

Tear strength is not particularly good in cast polypropylene, partly because of the orientation effects produced in manufacturing. Tear strength is, classically, extremely high in the transverse direction (across the web) and disappointingly low in the machine direction. The result of this combination is mostly bad: tear resistance to general perforation and snagging is not satisfactory for heavy bagged items, but at the same time knife cutoff on packaging machines is difficult because of the extreme toughness of the material in the transverse direction of the web. Special adaptations of many packaging machines by the addition of specially designed cutoff knives, or heated knives, have been necessary to permit operation with cast polypropylene.

4. Stiffness

Stiffness is excellent on cast polypropylene and is approximately three to four times that of polyethylene. In many applications, in which stiffness controls the required film thickness, the use of polypropylene permits the economic advantage of gauge reduction. The absolute stiffness of a sheet of material varies approximately as the third power of the thickness. Polypropylene, being ordinarily at least three times as stiff as polyethylene, can generally be used at a thickness ratio to polyethylene equivalent to $(1/3)^{1/3}$ or approximately 60% of the thickness of the corresponding polyethylene.

Stiffness is generally estimated by the measurement of the elastic modulus, which is satisfactory if the assumption is correct that the film is homogeneous in structure from one side to the other. This is not strictly true, but for thin cast films it is acceptable. Control of stiffness during the manufacture of cast polypropylene is complicated by the fact that the film ages rather slowly and stiffness will usually increase about 50% during the first three weeks from the time of manufacture.

5. Impact Resistance

Cast polypropylene is comparatively poor in impact resistance, particularly in the transverse direction. Orientation effects developed during manufacturing cause serious deterioration of this property. For this reason cast polypropylenes have had difficulty in going into bag markets that require heavy loadings. They can perform satisfactorily in lightweight bag applications, such as in small textiles and paper goods. The impact resistance is directional and, like tear strength, has tended to be a disadvantage in the cutoff sections of packaging machines, particularly those of the breadwapping type that use a fly-knife cutoff rather than a scissors or shearing action. Again, this problem has been solved generally by the use of specially designed cutoff knives, sometimes heated.

6. Low-Temperature Strength

The tendency of polypropylene to get brittle at temperatures below 40°F has been one of its most serious drawbacks in the film area. Many attempts have been made to solve this problem by the development of specially plasticized "winter-grade" materials. In every case in which a worthwhile reduction in the useful operating temperature has been achieved a serious loss of stiffness has also been encountered. The optical properties remain good, however. Though blending and random copolymerization are unsatisfactory solutions, stereoblock copolymers such as Eastman's polyallomers (Section II–E) largely achieve the desired balance of properties.

7. Temperature Resistance

The high melting point of isotactic polypropylene (168 to 171°C, 335 to 340°F) permits the polymer to be used in many applications that are closed to polyethylene. It withstands steam autoclave sterilization and can therefore be used for the packaging of surgical and antiseptic materials as well as for applications such as boil-in-bag and other food cooking.

Its high temperature resistance also permits polypropylene to function well in subsequent coating operations. Rather high drying temperatures can be used for applying polyvinylidene chloride, for instance. Polypropylene is substantially more resistant to solvent attack, generally, than is polyethylene. Polypropylene is also a good candidate for various laminates because of its temperature performance. Low creep and high strength-retention temperatures provide substantial advantages over either high- or low-density polyethylene. (See Table 9 for a comparison of some film properties [83].)

8. Heat-Sealing Range

Cast polypropylene has a surprisingly wide heat-sealing range. This property varies substantially with the materials of different vendors but will usually run between 30 and 50°F from the initial usable adhesion strength to burn-through. This compares with polyethylene at less than 10 Fahrenheit degrees. This feature, in addition to the excellent stiffness, has contributed heavily to the generally good machine operability of cast polypropylene in mechanized packaging operations. The tendency of polypropylene to block-seal at temperatures well below its

TABLE 9

Comparative Data on Some Transparent Packaging Films

Property	[a]Propafilm O 60 gauge[b]	[a]Propafilm C 70 gauge[b]	[a]Propafilm C 90 gauge[b]	Cast polypropylene 100 gauge[b]	Low-density polyethylene 100 gauge[b]	Cellulose films 300 MSAT	Cellulose films 300 MXXT
Density, g/cm$_3$	0.91	(a)	(a)	0.90	0.92	(c)	(c)
Yield, in.^{2}/lb	51,500	37,600	30,240	31,000	30,000	20,000	19,300
Tensile strength, psi MD^d × 10^{-3}	25–30	20–25	23–28	6.5	2.8	13–19	13–19
psi, TD^d × 10^{-3}	25–30	20–25	20–25	5.5	2.2	6–8	6–8
Elongation at break, % of original length, MD	50–85	50–75	80–110	750	300	20	20
TD	50–85	50–75	80–110	1000	650	60	60
Tear strength, g/0.001 in. MD	ca. 5	ca. 5	ca. 5	50	200	ca. 5	ca. 5
TD	ca. 5	ca. 5	ca. 5	80	150	8–12	8.12
Modulus of elasticity, psi × 10^{-5} MD	3.2–3.8	3.2–3.6	2.0–2.5	1.1–1.4	0.3–0.35	Depends on Grade	
psi × 10^{-5} TD	3.2–3.8	3.2–3.6	2.0–2.5	1.1–1.4	0.3–0.35	Depends on Grade	
Melting point, °C	168–171	168–171		168–171	107–112		
°F	335–340	335–340		335–340	225–234		

						Special grade necessary	
Low temperature flexibility, °C	−70	<−70	<−70	−10	−70		
°F	−94	<−94	<−94	+14	−94		
Permeability to water vapor, g/m²-24 hr at 90% rh at 38°C (100°F)	8–10	5–7	4–6	12	17	5–12	1–8 (uncreased)
Permeability, cm³/m²-24 hr at 25°C and 1 atm differential pressure							
oxygen	3000–3400	10–20	10–20	ca. 4500	ca. 9500	5–25	5–15
nitrogen	600–1200						
carbon dioxide	8000–10,000						

[a] *Propafilm C* is coated on both sides with a vinylidene chloride copolymer. *Propafilm O* is the base sheet, biaxially oriented and heat set.

[b] Gauge given in hundredths of a mil; i.e., 60-gauge film is nominally 0.0006 in. thick.

[c] See Chapter 7. MSAT: moistureproof, heat-sealing, anchored, transparent. MXXT: moistureproof, polyvinylidene-chloride coated, transparent.

[d] M.D., machine direction; T.D., transverse direction.

melting point can also give rise to what is sometimes referred to as "false seal." At the low end of its sealing temperature range adequate adhesion may be obtained when the seal is inspected directly from the packaging line. After aging for a few days, however, this adhesion, which is really little more than heavy blocking or matting, may disappear and leave an open package.

9. Barrier Properties

Cast polypropylene has roughly half the permeability of low-density polyethylene and a slightly lower permeability than high-density polyethylene. Resistance to hydrocarbons and general grease penetration is sufficiently good to permit use of this material in motor-oil packaging and as a grease barrier in packages containing fatty food material.

Polypropylene is an excellent water vapor barrier but in general has poor resistance to the light gases such as oxygen, nitrogen, and carbon dioxide (Table 9). On the other hand, it is too good a barrier to oxygen and carbon dioxide to permit its use as a breathing material in such areas as produce packaging. Resistance to aromatics and ester types of flavoring or aroma materials is usually poor, but because transmission of odors can vary widely specific tests should be made. Polypropylene exercises its protective properties primarily as a moisture barrier and as a grease or oil barrier.

10. Printability

Cast polypropylene can be treated and then printed in much the same fashion as cast polyethylene (Chapter 2). It is somewhat more difficult to treat, however, and care must be exercised to avoid overtreatment and blocking when dealing with high-gloss films. As with polyethylene, treatment is preferably done at the time of manufacture before the slip agents have had a chance to migrate to the surface. It is difficult to get adequate treatment on polypropylene after the slip agent has exuded.

Slip-containing films are also more prone to blocking and increase the requirement for careful control during the manufacturing process. Well-treated film which has been treated before slip exudation will maintain its ink-adhesion properties for at least six months. Some conflicting evidence that has been presented indicates loss of treatment with age. At least some of these anomalies are caused by the use of indirect tests for treatment, such as wettability or water sensitivity.

Although it is unquestionably true that the relative surface tension between water and the film surface will change with age, particularly when dealing with a slip-containing film, it is not necessarily an indication that ink will not adhere to it. Aging studies should be based strictly on ink adhesion and not on variations found in inferential tests.

It should also be noted that there is a considerable variation in the adhesion of different inks, even those from the same vendor and presumed to be of the same binder composition but of different colors. Inks used in the control of treatment for printing must be carefully standardized and reproduced, and it must be remembered that satisfactory exposure to the quality control test will not necessarily mean that every color and type of ink met with in the field will give satisfactory adhesion. Printing was a serious problem 10 years ago, but improved inks and treatment of base sheet have ameliorated this situation greatly. Standard inks for application to polyvinylidene chloride coatings can be printed beautifully on polymer-coated polypropylenes, such as ICI's Propafilm C (Table 9) or Olin's PCP grades (Table 10), all of which are approved for food-packaging applications.

11. Static Propensity

Polypropylene has a strong tendency to develop static. In most cases in which static is a problem in the use of the film on automatic packaging machines the problem can be substantially reduced or eliminated by the consistent use of static eliminators at every winding stage in the production operation, including printing. Static generation on wrapped packages and the associated attraction of dust and grime can be reduced only by the incorporation of antistatic agents in the resin or by application of such an agent as a surface coating. Satisfactory antistatic agents are available for most casting purposes from nearly all of the polypropylene vendors. Some antistatic agents tend to interfere with the development of the proper slip level and some act as slip agents as well as antistatic agents. All tend to increase blocking, particularly after surface treatment for printing.

12. Odor

Many of the early cast films had distinct and unpleasant odors which could be transferred to food products as an off-flavor or taste. The odors resulted from combinations of stabilizing additives plus residual solvents from the resin manufacturing process, aggravated by thermal

TABLE 10

Physical Properties of Olin Polyvinylidene-Chloride-Coated Polypropylene

Property	Film grades	
	280 PCP–11	220 PCP–11
Unit weight, g/m^2	25.0	32.0
Yield, in.^{2}/lb	28000	22000
Gauge, mil	0.9	1.15
Water vapor transmission rate,[a]		
TAPPI, g/100 in.^{2}/24 hr	0.32–0.36	0.24–0.28
Instantaneous permeability value,[b]		
g/100 m^2-day	35–45	30–40
Air permeability, ml/100 in.^{2}-24 hr	1.0–2.0	1.0–2.0
Tensile strength, psi, MD[c]	18000–22000	16000–20000
TD[c]	16000–20000	14000–18000
Elongation, %, MD	40–60	55–75
TD	45–65	60–80
Impact strength, Olin[d] kg-cm	15–25	20–30
Tear strength,[a] g, MD	4–6	6–8
TD	4–6	6–8
Stiffness,[a] g, MD	15–25	35–45
TD	15–25	35–45
Heat sealing range,[a] °F	210–275	210–275
Heat seal strength,[a] g/1.5 in.,		
275°F, 20 psi, and 0.5 sec	400–500	400–500
Coating anchorage,[a] min, time to fail in		
boiling water	30	30

[a] See Volume I, Chapter 13.

[b] Based upon a standard DuPont test: a relative humidity differential of 97–99% at 39.5°C is maintained on the two sides of the film.

[c] MD, machine direction; TD, transverse direction.

[d] Ninnemann impact tester, see Volume I, p. 580.

decomposition in casting. Most of the cast films still have distinctive and detectable odors, as have many polyethylenes. These odors are usually characteristic of the polymer source, and some operators claim to be able to identify the resins of different vendors by the smell. Odor is a property that shows substantial lot-to-lot variation. If a film is to be used to package any sensitive or delicately flavored food material, it is wise to make odor and taste panel tests on the product and, if necessary, to install a quality-control procedure for film odor on the product going to these customers.

13. Aging Properties

Cast polypropylene shows several aging effects which complicate its manufacture and quality control. These changes are caused by a slow recrystallization that takes place in the film following the rather severe quenching characteristic of the casting processes. They are often accompanied by the exudation of various additives, such as slip agents and antistatic agents, which bleed slowly to the surface and by their migration facilitate rearrangement and consolidation of the internal structure of the film. Film properties may change drastically in the first 72 hr after casting. In the case of modulus of elasticity full film stiffness may not develop for as long as three weeks. When initially measured, a film may have a modulus of 70,000 to 80,000 psi and ultimately attain a value as high as 150,000 psi. Impact and tear strength tend to decrease during the short-term aging period. Crystallization seems to continue at a progressively decreasing rate for as long as one to two years and may ultimately result in film with unsatisfactorily low impact resistance.

The instability of the internal structure of cast polypropylene also gives rise to a small but damaging amount of shrinkage, which can be disastrous if it occurs on a roll that is already tightly wound. This is a primary cause of crushed cores and stretched gauge bands in slit rolls of film. Allowance for this shrinkage tendency must be made during the manufacturing phases of the film. The shrinkage is not of sufficient magnitude to be noticeable on any packaging applications, and as far as package protection is concerned the appearance and barrier properties of the film will remain unchanged over many months of shelf-life.

Polypropylene is particularly susceptible to environmental oxidative degradation at moderately elevated temperatures and especially on exposure to sunlight. Special highly effective stabilizers have been developed for use under particularly difficult conditions, but none of these is acceptable in food-packaging applications and they are not used in most film extrusion polymers.

14. Extrusion Stability

Most film-grade polypropylenes with food-grade stabilizers undergo a distinct degradation at every extrusion step. Melt flow will generally increase about 50 % for each extrusion pass. This complicates the use of reclaim somewhat, but, on the other hand, re-extruded material tends to improve optical properties and gauge control when blended with new resin. The amount of thermal breakdown obtained during processing is, of course, widely variable from one type of polymer to the next, depending on the extent to which catalyst residues have been removed, the amount of stabilizers added, and also the particular extrusion conditions employed. If films are being extruded for nonfood use, such as in tapes or textile overwrap, more effectively stabilized resins can be employed and considerably less melt-flow change will be experienced.

C. ORIENTED FILMS

The great interest in oriented polypropylene arose from the discovery that the stiffness of the film increased phenomenally when it was stretched. Most other polymers can be strengthened substantially by orientation but the stiffness increases only nominally, if at all. The stiffness of oriented polypropylene film is approximately three times the stiffness of cast film and about eight times the stiffness of polyethylene films.

In many packaging applications the limitation on the film thickness, hence the cost of the package, develops from problems in handling the film and carrying out the wrapping process rather than from strength or protective requirements. The fact that the stiffness of a material varies as the third power of the thickness means that even moderate improvements in strength and protective properties are quickly nullified as the thickness is reduced; the film becomes so limp that it cannot be handled. The eightfold stiffness increase of oriented polypropylene held the promise that it might perform the same function as polyethylene in a package at half the thickness, that is, $(1/8)^{1/3}$, and could also compete on an equivalent thickness basis with cellophane.

The competitive potential has not yet been achieved, largely because of difficulties in achieving reasonable production economics. The performance potential is still valid, however.

1. Clarity and Gloss

Optical properties are generally superb in the thickness range of 0.5 to 0.8 mil. Above this thickness surface effects generated in the manufacturing process begin to produce haze. It is also easy to get a variety of pattern defects which are not identified by standard haze and gloss measurements but which must be controlled by over-all visual examination.

2. Tensile Strength

The tensile strength of well-oriented polypropylene is in the range of 20,000 psi. Very carefully produced film can go as high as 30,000 psi, and material oriented in one direction only can be produced up to 50,000 psi. The latter material fibrillates easily but finds use in applications that require only one direction of strength, such as binder twine, strapping tape, and some adhesive tapes. Tensile strength is directional and will vary if the degree of orientation in the longitudinal and transverse directions varies or is unequal.

3. Tear Strength

The tear strength of oriented polypropylene is zero for all practical purposes. This is the case with most oriented films, unless some special reinforcement is provided. Attempts have been made to produce oriented film with good tear strength by laminating together two webs of longitudinally oriented film with their orientation axes at a 45° angle. The opposing tear planes act as reinforcement for each other. This was accomplished by cross-winding the webs to form a tube and then slitting the tube. This process has not been commercialized at this date.

4. Stiffness

Stiffness, as estimated by the modulus of elasticity, can approach 500,000 psi. Common commercial film ranges from 300,000 to 450,000 psi. Stiffness is also directional, and the measurement will vary according to the degree of orientation in different directions of the film.

5. Impact Strength

Impact strength is controlled by a combination of tensile strength and ability to yield. Very highly oriented films, with little yield elongation left in them, may be inferior in impact to less thoroughly stretched film which still retains the ability to give with the stress.

Impact strengths of 500 ft-lb/in. are reasonable with oriented polypropylene, as measured with a falling-ball, Monsanto dart, or tensile impact tester.

It should be understood that any impact test is a restricted and artificial measurement, and application of the resulting numbers to a real-life-performance situation must be made empirically. The stress rates involved in a package failure, for example, may be widely different from the rates being used with one of the standard test methods. Also, the puncture type of test, such as the falling-ball, Monsanto dart, or Elmendorf tester, fail to distinguish directional properties. Impact strength is highly sensitive to direction in oriented film and to variations in degree of orientation.

The impact strength of most oriented films is quite adequate for most overwrapping and lamination requirements. It is excellent in machine-direction-oriented films which are slit to fibers and converted to binder twine or sisal substitutes (as in replacement for burlap bags). The impact strength is hopeless for heavy bag loadings (e.g., for apple and produce bags), but the absence of tear strength eliminates these markets anyway.

6. Low-Temperature Strength

Orientation almost entirely eliminates the problem of low-temperature strength which is so devastating in cast polypropylene. Serious brittleness is not encountered until the region of $-50°F$ is reached. Performance in ambient winter conditions is satisfactory in all parts of the United States.

7. Temperature Resistance

Oriented film which has been heat-set retains strength and dimensional stability above boiling-water temperatures. Some puckering will set in around 225°F and shrinkage may become serious around 250°F. Nonheat-set film will begin to shrink in the neighborhood of its last processing temperature but retains strength up to about 250°F.

8. Heat-Sealing Range

Oriented polypropylene is difficult to seal. The material begins to shrink and disorient before it reaches its melting point and, even when restrained, tends to perforate in the seal area.   If seals can be accomplished, the result is still unsatisfactory because the orientation has been disrupted just adjacent to the seal area and a small area of material has been melted and allowed to cool relatively slowly.   This results in a line of film adjacent to the seal which has all of the unsatisfactory properties of air-cooled polypropylene: brittleness, poor impact strength, poor tensile strength, sensitivity to low temperatures, and so on.

Much of the effort expended in developing coatings for oriented polypropylene is inspired by the need to develop a heat-sealing material. Some of the coatings, such as polyvinylidene chloride, also contribute to protective properties, but the main reason is still heat sealing.   Other successful coatings have been based on polyethylene, polyvinyl acetate, and polystyrene copolymers.

Eastman's polyallomer again comes to the fore in this area.   Some of the polyallomers show sufficient stiffness improvement with orientation to be attractive and, in addition, exhibit a barely tolerable heat-sealing range of 10 to 20 Fahrenheit degrees.

9. Barrier Properties

The protective barrier properties of oriented polypropylene show an improvement of 50 to 100 % over cast film.   This is just about enough to compensate for the reduced thickness that can be utilized.   Permeability to light gases (e.g., O_2, CO_2) is still inadequate for protection of fatty materials against rancidity or other oxygen-barrier requirements. Solvents with a plasticizing effect on polypropylene can cause disorientation and puckering of oriented film.   Even so, oriented polypropylene stands up well to solvent-coating applications, such as the manufacture of adhesive tape and typewriter ribbon.

10. Shrink Wrap

Oriented polypropylene looks like a good shrink-wrap candidate, since it has a shrink energy roughly equivalent to that of oriented polyvinyl chloride (Reynolon).   It has, in fact, found only limited use in this area.   The reason appears to be the high, and rather narrow, temperature range over which shrinking will occur.   Although polyvinyl chloride has no greater shrinking force, it begins to develop appreciable shrink tension as low as 130°F and increases it steadily up to around 180°F.   The result is that even though parts of the film are in direct contact with the package contents and do not get well heated in the shrink-tunnel nearly all areas of the wrap get hot enough to do some shrinking and a smooth package results.

Oriented polypropylene (nonheat-set) will begin shrinking rather abruptly in the region of 200°F. Portions of the film in contact with package contents (e.g., a tray of apples) will never get hot enough to shrink without damaging the contents because of excessive heat, and the film is too strong to yield under the stress of the shrinking areas. The result is a tightly shrunk package with a severe case of puckers at the boundaries of unshrunk spots. The appearance is unacceptable.

Some specialized shrink-wrap applications, such as record-album covers, have been successfully exploited. These uses involve situations in which there are no points of spot contact with a good heat absorber or the package contents can stand momentary attainment of temperatures in excess of 200°F.

11. Printability

Oriented polypropylene prints well, and treatment problems are about the same as those of cast film. Treatment in conjunction with film manufacture, usually by means of corona discharge, is most common. Post-treatment, particularly of slip-containing films is difficult. Blocking of treated film is much less a problem with oriented film than with cast film. Overtreatment, nevertheless, may cause sufficient adhesion between layers to produce unsightly blotches and marring of the surface. Backside treatment can occur more easily because the oriented film is usually thinner than cast film and also has more tendency to entrain air between the film and back-up roll during treatment. In addition to marring and rollblocking, backside treatment causes ink transfer after printing and elicits much displeasure from customers.

Coated films accept inks according to the nature of the coating.

12. Static Propensity

Oriented film tends to develop static even more than cast film, if for no other reason than that oriented film is usually thinner and there are more layers on a roll. Each layer adds its own increment to static charge. The effect of static on subsequent machine handling is also severer for a thinner film. The bulk of the film is less and the same electrostatic force can shove it around more easily. Strict manufacturing discipline regarding static removal is necessary if field complaints are to be avoided. Essentially static-free film can be delivered to the customer if static removal is carried out at every winding operation in the manufacturing process, from both sides of the film, and if the effectiveness of the removal equipment is monitored constantly by measuring static levels on the roll film with an electrostatic voltmeter.

As in cast film, development of static charges by the handling of wrapped packages, and the consequent accumulation of dust, can be reduced only by the incorporation of one of several commercial antistatic agents on the market. Heavier loadings are required for oriented film because the film is thinner; hence there is more surface to be coated with the antistat per pound of resin.

13. Odor

That odor-causing ingredients are removed with much greater thoroughness in oriented film than in cast film is due to the thinner film and to the more extended and severer heat-treatment to which the film is subjected during manufacture.

14. Balance

Oriented polypropylene is available in two sharply differentiated forms: a nearly balanced material, produced by processes that orient the film simultaneously in both directions, and a highly unbalanced film, produced by sequential machine-direction (MD) and transverse-direction (TD) orienting steps.

The unbalanced film is sometimes referred to as uniaxially or mono-oriented film, as opposed to biaxially oriented material. This is not true. The unbalanced film has substantial orientation in both directions, but it is about three times greater in the transverse direction.

The balanced films usually match properties in any direction of measurement within about 10%. The unbalanced film shows its lack of symmetry in tensile strength (typically 25,000 psi TD versus 8000 psi MD), in elongation (50% TD versus 400% MD), and, most important, in stiffness (about twice as stiff in TD as in MD). Impact strength is usually greater in the machine direction in spite of the lower tensile strength caused by greater elongation.

As long as performance requirements do not strain the boundaries of any of the quality measurements, it is difficult to tell the difference between the two films. The unbalanced film can be stretched appreciably by hand in one direction and not in the other. Balanced film cannot be stretched in either direction.

Since the two types of film are generally manufactured by competing companies, we may hear tendentious arguments about the superiorities of one film over the other. In fact, there are few applications in which difference or lack of difference in directional properties is of much importance, and these instances are likely to involve only the stiffness. Other film properties usually exceed the requirements of the use so thoroughly that it is futile to worry about existing variations. The purchaser is probably best advised to look to his specific requirements and then to the economics in making a choice.

15. Aging Properties

Oriented polypropylene appears to be fully crystallized, as estimated by density measurements, by the time its manufacturing processes are complete. This might be expected, considering the lengthy high-temperature history of the material. Nevertheless, the substantial property changes that can occur on aging indicate further stabilization and consolidation of the crystalline phases.

Aging effects are much more gradual than in cast film. Exudation of additives, such as slip agents, may take three to six weeks. Over this same period stiffness will increase; impact strength, decrease; and elongation, decrease. Appreciable changes will continue for more than a year. Along with additive migration and changes in physical properties, a small amount of shrinkage takes place. On an individual package it is of no concern, probably amounting to only 1 or 2%. If the shrinkage takes place on a tightly wound roll, however, serious damage to film and roll core can result. Tremendous pressures can be generated in a cylindrical form by small shrinkages. Film surface damage and stretching can occur along areas of gauge bands; buckling and core collapse may take place, and pressure can be high enough to disrupt the crystal structure of the slip agent, thus causing an increase in coefficient of friction instead of the anticipated lowering. Accommodation for this shrinkage may be required in the manufacturing process.

The slow rate of aging makes quality control something of a predictive game. Premature measurements must be made and projected to an anticipated aged value. The timing of quality measurements must be precise or the forecast will be wrong. Additive concentrations must sometimes be pushed far higher than would ultimately be required to achieve a desired short- or medium-term performance level.

Printing treatment shows little evidence of deterioration, although indirect tests of wettability may show substantial changes with time. Slip or antistatic additives, if incorporated excessively, may eventually come to the surface to such an extent that they will interfere with printing.

16. Resin Stability

Polypropylene degrades to a greater extent in the orientation process than from ordinary film casting. Reuse of scrap is an essential economic part of the film-making process, and this constant decrease of molecular weight must be accounted for in resin formulation.

Molecular-weight reduction is aggravated beyond the extrusion step by the extensive heat- and air-exposure history which is part of the orientation process. In addition to extending the opportunity for oxidative degradation, the exposure of very thin films in high-temperature ovens results in the removal and deactivation of the antioxidant systems, which are not outstandingly effective in food-grade materials anyway. Recycled scrap material is therefore deficient in protection against further degradation. It acts as a diluent on the antioxidant system in the virgin resin.

Thermal and oxidative degradation is not completely without merit. Blending back a material with the molecular-weight distribution shortened on the high-weight end produces a mixture with a broader distribution and easier processing properties. A significant portion of reclaim material blended with an appropriately selected virgin resin will often produce film with better gage uniformity and better optical properties than can be obtained from virgin polymer.

IV. Film Manufacturing Techniques

Processing of polypropylene into film is governed mainly by the rapid crystallization characteristics of the resin. In many respects it is similar to high-density polyethylene in this feature. The rate of crystallization apparently coincides so precisely with thermal conductivity rates that no quenching medium short of liquid nitrogen can halt crystallization below the 50 % level. The difference between water quenching and air quenching allows the development of nearly maximum annealed crystal-linity, 75 to 80 % in air-quenched films. The consequences of the formation of crystalline structure in the very nascence of film formation dictate the constraints of the processing systems and the properties of the product. The crystals become inextricably involved in the polymer flows which are taking place just at the moment of quenching, and consequently the macrostructure of the crystalline domains is distorted and oriented by the shaping forces of the process.

A. TUBULAR FILM

The tubing process, as conventionally used for manufacture of poly-ethylene film, has been unsuccessful in producing a useful polypropylene film. Several attempts have been made to modify the resin to permit tubing manufacture. These attempts center around efforts to control the crystallization in such a way that small crystals and small spherulites will be formed, thus eliminating the brittleness and poor physical pro-perties that are fatal characteristics of blown polypropylene tubing. It has been possible to make process modifications and produce film of acceptable appearance with good haze and gloss characteristics, but the weakness inherent in the slow air quench has remained.

It should be possible to produce a satisfactory polypropylene tubing by using one of the internal or external mandrel techniques or the liquid quench system as applied to Saran (see Volume I, Chapter 8). Good films have been produced experimentally with water-bathed sleeves or annular cascades as the cooling medium. The incentive to push these developments to commercial status has apparently been lacking up to now, possibly because of the poor low-temperature performance of unoriented polypropylene.

B. WATER-BATH QUENCHING

A beautiful tough film can be made by casting into a water bath, but the process is considerably more sensitive than when polyethylene is used. Water-bath casting of polypropylene requires some of the same precautions that must be taken when a film of high-density polyethylene is made. The melt is more sensitive, it appears to show a greater degree of plastic flow, and goes through a greater change in viscosity from extrusion temperature to solidification point than does low-density polyethylene. The result of the sharply changing viscosity, plus the increased sensitivity of viscosity to shear rate or flow rate, causes a good deal of instability in the molten web if a long drawdown is attempted. The water bath, therefore, must be mounted very close to the die lips. Air gap distances in the region of 1/4 in. must be maintained for smooth operation with thin-gage films. With heavier webs, about 5 mils and thicker, air-gap distances of as much as 1 to 1.5 in. can be used satisfactorily.

With very short air-gap distances between die and water it is apparent that a great deal of the drawdown of the web, from the initial die opening to the finished film thickness, must take place after the molten web has crossed the air-water interface. This is the main reason why waterbath quenching tends to develop stronger orientation effects than chill roll. A great deal more of the drawing down of the molten polymer takes place just at the freezing point, so that the streaming orientation, generated by the sharp shear effects, is frozen in place almost as it is generated. In fact, almost no permanent orientation occurs as a result of flow in the die or drawdown in the molten portion of the web. The flow orientation that may occur, momentarily, is quickly dissipated by thermal motion and the lack of any restriction toward movement in the fluid matrix. (Further discussion of these factors may be found in Volume I, Chapter 8.)

When the melt first touches the water, the outer layers of the polymer freeze instantly and accelerate to full line speed, telescoping off the internal molten layer and, in turn, exposing it to the quenching effect of the water. The film thus draws down with extreme rapidity once it has touched the water layer, in the manner of an expanding telescope, and generates a very high rate of shear at the interface between the solid and molten material. The shear-induced molecular orientation freezes in place before the molecules can again disorient themselves because of the rapidity of heat transfer at this point. Many orientation effects attributed to the setting of the die gap or to the velocity of flow through the die are, in fact, caused only indirectly by the gap setting. The true initiating factor is the change in percentage of drawdown that occurs at the quench point.

A smooth surface on the water bath is of critical importance in waterbath casting. With only 1/4 in. of air gap to work with and the very powerful orientation effects that occur at the moment of quenching, the effect of a 1/16-in. ripple or wave can be profound indeed.

Extreme care must be taken to baffle the tank thoroughly, so that splashback and ripples at the film exit are not propagated across the surface of the water, and to damp flow movement within the tank by means of porous baffles, so that surging does not take place from one end or side of the tank to the other.   Water inlets to the tanks should be designed so that no turbulence or local rippling results from the entrance of the water.

Temperature uniformity at the quenching point is also extremely important.   The physical properties of the film are strongly dependent on the internal geometry developed at the quench point.   Comparatively small temperature variations can bring about appreciable differences in quench rates and can modify gauge profile and the pattern of orientation.   Water temperature at the line of entry of the film should be continuously uniform within 1 to 2°F across the web.   This degree of uniformity in a circulating water bath calls for careful thought and effort in designing the water entries and circulation paths.

Water removal is one of the more difficult aspects of the water-bath casting process.   If high extrusion speeds are to be used (i.e., more than 500 fpm), removal of the water and complete drying of the film becomes a formidable problem.   The use of a hot-water bath, following the quench tank, helps considerably in water removal.   The viscosity of entrained water is lowered substantially, and such devices as high-pressure air knives or squeegee rolls are enabled to do their part of the removal job.   Retained heat in the film helps to evaporate the residual water left on the film.   The problem is aggravated severely by the use of various additives.   Incorporation of a slip agent such as oleamide or erucamide seems to increase water carryover two- or threefold

Air knives, squeegees, and pinch rolls reach their limit of effectiveness in the range of 300 to 400 fpm.   These devices are essentially pressure-actuated and remove water by developing enough backward pressure against the liquid film to force it back and off the moving substrate. Unfortunately the force required to resist the forward movement of the water increases as the square of the velocity of the film.   Thus as running speed increases, the ever-increasing nip pressures or air pressures required soon pass beyond the realm of practicable operation and a thin layer of water begins to pass, although the bulk of the carryover will still be removed.   At this point, multiple drying stages may be used to get further increase in speed, or additional removal techniques must be resorted to, such as oven drying or blotting with felt belts, a technique practiced in the paper industry.

C. CHILL-ROLL QUENCHING

Chill-roll quenching is by far the most popular method of producing cast polypropylene films. The quench is sufficiently rapid to produce a good looking film with generally satisfactory physical properties. Orientation effects, although still pronounced, are not so severe as in water-bath casting. High production speeds can be attained with properly designed chill-roll equipment. Speeds up to 1000 fpm have been reported in casting 1-mil and thinner films.

Low temperatures are required to obtain the rapid quenching necessary for good cast polypropylene properties. Water temperatures below 10°C are desirable. Chill-roll systems must therefore be set up with powerful circulation and special crossflow roll design to minimize the temperature differential from one side of the roll to the other and with good auxiliary cooling systems. Refrigeration is necessary in most areas throughout the year.

Cleanliness of the chill roll is of prime importance in producing good film of consistent quality. Additives from the resin bleed out and accumulate on the chill-roll surface. In addition, in humid weather the low chill-roll operating temperature will frequently cause condensation. Some cleaning procedure must be installed if unmarred film is to be produced.

A brief consideration of the mechanism involved in the formation of the final film is helpful in understanding the effects of different variables, including resin quality, on ultimate film properties. The molten web of polymer will leave the die lips downward at a thickness of 15 to 20 mils. It may move from 4 to 18 in. before it touches the chill roll, depending on the geometry of the die and the chill-roll arrangement. Because of the pull of the chill roll and the tension transferred back through the melt, the molten web will draw down and accelerate to perhaps two to three times the speed at which it left the die lips. There will be a concomitant thinning of the molten web so that it will be from 5 to 10 mils thick when it first makes contact with the chill roll. At the point of contact, the skin of polypropylene which is first quenched jumps instantly to full line speed and is dragged rapidly away from the still molten portion of the web, thus causing a strong shearing, peeling off action. The molten side of the web also accelerates but lags behind the solidified portion, thus accomplishing the final drawdown to the finished film thickness. Flow action continues as the film moves around the chill roll, with the solidified portion growing thicker, until finally it reaches the outside layer and all further flow and shear action cease.

Three elements of this dynamic stage in the formation of the film attract attention. The first is a large relative motion between one side of the film and the other during the quenching process. A drawdown between 500 and 1000% occurs at this point and takes place in just a few inches of travel. The second is that there is a distinct difference in quenching rate from one side of the film to the other, which is enough to alter the basic crystalline structure, produce a variation in optical properties from one side to the other, and generate curl in the film because of slight density differences. The third is that solidification of consecutive layers of polymer actually takes place under conditions of high strain. This generates the strong orientation effects characteristic of cast polypropylene.

Considering the micromechanism of film formation, the effect of specks or solid particles in the melt becomes obvious. A speck or a portion of the melt that will not flow becomes entrapped in the first layer of solidification to take place on contact with the chill roll. The particle must immediately accelerate to full line speed and then finds itself being dragged at a rapid rate through layers of still molten material, which are under melt tension and which are trying to slip backward over the solidified surface. A severe flow disruption, of course, is established and subsequently frozen into the film as further quenching proceeds, and forms an optical defect or blotch.

Operating procedures which tend to minimize the size and undesirable appearance of blotches concentrate on mechanisms for reducing the violence of the drawdown at the quenching point. Reducing the die gap will reduce the drawdown ratio. Increasing the distance from die to chill roll will permit more of the drawdown to take place in the molten state and allow flow patterns to smooth out before the quenching point is reached. Higher melt temperatures tend to increase the proportion of drawdown that takes place in the air gap and also, because of the reduced melt viscosity, tend to minimize the size of the flow patterns set up around the particles at the quench point.

Most of the operational steps that minimize blotch size also tend to minimize quench orientation and produce a more balanced film. This is usually desirable.

The effect of chill-roll deposits can be appreciated more fully in view of the sensitive and dynamic mechanism of film formation. Surface roughness can cause premature contact of portions of the web in relation to adjacent portions, thus giving rise to distortions in the film surface. Slight temperature differences caused by insulation effects of uneven surface deposits can also cause distinct differences in cooling rates of adjacent portions of film. Film deformities and local variation in properties are the result. These effects are in addition to the obvious marring of the surface that can take place.

It can also be seen from the mechanism of formation that there is likely to be a gradient of crystalline structure from one side of the film to the other so that additives incorporated in the film will find greater mobility toward one side of the film than the other. This gives rise to the phenomenon of different slip levels on the two sides of the film.

D. BIAXIAL ORIENTATION

Oriented polypropylene film has high strength, excellent appearance, and a crisp hand, which, combined with its low water-vapor permeability (Table 9), make it a preferred film for many packaging applications.

The basic principles involved in orientation of polymer films and a discussion of equipment used for this purpose are included in Volume I, Chapter 10.

The oriented films may be given the same degree of orientation in both longitudinal and transverse direction (the commonest practice) or orientation may be accentuated in the transverse direction. The first technique leads to a film with balanced properties, which, if not heat set (cf. Volume I, p. 462), may be used in skintight and shrink-wrap packaging. Cryovac Y-900 is such a biaxially oriented shrinkable polypropylene (properties shown in Table 11 [84]). Cast polypropylene film has no appreciable shrinkage to exploit in this kind of application.

The heat-set, balanced, oriented film possesses much lower permeability, much higher stiffness, and better low-temperature performance than unoriented cast film (Table 9).

Heat-set, stabilized, oriented films are often coated on both sides with vinylidene chloride copolymers (Tables 9, 10) to secure better heat sealing and barrier properties with good dimensional stability.

The same base sheet is also used in lamination (e.g., with cellophane) or is extrusion-coated (Crown-Zellerbach's DSF and FLS–11 films consist of oriented polypropylene base sheet coated on both sides with polyethylene) to secure better durability and protection in combination with the ability to heat-seal and run on standard packaging machines.

TABLE 11

Properties of Cryovac Y-900 Polypropylene Film

Gauges available, in $\times 10^5$	50, 65, 80
Density, g/cc at 23°C	0.90
Yield, in.^{2}/lb-mil	30,500
Yield, in.^{2}/lb, 50 gauge	61,000
Haze, %	1.2
Tensile strength, psi, 73°F	21,000–27,000
Elongation, % at break, 73°F	45–85
Tear initiation	Difficult
Tear propagation	Easy
Modulus of elasticity, psi, 73°F, $\times 10^{-3}$	350
Heat-sealing range, °F	320–330
Shrink temperature, air, °F	400–425
Unrestrained shrinkage, %, 310°F	55
Shrink tension, psi, 310°F	500–700
Maximum use temperature, °F	310–340 (melts)

Balanced and unbalanced films are available on the commercial market. The balanced films are produced by simultaneous orientation processes, in which the film is stretched equally in machine direction and transverse direction at the same time. The unbalanced films are produced by sequential processes, being stretched first in one direction and then in the other. These films are not completely lacking in orientation in the less well-oriented direction, as is often implied, but the ratio of orientation is generally about three to one. The machine-direction tensile strength of a typical "uniaxial" film, for example, is about twice the tensile strength of a cast film and about one-third the transverse-direction tensile measurement.

Orientation is accomplished close to the melting point of the polypropylene, and the severest quench that is practical should be used in preparing the film entering the orientation step. It is extremely difficult to orient a slow-quenched base film without splitting.

1. Line Drawing

Polypropylene is one of the polymers that "line-draws." This means that, when stretched, the entire area under stress does not extend in a uniform manner. Instead a portion of film yields abruptly and thins down in one step to a fraction of the initial thickness, thus forming a "step" between stretched and unstretched film. With continued application of force the polymer continues to stretch through the step, or line, as the line moves across the unstretched portion. There is little or no material between the stretched and not-stretched conditions; hence there is no choice or control of intermediate thicknesses between the original material and the stretched material. A line-drawing polymer has a natural stretch ratio that is characteristic of the material. For polypropylene the inherent stretch ratio is about 7:1.

The drawdown ratio applies in both directions, and for a balanced film the over-all area expansion ratio during the orientation step is about 50:1. To obtain a 1-mil oriented film we must start with about 50-mil base stock. In sequentially oriented film an initial machine-direction orientation is established in the casting step and subsequent machine-direction stretching of about 3:1 can be accomplished without line drawing. The transverse draw is a line draw, however; it is set at about 7:1.

The inherent draw ratio can change about 10% with different resin lots and with different vendors, but outside the inherent range either breakage or severe gauge variations will result. It is sometimes stated that a simultaneous orientation process can produce nothing but a balanced film. In other words, there is little leeway in tailoring stretch ratios to produce special directional film properties.

Some features of orientation are common to all production systems. Liquid quench is generally required to keep crystallinity near the lower level of 50% and eliminate extreme break sensitivity. Water removal before reheating for stretching is quite important, although it may sound trivial. At commercial production speeds a real effort is required. Scratches incurred during the preorientation steps are magnified and preserved in the finished product. Handling the heavier materials (25 to 50 mils) in the case of simultaneous processes without sliding contact on anything requires great care.

Uniformity of initial quench temperatures and subsequent reheating temperatures is critical. Minor variations in thickness or crystalline structure will be magnified as gauge defects in the oriented film. Gauge control itself is more demanding than in film casting. Gauge variations in the base stock show a tendency to carry through as absolute variations rather than as percentage variations. In other words, a gauge variation of 1 mil on the 50-mil base stock will often show a variation of 1/4 to 1/2 mil in the finished film, in spite of the 50/1 drawdown that might be expected to reduce it to a 2% variation.

The line-drawing characteristic of polypropylene places an additional premium on uniformity and temperature control in the stretching step, regardless of process. When line drawing occurs, stretching begins at one or more independent points, drawing down to the automatic ratio at those points and then spreading across the sheet. If drawing can be initiated at only one point, the stretch will spread uniformly, either to the edges or till it meets itself at the other side of the tube. If multiple stretch points are initiated, the drawlines will move across the web until two stretching areas meet each other. At the intersection an excess of gauge can occur.

Line drawing will start at the weakest points in the web; that is, thin spots, hot spots, areas of low molecular weight (a result of poor mixing in the extruder), and areas of lower crystallinity. Since weak spots frequently occur as a result of a pattern of nonuniformity somewhere in the process, multiple initiation points tend to repeat, as do the intersections of stretching areas. The result is a fairly consistent point of high gauge which can result in a severe gauge band on the finished roll.

There has been a great deal of heated discussion on the opposing merits of flat and tubular orientation. Some of the specific advantages and weaknesses of both processes bear discussion. The argument that a tubing process is greatly lower in capital cost should be dismissed at once, appealing though it may be. The equipment that can be compared for this judgment consists only of that involved in the orientation step itself and constitutes less than 25% of the total investment required for a film-producing plant. All auxiliary equipment (resin handling, storage, slitting, packaging, etc.) is essentially the same for installations of equal capacity. Even if the tubular equipment cost were half that of flat-processing equipment, the over-all difference in plant investment would not be more than 10%. If an additional heat-treating step is added to tubular orientation, little predictable margin is left, one way or the other. Commercial experience indicates no major economic advantage for either type of extrusion in full-scale production. A choice must be based on other factors.

2. Tubing Orientation

Orientation of tubing has immediate appeal because of the logical simplicity of the concept. Unfortunately, when the development of oriented polypropylene began, there was almost no appropriate prior art to draw on. Saran, the only material already oriented commercially by a tubing process, is profoundly different in characteristics from polypropylene, particularly in that it is blown in the amorphous state.

Production technology of light-gauge polyethylene tubing was largely inapplicable, since liquid quenching is required for polypropylene. So techniques for production were developed pretty much from a zero start.

In the production of base stock the initial problem lies in the difficulty of handling the extruded tube in anything but a straight line. The tube wall is so stiff that collapsing the tube or trying to turn it results in creasing the wall and establishing a point of weakness during the subsequent expansion step.

Quenching is similar to pipe extrusion. The tube can be run up, down, or sideways. In any case, seals and wipers to hold a water reservoir against the tubing and then to get the water off again must be used. Water-wet mandrels, internal or external, can also be considered. Uniformity is a major problem and avoidance of scratching at this stage is most difficult.

The orientation step is accomplished by heating the tube back to within 20° of the melting point and forcing in gas at several pounds pressure. Some patents teach the use of an internal liquid medium to carry out the tube expansion or an external liquid bath to do the reheating. These methods pose obvious problems unless straight-line handling of the tube can be eliminated. The more popular commercial processes use gas expansion and electric or hot-air heating.

The extrusion point must be isolated from the high-pressure gas in the expansion step, and, if the tube is not pinched, an internal plug or mandrel which fits fairly closely must be used. This is a convenient point for introducing the expansion air, but it must extend well past the quench point, making construction a bit awkward, and it is a critical point for scratching.

The flattening of the finished tube, following expansion and orientation, is a more formidable problem than those familiar with polyethylene tubing may realize. First, the tube is large, and finished film widths of more than 60 in. are desirable for acceptable slitting yields. Second, the oriented film is stiff and nonyielding, and finally it is inflated at several pounds pressure; so it does not change shape easily.

Geometrically it would be impossible to collapse and flatten continuously a tube of completely inelastic material by the use of flat collapsing guides.   The linear machine-direction elements of a tube being flattened travel different distances as the tube is deformed to a flat plane.   With an inelastic material the differential travel between adjacent elements can be compensated only by wrinkling and folding.   In the case of polyethylene the material is soft and stretchable and the restraining pressures in the tube are low; so satisfactory tubing collapse can usually be attained within 4 to 5 tube diameters.   In the case of polypropylene, much longer collapsing ratios must be used to minimize the percentage of tube distortion required to escape wrinkling or especially shaped collapsing shields must be fabricated to make all parts of the film path the same length.   As a compromise, an inch or two of wrinkling at the edges of the collapsed tube may be tolerated and trimmed off later before windup.

If shrink film is not desired, it is necessary to pass the film through a second heating stage to heat-set the stretched material.   The film must be heated above the orientation temperature, but not to the melting point, for 5 to 10 sec and must be restrained in both directions until it is cooled in order to prevent shrinkage and puckering.   Heat-setting may be done in various ways.   A light-duty tenter frame may be used to carry the film through an oven, the tube may be slid along an internal supporting frame or it may be reblown at low pressure, passed through an oven, and then collapsed again for slitting.   Heat setting may be done in line or as a separate step, working from mill rolls.

Size changes in manufacture of the tubing are not easy because of the critical nature of the components at the point of extrusion; so equipment is usually set for maximum size and customer variations are met by adjustment of waste trim in slitting.

Circular dies are not easily designed for fine point adjustment and indirect means must be used for making gauge adjustments, such as spot heating or control of the quenching medium.   It may be necessary to machine corrections into the die itself on a trial-and-error basis.

Scrap utilization is as much a part of tubular orientation as any other film process.   In addition to accumulations from breaks, start-up, etc., slitting trim of 5 to 20% must be accommodated, and 1 to 2 in., at least, must be recycled from the edges of the collapsed oriented tube.   The poundage to be recycled is not so high as in a flat process, but the problems involving reuse of light-gauge film are equally severe.   Under no circumstances can scrap utilization be ignored.   As in most film processes, the touchstone is not the quantitative amount of scrap to be handled but the determination and dedication with which the task is approached.

Tubing processes offer the invaluable potential of being able to rotate some part of the equipment such that film variations can be randomized at the windup.   To the extent that this trick can be exploited, the tubing process is a more tolerant procedure than flat orientation.   Tubing modules are practical in smaller capacities than for flat orientation so that initial investments can be minimized.   For equivalent capacity there is the opportunity for a certain amount of variation in production. Different gauges, or different film types, can be run on different machines if multiple units are available.   Balanced film of excellent physical properties can be produced.

3. Flat Orientation

Flat orientation is generally sequential, which produces an unbalanced film.   It is possible, though not commercial at this time, to produce a balanced film by sequential stretching steps.   Flat stretching machines capable of stretching both ways simultaneously have been built, and models have been put into commercial operation on vinyl film.   These machines have operated successfully, but not commercially, on poly-propylene.   These "two-way tenter frames," which are exceedingly complex mechanically, use independent clips that are guided along tracks at controlled rates to stretch the film in the machine direction as well as transversely.   The main problem is reliability, and up to now not enough effort has been put into the development of this equipment to achieve commercially feasible economics.   Kampf, in Germany, is the only supplier of this machine at present.

The Dow octopus deserves mention, if only because of its ingenuity. Not readily classifiable, this machine starts with a tubular extrusion, opens it up like a trumpet bell and pulls the film radially in a flat plane to eight separate winders arranged in a circle around the extruder. The radial stretching produces simultaneous orientation similar to the expansion of a tube.   A working model of this machine was built but it has never been commercialized.

In sequential orientation the film is extruded into a precisely built but conventional water bath, passed through a machine-direction orienta-tion step, then through a specially modified tenter frame for transverse stretching and heat setting, and is finally gauged, treated for printing (if desired), and wound.   The first commercial oriented fims were produced on this type of equipment.   It was chosen for development initially over tubing equipment because the technology of the equipment, namely tenter frames, is more than 100 years old, thus ensuring a reasonable level of reliability.   Also, the published work on orientation of polyester films provided a good starting point for further process development. (Volume I, Chapter 10, p. 466.)

Quench system design is critical, as discussed above (cf. "cast films"). Absolute surface placidity must be maintained along with vigorous circulation and rapid water turnover.   For best results refrigerated and deaerated water must be used.   Web speeds are high, compared with tubing, and water removal is correspondingly more difficult.   The base film must be absolutely dry before it reaches the heating rolls of the machine-direction orienter.

The orienter in the machine direction usually consists of two sets of rolls running at different speeds, according to the draw desired.   The initial roll set is heated electrically or with circulating liquid to bring the film to stretching temperature.   The second set of rolls is water-cooled.   The gap between the slow and fast rolls is kept as small as possible to restrict the area in which drawing occurs and minimize necking.   Additional refinements may be used to localize the stretching area, such as localized heating with infrared or hot gas heaters; or edge support devices that either prevent necking or carry the film close to the fast-roll surface before allowing stretching, may be used.   Nipping against the last heated slow roll may be required to prevent the film from stretching back over the surface of the roll.

After being chilled sufficiently to stop further machine-direction stretching the film is fed to the entrance of a tenter frame equipped with special film-holding clips.   The clips are designed with a wedging action on closing so that the harder the sidewise pull, the tighter the clips grip. Nevertheless, it is usually necessary to provide some additional security in the form of spring loading the clips or by beading the film edge so that a positive grip is obtained.   The clips, which are carried on a traveling chain, get quite hot as they circulate constantly through the tenter ovens, and to avoid overheating film in the clip jaws cooling tunnels are usually provided over the return path of the chain.

Several temperature zones are provided in the tenter housing.   The initial zone preheats the film before stretching begins.   This may be augmented with infrared or direct-contact gas heaters.   Two or more zones are provided during the stretching step, in which the film is drawn sidewise about 7:1. Temperatures are graded downward through the stretching zones so that the film is being cooled as it is being oriented. This convection cooling is preferential for the thinner areas so that thicker sections maintain a higher temperature and tend to stretch more easily, thus equalizing gauge variations. The final zones are used for heat-setting during which 5 to 10% transverse shrinkage is set on the tenter frame to permit maximum film relaxation and stabilization. An open zone is last to air-cool the film to room temperature.

After disengagement from the tenter frame the unstretched edge which has been held in the clips is trimmed off, ground, and fed directly back to the extruder or to a storage area for reclaimed material. The trimmed film is then wound into mill rolls.

In two-way tenter frames for producing simultaneous orientation no machine-direction orienter is used. The machine-direction stretch ratio does not have to be quite up to 7:1, as does the transverse stretch, because machine-direction orientation equivalent to approximately 2:1 stretch is developed in the initial casting step. The film coming from the orienter has a scalloped edge, since the clips are several inches apart by the time they leave the machine. Surprisingly, neck-in between the clips is not a serious problem. The scallop extends only about 1 in., and disruption of orientation extends only another inch or two from the film edge.

To obtain more desirable film properties oriented film from a tenter frame may be given a second stretch in the machine direction. Line drawing does not occur at this point and degree of stretch can be chosen, within limits of breakage. Another alternative, which has been technically successful but not used commercially, is to drive the initial machine-direction orienter to stretch ratios of about 6:1. If this is attempted gradually, film rupture will usually occur, but if the stretch ratio is jumped in one step, sustaining operation can sometimes be achieved. After transverse orientation this film is near balance. Handling in the tenter frame is more delicate than with the usual unbalanced product.

If film is being produced with a heavy additive content, such as slip agents, it may be necessary to age the mill rolls for several weeks before slitting into customer rolls. The slow dimensional change which accompanies the exudation of the slip agent is allowed to take place under relatively low stress, and the customer rolls subsequently produced can be smoothly and tightly wound with safety. This is similarly true of simultaneously stretched film. The sequential process will not produce a satisfactory shrink film.

Sequential tentering equipment can be run consistently at 200 to 300 yd/min. The basic equipment has the reliability of long commercial use. Equipment suitable for film handling can be obtained from commercial manufacturers in the United States and Europe. Several vendors of tenter frames have formed associations with extruder manufacturers and oven builders to offer complete packaged orientation units. Equipment costs are principally affected by speed and length but not by width. Hence economics usually dictates a wide machine of high-output capacity. The tenter frame can be operated in a narrow setting if desired, but little extra cost is involved in building it wide. Widths of 10 ft are in common use, and there are no obvious limitations to even larger machines. Wider equipment diminishes the percentage of edge trim but develops other handling problems with regard to mill rolls.

The film produced is satisfactory for most uses and has the critical elements of stiffness and low-temperature strength as well as enhanced protective properties over cast film.

V. Extrusion Equipment

A. Extruder Design

Polypropylene is somewhat more difficult to handle in extrusion equipment than polyethylene. The melt viscosity decreases much more rapidly with increasing temperature and also appears to have a substantially greater degree of plastic flow or change of viscosity with shear rate. The result is that polypropylene is much more sensitive to nonuniformities in the extrusion system. Local temperature variations in the melt can cause appreciable flow variations, which in turn result in substantial gauge variations in finished film.

The shear sensitivity tends to set up two types of flow situation because of its positive feed-back characteristic. Once an increased flow has been triggered in one portion of the die, for example, by a gobbet of material at high temperature, the reduction in viscosity due to the increased shear rate tends to increase the flow through this channel. This, in turn, reduces the viscosity further and generates yet a higher flow rate until increased flow friction finally balances the stream at a disproportionately high flow rate. This flow channel can remain in metastable equilibrium for a long period of time.

A second result which can easily occur is establishment of an oscillation between flows in different parts of the system. An uneven flow rate can build up through one portion of the die, as described, reach its peak, and then, at a rather minor interference, reverse the whole procedure. A slight pressure pulse may cause a momentary slowing of the flow, followed by a sharp increase in viscosity, which slows the flow further and acts almost as though a valve were closed in the channel. The sudden decrease in flow rate produces a pressure increase and formation of a channel in some other part of the flow path. This flow channel will then build up to a peak, close itself off, and return an oscillation to the first channel; or, depending on the capacities of the various parts of the flow system, alternate pressure increases and decreases can be generated by the positive feed-back action of the plastic flow mechanism, thus causing an over-all oscillation in flow rate based on cycling pressure changes and the compressibility of the fluid in the various flow chambers in the extrusion system.

Extruder design centers around the same principles controlling design of extrusion equipment for polyethylene (cf. Volume I, Chapter 8). High ratios of length to diameter of barrel, that is, 20 to 1 and 24 to 1, are favored. Screw design is very much a matter of personal opinion. Compression ratios in the region of 4 to 1 are generally accepted, but a certain amount of controversy exists over the relative lengths of the three functional sections of the extruder screw: the feed section, the compression section, and the metering section. Comparatively long metering sections are favored for the extrusion of polypropylene; they usually run from 30 to 50% of the total screw length. The long metering section has the advantage of improving temperature uniformity in the melt and of damping pressure pulsations which are initiated in the compression area. The disadvantage of the long metering section lies in the excessive shear working of the polymer and the generation of heat which makes melt-temperature control difficult at higher extrusion rates. Relatively long compression sections, running around 30% of screw length, are generally used, although some evidence has been presented that lengthening the feed section at the expense of the compression section gives a more uniform melt flow.

Extrusion dies have undergone considerable improvement in the last

10 years. Polyethylene and polypropylene designs seem to be interchangeable and a die giving good performance on one will generally be comparably good on the other. Long smooth flow passages, with particular attention to the elimination of hard edges and sudden direction changes, are required. Very long land lengths in the lips are a necessity for adequate polypropylene extrusion, and choker-bar dies for additional damping and back pressure control are helpful.

Die-lip adjustment for control of gauge is a delicate and artistic operation, and every effort should be made to provide equipment that will help the operator. Adjustment studs should be incorporated on the greatest possible frequency along the die, and combination push-pull adjusting bolts should be used at each point to provide greater latitude than exists with alternate pullers and pushers. Changes in flow through the die are highly disproportionate to the amount of adjustment made at any point in the die gap. Some of this effect results from the strong shear reaction of the melt and the sharp viscosity changes caused by initially minor variations in the flow rate. This gives rise to the phenomenon of bouncing gauge bands rather than the smoothing or elimination of the nonuniformities. Heavy-handed adjustment of the die gap at a point of high gauge will very often lead to the sharp disappearance of the band at that point and immediate appearance of another equally high gauge point in a distant part of the die. For this reason adjusting studs should be made with very fine thread movement, and preferably with differential threads, so that the operator can take a substantial turn on the screw without causing much movement of the die lip. Several minutes should be allowed after a die-lip adjustment to permit flow patterns within the die to readjust themselves to their new stable configuration. The entire effect of the adjustment must be in evidence before further changes are made.

B. Temperature Controls

The demands for high quality in film production have been progressively pushing extruder manufacturers to develop more sophisticated and effective methods of temperature control. Proportional controllers should be used on all heating circuits of the extruder. When on-off or cycling controls are used, it is usually easy to demonstrate a periodic gauge variation in the film that can be correlated directly with the operating cycle of any heater in the system. These effects will appear in bands when associated with the die heaters and as a periodic gauge variation in the machine direction when correlated with the barrel temperature controllers. For high rates of extrusion, in which continuous cooling of one or two zones of the metering section is required, proportional temperature control of the cooling fluid or boiling jacket controls should be used along with suitable melt-temperature control instruments.

C. Drive Systems

Polypropylene is very sensitive to pressure variations in the extrusion system, as has been discussed in detail. To minimize upsets and maintain uniform film tolerances the extrusion drive must be selected with as much thought and care as is given to any other part of the extruder.

The over-all power requirements are similar to those for polyethylene, and manufacturers' recommendations on horsepower for a given size of extruder can be safely accepted. Exceptional precision in speed control is desirable, however. Speed-control ratings on drive systems must be analyzed for meaning before application. Regulations of 1/2 or 0.1 % sound quite adequate, but one knows little about them unless the response time is also included. A drive system can recover very nicely from a load change and resume the original speed, but if 10 screw revolutions take place during the correction period serious film abnormalities can occur.

Polypropylene is generally susceptible to *backfiring* in the compression zone. It is apparent that with a 4:1 compression ratio, and an available compressibility from pellets to homogeneous solid of only 2:3, considerable backflow must occur in the compression section. This generates the extrusion pressure which hits its peak at the entrance to the metering zone and controls polymer flow through the metering zone, screen pack, and die. Pressure at this point is frequently twice the head pressure. Instead of maintaining a smooth and steady backflow, which produces a relatively constant driving pressure at the start of the metering zone, polypropylene will often build to a pressure peak and then discharge backward through the partially melted zone. This results in a momentary drop in pressure which is reflected forward through the metering zone and which also results in a momentary dropping of the motor load. The backfire does not last long, and pressure can be recovered within one or two screw revolutions, reloading the motor just as the regulation system has started to reduce current to accommodate the previous load drop.

Substantial screw-speed changes can thus occur within the space of one or two screw revolutions even with a precision-rated drive system. Such speed variations are not detectable with most speed-measuring systems. A tachometer-generator will give no clue that these short transients are occurring. The best way to detect them is by using stroboscopic techniques with which forward and backward shifts in even a fraction of a revolution can be observed.

The drive system selected should have a small range of regulation, but even more important it should have a short response time compatible with anticipated screw speeds. In order to minimize speed changes within the response limit of the control system, high-inertia drives should be selected. Ward-Leonard systems, which have a heavy dc motor armature directly connected to the drive system are preferred over eddy current drives which have equivalent control performance. If an eddy-current drive is used, it may be helpful to install a flywheel on the end of the screw drive to damp out short-term changes.

Backlash in the gearing and drive belts must also be held to a minimum. There can be enough backlash in an ordinary gearbox and belt drive to accommodate one eighth of a screw revolution.

D. RESIN FEED SYSTEMS

Thorough mixing and blending of the input resins to the extruder are vital to consistent film quality. As in all film processes, recovery of scrap and off-grade materials is an essential element of economical operation. This means that reclaim material from a variety of sources must be fed. Various additives must also be accommodated, often by the use of master-batch additions for which adjustment must be made in the recycled streams.

Preblending should be done as much as possible to minimize heterogeneous distributions at the feed point of the extruder. The problem can be extremely serious if ground film or fluff is being recycled directly. Local bulk-density variations in the feed hopper result in varying polymer feed rates at the screw flights. As these density variations arrive at the compression section, they permit wild pressure fluctuations, with consequent defects in the film tolerances.

Feed hoppers must be carefully designed for thorough mixing and remixing right down to the point of entry to the feed flights of the screw.

When pelletized mixtures are being fed, care should be exercised to keep pellets of the same size and geometry. Pellets of varying sizes and shapes segregate in the feed hopper as the mix drops; for example, if reclaim is being re-extruded and diced (a cubical pellet) and then blended with a virgin resin in the form of spherical pellets (underwater pelletizer), substantial separation and segregation can occur. These conditions will result in consecutive volume increments of sharply differing melt-flow number following one another into the die, with the attendant fluctuations in flow characteristics. Flow stability within the die will become completely disrupted and gauge control becomes almost impossible.

When different streams are being fed to the hopper, and we are relying on mixing during feed for homogeneity, care must be taken in the design of the feed systems so that the proportionality of the feed streams remains constant at all rates. A momentary shutdown of the extruder, as for a film break, can otherwise result in overloading the hopper with one component. When this abnormal feed layer works down through the hopper and hits the extruder, all control of film properties may be lost for several minutes. Intermittent feed devices should be avoided in favor of continuous interlocked systems.

VI. Markets and Applications

Properties on which markets for polypropylene are based are discussed in some detail above.

Properties most exploited are high clarity, high gloss, low haze, and (for the tensilized material) good dimensional stability, high tensile strength, good stiffness, good barrier properties against water vapor, oils, and greases, and good low-temperature strength.

This polymer has become a factor in the film business only within the last 10 years (Table 12), and as the technical problems associated with tensilization to preserve strength properties (particularly at low temperature) are solved costs will undoubtedly fall and the use of polypropylene either as self-supported film or as an element in composite films will continue to grow rapidly. The current cost of the polymer ($0.18 to 0.30/lb) is not sufficiently different from that of polyethylene ($0.14 to 0.20/lb) and oriented polystyrene ($0.60 to 0.63/lb) to deny it such large markets as overwraps for bakery products, snack packaging, shrink packaging of durable goods that should withstand rough handling and

TABLE 12

Polypropylene in the Packaging Film Market[a]

Film	1960	1961	1962	1963	1964	1965	1966	1967	1968
Cellophane	439	423	404	405	410	405	395	385	360
Polyethylene	275	370	400	455	500	615	730	735	795
Vinyl	18	20	22	24	30	40	40	70	90
Polypropylene		3	15	25	28	40	45	50	65
Saran	17	19	21	23	26	30	30	30	30
Polystyrene	3	5	7	11	15	15	15	15	15

[a] Estimates from several sources, including [85].

TABLE 13

Comparative Film Costs of Polypropylene and Some Competitors[a]

Film	Cost ($/lb)	Coverage (in.^{2}/lb)	Cost (cents/10^3 in.^{2})
Cellophane			
MS, 195[b]	0.67	19,500	3.4
MS, 220	0.67	22,000	3.0
Saran-coated, 140	0.81	14,000	5.8
Saran-coated, 195	0.81	19,500	4.2
Saran-coated, 250	0.75	25,000	3.0
Polyethylene-coated 182	0.82	18,250	4.5
Polypropylene, cast, 1 mil	0.62	31,000	2.0
Balanced, uncoated, 1 mil	0.90	31,000	2.9
Balanced, Saran-coated, 1 mil	1.20	28,000	4.3
Polystyrene, oriented, 1 mil	0.67	26,300	2.5
Polyethylene			
Low density, 1 mil	0.27	30,000	0.9
Medium density, 1 mil	0.31	30,000	1.0
High density, 1 mil	0.36	29,000	1.2
Polyethylene/cellophane, 1 mil/195 MS	1.05	11,800	8.9
Polyethylene/polypropylene/polyethylene, 1 mil	0.47	30,700	1.5

[a] Most of these data from [85], *Modern Packaging Encyclopedia*, July 1970, p. 20 (cf. Volume I, p. 13).

[b] Moistureproof, heat sealing; numbers are standard cellophane gauges (col. 3).

extended storage without embrittlement or yellowing, and form-and-fill wraps for dried fruits and rice.

Polypropylene film can be printed beautifully by gravure or flexographic process at speeds about equal to those used for cellophane. The film can be hot-melt-, solvent-, or dispersion-coated with a variety of materials ranging from styrene-butadiene, polyethylene, polyvinyl acetate, and polyvinyl chloride to polyvinylidene copolymers (sarans) as the preferred coating for the heat-set, tensilized, balanced base sheet. This capability has enormously enlarged the range of possible applications, for printing and overwrapping can be done at high speeds with excellent results.

On a cost-coverage basis polypropylene is two to three times as expensive as polyethylene and about the same as the cellophanes (which are competitors but do not function in the same way) and oriented polystyrene. Table 13 shows these data, taken from [85] and other sources, for polypropylene and several competitive films and film combinations.

References

1. K. Ziegler, *Brennstoff-Chem.*, **35**, 321 (1954).
2. K. Ziegler, Belgian Patent 533,362 (May 16, 1955).
3. K. Ziegler, Belgian Patent 534,792 (Jan. 31, 1955).
4. K. Ziegler, Belgian Patent 534,888 (Jan. 31, 1955).
5. K. Ziegler, Belgian Patent 542,658 (May 8, 1956).
6. K. Ziegler and H. G. Gellert, U.S. Patent 2,695,327 (Nov. 23, 1954); German Patents 878,560 and 889,299.
7. K. Ziegler and H. G. Gellert, U.S. Patent 2,699,457; German Patents 878,560 and 889,229.
8. K. Ziegler, E. Holzkamp, H. Breil, and H. Martin, *Angew. Chem.*, **67**, 426 (1955).
9. K. Ziegler, E. Holzkamp, H. Breil, and H. Martin, *Angew. Chem.*, **67**, 541 (1955); *Chim. Ind.* (*Milan*), **37**, 881 (1955).
10. G. Natta, P. Pino, P. Corradini, F. Danusso, E. Mantica, G. Mazzanti, and G. Moraglio, *J. Am. Chem. Soc.*, **77**, 1708 (1955).
11. G. Natta, *J. Polymer Sci.*, **16**, 143 (1955).
12. G. Natta, P. Pino, and M. Farina, *Simposio Internazionale di Chimica Macromolecolare, Ricerca sci.*, **25**, Suppl. A, 120 (1955).
13. G. Natta and P. Corradini, *Simposio Internazionale di Chimica Macromolecolare, Ricerca sci.*, **25**, Suppl. A, 695 (1955).
14. G. Natta, P. Corradini, and M. Cesari, *Atti. Accad. Nazl. Lincei, Mem. Classe Sci. Fis. Mat. Nat. Sez. II*[a], **21**, 365 (1956).
15. G. Natta, *Chem. Ind.* (*London*), **47**, 1520 (1957).
16. G. Natta, I. Pasquon, and E. Giachetti, *Angew. Chem.*, **69**, 213 (1957).

17. G. Natta, G. Mazzanti, G. Crespi, and G. Moraglio, *Chim. Ind. (Milan)*, **39**, 275 (1957).
18. G. Natta and P. Corradini, *J. Polymer Sci.*, **39**, 29 (1959).
19. G. Natta, *J. Polymer Sci.*, **34**, 21 (1959).
20. G. Natta, *Chim. Ind. (Milan)*, **42**, 1091, 1207 (1960).
21. G. Natta, I. Pasquon, A. Zambelli, and G. Gatti, *J. Polymer Sci.*, **51**, 387 (1961).
22. G. Natta, P. Corradini, and G. Allegra, *J. Polymer Sci.*, **51**, 399 (1961).
23. G. Natta, I. Pasquon, and A. Zambelli, *J. Am. Chem. Soc.*, **84**, 1488 (1962).
24. A. Zambelli, G. Natta, and I. Pasquon, *J. Polymer Sci.*, **C4**, 411 (1964).
25. G. Natta and I. Pasquon, *Advances in Catalysis*, Vol. 11, Academic, New York, 1959, pp. 1–66.
26. S. L. Aggarwal and O. J. Sweeting, *Chem. Rev.*, **57**, 665 (1957); *Usp. Khim.*, **27**, 1115 (1958).
27. J. K. Stille, *Chem. Rev.*, **58**, 541 (1958).
28. N. G. Gaylord and H. F. Mark, *Linear and Stereoregular Addition Polymers*, Wiley-Interscience, New York, 1959.
29. R. A. Raff and K. W. Doak, Eds., *Crystalline Olefin Polymers*, Wiley-Interscience, New York, 1965.
30. Phillips Petroleum Co., Belgian Patent 530,617 (Jan. 24, 1955); Australian Patent application 6365/55.
31. A. Clark, J. P. Hogan, R. L. Banks, and W. C. Lanning, *Ind. Eng. Chem.*, **48**, 1152 (1956).
32. Standard Oil Company (Indiana), British Patent 721,046.
33. E. Field and M. Feller, U.S. Patent 2,691,647 (Oct. 12, 1954).
34. E. F. Peters, U.S. Patent 2,700,663 (Jan. 25, 1955).
35. M. Feller and E. Field, U.S. Patents 2,717,888 and 2,717,889 (Sept. 13, 1955).
36. H. S. Seelig, U.S. Patent 2,710,854 (June 14, 1955).
37. R. A. Mosher, U.S. Patent 2,725,374 (Nov. 29, 1955).
38. E. Field and M. Feller, U.S. Patent 2,726,231 (December 6, 1955).
39. B. L. Evering, A. K. Roebuck, and A. Zletz, U.S. Patent 2,727,023 (Dec. 13, 1955).
40. E. Field, and M. Feller, U.S. Patent 2,727,024 (Dec. 13, 1955).
41. B. L. Evering and E. F. Peters, U.S. Patent 2,728,754 (Dec. 27, 1955).
42. E. Field and M. Feller, U.S. Patent 2,728,757 (Dec. 27, 1955).
43. E. Field and M. Feller, U.S. Patent 2,728,758 (Dec. 27, 1955).
44. E. Field and M. Feller, U.S. Patent 2,731,452 (Jan. 17, 1956).
45. E. Field and M. Feller, U.S. Patent 2,731,453 (Jan. 17, 1956).
46. G. Natta, P. Pino, and G. Mazzanti, U.S. Patent 3,112,300 (Nov. 26, 1963).
47. G. Natta, P. Pino, and G. Mazzanti, U.S. Patent 3,112,301 (Nov. 26, 1963).
48. K. Ziegler, H. Breil, E. Holzkamp, and H. Martin, U.S. Patent 3,257,332 (June 21, 1966); U.S. Patent 3,113,115 (Dec. 3, 1963).
49. K. Veselý, J. Ambrož, R. Vilím, and O. Hamřík, *J. Polymer Sci.*, **55**, 25 (1961).
50. R. A. Labine, Ed., *Chem. Eng.*, **67**, No. 11, 78 (May 30, 1960).
51. M. Hirooka, *Polymer Rept. (Japan)*, **94**, 22 (1966).
52. R. E. Dietz, U.S. Patent 3,219,647 (Nov. 23, 1965).
53. E. J. Vandenberg, U.S. Patent 3,051,680, (Aug. 28, 1962).
54. G. Natta, *Chem. Ind. (Milan)*, **42**, 1091 (1960).

55. G. Bier, *Kunststoffe*, **48**, 354 (1958).
56. *Chem. Eng.*, **67**, No. 13, 96 (June 27, 1960); H. W. Haines, Jr., *Ind. Eng. Chem.*, **55**, No. 2, 30 (1963).
57. *Modern Plastics*, **44**, No. 5, 118 (January 1967).
58. *Modern Plastics*, **44**, No. 8, 86 (April 1967).
59. J. Boor, Jr., and E. A. Youngman, *J. Polymer Sci.*, **A1, 4,** 1861 (1966).
60. R. L. Miller, "Crystalline and Spherulitic Properties", Ref. 29, Part I, Ch. 12.
61. P. H. Geil, *Polymer Single Crystals*, Wiley-Interscience, New York, 1963, pp. 57–63.
62. G. Natta and P. Corradini, *Nuovo cimento*, **Suppl.** to Vol. **15**, 1, 40 (1960).
63. Ref. 28, pp. 54–60, 65.
64. L. Marker, P. M. Hay, G. P. Tilley, R. M. Early, and O. J. Sweeting, *J. Polymer Sci.*, **38**, 33 (1959).
65. F. J. Padden, Jr., and H. D. Keith, *J. Appl. Phys.*, **30**, 1479 (1959).
66. Ref. 61, pp. 266–274.
67. G. W. Schael, *J. Appl. Polymer Sci.*, **10**, 901 (1966).
68. W. P. Conner and G. L. Schertz, *Soc. Plastics Engrs. Trans.*, **3**, 186 (1963).
69. K. D. Pae, D. R. Morrow, and J. A. Sauer, *Nature*, **211**, 514 (1966).
70. H. W. Wyckoff, *J. Polymer Sci.*, **62**, 83 (1962).
71. R. J. Samuels, *J. Polymer Sci.*, **A3**, 1741 (1965).
72. M. Peraldo, *Gazz. Chim. Ital.*, **89**, 798 (1959).
73. C. T. Liang and F. G. Pearson, *J. Mol. Spectry.*, **5**, 290 (1960).
74. R. L. Miller, *Polymer*, **1**, 135 (1960).
75. H. Tadokoro, M. Kobayashi, M. Ukita, K. Yasufuku, S. Murahashi, and T. Torii, *J. Chem. Phys.*, **42**, 1432 (1965).
76. M. P. McDonald and I. M. Ward, *Polymer*, **2**, 341 (1961).
77. T. Miyazawa, *J. Polymer Sci.*, **B2**, 847 (1964).
78. B. C. Repka, Jr., "Polypropylene", in Kirk-Othmer *Encyclopedia of Chemical Technology*, Ed. 2, Wiley-Interscience, New York, 1967, p. 287–288.
79. American Society for Testing and Materials, Philadelphia, *1966 Standards*, "Plastics", Parts 26 and 27.
80. *Properties and Characteristics of* Tenite *Polyallomer*, Materials Bulletin 16C, Eastman Chemical Products, Kingsport, Tenn. 37662, 1968.
81. *The Effect of Various Chemicals on* Tenite *Polypropylene*, Report TR–19A, Eastman Chemical Products, Kingsport, Tenn., 1968.
82. *Moplen Propylene Resin*, Montecatini technical bulletin, Novamont Corp., 100 East 42d St., New York 10017, 1968.
83. Propafilm *Polypropylene Film for Packaging*, Bulletin PF8, Films Group, ICI Plastics Division, Welwyn Garden City, Herts, April 1968.
84. *Cryovac Y-900 Film*, Cryovac Division, W. R. Grace & Co., Duncan, S.C., 1968.
85. *Modern Packaging Encyclopedia*, **43**, No. 7A (July 1970).

CHAPTER 4

POLYSTYRENE

JULES PINSKY†

Mearl Corporation, Ossining, New York

I. Manufacture of Polymer

The chemistry of styrene and styrene polymers has been so thoroughly treated elsewhere [1–3] that it is not necessary to give any extensive discussion of fundamentals here. Styrene was isolated from storax balsam in 1831 and polymerized by Simon in 1839. Synthetic styrene was prepared by Berthelot in 1869. The first patent on the polymers was issued in London in 1911. Thus polystyrene is clearly one of the oldest synthetic polymers. The first production of a pure monomer and a clear transparent polymer of commercial importance in the United States was made by The Dow Chemical Company in 1937.

Styrene polymerizes slowly simply by heating in bulk at 100°C., but it may also be polymerized readily by free-radical initiators in several ways. Commercially the polymer is made by either a batch or continuous process in greater than 95% conversion. Styrene polymerizes in a head-to-tail sequence in atactic configuration when free-radical initiators are used. Ziegler catalysts produce an unbranched, stereoregular, head-to-tail polymer.

$$CH_2{=}CH \longrightarrow \sim CH_2{-}CH{-}CH_2{-}CH\sim$$
$$\hspace{0.8cm} \underset{\displaystyle C_6H_5}{|} \hspace{2.5cm} \underset{\displaystyle C_6H_5}{|} \hspace{1cm} \underset{\displaystyle C_6H_5}{|}$$

† Formerly Director of Research, Monsanto Company Packaging Division, Hartford, Connecticut.

II. Polystyrene Films

Clear polystyrene film has the potential to become a prime mass-market packaging film [4, 5]. Whether it achieves such a level is contingent on three factors: (a) effective low-cost coatings for barrier properties, such as those that were developed for cellophane; (b) price, so that on a stiffness basis it becomes one of the lowest cost plastic packaging films; (c) development of machinery and techniques to use it at high speeds. For these reasons we deal chiefly here with clear (biaxially oriented) polystyrene film, but passing notice is given to styrene-acrylonitrile, acrylonitrile-butadiene-styrene, butadiene-rubber-styrene, and foamed-styrene films.

Flexibility of the polystyrene homopolymer and copolymer films is imparted by biaxial orientation. Therefore the theory and practice of biaxial orientation is examined closely.

A. Biaxial Orientation

The first work in this country on biaxial orientation was done by James Bailey, then research director and vice-president of Plax Corporation [6–12]. Bailey, who was later to receive the Hyatt Award and the Franklin Institute Medal for his pioneering work in blow-molding, was both an ingenious inventor and a fundamentalist. In his approach to the best means of formation of plastic containers he cast sheet, film, rod, and tubing from solvent solution, extruded into the same forms from all available commercial thermoplastics, and then stretched, compressed, and otherwise formed these shapes to determine the effects on physical properties.

One of the results that interested Bailey was the effect of stretching [13–15] on polystyrene, particularly with regard to elongation and tensile strength. He found, with uniaxial orientation of polystyrene, the following remarkable results.

1. Tensile strength of sheet (or film) was increased from 5000 to about 12,000 psi in the direction of orientation. Tensile strength of 2.5-mil fibers was increased to about 35,000 psi, with indications that 1-mil fibers would have considerably higher strength.

2. Elongation of polystyrene film (or sheet) was increased from 1 to 3% (theoretical) to about 60%, with fibers showing elongation of 85% and more.

3. Crazing strength thresholds were increased from 2000 psi to the ultimate breaking strength (as high as 20,000 psi).

4. Tensile strength and elongation values for sheet and film approached maximum limits at 300 % of stretch.

5. The increase in tensile strength in the orientation direction is accompanied by a sizeable decrease in tensile strength at right angles to the direction of orientation. In a typical case the tensile strength in the orientation direction increased to 11,000 psi, and a decrease to about 1500 psi was noted in the direction perpendicular to the orientation.

Further work established that if the polystyrene sheet or film is stretched sequentially longitudinally and then laterally, the tensile strength is increased in both surface directions but decreased in the thickness direction. With most polystyrene films and sheets the intrinsic thinness makes it difficult to measure the decrease in tensile properties in this plane. It has been demonstrated, however, that when biaxial stretching is performed at a maximum rate the sheet can show a tendency to delaminate; for example, when the sheet is roll-worked, punched, or hammered on, it separates into individual strata in the manner of mica.

The work of Bailey and his colleagues also established that molecular orientation occurred in every extrusion operation—in fact, in every fabricating process in which heat and pressure were used to achieve a shape. Thus extrusion through a cylindrical die theoretically causes orientation of the molecules parallel to the wall, and compression molding introduces a biaxial orientation similar to the two-step, two-direction orientation already described.

In actual practice it was found that the temperatures practiced in extrusion or compression molding allowed for a relaxation of some of the orientation effects. Thus total orientation was not achieved, but instead a compromise, which is a function of the temperatures and shear stresses employed. This explained why "spider" dies—used before the plastic achieves its final nozzle temperature and zero shear stress—introduced directional weaknesses at right angles to the extrusion direction (cf Volume I, p. 410 ff).

It is interesting to examine the theoretical considerations of the orientation of polystyrene from the experimental approach of Bailey and the mathematical evaluation of Samuels and Stein (Volume I, Chapter 7).

Bailey pictured the polystyrene molecule as an extended cylinder with a helical configuration that reduces the molecular length-to-width ratio 20–200-fold. In his analysis these cylinders were projected parallel with the plane of stretching when it occurred.

A cube of polystyrene, with the appropriate x-, y-, and z-coordinates is taken as the model, as shown in Figure 1. This material is to be

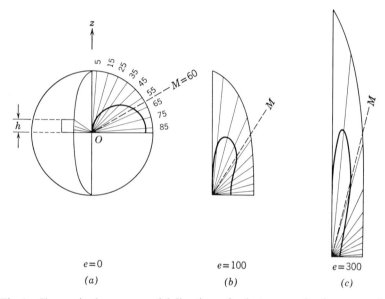

$e = 0$ $e = 100$ $e = 300$

(a) (b) (c)

Fig. 1   Changes in the average axial directions of polystyrene molecules on stretching in one direction.

stretched along the z-axis and the angles of the molecules, with respect to the coordinates, are to be determined. As in Figure 1a, the direction of a molecule is determined, its angle measured, and an extension of the direction is projected for all of the molecules in the original curve through a point on the sphere surface. By locating this point from any axis it is apparent that if the original structure were completely random each unit of area would contain an equivalent random number of points.

In the sphere the angle of any molecule with the stretching direction (O_z) is important. The directions of the molecules in the zone of height h, depicted, can be taken as a unit. This simplified model allows for one common radius passing through the half-height of the zone, with the number of molecules being directly proportional to the zone area. When the surface is arbitrarily divided into 10-degree zones and radii are drawn from 5 to 85°, each radius centers a 10-degree zone and the number of molecules in any zone is proportioned to the number of molecules. Taking radial sectors as indexes of the zone areas, their length gives the average angle. Thus in the unstretched stage this angle is 60° from the O_z-axis.

Now, when the stretch is 100% (the cube and the sphere are extended to twice their original height), the stretch ratio is $2X$. Such a stretch results in Figure 1b, with the directions of the molecules in the cube remaining parallel with the directions in the stretched sphere. The average molecular direction changes from 60° to the O_z phase to 31.5° with a stretch ratio of $2X$.

When the stretch ratio is $3X$, as shown in Figure 1c, with an elongation of 200%, the average molecular direction if 18.4°. Theoretically, a $4X$ stretch changes the angular direction to 12.2°. In practice, however, when stretching is done under low temperature or under mechanically restrained conditions that do not allow for relaxation, a $3X$ stretch ratio, for all practical purposes, produces the maximum of beneficial orientation effects.†

A diagrammatic representation of the statistical distribution of molecular direction, based on the preceding scheme, is shown in Figure 2.

When biaxial orientation is performed, a statistical change in the directional arrangement of the molecules can be inferred. This assumes that the stretching is done with equal force and temperature, in each of the two stretchings, and that there is no thermal relaxation. The average change in molecular direction with location to the face of the film is shown in Table 1.

Undoubtedly the arrangement of the long axes of the polystyrene in the plane of the film causes biaxially oriented polystyrene to become a flexible material of high tensile strength, as contrasted with the brittle lower tensile form in which so many of the chains are susceptible to separation and cleavage because their main tenacity results from Van der Waals forces.

A quite different approach to the problem of biaxial orientation was

---

† This is not valid, however, for films less than one mil in thickness.

RANDOM STRUCTURE

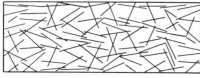

STRETCHED 100%

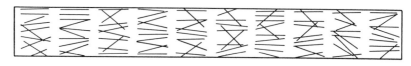

STRETCHED 300%

Fig. 2   Representation of molecular angular changes on stretching.

TABLE 1

Angular Change on Stretching

e^a	X^b	Angle A^c
0	1	60°
100	2	12°+
200	3	3°+
300	4	1.5°+

[a] e is the percentage increase in length and in width.

[b] X is the ratio between stretched length or width and original length or width, respectively.

[c] A is the minimum angle between the axis of the molecule and the face of the plate.

that of Stein [16–18]. He investigated the phenomenon from a mathematical thermodynamic analysis to explain the molecular displacement. Using equations developed by Joule [19], he established a relationship for the change in entropy S and internal energy E, with a correction term introduced to adjust for experimentation at constant pressure:

$$\left(\frac{\delta E}{\delta l}\right)_{V,T} \sim f - T\left(\frac{\delta f}{\delta T}\right)_{P,E} + \alpha l T\left(\frac{\delta f}{\delta l}\right)_{P,T},$$

where α is the coefficient of linear expansion of the unstretched material, l is the film plane dimension, f is the force applied, T is the absolute temperature, and V is the volume.

Then, by extension of this analysis to a two-dimensional system, the internal energy change of a film is $\alpha E = Q - W$, where Q is the heat absorbed, and W is the work done by the system. Whence,

$$\alpha E = T \, \partial S - P \, \partial V + f_1 \, \partial l_1 + f_2 \, \partial l_2,$$

where l_1 and l_2 are the plane dimensions of a film and f_1 and f_2 are forces (assumed to follow the principal stress directions l_1 and l_2).

The internal energy and entropy changes in the two directions of orientation and how these changes compare with a single orientation were calculated. The effects of thermodynamic changes in one direction or a second direction were also determined.

The internal energy and entropy per unit were defined as follows:

$$\bar{E}_{T_1} = \frac{E}{V_{0T_1}} = \frac{E}{l_{10}\, l_{20}\, l_{30}},$$

$$\bar{S}_{T_1} = \frac{S}{V_{0T_1}} = \frac{S}{l_{10}\, l_{20}\, l_{30}},$$

where T = temperature of measurement, V_{0T_1} = volume at zero deformation at temperature T_1, and l_{10}, l_{20}, l_{30} are the three principal dimensions.

Calculations from such a state would require maintenance of a constant thickness. Since this is most difficult, the equations were made to fit the conditions of constant pressure.

The relationship between constant-volume and constant-pressure quantities and their effect on configurational changes on extension and volume changes with pressure were then mathematically examined. The volume changes on extension to be of consequence were found to be the following:

1. The hydrostatic component of tension.
2. Phase changes involving a volume change.
3. Void formation or cavitation.

The equations at constant pressure were then transformed into relations at constant volume.

B. FILM MANUFACTURE

The early work of Bailey and his associates produced oriented polystyrene films with a rough surface. Such was the roughness that the films were not considered for packaging applications.

The rough surface was a characteristic of the polystyrene of high molecular weight then available and its wide molecular-weight distribution. The Plax sheeting made from such polystyrene was pressure-formed into toys and Christmas-tree ornaments.

One of the early devices built to accomplish pressure forming was the Strauch machine, an automatic continuous pressure-forming apparatus that was 20 years ahead of its time. The Strauch machine, used principally with cellulose acetate, but also workable with polystyrene, was a precursor of the automatic sheet- and film-forming machines of today. It was designed so that it could be coupled to an extruder, but in practice it was usually fed from roll film stock. The film was held by edge grippers and heated from above by "Calrod" heaters in front of aluminum reflecting plates. A series of water-cooled molds moved with the film at the film travel speed. There was a male metal plug assist, which was also used to cool the objects after they had been formed.

Another interesting application of the early Plax *Polyflex* was the formation of radomes. Layers of *Polyflex* were overlap-wound on each other over a male form and then introduced into an oven to heat-shrink and seal the Polyflex into the desired radome foil.

Similarly, sheets of Polyflex were stacked in a compression molding press and blocks formed up to 4 in. in thickness. These blocks were tougher and more craze resistant than if they had been compression-molded directly from molding powder or if they had been solvent-cast.

These two adaptations of Polyflex permitted fast and inexpensive building of radome models and accelerated the United States' development of radar during the war.

Polyflex also was used in cable wrapping because of its outstanding electrical properties and played a part in establishing the postwar Distant Early Warning line.

The Germans had also developed a biaxially oriented polystyrene film which they used, and still do, for cable winding. Their oriented polystyrene was produced on a much narrower, low-output "wishbone" machine and was called Styroflex. Styroflex was manufactured during the war in quantities of 20 tons a month by Nordeutsche Seekabelwerke [20], mainly in foil, ranging from 0.0004 to 0.0006 in. in thickness. Most of the Styroflex was wound on wire for coaxial cable insulation, but it was also used experimentally for capacitors, food packaging, and collapsible tubes.

The German process involved a large extruder with a barrel 7 in. in diameter which was run at a very low speed: 5 rpm. With stock temperature of approximately 390°F, a tube about 2.4 in. in diameter with a wall of 0.040 to 0.080 in. was extruded at a rate of about 5 to 10 lb/hr on the apparatus shown in Figure 3. The tube then passed onto a parabolic stretching frame, as shown in Figure 4. This frame had a series of small rollers over which the tube passed as it was being oriented. Small air jets were located along the two edges for cooling and for providing a more solid area for stretching. At the end of the parabola the film was slit and then passed onto the windup rolls. This method of low-output manufacture of biaxially oriented polystyrene in gauges of less than 1 mil is still being used at Nordeutsche Seekabelwerke as well as in the United States.

The Plax method, as described in U.S. patent 2,412,187 [9] was the first method for making biaxially oriented polystyrene film and sheet in commercial quantities. A sketch of the stretching apparatus is shown in Volume I, p. 473.

In the Plax process a polystyrene continuous ribbon (approximately nine times the ultimate film thickness) is extruded from a flat die at approximately 370°F. This ribbon is then threaded through a set of cooperative stretching rolls (which are internally cooled to regulate a temperature of approximately 200°F). The difference in linear speed on these rolls from the exit speed at the extruder nozzle accounts for the extrusion direction stretch. The temperature during the entire longitudinal stretching is carefully controlled by ovens and other regulating heat exchangers at approximately 255°F. In this process the edge sections are chilled and thickened so that they serve as a good

Fig. 3   Styroflex extruder nozzle viewed from takeoff ([20], p. 368).

gripping area. Then several inches of this edge are automatically trimmed, run through a shrink-back tunnel, ground to a selected size, and reintroduced into the hopper, along with new pelleted molding powder.

The optimum orientation is achieved when a 3:1 stretch ratio is imparted in each direction at the same temperature. This produces a film with good strength in both directions.

As the Plax method of achieving biaxial orientation came to the end of its patent life, various commercial machines which simulated the proprietary equipment were introduced. The approach was to use high-speed tenter-frame equipment, which had been originally designed for stretching fabrics (Volume I, Chapter 10). Commercial biaxial orientation machines have been developed by Waldron-Hartig, Bruckner Machinery, and others [11, 21].

In recent years experimentation and manufacture of thin film has been accomplished by the blown-film process, similar to the method of

Fig. 4   Parabolic stretcher for Styroflex, viewed from above ([20], p. 369).

manufacture of tubular polyethylene film.   A normal tubing-extrusion process is employed (usually extruding vertically upward).   As the tube leaves the die, air is introduced to accomplish the radial orientation, and the puller rolls ratio accomplishes the machine direction orientation. The tubing is usually slit into two or more flat films after it has passed through the initial set of puller rolls.

Difficulties with the tubular extrusion process have been the presence of gauge bands, the inability to impart higher balanced levels of orientation to thicker film and sheeting, and the fact that the film does not lie completely flat.   Gauge-band elimination has been overcome in some instances by the use of rotating dies.   Other technological advances will probably cure the remaining physical deficiencies found at present in film made by this process.

An exotic method of manufacture has been vaguely and only generally described.   In such an operation a tube is extruded upward.   At its orifice it is slit into seven segments which are then stretched upward and outward to impart biaxial orientation. From a casual examination, and without more factual acquaintance than a brief generalized description, start-up and thickness control problems loom large in production with this apparatus.

C. FILM PROPERTIES [22–24]

Two principal ASTM specifications concern the film under discussion: *D 1463*, "Biaxially Oriented Styrene Plastics Sheet" [22]; and *D 1504*, "Orientation Release Stress of Plastics Sheeting" [23]. D 1463 covers two types of biaxially oriented film:

Type I. A general purpose sheet made from polystyrene with or without the addition of colorants and small amounts of lubricants.

Type II. A sheet made from a styrene-acrylonitrile copolymer with or without the addition of colorants and small amounts of lubricants. The detailed requirements are shown in Table 2.

TABLE 2

Detailed Requirements for Biaxially Oriented
Polystyrene and Styrene-Acrylonitrile Sheet

	Type 1	Type II
Specific gravity	1.045–1.07	1.065–1.085
Orientation release stress, minimum, psi	75	75
Orientation ratio, each direction, minimum	2:1	2:1
Tensile strength, minimum, psi [24]	8000	10,000
Dimensional stability, maximum, change, %		
One month at $23 \pm 1°C$ ($73.4 \pm 1.8°F$),		
$50 \pm 2\%$ relative humidity	0.05	0.1
One month at $71 \pm 1°C$ ($160 \pm 1.8°F$)	0.8	1.6

TABLE 3

Properties of Biaxially Oriented Polystyrene Film

General	
Forms available	Sheets, rolls, slit to width
Film and sheet gages available	0.001 to 0.020 in.
Standard roll widths	21 and 42 in.
Yield (theoretical in.^{2}/mil)	26,300
Clarity and sparkle	Excellent
Visible light transmission	92%

TABLE 3-*continued*

Properties of Biaxially Oriented Polystyrene Film

Aging properties	
indoors	Outstanding
outdoors	Limited
Mechanical	
Tensile strength,[a] psi	9000–12,000
Modulus of elasticity in tension[b]	450,000
Heat-sealing range (1 mil)[c]	250–300°F
Adhesive bonding[d]	Fast sealing
Chemical	
Water-vapor transmission rate[e] [29, 30]	
0.001 in.	4.4
0.005 in.	1.3
0.010 in.	0.7
0.015 in.	0.53
Water absorption, % weight gain in 24 hr	Very low (0.04–0.06)
Gas permeability[f]	
O_2	213–300
N_2	42–65
CO_2	790–926
Air	62–90
Resistance to weak acids, alkalies	Excellent
Resistance to strong acids, alkalies[g]	Good
Resistance to organic solvents	Good to poor
Resistance to alcohols	Good
Resistance to greases and oils[h]	Good to excellent
Thermal	
Minimum use temperature[i]	Excellent, subzero
Maximum use temperature,	
brief exposure	Up to 200°F
continuous exposure	175°F
Flammability	Slow burning

[a] 1–10 mil, both directions, D 882–64T [24].

[b] Psi, D 882–64T [24].

[c] Heavier gauges, 250° and up, depending on application. Seals by contact method, not electronic.

[d] Variety of solvents, emulsion, and latex adhesives. Coated board available for blister-packs.

[e] g/100 in.^{2}-24 hr at 100°F, 95% relative humidity.

[f] ml of gas/100 in.^{2}-24 hr-mil at 73°F and 1 atm, D 1434–63 [30] (1–5 mil range).

[g] Attacked somewhat by oxidizing acids (chromic, concentrated nitric).

[h] Lubricating and vegetable oils generally good; mineral oil generally excellent; some exceptions.

[i] 0.010-in. gauge can be bent around 1/4-in. diameter mandrel at −80°F without breaking.

The orientation release stress test method, as delineated in ASTM Method D 1504, is the parameter of prime significance in measuring the degree of effective orientation. This test should be done in both directions of the film and must be coordinated with the amount of orientation, thickness, and other qualities of the film in judging over-all quality.

The physical properties of oriented polystyrene film are different from those of many other packaging films, especially in stiffness [25–28]. A summary of properties is made in Table 3.

III. Commercial Applications of Film

The commercial large-scale use of polystyrene film and sheeting has been stayed because it differs from conventional materials. When converters in Taber equipment attempted to fold or bead it, its elastic memory caused it to shrink and jam the dies. It was virtually the first shrink material at a time when there was no commercial recognition of the merit of this property.

It did not heat-seal in conventional equipment and required special inks for printing. Therefore its first uses were for toys and novelties, shelf edging, and place mats and as a base for photographic film for color-separation plates. Polystyrene replaced cellulose acetate propionate for this application because of its superior dimensional stability. The polystyrene film absorbed virtually no water vapor and therefore was superior for registering colors in multicolor printing plate applications.

When vacuum-forming equipment became available in the 1950s, several attempts were made to produce packages and trays from biaxially oriented polystyrene sheet. Usually these attempts ended in frustration for one of two reasons. As the radiant heat from the source was absorbed in the polystyrene, the sheet heated up in zones, rather than uniformly, and started to shrink and thin out. The use of baffle plates and other devices served to make the heat pattern more uniform, but the retractive force of the polystyrene film (a 200 psi orientation release stress on a 10-mil sheet causes a 2-lb force per linear inch) pulled the polystyrene from the clamping. The use of pins and special clamping cushions overcame this condition, but at about this time the Emhart Manufacturing Company offered their pressure-forming apparatus. The Emhart machines employed a heated platen against which the polystyrene sheet was hydraulically held. When the sheet was heated to forming temperature, it was blown by pressure (which could be many times the limiting 15 psi of the vacuum-forming system) into the mold. The first machine offered by Emhart, designated EM–132 formed lids; the next machine, called the EM–134, formed trays, cottage-cheese cups, etc. Other pressure-forming machinery is produced by Tronomatic, Thermatrol, Kirkoff, Atlas-Vac, and others.

Biaxially oriented polystyrene in thin form can be heat-sealed on Packaging Machinery's FA and most other automatic heat-sealing machines if the orientation release stress is not too high (maximum about 275 psi) and if it is balanced in both stretching directions.

Various automatic machinery for processing polystyrene film and sheeting has been worked on over the years. One such machine which die-cuts blanks and then folds the sheet into transparent boxes is built by Palmer-Jones. Many other mechanisms for continuously processing polystyrene film are in the development or prototype stage.

Early polystyrene film was probably the first shrink film. It was tried for bulk-grouping of beer, fruit, and other commodities, but because of the lack of good shrink tunnels it was not until recent years that oriented polystyrene film found mass markets for shrink application. One present application is for packaging lettuce in the field. Machinery has been developed to wash the lettuce in the field and automatically wrap it in the polystyrene film, which is then heat-shrunk to the contour of the lettuce. Not only does polystyrene provide a sparkling, sanitary, transparent wrapper but the oxygen and carbon dioxide transmission rates are such that the lettuce can continue its normal respiration and hold most of its water content, thus keeping spoilage at a minimum. Figure 5 shows a field operation wrapping lettuce in oriented polystyrene film. Figure 6 shows a typical shrinkwrap package of apples, and Figure 7 illustrates a similar overwrap for tomatoes.

The overwrap consumption for meat and produce is about 9 million lb of all films a year. Oriented polystyrene has just started to penetrate this market, which is still growing. There are, in addition, many miscellaneous overwrap applications. In these uses cardboard cartons are overwrapped for items such as candles, models, games, and soft goods.

This clear film is being used also as windows for envelopes and cartons, where its clarity and dimensional stability against high humidity are proved advantages. It is also being employed for laminations and silk-screen signs like those applied to supermarket windows.

As discussed earlier, the biaxially oriented sheet first found packaging applications as transparent lids for cottage cheese tubs and delicatessen

Fig. 5   Field-packaging of lettuce in polystyrene.

goods in general.   The base packages were initially on paperboard, but now many of the trays are being made from impact and foamed poly-styrene sheet.   Other lids are shown in Figure 8 for such diverse foods as strawberry shortcake, pizza, baked beans, pies, chicken, angel cake, jellies, and caramel corn.   Individual servings of pastry, frozen desserts, and ice cream (Figures 9, 10), for which the entire package is made of oriented polystyrene, are finding favor.

One rapidly growing application is in formed trays for cookies and candies (Figures 11, 12).   In these instances the formed container has lowered the packaging cost because of the formed recesses; the over-all packaging operation is more economical than when paper separators

Fig. 6   Apples shrunk-wrapped in polystyrene.

are used.   Another growing area is in packaging individual portions of food for automatic vending machines (Figure 13).

Incidentally, nonpackaging applications, such as document protectors, also represents a growing market.

In addition to unmodified polystyrene, film and sheet have been made from styrene-acrylonitrile, impact polystyrene of various grades, and foamed polystyrene.   In general these materials have been used in the same applications as crystal polystyrene sheet; namely, disposable plates and soup bowls, cups for hot and cold drinks, meat and produce trays, interior-formed and embossed sheet for packaging, egg cartons, apple trays, greeting cards, and printed advertising.   Some of these applications are shown in Figure 14.   Currently the largest single

Fig. 7   Tomatoes overwrapped in shrinkable polystyrene.

application for oriented polystyrene trays is for packaging meat. Legislation in New York City requires that both top and bottom trays be transparent. This application alone may reach a level of 4 billion trays in the current year (Figure 15).

It is only within the last several years that the market for polystyrene film and sheeting has grown to major status. The estimate of about 30 million lb in 1969 [5] was probably exceeded. The major markets for food packaging are scheduled for the future when suitable coatings and methods of application have been worked out [31]. The base film is cheaper than cellophane and superior in dimensional stability and

Fig. 8   Foods packaged with formed lids and covers of molded polystyrene.

297

Fig. 9   Individual servings of desserts.   The outer container is of molded
polystyrene.

Fig. 10   Entire ice cream package of molded polystyrene.

Fig. 11   Candy packed in molded polystyrene tray.

Fig. 12   Design of tray and overwrap of polystyrene for cookies.

Fig. 13 Automatic vending-machine tray.

Fig. 14 Illustrative applications of high-impact polystyrene and copolymers.

300

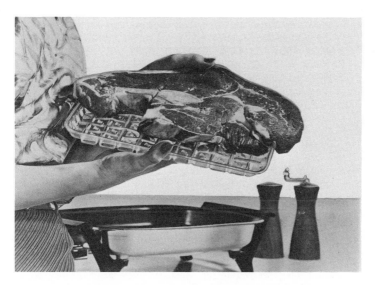

Fig. 15   For packaging fresh meat in transparent trays (both top and
bottom), oriented polystyrene finds extensive use.

stiffness modulus. For these reasons, when barrier coatings now in
laboratory and pilot-plant stage are placed in full production, coated
oriented polystyrene film may replace cellophane in some markets.

When the coatings become commercial, oriented polystyrene in sheet
thickness could replace opaque packages such as paperboard and aluminum foil for food containers.

References

1. R. H. Boundy, R. F. Boyer, and S. M. Stoesser, *Styrene, Its Polymers, Copolymers, and Derivatives*, Reinhold, New York, 1952.
2. W. C. Teach and G. C. Kiessling, *Polystyrene*, Reinhold, New York, 1960.
3. W. Kern, H. Cherdron, and H. Logemann, "Polymerization of Styrene," in
   Houben-Weyl *Methoden der Organischen Chemie*, Eugen Müller, Ed., Vol. 14,
   Part 1, Thieme, Stuttgart, 1962, p. 753 ff.
4. *Plastics*, **7**, No. 2, 34–36, 58 (1947); *Rubber World*, **131** (January 1954); *Chem.
   Week*, **82**, No. 22, 23 (May 31, 1958); *Mod. Packaging*, **32**, No. 1, 122 (September
   (1958); *Oil, Paint, Drug Reptr.*, May 26, 1958; *Mod. Plastics*, **39**, No. 6, 82
   (Feb. 1962); *Chem. Week*, **92**, No. 19, (May 11, 1963); *Mod. Plastics*, **40**, No. 10,
   80 (June 1963); *Mod. Plastics*, **41**, No. 5, 95 (January 1964); *Oil, Paint, Drug
   Reptr.*, March 16, 1964.

5. R. B. Bishop, *Plastics World*, **26**, No. 9, 22 (September 1968).
6. J. Bailey, *Mod. Plastics*, **21**, No. 4, 90 (December 1943).
7. J. Bailey, *India Rubber World*, **118**, No. 2, 225 (May 1948).
8. J. Bailey, U.S. Patent 2,545,868 (March 20, 1951).
9. F. E. Wiley, R. W. Canfield, R. S. Jesionowski, and J. Bailey, U.S. Patent 2,412,187 (Dec. 3, 1946).
10. R. W. Canfield, U.S. Patent 2,541,203 (Feb. 13, 1951).
11. *Plastics Technol.*, **7**, 60 (Jan. 1961).
12. *Soc. Plastics Engr., Trans.*, **1**, 164 (1961).
13. L. E. Nielsen and R. Buchdahl, *J. Appl. Phys.*, **21**, 488 (1950).
14. L. E. Nielsen and R. Buchdahl, *J. Chem. Phys.*, **17**, 839 (1949).
15. E. Merz, L. E. Nielsen, and R. Buchdahl, *J. Polymer Sci.*, **4**, 605 (1949).
16. R. J. Stein, *J. Polymer Sci.*, **50**, 339 (1961).
17. S. Hoshino, J. Powers, D. G. LeGrand, H. Kawai, and R. S. Stein, *J. Polymer Sci.*, **58**, 185 (1962), and previous Stein papers.
18. R. S. Stein and R. R. Wilson, *J. Appl. Phys.*, **33**, 1914 (1962).
19. J. P. Joule, *Phil. Trans. Roy. Soc.*, **149**, 91 (1859).
20. J. M. DeBell, W. C. Goggin, and W. E. Gloor, *German Plastics Practice*, DeBell and Richardson, Springfield, Mass., 1946, pp. 365–471.
21. *British Plastics*, **35**, 429 (Aug. 1962).
22. American Society for Testing and Materials, *1966 Standards*, Part 26, p. 103; Method D 1463–63. Parts 26 and 27 of the *Standards* contain references to standards of manufacture of polystyrene.
23. American Society for Testing and Materials, *1966 Standards*, Part 26, p. 115; Method D 1504–61.
24. "Tensile Properties of Thin Plastic Sheeting," American Society for Testing and Materials, *1966 Standards*, Part 27, p. 366; Method D 882–64T.
25. *Modern Packaging Encyclopedia*, **43**, No. 7A (July 1970), McGraw-Hill, New York.
26. *Modern Plastics Encyclopedia*, **47**, No. 14A (October 1970), McGraw-Hill, New York.
27. M. C. Slone and F. W. Reinhart, *Mod. Plastics*, **31**, No. 10, 203 (June 1954).
28. F. C. Dulmage, *Mod. Packaging*, **32**, No. 1, 154 (September 1958).
29. H. A. Scopp and A. Adakonis, *Mod. Packaging*, 32, No. 4, 123 (December 1958).
30. "Gas Transmission Rate of Plastic Film and Sheeting", American Society for Testing and Materials, *1966 Standards*, Part 27, p. 494, Method D 1434–63. See also Chapter 1.
31. H. A. Scopp and S. Black, *Mod. Packaging*, **32**, No. 11, 116 (July 1959).

CHAPTER 5

POLYVINYL CHLORIDE

ORVILLE J. SWEETING*

Yale University, New Haven, Connecticut

I. Introduction

Vinyl chloride and its polymers are among the very oldest pure organic substances isolated and characterized. The monomer was first obtained by Liebig in 1835 and characterized by his student, Regnault [1].

Properties of the monomer, techniques of polymerization, and polymer properties and applications have been treated in several monographs on films and fibers, which should be consulted for further information [2–4]. Though these volumes suffer to some extent from age, they are still useful. Developments in Germany during World War II have been described [5–6]; they were of little significance.

The best recent comprehensive summary of the state of the art of polymerizing vinyl chloride appears in Houben-Weyl [7].

* Formerly Associate Director of Research and Development, Olin Film Division, Olin Corp., New Haven, Connecticut. Present address. Quinnipiac College, New Haven 06518.

II. Preparation of Vinyl Chloride

Vinyl chloride monomer is a gas, boiling at $-13.9°C$; it has a refractive index of n_D^{15} 1.38, a density of 0.992 at 25°C, and a heat of polymerization $\Delta H = -17.0$ kcal/mole. Until quite recently it has usually been prepared industrially by passing a mixture of acetylene and hydrogen chloride over suitable catalysts such as mercuric chloride on silica gel at 20 to 30°C or activated carbon at 200°C.

$$CH \equiv CH + HCl \xrightarrow{\text{H}_g\text{Cl}_2} CH_2 = CHCl \qquad (1)$$

It can also be prepared from ethylene dichloride, and this was formerly the preferred commercial method in the United States. (This is the method first used by Liebig and Regnault and is still the most convenient

$$CH_2ClCH_2Cl + OH^- (\text{at } 60° \text{ in } CH_3OH) \longrightarrow$$
$$CH_2 = CHCl + Cl^- + H_2O$$

laboratory synthesis.) The same result can be achieved by cracking ethylene dichloride at 300 to 600°C in a hot tube or over a contact catalyst, and in this case the hydrogen chloride liberated can be used to make vinyl chloride by action on acetylene. This cracking process is also used to some extent industrially.

In recent years vinyl chloride monomer has been most economically produced [8–9] by a combination of reaction (1) with the chlorination (2) or oxychlorination (3) of ethylene, followed by cracking the dichloroethane produced (4). Thus HCl and Cl_2 may be used in such

$$CH_2 = CH_2 + Cl_2 \longrightarrow CH_2Cl - CH_2Cl \qquad (2)$$

$$CH_2 = CH_2 + Cl_2 + 1/2\ O_2 \longrightarrow CH_2Cl - CH_2Cl + H_2O \quad (3)$$

$$CH_2Cl - CH_2Cl \longrightarrow CH_2 = CHCl + HCl \qquad (4)$$

a balanced process that about equal numbers of moles of ethylene and acetylene result in vinyl chloride in highly efficient use of the chlorine charged and very little hydrogen chloride by-product [since it may be consumed in (1) at about the rate of formation]. The kinetics are favorable toward a high yield of vinyl chloride as a result of simultaneous operations of reactions (1), (2), and (4).

III. Polymerization of Vinyl Chloride

Polyvinyl chloride polymers are almost exclusively produced by emulsion or suspension polymerization rather than by bulk polymerization. The high heat of polymerization ($\Delta H = -17.0$ kcal) makes the latter difficult to control and the polymer has a tendency to develop hot spots, decompose by loss of HCl, and discolor (Section IV-C].

Although the term *vinyl polymers* includes a large family, it has come to be used to designate polymers of vinyl chloride or copolymers largely composed of vinyl chloride. Production in the United States is now well over 2 billion lb a year; of this total production, however, only about 4 % finds its way into film. Table 7, Section VI, shows the growth of world production of these polymers during the last 20 years.

Vinyl chloride is usually polymerized rapidly at a temperature as low as possible, since in this way a minimum discoloration results. Therefore polymerization in suspension or emulsion in water, or in solution in organic solvents such as butane, is preferred to polymerization in bulk. The recent development of emulsion polymerization results in very small dispersed globules, and reaction is readily controlled. Such problems as heat transfer of the heat of polymerization and control of particle size in the resulting polymer are minimized.

Vinyl chloride is polymerized by mechanisms similar to those used in the polymerization of ethylene, vinyl acetate, or a variety of other monomers containing a structure that can form π-complexes with free-radicals. The free-radical initiators vary greatly and include peroxides such as dilauroyl peroxide, or dibenzoyl peroxide, and others such as α,α'-azoisobutyryldinitrile. The mechanisms of these polymerizations have been studied for many years [2, 7, 10–12], and though some details of the emulsion polymerization mechanisms have not been well worked out the general principles are clear.

The process can be divided into three steps, initiation, propagation, and termination [13–15].

Initiation
$$\text{initiator} \longrightarrow \text{R·}$$

Propagation
$$\text{R·} + n\text{CH}_2\text{—CHCl} \longrightarrow \text{R—(CH}_2\text{—CHCl·)}_n$$

Termination

(a) by coupling:
$$R-(CH_2-CHCl)_n-CH_2-CHCl\cdot$$
$$+ R-(CH_2-CHCl)_m-CH_2-CHCl\cdot \longrightarrow$$
$$R-(CH_2-CHCl)_{n+1}-(CHCl-CH_2)_{m+1}-R$$

(b) by adding a simple free radical:
$$R-(CH_2-CHCl)_n-CH_2-CHCl\cdot + R'\cdot \longrightarrow$$
$$R-(CH_2-CHCl)_{n+1}-R'$$

(c) by disproportionation:
$$2R-(CH_2-CHCl)_n-CH_2-CHCl\cdot \longrightarrow$$
$$R-(CH_2-CHCl)_nCH=CHCl + R-(CH_2-CHCl)_n-CH_2CH_2Cl$$

Chain transfer can also occur, resulting in branching and the new radical can continue to grow. This reaction does not occur to an appreciable extent in emulsion polymerization, however.

Chain Transfer
$$R-(CH_2-CHCl)_n-CH_2-CHCl\cdot + \sim CH_2-CHCl \sim \longrightarrow$$
$$R-(CH_2-CHCl)_nCH_2-CHCl-\overset{\cdot}{C}H-CHCl \sim$$

There are two possible arrangements of the vinyl groups: either a head-to-head or head-to-tail.

Head-to-Head

$$2n\ CH_2=CHCl \longrightarrow (-CH_2-CHCl-CHCl-CH_2-)_n$$

Head-to-Tail

$$2n\ CH_2=CHCl \longrightarrow (-CH_2-CHCl-CH_2-CHCl-)_n$$

The structure of polyvinyl chloride has been much studied, and all evidence is consistent with a view that polymerization takes place largely in the head-to-tail fashion. There is some evidence to indicate that certain vinyl chloride polymers have branched chains. Branching apparently occurs more frequently if the polymerization temperature is high, which brings about chain transfer. Bulk polymerization or concentrated mixtures in heterogeneous polymerizations also favor chain transfer.

Vinyl chloride and vinylidene chloride polymerizations differ from most of the other vinyl polymerization systems (e.g., styrene, acrylate esters) in that the polymer is not soluble in the monomer and therefore separates from the system as fast as it forms. This causes some complications in theoretical discussions of emulsion polymerization. Harkins's general theory [16, 17] has had to be modified in recent years, to account for differences in mechanism that appear to result from the low solubility of polymer in monomer and vice versa [18–21].

Vinyl chloride can also be polymerized with the help of Ziegler catalysts [22] to yield more highly crystalline vinyl polymers than hitherto known. Except for the x-ray pattern, the properties found were not greatly different from those listed in Table 1.

TABLE 1

Properties of Vinyl Chloride, Vinylidene Chloride,
Styrene, and Methyl Methacrylate Polymers

	Polyvinyl chloride	Polyvinylidene chloride	Polystyrene	Polymethyl methacrylate
Specific gravity	1.35–1.55	1.65–1.72	1.04–1.06	1.18–1.19
Refractive index, n_D^{20}	1.54–1.55	1.60–1.63	1.59–1.60	1.49
Tensile strength, psi	5–9000	3–5000	5–9000	8700–9500
Elongation, %	5–15	to 250	1.0–3.6	3–10
Specific heat, cal/°C-g	0.20–0.28	0.32	0.32	0.35
Thermal expansion, $\times 10^5/$°C	5–6	19	6–8	9
Heat-distortion temperature, °C	50–60	55–65	70–100	70–90
Softening temperature, °C	70–80	185–200	>100	>100
Volume resistivity, Ω-cm (50% RH, 23°C)	>10^{16}	10^{14}–10^{16}	10^{17}–10^{19}	>10^{14}
Dielectric strength, short time, 1/8 in. thick, volts/mil	7–1300	350	5–700	450–500
Dielectric constant, 60 cylces	3.4–3.6	4.5–6.0	2.45–2.65	3.5–4.5
10^3 cycles	3.0–3.3	3.5–5.0	2.40–2.65	3.0–3.5
10^6 cycles	2.8–3.0	3.0–4.0	2.40–2.65	2.7–3.2
Burning rate	Self-extinguishing	Self-extinguishing	Slow	Slow
Effect of sunlight	Darkens	Darkens	Very slight	Slight yellowing

IV. Properties of Polyvinyl Chloride and Copolymers

A. POLYVINYL CHLORIDE

In common with most vinyl products, polyvinyl chloride is characterized by extreme flexibility, resiliency, and toughness; but above all it has excellent resistance to water and a low inflammability. Polyvinyl and polyvinylidene chloride are clearly superior to other vinyl polymers in these two properties. Polyvinyl chloride is resistant to swelling and distortion by mineral and vegetable oils and also highly resistant to aging and weathering, though the action of light causes the slow liberation of hydrogen chloride, with consequent degradation of the polymer. Tubing made from plasticized polyvinyl chloride is chemically resistant and widely used in handling liquid and gaseous corrosive chemicals. In most of these applications it is superior to natural rubber.

The chief uses of polyvinyl chloride include coatings, pipe, automobile upholstery, tubing, gasket material, floor covering, chemically resistant finishes, toilet articles, shower curtains, raincoats, packaging films, and other applications in which a moisture barrier is desired.

Polyvinyl chloride is usually plasticized, though in Germany during World War II unplasticized calendered sheet was made in thicknesses down to 1 mil [5]. The amount of plasticizer used is ordinarily much greater than in most polymers, and certain types may contain as much as 85%. The effectiveness of a plasticizer is a complicated business, but one factor on which it depends is certainly its solubility in the resin. The low solubility of the common plasticizers in polyvinyl chloride at room temperature (compared with other polymers) is not necessarily a fatal disadvantage, for many of these materials can be incorporated with the polymer at elevated temperatures (e.g., 150°C with tricresyl phosphate) to give a material that remains homogeneous at room temperature and can be molded, cast, or extruded. Useful plasticizers include phthalate, adipate, and sebacate esters, trioctyl phosphate, polypropylene glycol sebacates and adipates, and certain copolymers of butadiene and acrylonitrile of low molecular weight.

For the general applications such as molding, extrusion, and sheeting out, the amount of plasticizer depends on the heat resistance and flexibility required. Films and molded materials of all kinds may be altered in various ways by changing the kind and amount of plasticizer. The plasticized polymer may be further compounded by the addition of dyes, pigments, lubricants, diluents, and fillers as required.

Polyvinyl chloride fabrication is generally by molding or extrusion. Compression molding conditions as a rule require pressures of 1500 to 2000 psi and temperatures of 175 to 200°C. The temperature must be kept as low as possible, however, since polyvinyl chloride shows a marked tendency to undergo thermal decomposition; this decomposition is self-propagating by a self-catalysis that results in an ever-progressive decomposition of the polymer, with charring and liberation of hydrogen chloride. Various heat stabilizers of a wide variety have been used [Section IV-C], but the results are not entirely satisfactory, since most of these compounds are carbonates, hydroxides, oxides, or similar basic substances, the purpose of which is to absorb hydrogen chloride. Their basic nature tends to bring about splitting off hydrogen chloride from the polymer, a process that breaks it down still further.

Table 1 lists some of the properties of polyvinyl chloride, as well as the corresponding properties of polyvinylidene chloride, polystyrene, and polymethyl methacrylate for comparison [23].

A large number of copolymers of vinyl chloride has been described in the literature [24]. Here we discuss only one large group, those with vinyl acetate.

The copolymers with propylene seem to be filling a growing need for applications that require greater film stiffness than found with polyvinyl chloride alone, coupled with higher softening temperatures and better barrier properties to oils and greases.

B. Copolymers of Vinyl Chloride and Acetate

Attempts have been made to obtain a combination of the properties of polyvinyl chloride and polyvinyl acetate by physical mixtures of the two polymerized materials, but they show poor compatibility and no worthwhile properties which have led to any important industrial applications. It has been necessary to obtain the desired intermediate properties by copolymerizing vinyl chloride and vinyl acetate. In general, the useful copolymers have been those that contain 80 to 95% vinyl chloride. The degree of polymerization may be altered by varying experimental conditions such as temperature, solvent dilution, and catalyst. In this way physical properties such as tensile strength, elongation, impact strength, and solubility may be changed over a wide range.

In general, less plasticizer is required with copolymers of polyvinyl chloride than is needed with the pure polymer, presumably because of the internal plasticization brought about by the vinyl acetate units in the chain. The type of plasticizers used with the copolymers is approximately the same as those used with polyvinyl acetate. Typical examples of suitable plasticizing agents include dioctyl and dibutyl phthalates, dibutyl cellosolve phthalate, cellosolve ricinoleates, glyceryl esters, tricresyl phosphate, camphor, and triacetin. Pigments, dyes, mold lubricants, and fillers are often added to change the properties. Heat stabilizers (normally basic organic or inorganic compounds) may be added in 1 to 2% to render the polymer more stable toward heat.

For injection molding, pressures of 15 to 20,000 psi and temperatures of 150 to 165° are employed. A molecular weight of 9500 to 10,500 is ordinarily considered optimum. The copolymers can also be extruded in the form of rods, tubing, sheeting, and film. Sheet and film are available commercially in plasticized or unplasticized forms, colorless or colored, transparent, translucent, or opaque.

Vinyl acetate-chloride copolymers show excellent resistance to water, salt solutions, acids, alcohols, alkalis, and oils of all kinds. Poor resistance is shown toward aqueous ammonia, ketones, esters, aldehydes, aromatic hydrocarbons, and some organic acids. Absorption of water may be as low as 0.05% when the polymer sheet is immersed in water at 25°C for one week. The low water absorption results in excellent dimensional stability under all sorts of weather conditions and accounts for the use of unplasticized sheeting in watch crystals, drafting instruments, navigation instruments, radio dials, aircraft windows, phonograph records, bindings for books, and various other products.

Sheets obtained from the highly plasticized copolymer resins are tough, flexible, highly elastic, and resistant to tearing, scuffing, and abrasion. Their crease resistance is excellent, and some of them can be elongated up to 300%. These films make excellent flexible packaging for food as well as a great variety of other materials. Some of their properties are presented in Table 2 [23].

The copolymer is also used extensively in coating paper. This is done by calendering a 2-mil layer of plasticized film onto a thin sheet of paper. The resin becomes firmly bonded to the paper, and this product is used as a liner for closures for all sorts of products, including foods. These uses are possible, since the copolymers are nontoxic, tasteless, and odorless and possess excellent water resistance.

TABLE 2

Properties of Vinyl Chloride-Acetate Polymers

Specific gravity	1.34–1.45
Refractive index, n_D^{20}	1.52–1.53
Tensile strength, psi	6000–7000
Specific heat, cal/°C-g	0.23
Heat distortion temperature, °C	60–65
Softening temperature, °C	70–125
Volume resistivity, 50% and 23°C	10^{16} Ω-cm
Dielectric strength, short time, 1/8 in. thick	425 V/mil
Dielectric constant, 60 cycles	3.2–3.3
10³ cycles	3.1–3.2
10⁶ cycles	3.0–3.1
Water absorption, 24 hr, 25°C, 1/8 in. thick, %	0.07–0.10
Burning rate	Self-extinguishing
Effect of sunlight	Darkens on prolonged exposure

C. Degradation of Vinyl Polymers

In the case of vinyl chloride, vinylidene chloride, or copolymers of vinyl chloride and vinyl acetate the decomposition that occurs during exposure to heat and to light is the result of rather complex chemical changes [25, 26]. The first step probably involves the splitting out of the elements of hydrogen chloride and acetic acid. Following this process, double bonds form, the polymer becomes crosslinked, or the active center produced may attack a dead polymer chain and produce branching. Oxidation and splitting of polymer chains may also occur. It is not surprising that no single method to prevent this complicated series of chemical processes has yet been found.

The evolution of hydrochloric and acetic acids has been studied quantitatively, and it has been found that under both heat and light this reaction occurs [27]. A small amount of these acids is released from the resin at the rate of about 0.01%/hr for unstabilized polymer, accompanied by slow discoloration, but heat aging splits them from the vinyl resins at a much more rapid rate. Since the formation of double bonds in the polymer renders an adjacent chlorine atom allylic and therefore subject to easy removal, the formation of conjugated double bonds occurs with considerable ease. Indeed, under the influence of heat or light a chain of conjugated double bonds may form, and these polyunsaturated conjugated chains may produce highly colored polymer.

The stabilization of polyvinyl or vinylidene halides has been approached from a standpoint of adding some substance that will bind the acids produced in the decomposition, since accumulation of hydrogen chloride in the presence of catalytic materials greatly accelerates the rate of thermal decomposition. This is not a complete answer to the problem, of course, since such basic materials also increase the rate at which the acid elements are lost by the polymer. Nevertheless, a large number of compounds has been used as thermal stabilizers, nearly 100 such patents having been issued since 1940. Most of these materials are weak basic substances, preferably a mixture containing some buffering material. The more useful of the inorganic salts include lead and calcium salts (e.g., lead oleate, lead stearate, calcium oleate, and calcium stearate), as well as the lead and calcium salts of the lower fatty acids, such as the acetates and butyrates. Most of these basic stabilizers are by their very nature prohibited from use in packaging films that would come in contact with foods, but another disadvantage appears in the fact that certain stabilizers, particularly the compounds of lead, become cloudy on prolonged heating, even at moderate temperatures. This clouding is probably caused by the formation of lead chloride or a basic lead chloride, and the high refractive index and insolubility of these compounds are assumed to be the cause of the pronounced cloudiness. Furthermore, most of the basic substances which have been used as stabilizers themselves develop yellow colors even at slightly above room temperature, though darkening of the resin may be prevented for a long time.

The quantity of stabilizer used is in general between 0.1 and 1% by weight, since larger amounts may detract seriously from the desired physical characteristics of the polymer. Thus it is important that care be exercised in selecting both the type and amount of stabilizer to be used. In general, it is inadvisable to exceed 5% of the resin content for any reason. The adverse effect of stabilizers may be any of the following: cloudiness or color, degradation of impact strength, embrittlement as a result of inhomogeneity, and a change in extrusion characteristics.

Two methods are in common use for evaluating the effectiveness of a stabilizer, namely, evolution of acid and color development. For the evolution of acid, samples of the resin are heated in a closed system or subjected to irradiation by ultraviolet light. The acid evolved is determined by titration. Measurements of color development are usually made on samples heated at constant temperature for varying lengths of time.

As pointed out above, elimination of hydrogen chloride and acetic acid may form double bonds, and it has also been shown that if a chain of four or more conjugated double bonds occurs in a molecule absorption of blue light occurs and yellowing of the resin results. An interesting kind of stabilizer is one that would prevent the formation of a series of conjugated double bonds and therefore prevent discoloration. It has been found that maleic acid and some of its derivatives readily add to conjugated double-bond systems and are therefore vinyl stabilizers. In fact, these maleates have been found to bleach an already discolored resin. Thus it happens that di-2-ethylhexyl maleate, which has little or no stabilizing value alone, will stabilize vinyl quite satisfactorily when used in combination with dibutyltin oxide.

It is also possible to protect vinyl polymers against decomposition by light by the use of compounds that absorb ultraviolet light. Many organic molecules such as hydroquinone, resorcinol, and the salicylates absorb strongly in the ultraviolet. Usually esters of these compounds are used. Resorcinol alone discolors vinylite resins, as does resorcinol monoacetate, but resorcinol dibenzoate provides stabilizing action and resorcinol disalicylate is a useful stabilizer.

Table 3 lists those compounds that have been found to be effective stabilizers against light. They appear in the approximate order of decreasing effectiveness. It should be mentioned that many of the cutoff values were not sharply defined, and in some cases absorption bands above these values were observed. Among the compounds listed that improve the light stability of vinyl resins only two have low cutoff values. It is known that resorcinol dibenzoate absorbs wavelengths above the cutoff value and the same is probably true of o-dichloro-benzene. Most of the compounds that were of no effect in stabilizing vinyl resins had low cutoff values, that is, below about 3000 Å. It will be noted that all derivatives of salicylic acid listed have nearly the same values. It is expected, therefore, that these salicylates would show little or no difference in effectiveness as light stabilizers if this effectiveness were principally a function of ultraviolet transmission. This assumption appears to be inaccurate, for the salicylates do exhibit differences in effectiveness, probably resulting, in part at least, from differences in

ORVILLE J. SWEETING

TABLE 3

Light Stabilizers Arranged in Order of Decreasing Effectiveness

	Ultraviolet cutoff value Å[a]
Resorcinol disalicylate	3500
Butyl carbitol salicylate	3400
Menthyl salicylate	3370
Phenyl salicylate	3500
Butyl cellosolve salicylate	3400
Calcium salicylate	3360
Glyceryl salicylate	3440
Isopropyl salicylate	3370
Benzyl salicylate	3410
Acetylsalicylic acid	3040
Salicylic acid	3440
Resorcinol dibenzoate	2930
o-Nitrophenol	4100
2,6-Dinitro-4-chlorophenol	4100
o-Nitroacetanilide	3830
2,5-Dichloroaniline	3330
o-Nitrochlorobenzene	3730
2,5-Dichloronitrobenzene	3430
2,4-Dichloroaniline	3330
2,4-Dinitrophenol	3880
o-Dichlorobenzene	2850
2,4-Dinitro-4'-hydroxydiphenylamine	4800

[a] Wavelength at which a 0.1% solution in chloroform gives 50% transmission; solution depth 1 cm.

boiling points, since materials of low boiling point are lost more rapidly in hot compounding.

Phenyl salicylate has been widely used in polyvinyl chlorides and its value as a light stabilizer has been argued. Investigators have found good and poor results, even with the same formulation. Current opinion is that phenyl salicylate is of some benefit in plasticized compounds and quite effective in rigid formulations. Acetylsalicylic acid and salicylic acid are both of little value in plasticized vinyls and are detrimental to rigid vinyls. Resorcinol disalicylate is excellent. Its high effectiveness does not seem to be simply related to its ultraviolet cutoff value, which is very close to that of the other salicylic ester.

V. Polyvinyl Chloride Films

A. PRODUCTION

Vinyl films for packaging range in stiffness from films similar to cellulose acetate to soft flexible films similar to polyethylene. Film can be made by solvent casting, extrusion, or calendering, but the preferred and cheapest method is by extrusion.

Extruded film can be made by standard methods (Volume I, Chapter 8), in thicknesses of 0.5 to 3 mils or heavier. Nearly all extrusion polymers for film contain plasticizer and stabilizing agents, the former to control film properties and the latter to permit manufacture without serious decomposition. Attention must be given to extruder design and die flow to avoid holdup and overheating within the barrel of the extruder.

Plasticizers and stabilizers were discussed in Section IV-C. The amounts of plasticizer used vary widely, depending on the end use to which the film may be put.

Most extrusion resins are polyvinyl chloride or copolymers of vinyl chloride and vinyl acetate. The newer polypropylene and vinyl chloride copolymers, however, are rapidly becoming an important factor in the commercial market.

Oriented vinyl films for shrink-wrap applications are produced in the thinner gauges by mechanical tensilization or by bubble technique (Volume I, Chapter 10).

B. FILM PROPERTIES

Table 4 presents the generalized film properties of polyvinyl chloride films in comparison with chief competitors, polyethylene, polypropylene, polyvinylidene copolymers (Sarans), and polystyrene [28]. Values vary widely, depending on formulation—in particular, the amount of plasticizer present. In Table 5 we have presented the experimental values for a small number of typical but specific films [30, 31].

Table 6 lists some of the comparative data for copolymer films from vinyl chloride and propylene [28].

VI. Markets and Applications

Of all the polymers produced, polyvinyl chloride polymers are the second largest group in the United States (polyolefins, which by loose construction can be classified as vinyls, though not usually so considered.

TABLE 4

Properties of Polyvinyl Chloride Films and Their Chief Competitors

Property	Polyvinyl chloride	Polyeth-ylene	Polyvinyl-idene chloride (Sarans)	Polypro-pylene (unoriented)	Polystyrene (oriented)
Coverage, in.^{2}/lb	19–22,000	30,000	16,300	30,300	26,300
Density, g/cc	1.23–1.35	0.915	1.68	0.89	1.05
Tensile strength, psi[a]	2–19,000	2500	14,000	4500	10,500
Impact strength, pendulum[b]	12–20	9	12	2	3
Tear strength, g/mil[c]	15–100	250	15	200	15
Elongation, %[b]	5–500	400	60	350	40
Stiffness, Handle-O-Meter, g MD[c]	7.5–40	3	10	20	50
TD	10–45	5	15	20	50
Heat-sealing range, °F	200–350	250–350	280–300	325–400	250–325
Maximum use temperature, °F	ca. 200	150	290	250	185 (shrinks)
Minimum use temperature, °F	Depends on plasticizer	−60	≪0	Not recom-mended	<0
Dimensional change at high rh, %	None	None	None	None	None
Inflammability	Self-extinguish-ing	Slow burning	Self-extinguish-ing	Slow burning	Slow burning
Permeability					
H$_2$O,g/24hr-m^2 at 90%rh,100°F[d]	25+	18	3	9	100+
O$_2$, ml/mil-m^2-24 hr-1 atm at 0% rh, 73°F[e]	77–7500	8000	15	4000	5000
CO$_2$, same conditions[e]	770–55,000	50,000	100	15,000	15,000

[a] ASTM D 882–64T.
[b] [29].
[c] Elmendorf, ASTM D 1922–61T.
[d] ASTM E 96–63T Procedure El.
[e] ASTM D 1434–63.

are first). In 1966 production was more than 2 billion lb in this country, but less than 4% of it went into film. Production since 1955 has grown at a rate of 15% a year in the United States. Huge volumes of vinyl chloride polymers are produced both here and abroad; new plants in the United States recently completed have a capacity of 500 million lb a year each (or greater) for economy.

Properties of Polyvinyl Chloride Films

A. Reynolon General-Purpose Oriented Film[a]

Property	Value							Method
	50 Gauge	60 Gauge	75 Gauge	100 Gauge	125 Gauge	150 Gauge	200 Gauge	
Coverage, in.^{2}/lb	40,600	34,900	28,700	22,300	17,600	15,300	11,200	ASTM D 646–63T
Density, g/cc[b]	1.322							ASTM D 792–64T
Yield strength, psi[b]	4500							ASTM D 882–64T
Tensile strength, psi[b]	12,000							ASTM D 882–64T
Elongation, %	120	130	150	130	140	150	180	ASTM D 882–64T
Tear strength, g/mil[b]	18							ASTM D 1922–61T
Impact resistance, at 70°F and 30% shrinkage	105	115	125	225	250	300	350	Spencer
Slip-coefficient of friction	5	5	1	1	<1	<1	<1	ASTM D 1894–63
Sealing range, °F (Sentinel, 1.0 sec/30 psig)	305–370							
Shrinkage (film temperature), %[b]								ASTM D 1204–54
at 100°F	1							
at 150°F	12							
at 200°F	28							
at 250°F	45							
at 300°F	60							
Water vapor transmission rate g/100 in.^{2}-24 hr at 100°F and 90% rh	15.0	12.0	8.0	5.0	4.5	4.0	3.0	ASTM E 96–63T (Procedure E)
Gas permeability, ml/100 in.^{2}-24 hr-mil at 77°F and 1.0 atm[b]								ASTM D 1434–63
oxygen	75							
nitrogen	20							
carbon dioxide	425							

[a] Formulation 5155, the most versatile Reynolds Metals general-purpose heat-shrinkable film; medium soft; biaxially oriented; prints and machines well; not acceptable to Food and Drug Administration for contact with foods.
[b] Value applies to all gauges.

TABLE 5—*continued*

B. Reynolon Biaxially Oriented Semirigid Film[c]

Property	Value 50 Gauge	60 Gauge	75 Gauge	Method
Coverage, in.^{2}/lb	41,000	34,600	28,700	ASTM D 646–63T
Density, g/cc[b]	1352			ASTM D 792–64T
Yield strength, psi[b]	8000			ASTM D 882–64T
Tensile strength, psi[b]	14,000			ASTM D 882–64T
Elongation, %	110	120	140	ASTM D 882–64T
Tear strength, g/mil[b]	12			ASTM D 1922–61T
Impact resistance, at 70°F and 30% shrinkage	60	65	70	Spencer
Slip-coefficient of friction	<1	<1	<1	ASTM D 1894–63
Sealing range, °F (Sentinel 1.0 sec/30 psig)	305–370			
Shrinkage (film temperature), %[b]				ASTM D 1204–54
at 100°F	1			
at 150°F	12			
at 200°F	30			
at 250°F	50			
at 300°F	62			
Water vapor transmission rate, g/100 in.^{2}-24 hr at 100°F and 90% rh	9.0	8.0	6.0	ASTM E 96–63T (Procedure E)
Gas permeability, ml/100 in.^{2}-24 hr-mil at 77°F and 1.0 atm[b]				ASTM D 1434–63
oxygen	25			
nitrogen	5			
carbon dioxide	100			

[b] Value applies to all gauges.

[c] Formulation 3055, a general-purpose heat-shrinkable, biaxially oriented, semirigid film with high slip for high speed wrapping on automatic packing machines; approved for food packaging by Food and Drug Administration.

TABLE 5—*continued*

C. REYNOLON UNIAXIALLY ORIENTED SHRINKABLE FILM[d, e]

Property		50 Gauge	25 Gauge	100 Gauge	Method
Coverage, in.^{2}/lb		41,300	29,900	21,500	ASTM D 646–63T
Density, g/cc[b]		1.281			ASTM D 792–64T
Yield strength, psi	MD	3100	2700	2800	ASTM D 882–64T
	TD	1700	1600	1800	
Tensile strength, psi	MD	13,000	13,000	13,000	ASTM D 882–64T
	TD	5000	5000	5000	
Elongation, %	MD	100	135	125	ASTM D 882–64T
	TD	325	325	325	
Tear strength, g/mil	MD	[f]	[f]	[f]	ASTM D 1922–61T
	TD	28	30	46	
Impact resistance at 70°F and 30% shrinkage		70	72	76	Spencer
Slip-coefficient of friction		Nonslip	Nonslip	Nonslip	ASTM D 1894–63
Sealing range, °F (Sentinel 1.0 sec/30 psig)		300–350	300–350	300–350	
Shrinkage (film temperature), %					ASTM D 1204–54
at 100°F	MD	1	1	1	
	TD	0	0	0	
at 150°F	MD	33	30	25	
	TD	1	2	1	
at 200°F	MD	40	35	35	
	TD	5	6	5	
at 250°F	MD	58	53	54	
	TD	8	8	7	
at 300°F	MD	60	55	55	
	TD	12	12	12	
Water vapor transmission rate, g/100 in.^{2}-24 hr at 100°F and 90% rh		16	13	8	ASTM E 96–63T (Procedure E)
Gas permeability, ml/100 in.^{2}-24 hr-mil at 77°F and 1.0 atm[b]					ASTM D 1434–63
oxygen		180			
nitrogen		50			
carbon dioxide		1150			

[b] Value applies to all gauges.

[d] Formulation 8761, a soft uniaxially oriented film for economical sleeve over-wrapping; sufficient orientation to give a heat-shrinkable film for tight packaging with high clarity and gloss; approved for food packaging by Food and Drug Administration.

[e] MD and TD refer to the machine and transverse direction of manufacture, respectively.

[f] Not measurable in the machine direction.

TABLE 5—*continued*

D. BAKELITE RIGID GENERAL-PURPOSE FOOD-GRADE FILM[e, g]

Property		Value	Method
Coverage, in.^{2}/lb		21,000	
Specific gravity		1.32	D 1505–57T
Tensile strength, psi	MD	13,500	D 882–64T
	TD	11,500	
Tensile modulus, psi		400,000	D 882–64T
Elongation, %	MD	120	D 882–64T
	TD	130	
Tear resistance, g/mil	MD	14	D 689–44
	TD	16	
Dart drop, g/mil		30	D 1709–62T
Coefficient of friction		0.30	D 1894–61T
Strain release, °F		135	
Inflammability, sec		Self-extinguishing	D 568
Specular gloss, gloss/mil 45°		90	D 523–53T
60°		190	
Haze, %		1.8	D 1003–52
Specular light transmission, %		89	
Water vapor transmission rate, g/100 in.^{2}-24 hr at 100°F and 90% rh		3–4	E 96–63T–E
Gas permeability, ml/100 in.^{2}-24 hr (1 atm at 73°F)			D 1434–56T
oxygen		15–22	
carbon dioxide		30–40	

[e] MD and TD refer to the machine and transverse direction of manufacture, respectively.

[g] Union Carbide VGA–0403 NT. 7. All tests on 1-mil film.

The rate of growth has been even greater abroad (Table 7).

In recent years the vinyl polymers market has been somewhat unsettled. Price cutting of film-grade resins resulting from overproduction began in this country about 1955 [32, 33], and since then the average annual decrease has been 7 to 8% [34]. It has been estimated that production of vinyl chloride polymers in North America will be 2.9 to 3.3 billion lb by 1970 and 4.2 to 5.1 billion lb by 1975 (Table 8) [34]. This would be reflected in film-resin prices of about 7 cents/lb (compared with 31 cents in 1955).

TABLE 6

Some Properties of Copolymer Films of Vinyl Chloride and Propylene

Property	Value
Density, g/cc	1.37
Area coverage of 1 mil film, in.^{2}/lb	19,500
Tensile strength, 75°F, psi	7,500
Elongation to break, 75°F, %	100
Izod impact strength, 75°F	65
−40°	0.4
Heat distortion temperature, °F, 66 psi	153
Stiffness, psi	420,000
Permeability	
H_2O, g/m^2-24 hr-mil at 95% rh and 100°F	4.3
O_2, ml/m^2-24 hr-ml-atm at 25°C	45

TABLE 7

World Production of Polyvinyl Chloride Polymers

	Pounds, $\times 10^{-6}$	Average Annual Increase, %
1945	110	
1950	480	35
1955	1100	19
1960	3200	23
1965	6800	16

The uses of vinyls in the United States are summarized in Table 9 [10].

Packaging film totaled approximately 90 million lb in 1968, led only by the polyethylenes and cellophanes (Table 10).

Polyvinyl chloride film has become a major factor in food packaging since the problems of acceptable plasticizers and stabilizers were solved. Nearly all fresh meat in supermarkets is now wrapped in vinyl film.

TABLE 8

Estimated Production of Polyvinyl Chloride Polymers 1970–1975

	Pounds, $\times 10^{-6}$	
	1970	1975
North America, high	3.3	5.1
low	2.9	4.2
Western Europe, high	5.1	7.5
low	4.4	5.9
Japan, high	1.8	2.6
low	1.5	2.0
World Free Economy, high	11.0	17.0
low	9.2	13.0

TABLE 9

United States Uses of Vinyls in 1965

	%
Flexible Products	
Cable and wire coating	12
Film	4
Sheeting	14
Floor covering	17
Coatings	15
Other extruded products	14
Miscellaneous	14
	90
Rigid Products	
Pipe and fittings	5
Records, containers, etc.	3
Film and sheeting	2
	10

TABLE 10

Consumption of Films in Packaging, 1960–1968
(millions of pounds)

	1960	1961	1962	1963	1964	1965	1966	1967	1968
Polyethylenes	275	370	400	455	500	615	730	735	795
Cellophanes	439	423	404	405	410	405	395	385	360
Vinyls	18	20	22	24	30	40	40	70	90
Polypropylenes		3	15	25	28	40	45	50	65
Sarans	17	19	21	23	26	30	30	30	32
Polyesters	8	8	10	10	10	20	20	20	20
Cellulose acetates	10	10	11	12	13	15	15	15	15
Polystyrenes	3	5	7	11	15	15	15	15	15
Pliofilm	12	13	14	15	15	15	15	10	7
Miscellaneous	2	2	3	3	4	4	5	8	10
Total packaging uses	784	873	907	983	1051	1199	1310	1338	1409

compared with only 10% about five years ago when most meat was packaged in cellophane (a small part in rubber hydrochloride, Pliofilm). About 75% of fresh meat is now packaged in the store, and this operation accounts for a major portion of the 90 million lb consumed annually.

Polyvinyl chloride films now have a portion of the skin-packaging market for meats, and if this is extended an additional 10 million lb or more may be gained.

Produce such as fruit, vegetables, and lettuce now consumes about 15 million lb of vinyls a year, sharing the market with polystyrene and polyethylene.

Blister packaging and various kinds of package overwraps constitute the rest of the current market, most of which exploits the shrink properties of oriented vinyl films.

Besides fruit, produce, and central meat packaging, tremendous growth is expected for vinyls within the next few years in a variety of other bulk packaging applications: carton overwrapping, almost unlimited (600,000 tons of corrugated paper); vacuum forming, 100 million lb, and toys, hardware, office supplies, 50 million lb [35, 36].

References

1. V. Regnault, *Ann.*, **14**, 22 (1835).
2. Franz Kainer, *Polyvinylchlorid und Vinylchlorid-Mischpolymerisate*, Springer, Berlin, 1951.
3. C. E. Schildknecht, *Vinyl and Related Polymers*, Wiley, New York, 1952.
4. Rudolf Pummerer (Ed.), *Chemische Textilfasern, Filme, und Folien*, Ferdinand Enke, Stuttgart, 1953.
5. J. M. de Bell, W. C. Goggin, and W. E. Gloor, *German Plastics Practice,* Debell and Richardson, Springfield, Mass., 1946.
6. L. H. Smith (Ed.), *Synthetic Fiber Developments in Germany*, Textile Research Institute, New York, 1946 (P.B. 7416).
7. H. Bartl, *"Polymerisation des Vinylchlorids,"* in Houben-Weyl *Methoden der Organischen Chemie*, 4th ed., Eugen Müller, Ed., Vol. 14/1, George Thieme, Stuttgart, 1961. *Added in proof*: The most recent comprehensive work is by George E. Ham, Ed., *Vinyl Polymerization*, Dekker, New York, 1969.
8. L. F. Albright, *Chem. Eng.*, **74**, No. 7, 123 (March 27, 1967).
9. L. F. Albright, *Chem. Eng.*, **74**, No. 8, 219 (April 10, 1967).
10. L. F. Albright, *Chem. Eng.*, **74**, No. 10, 151 (May 8, 1967).
11. L. F. Albright, *Chem. Eng.*, **74**, No. 11, 145 (June 5, 1967).
12. L. F. Allbright, *Chem. Eng.*, **74**, No. 14, 85 (July 3, 1967).
13. P. J. Flory, *Principles of Polymer Chemistry*, Cornell University Press, Ithaca, N.Y., 1953.
14. C. H. Bamford, W. G. Barb, A. D. Jenkins, and P. F. Onyon, *The Kinetics of Vinyl Polymerizations by Radical Mechanisms*, Academic, New York, 1958.
15. F. W. Billmeyer, Jr., *Textbook of Polymer Science*, Wiley-Interscience, New York, 1962.
16. W. D. Harkins, *J. Am. Chem. Soc.*, **69**, 1428 (1947).
17. W. D. Harkins, *J. Polymer Sci.*, **5**, 217 (1950).
18. E. Peggion, F. Testa, and G. Talamin, *Makromol. Chem.*, **71**, 173 (1964).
19. P. M. Hay, J. C. Light, L. Marker, R. W. Murray, A. T. Santonicola, O. J. Sweeting, and J. G. Wepsic, *J. Appl. Polymer Sci.*, **5**, 23 (1961). See also George E. Ham, *Vinyl Polymerization*, Dekker, New York, 1969, Part II, pp. 28ff.
20. J. C. Light, L. Marker, A. T. Santonicola, and O. J. Sweeting, *J. Appl. Polymer Sci.*, **5**, 31 (1961).
21. C. P. Evans, P. M. Hay, L. Marker, R. W. Murray, and O. J. Sweeting, *J. Appl. Polymer Sci.*, **5**, 93 (1961).
22. N. G. Gaylord, *Linear and Stereoregular Addition Polymers*, Interscience, New York, 1959.
23. *Modern Plastics Encyclopedia*, 1969 (*Mod. Plastics*, **46**, No. 14A, Oct. 1969), and other sources.
24. G. E. Ham, *Copolymerization*, Wiley-Interscience, New York, 1964, pp. 812–815.
25. *Polymer Degradation Mechanisms*, Nat. Bur. Std. (U.S.), Circ. 525, Nov. 16, 1953, especially Ch. 1, 4–7.
26. H. H. G. Jellinek, *Degradation of Vinyl Polymers*, Academic, New York, 1955.
27. W. M. Gearhart and L. W. A. Meyer, *Ind. Eng. Chem.*, **43**, 1585 (1951).

28. *Modern Packaging Encyclopedia*, 1969 (*Mod. Packaging*, **42**, No. 7A, July 1969), and other sources.
29. K. W. Ninnemann, *Mod. Packaging*, **30**, No. 3, 163 (November 1956).
30. Technical Data Sheets, Reynolon films, Reynolds Metals Co., Richmond, Va, 1969.
31. Bakelite Vinyl Technical Film Data, Union Carbide Corp., Plastics Division, Film Department, 270 Park Ave., New York, 1969.
32. *Chem. Week*, **100**, No. 1, 32, (January 7, 1967).
33. *Chem. Eng. News*, **46**, No. 26, 28 (June 17, 1968).
34. G. Olivier, *Hydrocarbon Processing*, **45**, No. 9, 281 (1966).
35. *Mod. Plastics*, **43**, No. 12, 94 (August 1966).
36. *Chem. Eng. News*, **45**, No. 7, 20 (February 13, 1967).

CHAPTER 6

VINYLIDENE CHLORIDE POLYMERS
AND COPOLYMERS

A. T. WIDIGER and R. L. BUTLER

The Dow Chemical Company, Midland, Michigan

I. Process

A. POLYMERIZATION AND POLYMER PROPERTIES

Polyvinylidene chloride is a white porous powder with a softening range of 185 to 200°C and a decomposition temperature of about 225°C. Heated to the melting point and then cooled, the polymer is colorless and nearly transparent. A fundamental property of polyvinylidene chloride is its crystallinity, as determined by x-ray diagrams (Figure 1). The polymer has been described as a head-to-tail configuration and the crystallinity is attributed to its regular molecular structure (Table 1).

TABLE 1

Properties of Polyvinylidene Chloride

Molecular formula	$(-CH_2-CCl_2)_n$
Molecular weight	10,000 to 100,000
Softening temperature	185–200°C
Decomposition temperature	210–225°C
Density	1.875 g/cc at 30°C
Index of refraction	$n_D^{20} = 1.63$

Measurements of solution viscosity and osmotic pressure indicate a range of chain lengths of 100 to 1000 monomer units. The crystalline polymer in the fused state is hard, tough, and similar to ordinary plastic materials in many aspects. Amorphous polyvinylidene chloride is soft, rubbery, capable of mechanical working, and tends to crystallize on aging. When oriented into a filmlike product, the polymer is strong and highly flexible.

Vinylidene chloride will form copolymers with many substituted ethylenes (particularly the common vinyl compounds), with dienes and their derivatives, and with a number of other unsaturated compounds. Important copolymers of vinylidene chloride are those with vinyl chloride, vinyl acetate, styrene, esters of acrylic and methacrylic acid, and vinyl cyanide [1].

The most useful vinylidene chloride copolymers for fabrication into films are the vinyl chloride copolymer compositions; their greatest utilization is with 10 to 27% vinyl chloride. Copolymerization introduces units in the polymer chain that tend to destroy its regularity and

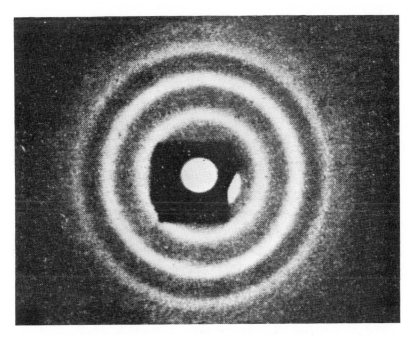

Fig. 1   X-Ray pattern of polyvinylidene chloride powder.

consequently its ability to crystallize, although the introduction of small amounts of other monomers merely results in minor discontinuities in the crystalline regions.   In general copolymers that contain less than about 70% vinylidene chloride are essentially noncrystalline.   Other effects of copolymerization related to the decreased crystallinity are reduced softening temperature and increased solubility in organic solvents.

Monomeric vinylidene chloride, 1,1-dichloroethylene, is a colorless liquid with a mild characteristic odor.   Important properties are shown in Table 2.   The most convenient source of vinylidene chloride is the reaction of 1,1,2-trichloroethane with aqueous alkali.   1,1,2-Tri-chloroethane may be made from petroleum and brine by well-known reactions involving ethylene and chlorine.

Vinylidene chloride polymerizes readily at temperatures above 0°C to form a polymer that is insoluble in the monomer and precipitates as a

TABLE 2

Properties of Vinylidene Chloride

Molecular formula	$\begin{array}{c} Cl \qquad\qquad H \\ \diagdown\;\;\;\;\diagup \\ C = C \\ \diagup\;\;\;\;\diagdown \\ Cl \qquad\qquad H \end{array}$
Molecular weight	96.95
Boiling point	31.7 at 760 mm
Freezing point	$-122.1\,^\circ C$
Density	$d_4^{20} = 1.2129$; $d_{25}^{25} = 1.2085$
Index of refraction	$n_D^{20} = 1.4249$

white powder. The presence of dissolved oxygen, which reacts to form acid chlorides and peroxides, catalyzes the polymerization. Although vinylidene chloride is not a vinyl compound, much of the technique of vinyl chloride polymerization cure is applicable to vinylidene chloride. The many chemical catalysts for polymerization which have been used successfully may be classified into five groups: organic and inorganic peroxygen compounds, organometallic compounds, organic carbonyl compounds, inorganic salts, and inorganic acids. Benzoyl peroxide in concentrations of 0.05 to 2.0% has been used frequently in laboratory work, and a typical polymerization curve is shown in Figure 2 (cf p. 307).

A very narrow softening range is typical of the crystalline polymers and copolymers of vinylidene chloride. A few degrees above the softening range these polymers have a sharp, reproducible crystalline melting point which probably corresponds to the melting of the most stable crystalline regions. At higher temperatures the polymer may be quite fluid. The sharp softening point of a crystalline vinylidene chloride-vinyl chloride polymer is shown in Figure 3.

During the film-forming processes, when the normally crystalline polymer is heated to a temperature sufficient to melt the crystalline portion and then cooled below the crystalline melting point so that no recrystallization occurs, it is amorphous and said to be supercooled. Under such conditions the ability to crystallize is a function of many variables, the most important of which are the copolymer composition, the time and temperature of storage, and the presence of additives such as plasticizers. Knowledge of the theory of crystallinity and the factors

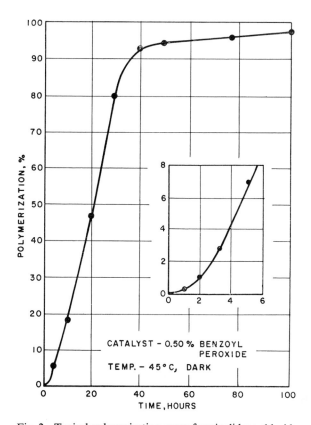

Fig. 2   Typical polymerization curve for vinylidene chloride.

controlling the degree of crystallinity are essential to proper fabrication of vinylidene chloride-vinyl chloride films.   Figure 4 shows the effect of temperature on the induction time that elapses before recrystallization begins.

Vinylidene chloride polymers and copolymers exhibit orientation when either the crystalline or amorphous modifications are mechanically worked, but the degree of orientation may vary, depending on the polymer composition and treatment.   The films may be stretched to an elongation of 350 to 400%, depending on previous treatment.

The family of vinylidene chloride polymers exhibits low thermal

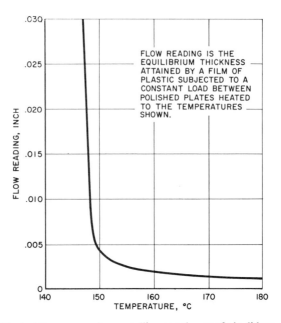

Fig. 3   Typical flow curve of a crystalline copolymer of vinylidene chloride and vinyl chloride.

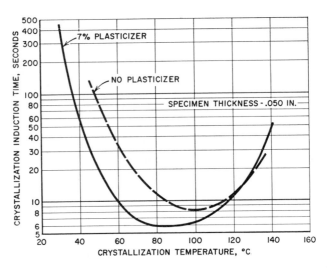

Fig. 4   Typical crystallization induction period curves for a normally crystalline copolymer of vinylidene chloride and vinyl chloride.

332

stability and various stabilizing and plasticizing additives are usually utilized to minimize thermal breakdown and optimize melt flow in fabrication equipment.   The low compatibility of crystalline vinylidene chloride polymers and copolymers with other materials limits the number of plasticizers that are functional.   Heat and light stabilizers, dyes, pigments, and lubricants have been developed specifically for many applications.

B. Extrusion

The current commercial Saran† resins at the extrusion temperature are heat-sensitive and exhibit a limited thermal life which is a function of time with respect to temperature.   Special machine-design features are necessary to accomplish successful extrusion.   Corrosion-resistant metals are necessary throughout the working parts of the extruder to avoid catalyzing polymer degradation.   The sharp melting point of Saran requires special operating techniques other than those employed by conventional extrusion techniques.

The metals that have proved to be satisfactory with Saran are nickel and cobalt alloys.   These alloys are not affected by the hydrogen chloride normally present in Saran extrusion and have no appreciable catalytic effect.   With care they can be machined, are hard enough for normal usage, and have a coefficient of expansion similar to that of steel.

The most important part of the extruder is the screw, and the quality and output of the extrusion operation is dependent on the screw design. As the screw forwards the polymer from the feed hopper to the die, the polymer undergoes considerable temperature change, is compressed, and the air must be ejected back through the feed hopper.   These functions must be performed properly to give satisfactory extrusion operation. (See Volume I, Chapter 8.)

A variety of screw designs has been proposed, fabricated, and tested. Among the types tested were constant- and variable-depth screws, single- and multiple-flight screws, constant- and variable-pitch screws, screws over a wide range of land widths and clearances, and screws over a wide range of pitch to diameter ratio.   Most commonly employed is a screw design similar to the one shown in Figure 5.

As the pitch of a screw decreases, the screw efficiency increases, but the

† Trademark of The Dow Chemical Company abroad.

Fig. 5   Typical extruder screw design for Saran resins.

volumetric capacity decreases. On the other hand, as the pitch increases, the capacity increases, but the efficiency decreases. Therefore compromise should be made between the low-efficiency, high-capacity screw on the one hand and a high-efficiency, low-capacity screw on the other.

Figure 6 indicates the effect of the ratio of screw pitch to diameter on the volumetric efficiency (E_v) and the Q_c ratio (pounds per hour per rpm) of a 3-1/2 in. extruder. As shown on the curve, the points that compromise between the low and the high volumetric efficiency and the highest Q_c ratio are pitch-to-diameter ratios of 0.8 to 1.0. Therefore a 2-in. screw might have a pitch of 1.6 to 2.0 in. It has been found that the optimum ratio of hopper opening to screw diameter is 1 : 1. Circular hopper openings are preferred to rectangular openings, since the latter tend to give polymer holdup in the corners.

For proper feeding at the hopper the flight depth at the hopper is usually $\frac{3}{8}$ to $\frac{1}{2}$ in. Depending on the resin and formulation employed, the compression ratio is in the range of 3.5 to 5.0 to 1. Screw land widths of $\frac{1}{4}$ in. have been utilized to prevent leakage flow over the land and provide adequate bearing surface. With the clearance between the flight lands and the cylinder liner, experience has indicated that a clearance of 0.005 to 0.010 on the diameter is desirable. For 2- to $3\frac{1}{2}$-in. diameter screws, radii of 0.125 to 0.30 in. on the pressure edge of the flight and 0.30 to 0.50 in. on the trailing edge give the most satisfactory results.

The heat sensitivity of the Saran resins requires short residence time in the extruder. Consequently screws are short compared with other conventional thermoplastic extruders. Ratios of screw length to screw diameter of 9 or 10 have proved most satisfactory for extruders.

The problem involved in torpedo design is to provide for the rapid and thorough melting of the material and at the same time keep the pressure rise across the torpedo low enough so that it will not be objectionable from the standpoint of raising the melting point and causing decomposition. It has been found that torpedo lengths of 3 to

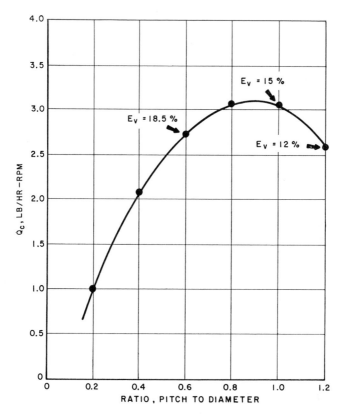

Fig. 6   Effect of ratio of screw pitch to diameter on volumetric efficiency
(E_v) and Q_c ratio; 3.5-in. extruder.

5 in. and torpedo clearances of 0.060 to 0.125 in. are a good compromise between the pressure drop and the rapid and thorough melting of the polymer.

The die design for Saran films requires streamlining to prevent holdup and subsequent polymer degradation.   The basic crosshead circular film die commonly in use for other films is employed with redesign for the properties of the thermally sensitive Saran resins.   Sheet dies are used in special applications and require a design specific to Saran resins.

The most satisfactory means of heating extrusion equipment is steam. Moderate pressures are required (170 psi maximum) and very rapid heating and cooling may be readily accomplished. The utilization of the heat of vaporization in a steam system permits rapid heating and minimizes decomposition of Saran via slow preheating. Electrical heating has been limited because of its inability to achieve rapid temperature changes. Improvements in electrical heaters with internal cooling methods could lead to future use of electrical systems. Oil heating to date has not produced temperature changes that are rapid enough for extrusion heatup and cooling.

1. Bubble Process

Vinylidene chloride-vinyl chloride films are commonly extruded by the bubble process, as shown in Figure 7. The tube is extruded downward into a water bath for rapid quenching of the plastic melt in order to minimize crystallization of the polymer. The supercooling bath is controlled as a function of copolymer ratio, the formulation, the size of tube, and the extrusion rate. From the supercooling bath the film is passed into a warming bath in which the film is warmed before the subsequent orientation step of the process. The warming bath prepares the film for optimum blowup conditions in the bubble orientation step.

The bubble orientation is accomplished by trapping air in the tube, which subjects the film to biaxial tensions plus the longitudinal stress from the draw rolls that results in orientation of the film in both directions (Figure 8).

The oriented tube is collapsed via diverging rolls and wound into rolls as a continuous tube. The tube can be the final product or it may be slit into single- or double-layer film as needed. The winding quality is a function of the film gauge variation, the winding tension, the polymer

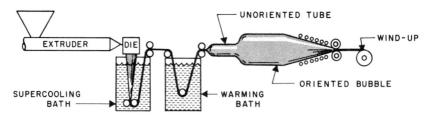

Fig. 7   Bubble process for extrusion of Saran film.

Fig. 8   Bubble orientation in extrusion of Saran film, before passage
into flattening rolls on the way to windup.

type, the film formulation, and other process variables.   The variables
are adjusted to meet the demands of the application.   In some cases the
oriented film is preshrunk via controlled heat application during a
rewinding operation.   The preshrinking temperature and film speed
determine the extent of preshrinking accomplished.   In order to control
the slip of the film for specialized applications, the film is dusted to
transform the high-cling Saran film into a slip-grade film.

   The extrusion rate, die diameter, die gap opening, drawdown of the
oriented tube, bath conditions, longitudinal orientation, bubble orientation, film take-away speeds, and other process variables are adjusted to
produce films in gauges of 25 up to 300 ($\frac{1}{4}$ to 3 mils) and in widths of
3 in. for heavy gauges up to more than 50 in. for thinner gauges.

   The bubble process for Saran film is divided into three major types
of product line.

Household Wrap. A special formulation designed for optimum cling, optics, and wrapping characteristics is extruded into large rolls of film. Slitting and winding operations convert the extruded tube into the household sized rolls.

Industrial Film. A family of specialized formulations developed for optimum properties of barrier, optics, machinability, and other product properties is extruded in the bubble process into large tubes of width and gauge required by the application. Slitting and rewinding convert the tube into the roll size and width needed by the application. Further treatment of the film for modification in slip, shrink, and other properties is conducted as necessary.

Shrink Film. Modifications in polymer, additives, and process conditions are utilized in extruding a family of high-shrink formulations utilized for shrink packaging with bags. The tube is sized to the application. The film is printed to customer order and cut into bags for product packaging. The Cryovac Division of W. R. Grace & Company was one of the leaders in pioneering this use of Saran films.

2. Cast-Film Process

Extrusion of Saran resins in the cast-film process has been limited to specialized applications in view of the die design problems with the heat sensitive resin and the success of the bubble process in producing a satisfactory film for most applications. A cast film die must be designed with streamlined flow throughout to avoid holdup of the vinylidene chloride copolymers and subsequent rapid decomposition of the resin with product discoloration or release of decomposed polymer particles. The problem becomes more complex as the die width increases and the product qualities (clarity, food approval applications, etc.) limit the number and type of stabilizing additives that may be utilized.

Saran resins are extruded by the cast-film process in specialized applications [2] that minimize the die-design problem and can utilize the product advantage of Saran resins. A commercial example is the system that utilizes Saranpac† developed by Oscar Mayer and Dow. This process is shown in Figures 9 and 10.

As shown in Figure 9, three different resins (two of Saran and one of polyvinyl chloride) are extruded into six flat films that form separate three-ply top and bottom laminates. The product is then loaded onto

† Trademark of The Dow Chemical Company.

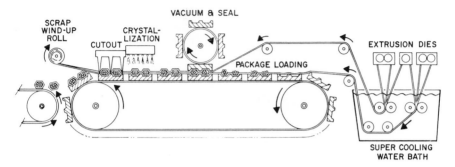

Fig. 9   Continuous extrusion and packaging process.

the bottom film, which is then joined by the top film at the vacuum-seal station to form a contoured package.   After radiant heaters convert the Saran to the crystalline state the packages are cut from the continuous film web.   The Saran resins are selected for their good sealing qualities and moisture and vapor impermeability; the polyvinyl chloride formulation gives toughness and low-temperature flexibility to the product.

The die design for the two types of Saran resin employed is aided by the relatively narrow product (15 in.) and the nature of the application and process.   The formulation and die design have been modified to meet the needs of the process satisfactorily by using Saranpac.

The key to the use of Saranpac is a method of sealing Saran to itself in an amorphous or supercooled state.   The Saran film is quenched in the supercooling bath and then, while still in the amorphous state, the tacky inner ply of Saran seals to itself without conventional use of adhesives, heat, or dielectric methods.   The film fuses at the joined edges to give a strong seal.

II. Applications

A. Introduction

The films from copolymers of vinylidene chloride and vinyl chloride are known and generally identified as Saran and are essentially odorless, tasteless, nontoxic, strong, tough, and flexible, and have high chemical resistance and low permeability to moisture and gases.   These films are

Fig. 10   Packages wrapped on the continuous extruding-packaging line.

being marketed under the trademarks of Saran Wrap,† Saran Wrap-S,† and Cryovac S.‡

The film properties, such as chemical resistance, water-vapor transmission rate, gas transmission rate, degree of shrinkage and heat sealability, can be controlled or altered by changing the ratio of vinylidene

† A trademark of The Dow Chemical Company.
‡ A trademark of W. R. Grace & Company.

chloride to vinyl chloride in the copolymer and by the amount of other blended additives.

Vinylidene chloride-vinyl chloride copolymer films are formulated to exacting end-use requirements. Generally the following property characteristics can be expected from films having a *high* vinylidene chloride and *low* vinyl chloride content.

Excellent chemical resistance.
Low water-vapor transmission rate.
Low gas transmission rate.
High heat-seal range (250–315°F).
Moderate shrinkage at elevated temperatures (212°F).

As the vinyl-chloride content is increased, the following film properties will be altered to some degree:

Somewhat poorer chemical resistance.
Slightly higher water-vapor transmission rate.
Increased gas transmission rate.
Lower sealing range (230–280°F).
Higher shrinkage at elevated temperatures (212°F).
Improved cold-temperature toughness.

B. History

Currently the accepted practice when introducing a new product or film chooses a name that by itself has no meaning and does not suggest reference to any other product.   On the other hand, some of the older products were named after places or people.   Saran film is one of these older products.   One of the stories is that the trademark "Saran" was the result of combining "Sarah" and "Ann," the names of the wife and daughter of one of the pioneers in the development of Saran films.

Initially the film was known as "Saran film" and was produced in one type only.   Type "M" Saran film was first manufactured during the early years of World War II and in tube form was widely used for the barrier packaging of machine guns, aircraft engines, and spare parts. The "M" type was a tough, crystal-clear, gas- and water-tight film with excellent chemical resistance.   It was ideally suited for the critical task of providing maximum protection for the packaging of many war-essential items.

Type "M" film was unsuitable for food packaging, for the additives used in its manufacture were not acceptable under regulations of the Food and Drug Administration.

Saran, with its many desirable properties, was an ideal beginning in the development of a family of food-grade packaging films. Near the end of World War II, work was well underway toward producing films that were acceptable for direct food contact.

Saran film 517, marketed in 1947, was the first of the food-grade Saran films to be produced. It was a water-clear film with excellent toughness, low water-vapor and gas transmission properties, and a high degree of cling. This high-cling property made the film suitable for packaging a wide range of food products, including almost all types and varieties of cheese and processed meat, both of which require the exclusion of oxygen and the retention of moisture to maximize package shelf life. In order to make the utilization of these films more diversified it was necessary to modify the films to a degree. This modification was accomplished by surface treatment and internal additives. The improved films could then be made into bags and pouches on automatic equipment, utilized on overwrap machines, made into packages on form, fill, and seal units, printed, and laminated.

One of the many interesting uses that developed for this water-clear film with the high specular gloss was the development of a thin-gauge film with a controlled degree of cling. This plastic film became known the world over as a household wrapping film, the first to be widely utilized for this purpose. It was designated Saran Wrap (Figure 11). Household rolls of Saran Wrap are produced in two convenient widths, 12 and 18 in. and are used by housewives for wrapping a wide variety of household items, including most types of food, for both refrigerator and deep-freeze storage. Its excellent barrier properties prevent flavor and aroma transfer as well as desiccation of food items in the refrigerator. In the deep freeze it affords more than adequate protection for the extended storage of meats, bakery items, and other foods.

Shortly after the marketing of household rolls of the film (early 1953) the trademark Saran Wrap was adopted for all Saran films by Dow.

C. Film Constructions

Vinylidene chloride-vinyl chloride films are supplied in thicknesses ranging from the thinnest, 40 gauge (0.0004 in.), to the thickest, 300 gauge (0.003 in.) and up to 40 in. wide. The films are supplied in single-ply

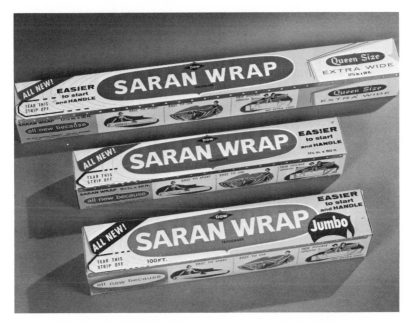

Fig. 11   Saran Wrap household rolls, first commercially available.

as well as multiple-ply thickness, centerfold sheeting, layflat tubing, and fabricated bags (Figure 12).

Single-thickness films (single-wound) are used for packaging luncheon meats, cheese, and candy, and for laminating to board stock for cap liners.

Double-ply films (double-wound) are used for packaging many food products, such as liver sausage, processed cheese, cookie dough, and candy.

Layflat tubing is utilized in the fabrication of drum liners for packaging various food and nonfood products. These liners are generally made with an electronic seal across the bottom and either an electronic seal for top closure or a mechanical closure: twist, fold, tie, or metal clip.

Centerfold sheeting is supplied in the liner-stock grade only and is used to obtain a maximum width sheet for fabrication into chemically resistant tarpaulins and large liners.

Multiple-ply films are supplied with different film types and thicknesses

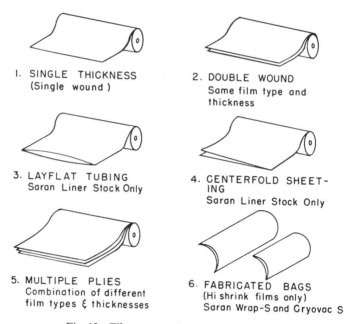

1. SINGLE THICKNESS
(Single wound)

2. DOUBLE WOUND
Same film type and
thickness

3. LAYFLAT TUBING
Saran Liner Stock Only

4. CENTERFOLD SHEET-
ING
Saran Liner Stock Only

5. MULTIPLE PLIES
Combination of different
film types & thicknesses

6. FABRICATED BAGS
(Hi shrink films only)
Saran Wrap-S and Cryovac S

Fig. 12 Film constructions available in Saran.

wound together. Films of this construction are utilized for improved barrier and strength at reduced temperatures. Candy bags, cheese packaging, and small-unit packages find this construction ideal.

Fabricated bags are supplied in a range of sizes and thicknesses. These bags are sealed electronically with a curved electrode to provide a rounded end and are either plain or flexographic printed. They are available in the Saran Wrap-S and Cryovac S high-shrink films and are utilized for packaging processed meat, cheese, and poultry.

Colored Saran films are available in a range of colors that is acceptable to the Food and Drug Administration and the Meat Inspection Division for packaging food items and for various industrial uses. These colors are red, flame, black, white, and aluminum.

D. Film Types

Saran Wrap films are specially formulated for packaging a host of food products as well as for a number of industrial nonfood applications.

TABLE 3

Comparative Saran Wrap Film Types

Series	General description	Type	Gauge	Construction	Characteristics	Typical applications
I	Lowest permeability to water vapor and other gases. High chemical resistance. Maximum protection.	3	60–75	Single	Clear, high-cling, wider seal range	Overwraps, cheese
		8	40–200	Single and double	Clear, high-cling	Overwraps for processed meats, cheese, fruit
		18	40–200	Single and double	Slip for machinability	Overwrap for processed meats, Kartridg Pak, laminations
		18L	75–200	Single	Preshrunk for laminating	Laminations to paper, foil, and other films
		11D	40–200	Single and double	Dusted for bags	Bags for candy, cookies, crackers, snacks
		11M	50–200	Single	For metallizing	Christmas decorations, industrial uses after metallizing
II	Improved low-temperature flexibility with greater permeability than Series I	27	50–200	Single and double	Clear, high-cling	Overwrap for processed meats, cheese, candied fruits
		36	50–200	Single and double	Slip for machinability	Kartridg Pak for cookie dough, other viscous products
III	Increased low-temperature flexibility, higher shrink, more elongation, easier heat sealing and greater permeability than Series I and II	41	50–200	Single and double	Clear, high-cling	Overwrap for processed meats, cheese, frozen foods
		42	50–200	Single and double	Slip for machinability	Kartridg Pak for cold or frozen products
IV	High-shrink series with maximum elongation, cold-temperature flexibility, and heat sealability. Greatest permeability of all series.	50	50–200	Single and double	Clear, high-cling	Overwrap for frozen foods and in combination with other types for cold-temperature pouching

Saran Wrap is well known for its unique properties: sparkling clarity, low moisture and gas transmission rates, high chemical resistance, negligible water absorption, and excellent aging and shrinkability.

Saran Wrap films are produced in four distinct series, each of which consists of one to several types.   The various types of film in each series differ in clarity and surface characteristics as well as in physical properties.

Saran Wrap film Series I has the highest chemical resistance and the lowest gas and water-vapor transmission rates.   Saran Wrap films Series II and III possess slightly less chemical resistance and higher gas and water-vapor transmission rates.   Saran Wrap Series IV and Saran Wrap-S films have less chemical resistance than Series I, II, and III. The gas and water-vapor transmission rates of Series IV and Saran Wrap-S films are higher than Series II and III films.   Cryovac S films compare generally with Saran Wrap films in Series III and IV.   Table 3 summarizes the four series.

E. PHYSICAL PROPERTIES

1. General Physical Properties of Saran Wrap, Saran Wrap-S, and Cryovac

The general physical properties of vinylidene chloride-vinyl chloride copolymer films produced by The Dow Chemical Company are shown in Tables 4 and 5.   Table 6 shows the general physical properties of vinylidene chloride-vinyl chloride films produced by W. R. Grace & Company.

2. Shrink Characteristics

Highly oriented films will shrink when exposed to elevated temperatures.   Although this property is desirable when the film is used as an overwrap, it can be a disadvantage in laminating.   Two grades of film are available: regular and preshrunk.   The preshrunk grade is designated with an "L" suffix.

Figures 13a, b, and c show the shrinkage characteristics of Series I, II, III, and IV Saran Wrap films.   Figure 13d shows the shrinkage curve of preshrunk film Series I (18L).

TABLE 4

Properties of Saran Wrap

Properties[a]	Series I	Series II	Series III	Series IV
Specific gravity (ASTM D 792–60T)	1.68	1.64	1.62	1.57
Water-vapor transmission g/100 in.²-24 hr at 100°F and 90% rh (ASTM E 96–53T–E)	50 gage 0.30 100 gage 0.20 200 gage 0.10	0.45 0.30 0.15	0.60	2.60 1.40
Water absorption (ASTM D 570–59T)	Negligible	Negligible	Negligible	Negligible
Gas transmission (ASTM D 1434–58T)	See Tables 8 and 9	See Table 8	See Tables 8 and 9	See Tables 8 and 9
Sealing temperature	250–315°F	240–300°F	230–290°F	230–280°F
Burning rate (ASTM D 568–61)	Self-extinguishing	Self-extinguishing	Self-extinguishing	Self-extinguishing
Dimensional stability	See Figure 13a and d	See Figure 13b	See Figure 13b	See Figure 13c
Tensile strength, psi (ASTM D 882–61T)	100 gage 8000–20,000	8000–20,000	8000–16,000	7000–12,000
Percent elongation (ASTM D 882–61T)	100 gage 40–80	35–70	40–90	50–110
Transmission of white light	90%	90%	90%	90%
Ultraviolet cutoff	3000 Å	3000 Å	3000 Å	3000 Å
Transmission of infrared	88%	88%	88%	88%
Resistance to sunlight	Good	Good	Good	Good
Dielectric strength, volts/mil (ASTM D 149–61)	50 gage 4000–7000 100 gage 3000–5000 200 gage 3000–4500	4000–3000 3000–5000 3000–4500		
Surface resistivity, Ω (ASTM D 257–61)	10^{12}–10^{15}	10^{12}–10^{15}		
Volume resistivity, Ω–cm (ASTM D 257–61)	10^{12}–10^{15}	10^{12}–10^{15}		

[a] Typical average values.

347

TABLE 5

General Physical Properties[a] of Saran Wrap-S
(Thickness 1 mil)

Gauge	0.001 in.
Specific gravity[b]	1.58
Area factor, approx. in.^{2}/lb	17,500
Water-vapor transmission, g/100 in.^{2}-24 hours at 100°F and 90% rh[c]	1.0–1.3
Water absorption[d]	Negligible
Burning rate[e]	Self-extinguishing
Greaseproofness	Excellent
Chemical resistance	Excellent
Percent shrinkage at 212°F	40–50
Tensile strength, psi[f]	6000–10,000
Percent elongation[g]	75–150
Bursting strength, psi–Mullen[h]	30
Drop height, kg/cm-mil-75°F	23.0
-0°F	16.0
Refractive index[i]	1.602
Transmission of white light	90%
Ultraviolet cutoff	3000 Å
Transmission of infrared	88%
Resistance to sunlight	Good
Resistance to	
acids	Excellent
alkalies (except NH_4OH)	Good
oils	Excellent
organic solvents	Excellent
alcohols	Good

[a] Typical average values.
[b] ASTM D 792–50.
[c] ASTM E 96–53T–E.
[d] ASTM D 570–54T.
[e] ASTM D 568–40.
[f] ASTM D 882–54T.
[g] ASTM 882–54T.
[h] TAPPI.
[i] ASTM D 542–42.

TABLE 6

General Physical Properties[a] of Cryovac S Film

[a]Property	Units	MS 750[b] and MS 755[c] Clear	Orange	S–900
Density	g/cc	1.68	1.61–1.64	1.61–1.64
Ultimate tensile	psi	10,000–15,000	7,000–14,000	7,000–14,000
Ultimate elongation	%	30–60	50–100	50–100
Modulus of elasticity				
at 73°F	psi	50,000–65,000	16,000–24,000	16,000–24,000
at 10°F	psi	700,000–800,000	450,000–600,000	
Unrestrained shrink				
at 200°F	%	25–35	40–50	40–50(205°F)
at 220°F	%	34–40		
at 240°F	%	40–50		
at 260°F	%	50–57		
Shrink tension at 200°F	psi	190–280	50–200	50–200(205°F)
at 220°F	psi	180–230		
at 240°F	psi	140–230		
at 260°F	psi	90–180		
Shrink temperature, Air	°F		225–260	225–260
Shrink temperature, Water	°F		195–205	195–205
Heat-seal range (overlap seal)	°F		225–260	225–262
Haze	%		1.5–6.0	1.5–50
Water-vapor transmission rate at 100°F, 100% rh	g/mil-100 in.²-24 hr	0.1–0.2	0.4	0.55
Oxygen permeability	ml/mil-100 in.²-24 hr-atm	1.0–1.6	4.5–10.	1.6—12.9
Resistance to acids and alkalis		Excellent Except to Ammonia		
Resistance to grease and oil		Excellent		
Yield	in.²/lb-mil			16,700–17,000
Water absorption				Negligible

[a] Typical average value.

[b] MS 750 is designed for hand-wrap applications.

[c] MS 755 is designed for machine applications.

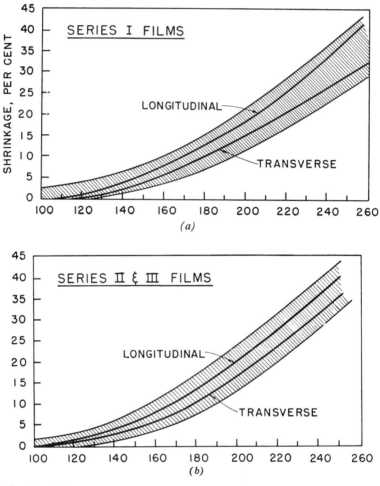

Fig. 13 Shrinkage characteristics of Saran Wrap as a function of temperature when exposed for 1 min in a liquid bath. (a) Series I films, (b) Series II and III.

350

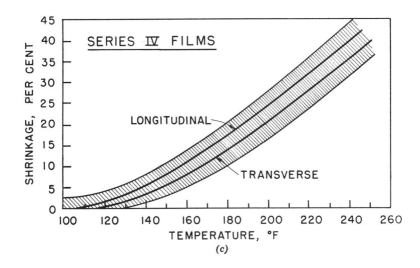

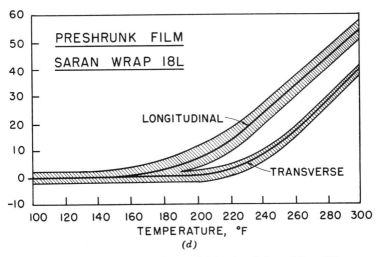

Fig. 13 (*continued*)   (*c*) Series IV, (*d*) Preshrunk Saran Wrap 18L.

3. Chemical Resistance

Saran Wrap films are combined with various substrates for numerous packaging applications to utilize their excellent chemical resistance and barrier properties. Combined with paperboard they are used as cap liners for the packaging of hard-to-hold items (Figure 14).

Saran Wrap is resistant to most organic and inorganic compounds except aqueous ammonia, cyclic ethers and ketones, and concentrated sulfuric or nitric acid (Table 7 shows the chemical resistance of Series I films).

Chemical resistance decreases with increases in temperature. The chemical resistance of Saran Wrap Series II, III, IV, and Saran Wrap-S and Cryovac S is generally less than that of the Saran Wrap film Series I; therefore the latter type is normally used in packaging uses in which chemical resistance is an important factor.

TABLE 7

Chemical Resistance, Series I, at 25°C

Dilute mineral acids	Excellent
Concentrated mineral acids (except H_2SO_4 and HNO_3)	Excellent
Organic acids	Excellent
Alkalies (except NH_4OH)	Good
Alcohols	Excellent
Aliphatic hydrocarbons	Excellent
Oils, fats, and waxes	Excellent
Pharmaceuticals and detergents	Excellent
Organic solvents (except cyclic ethers and cyclic ketones)	Good to fair

4. Permeability

Saran films are noted for their extremely low gas transmission rates. The permeability rate is inversely proportional to the film thickness.

The data listed in Table 8 indicate the gas transmission through 100-gauge (1-mil) films at 73°F. The permeability rate is also affected by the temperature: the lower the temperature, the lower the gas transmission rate. The data listed in Table 9 indicate the gas transmission through 100-gauge film at various temperatures.

Fig. 14   Saran Wrap is used as a cap liner for these and similar products
which tend to leak or lose flavor.

The very low permeability rate of the films is believed to be beneficial
in providing protection against insect infestation of packaged foodstuffs.   Laboratory studies of dried fruits such as figs and prunes
packaged in the film showed no insect penetration after two months'
exposure to adult stages of the saw-toothed grain beetle, Mediterranean
flour moth, and cigarette beetle.

The water-vapor transmission rate of Saran films is also dependent on
film thickness and exposure temperature.   Figure 15 relates the water-
vapor transmission rate of Series I films at various temperatures.

5. Optical Properties

The vinylidene chloride-vinyl chloride copolymer films are not recommended for outdoor applications in direct sunlight.   On direct exposure
to sunlight under glass there is only a slight decrease in the tensile
strength, elongation, bursting strength, and barrier properties.

TABLE 8

Gas Transmission Rates ml/100 in.^{2}-24 hr-atm at 73.4°F
adjusted to 1 mil thickness

Gas	Film Formulation			
	Series I	Series II	Series III	Series IV
O_2	0.8–1.1	2.2–4.9	4.4–6.9	18– 24
CO_2	3.8–6.0	7.0–29	24–44	120–150
N_2	0.12–0.16	0.3–1.6	1.0–1.5	6–10
Air	0.21–0.44	1.6–2.4	1.7–2.6	8–12

TABLE 9

Oxygen Transmission Rates ml./100 in.^{2}-24 hr.-
atm., adjusted to 1 mil thickness

Temp., °F.	Series I	Series III	Series IV
0	0.005	0.06	0.5
23	0.03	0.34	1.8
48	0.23	1.5	7.4
73	1.0	5.0	20.0

Although 90% of the white light is transmitted through the film, most of the ultraviolet light rays which cause product color fading are cut off. Figure 16 shows light transmission as a function of wave length for Series I and II films.

F. Fabrication

The full utilization of Saran films was hampered initially by the lack of proper fabricating equipment. One of the characteristics of the vinylidene chloride-vinyl chloride copolymer film is the extremely sharp, almost superimposed softening and melting points. The property of having a narrow temperature-sealing range (sealing temperature controlled ±5°F) made it difficult to seal the film by conventional methods.

It was learned that the film could be heat-sealed provided the sealing surface was carefully controlled within a narrow temperature range. In

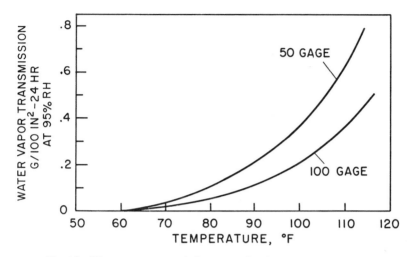

Fig. 15 Water-vapor transmission rates of Series I Saran Wrap as a
function of temperature.

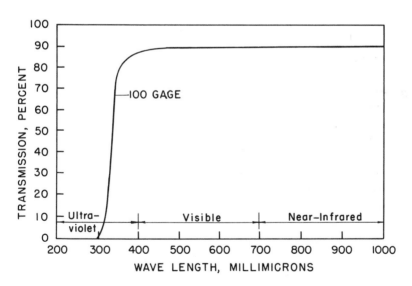

Fig. 16 Percentage of light transmitted by Series I and II Saran Wrap
films as a function of wavelength.

addition, the sealing surfaces were masked with a nonsticking surface (glasscloth coated with Teflon). This type of sealing was adequate for an overwrapped package of cheese or luncheon meat in which the film is shrunk around the product during the sealing sequence, but it was not suitable for the fabrication of bags or tubes because of the distortion in the seal area caused by thermal shrinkage.

Electronic or radio-frequency sealing is utilized for the fabrication of bags and on form, fill, and seal equipment. In this method of sealing the film is passed between cold electrodes with the high frequency current effecting a seal without causing excessive shrinkage of the film. Optimum seals are made at 27 Mc. The film can be sealed, however, at higher frequency ranges.

G. Packaging Machinery

Machinery manufacturers are now supplying equipment for the utilization of limp films. Almost all these machines can handle the vinylidene chloride-vinyl chloride copolymer films.

1. Machinery Manufacturers

The manufacturers of packaging machinery listed here, although not a complete list, are representative of the types of automatic and semiautomatic equipment suitable for packaging with Saran films.

Company	Type of Equipment
Alpma Machine Company Alpenland Maschinenbau Gesellschaft Hain & Company KG 8093 Rott/Inn, Postfach	Overwrap
Battle Creek Packaging Machines, Inc. Merrill Park Battle Creek, Michigan 49016	Overwrap
Crompton & Knowles Corporation Suffield Street Agawam, Massachusetts	Overwrap
Franklin Electric Company, Inc. Bluffton, Indiana	Overwrap

Company	Type of Equipment
Hayssen Manufacturing Company Highway 42 North Sheboygan, Wisconsin	Overwrap
The Kartridg Pak Company 807 West Kimberley Rd. Davenport, Iowa	Form, fill, and seal
Oliver Machinery Company 445 Sixth Street, N.W. Grand Rapids, Michigan 49502	Overwrap
Omori Machinery Company, Ltd. 7–4 Negishi 5–Chome Taito-Ku Tokyo, Japan	Form, fill, and seal
Corley-Miller, Inc. 18 South Clinton Street Chicago 6, Illinois	Semiautomatic packaging lines. Sheeters, hot plates, etc.
Great Lakes Corporation 2500 Irving Park Road Chicago, Illinois 60618	Semiautomatic packaging lines. Shrink tunnels, hot plates, etc.

H. Printing

Single-wound and double-wound films are printed for many packaging applications by both flexographic and rotogravure processes. The inks used are special Saran-type products specially formulated to bond to the highly chemically resistant film. Single-wound films are surface-printed and require an overlacquer to prevent offset. The overlacquer also provides a high-gloss printed surface and protects against scuffing.

The double-wound films are printed by surface-printing methods and by the trapped printing process, which is known as "Unilox" and described in detail in U.S. Patents 2,679,968 and 2,679,969. In the "Unilox" process the double-wound film is run through the printing process (flexographic or gravure) as a single web. During the subsequent slitting operation to produce the finished roll the plies are separated and one ply of the film is brought around the roll, trapping the printing between the plies, thus protecting the printing from scuffing during use and giving it a very glossy appearance.

I. Storage Considerations

Saran films will not support combustion and are self-extinguishing. They age well, their water absorption is negligible, and therefore the film is unaffected by changing humidity conditions. The films should always be stored at moderate temperatures, (e.g., at 73°F). Storage at elevated temperatures for prolonged periods is not recommended because possible shrinkage can deform the roll and cause difficulties in machining.

J. Food Uses

The vinylidene chloride-vinyl chloride copolymer films are being utilized for the packaging of a variety of items, and although the following is not a complete listing the prime packaging applications that utilize the combination of unique properties found in the Saran films are outlined.

1. Bakery Items

Single-ply film is used to machine and handwrap fresh-frozen bakery items. The high-transparency films are used to give maximum protection against moisture and flavor transfer or loss (Figure 17).

2. Fresh Ground Beef

Packaging in barrier films is a more recent development (early 1960), described in U.S. Patent 2,847,313, Reissue No. 24992, issued to Tee-Pak Incorporated. Double-ply, white, opaque films are utilized to package this product on form, fill, and seal machines that produce a chub-shaped package usually in the 1- and 3-lb size (Figure 18).

This package is unique in that a high barrier film is used to package fresh red-meat products. Barrier films are not utilized to package cuts of red meat because of the very undesirable dark color change which the meat undergoes shortly after packaging. This color change is caused by the exclusion of oxygen from the package.

With ground beef, however, a barrier film is desirable and necessary. The meat is ground and immediately packaged in chub-shaped airtight packages which are stored at refrigerator temperatures of 32 to 40°F. Film overwrapped trays of ground beef normally have a shelf life of

Fig. 17   Frozen baked goods packaged in Saran.

1 to 3 days.   This new Saran film package provides maximum protection to the product and extends its shelf life to six days.   The product on removal from the package regains its fresh red color appearance or *bloom* of fresh ground beef.

The chub packaged product can be frozen for extended storage.   The intimate contact of the high barrier film with the meat protects the product against freezer burn.

3. Candy

Single-, double-, and triple-ply films are utilized in a fabricated bag form to package all types of candy (Figure 19).   The highly transparent films give picture window display to the packaged product.   The high barrier property protects the candy against moisture loss or gain and retains the flavor and aroma by maintaining the essential oils and flavorings.

Fig. 18   Chub packages of freshly ground beef packaged in Saran Wrap.

4. Cheese

Cheese is overwrapped by machine or hand (heat-sealed) with single- and double-ply films in roll form.   These films are used as a curing block wrapper and for wrapping all types of consumer size cuts of natural and sliced processed cheese (Figure 20).   Both plain and printed films are available.   Processed cheese is also packaged in link-shaped packages on the Kartridg Pak chub machine, a form, fill, and seal unit. This machine utilizes electronic sealing for the side seams and aluminum wire clips for the end seals.   Fabricated bags made from high-shrink film are used to package irregularly shaped cheeses such as gouda, longhorn, and half moon.   The cheese item is placed in the high-shrink bag, a vacuum is applied to eliminate the air from the package, the top of the bag is twisted, and a metal clip seal is affixed as the final closure. The wrapped item is then dipped rather quickly in a hot-water tank (the

Fig. 19   Candy wraps of Saran.

film shrinkage is instantaneous) or passed through a steam or hot-air shrink tunnel.   Again the transparency achieves a fresh "just wrapped" appearance.   The low water-vapor transmission rate prevents weight loss and the low gas transmission rate protects the product from oxygen degradation.

5. Cookie Dough

Roll stock in double-ply thicknesses is used to package a variety of ready-to-bake products on the Kartridg Pak form, fill, and seal machines in chub form (Figure 21).   The prime requisites are protection from moisture, flavor loss, and oxidation.

6. Processed Fruits

Roll stock in double-ply thickness is utilized on the Kartridg Pak form, fill, and seal machines to package processed fruits in chub form. The cooked and heated product is packaged hot and further heat processed after the packages have been made (Figure 22).   Gas and moisture barriers as well as chemical inertness are features of this package.

Fig. 20   Cheese packaging is one of the largest uses for Saran films.

7. *Laminations*

Single-ply film is laminated to board stock, die-cut, and used as a cap liner for protection against flavor or moisture loss, oxidation, chemical action, or loss of essential oils in food and nonfood items such as salad dressings, mustard, catsup, wine, whiskey, shampoos, dentifrice, cream deodorants, hair oils, and various chemical compounds (Figure 14).

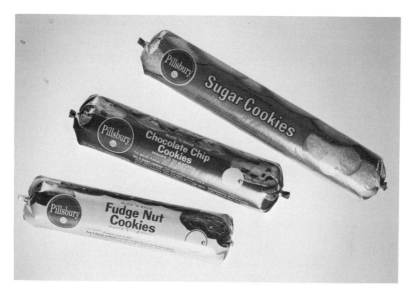

Fig. 21   Cookie dough packaged, ready for baking.

Fig. 22   A hot-processed product packaged in Saran.   Such products
have usually required a metal container.

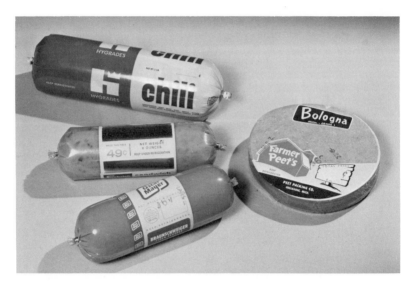

Fig. 23   Packages of processed meat products.

8. Processed Meats

Single-ply films are used for both overwrapping on Wrap King machines and hand wrapping of all kinds of sliced luncheon meat (heat-sealed packages).   Many pumpable products—liver sausage, bologna, sandwich spread, etc—are packaged on the Kartridg Pak form, fill, and seal machines (electronic seal) with double-ply films (Figure 23). Smoked hams, ham butts, and ring bologna are packaged in fabricated bags made of high-shrink films with the vacuum, metal clip, and shrink process.   Moisture and flavor loss, protection against oxidation, and color retention are prime reasons for the use of this film.

9. Poultry

Fabricated bags made from high-shrink films (Saran Wrap-S and Cryovac S brands) are used to package turkeys, ducks, and other poultry (Figure 24).   The process is similar to that of packaging irregularly shaped cheese.   After the film bag has been shrunk tightly around the item, the package is quickly frozen.   The poultry item is held at subfreezing temperatures during storage, shipment, and market-ing.   Both plain and printed bags are available.

Fig. 24   Poultry wrapped in shrink packages of vinylidene chloride
copolymers.

K. Industrial Uses Other Than Food

1. Protective Pipeline Overwrap

Single-ply film in the 200-gauge thickness protects buried pipelines from corrosion by soil acids, bacteria, and oxidation [3].   The film is applied to the pipe in the over-the-trench method with a specially formulated microcrystalline wax as the adhesive (Figure 25).   The wax is applied hot (365°F) to the unprimed pipe surface by a special coating and wrapping machine which wraps the film helically around the wax-coated pipe.   The hot wax shrinks the film and firmly bonds the film to the pipe [4].   An 80-lb Kraft paper is wrapped helically over the film which acts as a shield to protect the film from abrasion during placement in the trench and during backfilling of the soil.   Pipelines from 4 to 30 in. in diameter have been installed with this technique.

Fig. 25   Wrapping an oil pipeline with a resistant polyvinylidene
chloride film.

A survey after 10 years of underground service indicates that these
film-wax protected pipe-lines have stood the test of time [5].   Physical
property determinations were made on film specimens removed from
these buried lines and compared with the unexposed film.   The tensile
strength, percentage elongation, dielectric strength, and water-vapor
transmission rates were found virtually unchanged, as were current
requirements for cathodic protection.

2. Thin-Wall Insulation

One major appliance maker, Hotpoint, has used a lamination called
"Wonderwall" as the thermal insulator for a line of refrigerators.
"Wonderwall" was made up of Saran Wrap/polyester film/Kraft paper
and was used as an insulating envelope, filled with glass fiber, Freon, and
sealed.   The resulting package was an insulator equivalent to polyurethane foam [6].

3. Christmas Tree Icicles

Single-ply film of a preshrunk grade is coated very thinly with pure aluminum metal by the vacuum deposition process. The metallized film is then slit into the required narrow widths (3/32 in.). Saran Wrap, the pioneer in this market, was introduced in 1957 as the first plastic icicle. The soft drape and easily metallizable surface of Saran Wrap coupled with its self-extinguishing property makes it ideally suited to this purpose.

L. INTERNATIONAL ASPECTS

1. Film Manufacturers

Vinylidene chloride-vinyl chloride copolymer films are now being produced by a number of companies outside the United States.

Company	Location	Brand
Asahi-Dow Ltd.	Japan	Saran Film
W. R. Grace & Co.	Canada	Cryovac
W. R. Grace & Co.	Europe	Cryovac
Dow Chemical Europe S.A.	Europe	Saran Film
Kureha Chemical Industry Co.	Japan	Kurehalon

2. Canadian and European Markets

The dominant uses for Saran films in Canada and Europe are in the packaging of processed meats and cheese.

A unique use for the film in a number of European countries is in packaging sliced bread. Single-wound film in the 60-to-75-gauge thickness is used to overwrap several slices. The packages are then heat-sterilized to shrink and seal the film. The resulting package has several weeks of moldfree shelf life.

3. Far East Markets

In Japan the prime markets for these films are in packaging fish sausage, fish cakes, bean curd, ground beef, and processed meats. Packaging fish sausage is by far the greatest. This product is wrapped with double-wound films either semi-automatically in prefabricated tubes or automatically on form, fill, and seal machines. The Kartridg Pak or the Omori chub machines are being used.

M. FUTURE

Hermetically sealed Saran film packages can withstand the temperatures and pressures of autoclaving. Acceptable food packages have been made with the radiation sterilization process. As packaging technology advances and improves, the polyvinylidene chloride-vinyl chloride copolymer films appear in many new and unique packaging applications.

References

1. R. C. Reinhardt, *Ind. Eng. Chem.*, **35**, 422 (1943).
2. P. J. Meeks, *Modern Packaging*, **36**, No. 6, 130 (February 1963).
3. Ted Kennedy, Jr., *Corrosion*, **13**, No. 3, 56 (March 1957).
4. Ted Kennedy (to the Trenton Corp., Ann Arbor, Mich.) U.S. Patent 2,713,383 (July 19, 1955).
5. A. A. Brouwer, *Materials Protection*, **1**, No. 6, 16 (June 1962).
6. *Chem. Eng. News*, **43**, No. 30, 80 (July 26, 1965).

CHAPTER 7

CELLOPHANE

H. H. SINEATH and WALTER R. PAVELCHEK

Film Operations, American Viscose Division, FMC Corporation,
Marcus Hook, Pennsylvania

I. Introduction

A. Cellophane Defined

The name *Cellophane* was coined from the words cellulose and dia-phane by J. E. Brandenberger in 1911 [1] to apply to regenerated cellulose film obtained by acid coagulation and regeneration of cellulose xanthate (see Volume I, Chapter 4-B). Use of the name or term, however, was never restricted to the regenerated cellulose itself. Initi-ally, cellophane referred to the film composite of regenerated cellulose containing a minor amount, say 10 to 20%, of an absorbed polyol plasticizer or *softener* such as glycerol, an amount of absorbed moisture determined by equilibrium with the surrounding atmosphere, and an optional surface dusting of a powder to provide surface slip. When film-forming coatings were later added to the original cellophane, no distinction was made, so that now the name is given to all such regenerated cellulose films, whether they have been dyed, colored, primed for coating anchorage, coated continuously on one or both surfaces, or laminated—as long as all treatments are carried out by the original manufacturer. Indeed, even those regenerated cellulose films that have been extrusion-coated with polyethylene by the cellophane manufacturer are called cellophane, whereas a similar product prepared by a converter is customarily called a "coated cellophane."

Cellophane is still a registered trademark in many countries but not in the United States, where a 1936 U.S. court decision against E. I. duPont de Nemours & Company, declared the name to have become generic by usage [2]. Other recognized trademarks for cellophane throughout the world are Avisco, Cellglass, Clar-Apel, Heliozell, New-Wrap, Rayophane, Sidac, Sylphrap, Viscacelle, Transparit, and Cellophan [2, 3].

B. History

Present-day cellophane manufacture can trace its beginning to the discovery patented in England in 1892 by Cross, Bevan, and Beadle [4] that cellulose could be solubilized in a caustic solution by reaction with carbon disulfide, although E. Weston [5] had made regenerated cellulose 10 years earlier by the denitration of nitrocellulose. Cross also prepared laboratory samples of film [6].

The next historical milepost was the issuance of a British patent in 1898 to C. H. Stearn [7] which covered a process for making threads or sheets by coagulating the solution of cellulose, called viscose, in a solution of ammonium chloride. By 1904 M. J. Teillard's experimentation had led to the development of a process for preparing a glass substitute by spreading a layer of viscose on an inert substrate for support during coagulation and regeneration treatments [8]. Then in 1908 the results of work by J. E. Brandenberger began to appear in the public record.

Brandenberger's experiments began while he was an employee at the Blanchisserie et Teinture les Voges in Alsace, a French textile firm. The experiments were designed to form a regenerated cellulose film within the structure of a woven cloth to provide a high luster. Viscose was sprayed onto the cloth which was then immersed in acid solutions to regenerate the cellulose [2, 9, 10]. The cloth became smooth and glossy but, unfortunately, also stiff and brittle. Therefore he next directed his efforts toward first making a film of regenerated cellulose and then applying it to the cloth [11]. Brandenberger could not obtain satisfactory adherence of the film to the cloth but in the process designed a machine for producing cellophane for which he obtained German, French, British, and U.S. patents by 1912 [9, 10, 12–19].

The earliest cellophane prepared by Brandenberger on a machine built in 1908 was 8 to 10 mils (0.008 to 0.010 in.) thick and quite brittle. By 1912 he had succeeded in preparing more flexible cellophane as thin as 0.8 mil and a use in packaging had been developed for it [11, 20]. However, the entire output of this machine was commandeered by the French government during World War I to make surgical dressings and gas-mask eyepieces [2].

In 1913 La Cellophane Société Anonyme was formed in Paris [20]. With the financial backing of Comptoir des Textiles Artificiels, the first plant built in France specifically for the manufacture of cellophane was constructed in Bezons in 1920 [8]. This plant used a Brandenberger process machine which was capable of operation at 160 ft/min [21].

In 1923 the duPont Company obtained a license for the United States rights to the Brandenberger process from La Cellophane and started construction of the first cellophane plant in the western hemisphere at Buffalo, New York. Plain sheeting made for the first time in the spring of 1924 was sold as a luxury packaging material at $2.65/lb. Approximately 500,000 lb were sold [8]. At the same time interest in cellophane manufacture spread quickly. In 1925 manufacture was started by Wolff and Company in Bomlitz near Walsrode, Germany, with a drum-casting technique rather than the conventional Brandenberger method used by Kalle and Company, which in the same year also started production in Germany at Wiesbaden-Biebrich [3]. A group of technicians from La Cellophane also developed and patented a modified process in Belgium. They formed La Société Industrielle de la Cellulose (Sidac), built a plant, and started manufacture [20].

Sales in the United States rose to 2,000,000 lb in 1927 [22] when the first cellophane with a moistureproof coating developed by Charch and Prindle of duPont [22, 23] was sold commercially, thereby opening new markets such as wrapping baked goods [24]. For this development Charch was awarded the Schoellkopf Medal of the Buffalo Section of the American Chemical Society in 1932 [25].

In 1929 Feldmühle became the third German cellophane manufacturer at Odermünde [26], Wolff and Kalle formed a sales organization, and Transparit was founded at Wiesbaden [3].

The second DuPont plant, built in 1929 at Old Hickory, Tennessee, was expanded in the following year to twice its original capacity when the third DuPont plant was built at Richmond, Virginia [27].

Sylvania Industrial Corporation became the second United States manufacturer in 1930 with four cellophane spinning machines making "Sylphrap" at Fredericksburg, Virginia, under license from Sidac [2]. In the same year Langheck and Company started production of regenerated cellulose films and sheets at Wurtemberg, Germany, with one machine [3].

The next major development was the introduction of heat-sealing coatings by Sylvania in 1931 [28, 29], thereby greatly increasing the potential uses of cellophane so that annual United States sales reached 34 million lb in 1932 [24], the year duPont of Canada started production at Shawinigan, Quebec [30]. Nineteen thirty-one was also the year in which Courtaulds, Ltd., began manufacture and marketing of "Viscacelle" in competition with the English sales organization of La Cellophane [31]. J. P. Bemberg, A.-G., built one machine at Wuppertal-Oberbaumen in 1933, a second in 1933, and a third in 1937. Meanwhile, in 1934 additional markets were opened by the introduction of anchored coated cellophanes [24].

Courtaulds and La Cellophane joined forces in 1935 by setting up a joint subsidiary, British Cellophane, Ltd., which began manufacture of "Cellophane" at Bridgwater, Somerset, in 1937 [31].

Sales in the United States in 1937 were estimated at 70 million lb [32]. Between 1924 and 1939 the listed selling price of cellophane in the United States had been reduced 21 times, from $2.65 to $0.40/lb, but raw materials and wage rates had increased 118 and 227%, respectively, attesting to the increased operating efficiencies accomplished by the industry [24]. This was further borne out by the publication in 1940 of an estimate that the minimum economic plant size had reached 35 million lb/yr [33]. By 1942 total U.S. sales had exceeded 100 million lb/yr.

Before World War II cellophane production had reportedly been started in Poland, Holland, and Japan [20]. During the war, shortages of raw materials in Germany seriously curtailed production and machinery maintenance. These factors probably figured in cessation of operations by Langheck in 1942 [3]. In 1945 total German production capacity was reportedly increased by 2.4 million lb/yr to 15,336,000 lb/yr (Kalle—8,736,000; Wolff—2,640,000; and Feldmühle—3,960,000) [3, 34].

Postwar shortages due to rapidly increasing demands, particularly in the United States, initiated considerable expansion of production facilities during the following 13 years. American Viscose Corporation's absorption of Sylvania in 1946 was the forerunner of major additions to the Fredericksburg plant, starting in 1949, which increased its capacity to 100 million lb/yr and made it the largest cellophane plant in in the world [28, 35]. Also in 1946, duPont began a study of West Coast locations for a new plant and was considering locating it near Springfield, Oregon [36]. This—including the one under construction at Clinton, Iowa, which began production the following year [24]— would have been the fifth duPont plant. However, with the expiration of the moistureproof coating patents, the Anti-Trust Division of the Justice Department filed suit against duPont, alleging restraint of trade in cellophane manufacture and marketing so that duPont then undertook to set up a new competitor. This was accomplished by licensing Olin Industries, whose subsidiary, Ecusta Paper Co., built a plant at Pisgah Forest, North Carolina. This plant with its nine machines was rated at a 33-million-lb/yr capacity and began production in 1951 [35]. Meanwhile polymer-coated cellophane had been introduced commercially in 1949, thereby further increasing the potential applications for cellophanes [24].

Between 1939 and 1952 the price of cellophane had been forced up about 35% by major increases in labor and raw-materials costs, but annual poundage sales tripled. During the same period, selling prices of paper-based wrapping materials almost doubled [8].

Post-World War II cellophane-production capacity was increased around the world with new facilities in Sweden, Spain, Italy, Egypt, Czechoslovakia, and Brazil [20, 37]. At the same time total U.S. installed capacity was further increased to 570 million lb/yr by addition of a second 40-million-lb/yr Olin plant, at Olin, Indiana, near Danville, Illinois, in 1956, doubling the Olin Danville plant to 80 million lb/yr in the mid-1950's, by the addition of the second American Viscose plant rated at 50 million lb/yr at Marcus Hook, Pennsylvania, in 1958, and by the addition of a fifth duPont plant at Tecumseh, Kansas, in 1958 [24, 28, 38–40].

World production of cellophane for 1965 has been estimated in excess of 1 billion lb (1,080,000,000 lb). Europe and the United States accounted for about three quarters of a billion pounds, Canada produced about 45 million lb, Japan about 141 million, Central and South America about 50 million, and the remaining 62 million lb came from all other installations. With the generally increased rate of industrial and sociological development throughout the world, and the attendant increase in packaging, cellophane utilization should increase even more.

Annual cellophane sales in the United States have shown some decline since 1958 (Figure 1), to about 70 % of original installed capacity. Since the closing of duPont's Old Hickory plant in 1964, annual sales now represent about 85 % of the original installed capacity. This level-off of sales is attributable in part to the rapidly increasing use of plastic

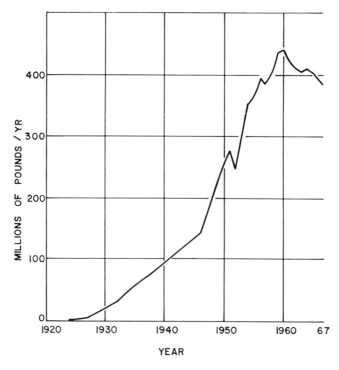

Fig. 1   Cellophane sales in the United States from 1922 to 1968 [24, 35, 166, 167].

films in markets formerly served entirely by cellophane. Another significant cause has been the introduction of higher yield cellophanes, starting about 1960, which provide an increased number of wraps per pound or the same number of wraps using fewer pounds of film. The steady stream of new and improved types of cellophane by the three U.S. manufacturers has been notably effective in increasing the number of applications and markets for cellophane rapidly enough to compensate for these factors and to achieve some degree of sales stability poundage-wise.

II. Present Commercial Status and Uses

Cellophane has been called the granddaddy of transparent flexible films and is worthy to bear this name because it has been in the forefront of transparent packaging materials since its inception in 1925, when only a few hundred thousand pounds were sold in the United States. One of the major reasons for this status has been the fact that cellophane is not one product, but a complex product line of more than 120 different types. Each type has been tailored to do a specific job, with the only thing in common being general appearance and a cellulosic substrate. For example, films developed for packaging fresh red meat are ineffective for packages requiring aroma retention. Conversely, films that do an outstanding job on aroma retention would turn meat dark in a few hours.

Of the more than 385 million lb of cellophane sold in the United States in 1967, about 90% was consumed for packaging food and tobacco. The approximate breakdown into the various categories of use is shown in Figure 2. Various recommended uses for cellophane for specific packaging purposes, including a partial list of approximate competing grades by the three major manufacturers in this country, are given in Table 1. The complexity of codes and types of cellophane manufactured in the United States is readily evident.

Until a few years ago cellophane types produced in the United States were coded according to the European practice of designating types by their weights in grams per square meter. Type 300 weighed 30 g/m^2; type 450 weighed 45 g/m^2, etc. Recently, this was changed to a more useful code system in which the first three numerals of the code, as shown in Table 2, represent coverage yield in square inches per pound.

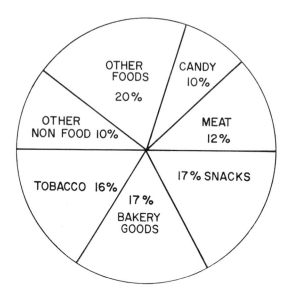

Fig. 2   Uses of cellophane in the United States.

Prices per pound for most cellophanes range from 62 to 79 cents, al-though several types have higher per pound prices.   The cost of cello-phanes per 1000 in.^{2} ranges from 2.8 to 8.2 cents, depending on type and yield.

Cellophane is sold in rolls and sheets of varying widths to meet specific packaging requirements in end-use applications.   This is re-quired by the various package sizes and packaging machines on which the material must be run in order to provide the customer with the packaged products.

Cellophane is distributed in the United States through complex distribution channels, including direct sales to customers by the manu-facturer, sales through distributors, and sales to converters who perform other operations on cellophane before it is used or before it is ready for use on the finished product [24].

Distributor sales are arranged by individual distributors who can quickly provide service to the customer from their warehouse stock. They generally supply users whose requirements are relatively small.

TABLE 1

Cellophane Grade Classifications

Designation[a]			Characteristics	Typical uses
Avisco	Du Pont	Olin		

Standard Performance

MOISTUREPROOF, HEAT-SEALING, NITROCELLULOSE-COATED

Avisco	Du Pont	Olin	Characteristics	Typical uses
MS 1	MSD 51	MST 51	Good general properties	Bread and some sweet doughs; carton overwraps, small products
MS 2	MSD 52	MST 52	Strong heat seal; extra rigidity	Direct wrap for baked goods; carton overwraps; salty, oily products
MS 3, 100	MSD 53	MST 53	Extra strong heat seal; extra toughness and durability, even at low temperatures	Bags, for which strong heat seal and durability are required
MS 400	MSD 54	MST 54, 44	Good durability and flexibility; extra strong heat seal	Carton overwraps and bags for greasy or oily products; laminations
MS 8	MSD 58	MST 58	Good general properties plus good heat seal	Wrap on cigars, cut tobacco, cigarettes, where heat seal is desired
MS-12	MSD 58T	MST 66	Improved rigidity for highspeed performance	Direct wrap on cigars
MS 10	MSD 60	MST 60	Improved dimensional stability	Bread, cake, sweet doughs; carton wraps
MSB 6	MSAD 86	MSAT 86	Superior strength; highly water-resistant	Wet products; bags for high humidity
MSB 7	MSAD 87	MSAT 87	Similar to types 6, 86, but for low-temperature use	Wraps on moist products and frozen foods

MOISTUREPROOF, NONHEAT-SEALING, NITROCELLULOSE-COATED

Avisco	Du Pont	Olin	Characteristics	Typical uses
M 1, 2	MD 31, 32	MT 31	Excellent moistureproofness, solvent-sealing	Types 1, 31—cigarettes; types 2, 32—cigars, cigar tubes; types 5, 35—twist wraps
M 5	MD 35	MT 35		

377

TABLE 1—continued

Avisco	Du Pont	Olin	Characteristics	Typical uses
M 6	MD 36	MT 33	Excellent moistureproofness; superior dead fold	Twist wrap, candies; direct wrap, caramels
M–300	MD 200		Superior moistureproofness	Cigarettes
MBO	MAD 10		Moistureproof non-heatsealing for polyethylene extrusion coating and adhesive laminations	Packaging food and pharmaceutical products

(Designation[a]: Avisco, Du Pont, Olin)

INTERMEDIATE MOISTURE PROTECTION, NITROCELLULOSE-COATED

Avisco	Du Pont	Olin	Characteristics	Typical uses
DS	LSD	LST	Decreased moistureproofness; good heat seal	Glazed doughnuts, hard crusted breads and some dry products
DSB	LSAD	LSAT	Decreased moistureproofness; strong heat seal; highly water-resistant	Wet produce; processed meats, items needing limited moisture protection
MSBO, AMO	MSAD 80	MSAT 80 OF 16	Moistureproof coating one side; strong, flexible, good wet strength; coated sides seal strongly	Fresh meat (uncoated side applied directly to meat)
MSBO 10	MSAD 10	MSAT 10	Moistureproof; water-resistant; coated on one side only; coated sides heat seal strongly	For lamination and extrusion coating

Premium Performance and Special Types

SARAN-TYPE COATING

Avisco	Du Pont	Olin	Characteristics	Typical uses
RS 7	K 207, 208, 307	V 3, 5	Optimum durability with excellent machining characteristics; improved release from seal jaws	Preformed bags and form-and-fill packaging of cookies, candies, snacks, etc.

TABLE 1—*continued*

	Designation[a]			
Avisco	Du Pont	Olin	Characteristics	Typical uses
RS-4	K 204	V-4	Saran-coated cellophanes designed for strong bonds	For extrusion coatings and laminations
RS 99			Specially formulated for use on push-feed packaging machines	Direct wrap on small cakes, overwraps on other baked goods
RSO			One-side-coated, uncoated side treated for improved anchorage	Direct and overwrap applications; printing; polyethylene extrusion and adhesive laminations
T-79			Two light-gauge cellophane sheets laminated with 1/4 mil of polyethylene and coated on both sides with polyvinylidene chloride; superior durability	Special applications in carton and direct overwrap, especially for snack items
POLYETHYLENE COATING—ONE SIDE				
REO	MSAD 90	OF 18	Superior durability and tensile strength; highly punctureproof; permits extremely tight contour wrapping	Fresh meat (uncoated side applied directly to meat); industrial applications
VINYL COPOLYMER COATINGS				
R-18			Good moistureproofness and gasproofness; strong heat seal; superior dimensional stability	Bread, sweet doughs, cookies, pretzels and candy
ANTIOXIDANT INGREDIENTS				
		OF 20	Nitrocellulose coating with antioxidant combinations	Potato chips, nut meats, cookies

TABLE 1—continued

	Designation[a]		Characteristics	Typical uses
Avisco	Du Pont	Olin		
UNCOATED				
P-1	PD		Protective wrap when moisture protection and heat seal are not required; air, dust, grease, and nonboring-insectproof	Wax products, roll leaf applications
P-7			Specially designed curl-resistant carrier film for the roll-leaf industry and release film for the low-pressure molding industry	Roll-leaf and low-pressure molding applications
PL			Gloss controlled	Low pressure mold release for decorative panels, etc.

[a] Cellophane designations

Type of Film	Avisco	Du Pont	Olin
Anchored water-resistant coating	B	A	A
Antioxidant coating			OF–20
Polyvinylidene chloride coating	RS	K	V and OX
Polyethylene coating	RE	90	OF–18
Intermediate moistureproofness	D	L	L
Moistureproof	M	M	M
Heat-sealing	S	S	S
Vinyl coating	R–18		
Uncoated	P, PL	PD	

380

TABLE 2

Cellophane Yield and Thickness

Typical Gauges	Square inches per pound	Thickness (inches)
250	25,000	0.0008
220	22,000	0.0009
210	21,000	0.0009
195	19,500	0.0010
180	18,000	0.0011
140	14,000	0.0014
116	11,600	0.0017

Between the manufacture of cellophane and its final use many operations may be performed on it by a converter. Converters are a major force in cellophane distribution channels (25 to 30% of total sales). They modify the product by extrusion, by hot melt, solvent or aqueous coating, and by combining cellophane with itself and other films to form an almost unlimited variety of materials, tailor-made for specific uses. One of the most important functions of the converter is printing. Printing on cellophane enables the maker of the packaged product to provide a package for his material that is informative and appealing to the eye. Printing can be done by a variety of methods including the flexographic or rotogravure techniques. Cellophane can be direct- or reverse-printed, depending on the application. Reverse printing leaves a glossy surface on the outside of the package for enhanced appearance.

Direct sales can be defined as those shipments that go from the manufacturing plant directly to the end user without subsequent processing. These shipments represent a significant share of the total poundage of cellophane shipped (60 to 65%) and can take the form of rolls or cut-to-size sheets. Cellophanes in roll form are used for a multiplicity of overwrap applications on automatic packaging equipment. Cut-to-size sheets are used for a variety of hand or semiautomatic operations.

With continuing increases in basic costs of packaging, cellophane users have been forced to higher efficiency in packaging operations. This has placed increasing demands on cellophane for ever-higher speeds of operation on a vast array of packaging machines. Over-wrapping operations at 200 to 400 wraps per minute are common today. Demands for extended shelf life of the packaged product have resulted in the development of new cellophane materials and new methods of packaging. Recently the developments of cellophane-cellophane laminations and cellophane-plastic film laminations have evolved rapidly. Such composite structures combine the advantages of one film with the advantages of another. Viewing the packaging scene against the backdrop of technological advances, it appears that composite structures of various types will find an ever-increasing application in the world of packaging. As a result it should be expected that an increasing portion of the cellophane sold in the United States will be distributed through the converting industry.

III. Manufacturing

A. OVER-ALL REQUIREMENTS

The entire process for manufacturing cellophane is illustrated in Figure 3. The three major operations within the over-all process are viscose preparation, film casting, and coating. These operation are discussed in detail in the following sections, but first let us review the requirements of the process and its position in the broader industrial picture. The most extensive analysis of requirements for cellophane manufacture was published in 1951 by McIndoe [35]. Monbiot et al. [3] gave a comprehensive description in 1946 of German operations before and during World War II, but it must be assumed that significant changes have been made since then and it is quite probable that requirements for operation on a larger scale, as in the United States, are somewhat different. Data published by Clunan [24] in 1960 are limited but, being relatively current, provide some basis for estimating current uses. By drawing primarily on these sources raw materials, utilities, and manpower requirements for a cellophane plant with a 50-million-lb/yr capacity have been estimated and are shown in Table 3.

As these estimates indicate, the cellophane industry provides multimillion-pound markets for the pulp and chemical industries, employs large numbers of people, and uses large amounts of utilities. In addition, cellophane manufacturers are large-volume chemical producers, for they convert most of the sodium hydroxide, carbon disulfide, sulfuric acid, and sodium sulfide to sodium sulfate, for which there has been an available market [41].

B. By-products and Wastes

The major waste material from the viscose preparation process is the hemicellulose removed from spent caustic. This material, which contains a small amount of caustic, is generally run to a settling basin before going to the sewer [8, 41]. Major reductions in the amount of this material have been effected by significant reductions in the amount of hemicellulose in the wood pulps used. Also, the effective use of a dialysis process to recover caustic from the spent steeping liquor has reduced the amount of caustic in the waste hemicellulose to a negligible level.

As already noted, sodium sulfate produced during the coagulation and regeneration steps is recovered and sold as a by-product. In this process essentially all of the unreacted sulfuric acid is recovered for reuse, thereby effecting a major process economy as well as avoiding a serious waste-disposal problem. Losses to the sewer from other treatment baths in the wet processing section of the casting machine are minimized by recirculation and replacement of consumed chemicals whenever possible. Also, wiping techniques have been developed to minimize carry-over of liquid on the film from one bath to the next. The wash water and tank overflow that does find its way into the sewer has a very low level of chemical wastes. Other techniques such as the use of a sodium sulfate solution in regenerating water softeners and sequential back flushing of viscose filters have also helped reduce chemical losses to the sewer.

Recirculation and make-up addition to the plasticizer bath is also practiced. In addition, "spent" plasticizer is recovered for reuse by evaporation, thereby eliminating what formerly constituted a major addition to biological oxygen demand on cellophane-plant liquid waste streams [41].

Carbon disulfide and hydrogen sulfide fumes generated during the viscose preparation and coagulation-regeneration processes are carefully collected, greatly diluted, and dispersed into the atmosphere [8]. Processes have been proposed for the recovery and reuse of carbon disulfide fumes, but it has not been revealed whether this is actually being done commercially.

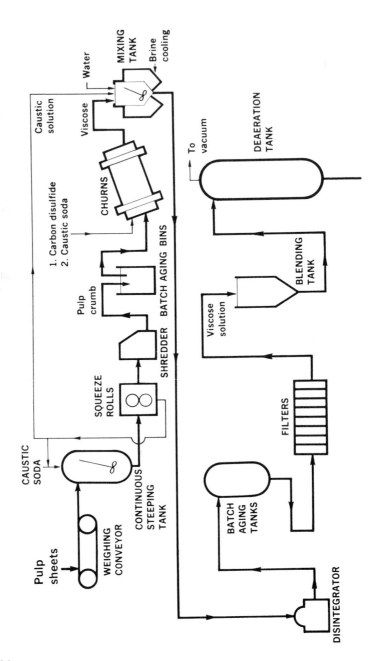

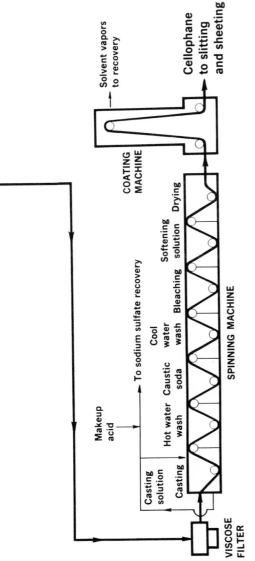

Fig. 3 Flow diagrams of a cellophane manufacturing plant.

TABLE 3

Estimated Requirements for Producing 50,000,000 Pounds of Coated
Cellophane a Year

CHEMICALS	
Wood pulp, MM lb/yr	47.5
Plasticizers, MM lb/yr	7.7–9.1
Coating components, MM lb/yr	4–6
Sodium hydroxide, MM lb/yr	35–60
Carbon disulfide, MM lb/yr	15.5
Sulfuric acid, MM lb/yr	50–70
Sodium sulfide, MM lb/yr	5–8
Chlorine, MM lb/yr	0.09–1.1
(dyes and colorants: 2–10% of film weight when used)	
UTILITIES	
Electricity, kWh	15,000
Water, MM gal/hr	1.2–2.8
Refrigeration, tons	850,000
Steam, lb/hr	250,000–400,000
MANPOWER	
Man-hours, MM	2–5

Use of recovery processes for coating solvents effectively keeps losses of solvent vapors to an extremely low level. Recovered solvent mixtures may be reused or shipped to a company specializing in their further rectification and recovery. Also, solvents that do not tend to form unwanted by-products during processing, which would create unusual recovery or disposal problems [41], are chosen for use.

Dry cellophane waste from any part of the operation is usually burned in a company-owned incinerator. It may, however, be treated to recover the plasticizer content, dried, chipped (flaked), and sold for various uses.

Wet cellophane, partially coagulated cellulose, and waste viscose are not easily combustible and are therefore usually disposed of as land fill in an area owned and maintained by the company [41]. This is also done with solid or semisolid residues from the plasticizer and solvent recovery operations.

C. VISCOSE PREPARATION

The wood pulp, caustic soda, and carbon disulfide raw materials required for cellophane manufacture via the viscose process are the same as those used in the manufacture of viscose rayon with one notable difference. Wood pulps for cellophane viscose are treated to give a highly reactive alkali cellulose which permits the minimum use of carbon disulfide and caustic in the viscose, yet gives a maximum filter press throughput [42, 43]. Pulp viscosity is also reduced below that for rayon pulp to reduce aging times. This is permitted, since cellophane is manufactured at lower degrees of polymerization (DP) levels than rayon (cf. Volume I, Chapter 4A).

In earlier years the manufacture of viscose was a batch process involving steeping the pulp in caustic soda, pressing it to a specific wetness-dryness ratio, then shredding it to produce what is called an alkali cellulose crumb. The shredders were discharged into aging vessels ("cans") which were then stored at controlled temperature and humidity for specific aging times. After aging, the alkali crumb was discharged into a baratte where carbon disulfide was introduced for xanthation. After completion of xanthation, this material was discharged into dissolving equipment and suspended in caustic soda by agitation.

In recent years much progress has been made in the automation of viscose manufacture in various cellophane plants. Batch processes have given way to what could be called batch continuous processes, since part of the processes is batch and part is continuous. Steeping and shredding operations have been replaced by processing in continuous steeping and shredding units, such as slurry steepers. Aging cans are being replaced by continuous aging devices or towers [44, 45]. Barattes are being replaced with continuous churns which permit xanthation and mixing to be carried out in a single vessel. Also, continuous xanthation processes have been suggested [46, 47] and are practiced in at least two installations, using moving belts.

Pulp is received for cellophane manufacture in the form of bales of sheets or continuous rolls. Regardless of the form received, the material is fed continuously or semi-continuously into the steeping process. During the steeping process the hemicelluloses in the pulp are removed with the steeping soda. A balance of hemi content and steeping soda is reached, depending on the draw-off rate of the "hemibleed caustic," which is subsequently used in the dissolving of the viscose or processing on a casting machine.

After the pulp is processed to its proper swollen condition it is pressed to a predescribed or predetermined wetness-dryness ratio, shredded, if the process is batch, and discharged into the aging cans or towers. The alkaline cellulose in the presence of sodium hydroxide undergoes chemical modifications by rearrangement in a process called aging. This is really a depolymerization of the cellulose molecule, and the control of this process contributes significantly to the properties of the finished viscose.

After aging to the desired degree the alkaline cellulose interacts with carbon disulfide to form cellulose xanthate. The amount of carbon disulfide added in this xanthate reaction and the conditions under which the process takes place significantly affect the quality of the viscose and the regenerated cellulose product made from it (cf. Volume I, Chapter 4-B, C). Suffice it to say that sulfur is added to the cellulose molecule in a manner that puts it in a state such that it can be suspended in caustic. Consequently viscose is a colloidal suspension of cellulose xanthate in sodium hydroxide solution, which by proper coagulation at a later stage in the process can be made into a sheet of cellophane.

From a consideration of cellulose and cellulose xanthate chemistry it is obvious that one can make a variety of viscoses from these raw materials, depending on the concentrations of materials and the time-temperature relationships used. The ripening of viscose (the aging process), which takes place normally somewhere during the deaeration step or between mixing and deaeration, is carried to a somewhat lower level of xanthate substitutions than for rayon-type viscoses, thus producing a cellophane of high clarity and gloss. However, if the level of xanthate substitution is permitted to reach too low a point, the resulting film will be weak in the wet state and difficult to process. Between these two limits there is a broad area of operation for making a satisfactory film with the properties required.

After the xanthated crumb is mixed with caustic soda to form a viscose by any of several dissolving methods, it is necessary to remove air or gases included in the viscose by a process called deaeration. During deaeration, whether it is done batchwise or continuously, under vacuum or pressure, the salt index of the viscose changes, thus indicating a change in D.P.*   Viscose is held in the various stages of this process

---

* The salt index is a measure of the precipitability of the cellulose, which in turn, depends on the degree of polymerization.   A salt index of 2, for instance, means that a 2% solution of sodium chloride will just cause precipitation of the cellulose xanthate (whereas it will not precipitate in salt solutions of less than 2%). Indexes between 1 and 3 are used in the industry, depending upon the character and internal composition of the viscose prepared and the intended use of the regenerated product.

until it reaches the correct salt index for spinning. It is then given a final filtration and transferred to the casting machine for casting into a sheet of regenerated cellulose.

D. Coagulation and Regeneration

There are several systems in use for pumping and filtering the viscose, each of which has been carefully designed to prevent pulsation or vibration as the material is extruded from the hopper or nozzle. The process by which viscose is converted into a cellulose film or sheet has been referred to as casting or spinning. It would be more logical to refer to this as extrusion, since the material is actually extruded from a die into a coagulating bath. The term casting, however, has been used to designate all of the steps carried out on a cellophane machine before coating (Fig. 3): extrusion, coagulation, regeneration, desulfurizing, bleaching, plasticizing, resin impregnation, drying, conditioning, and winding.

The extrusion die used is often referred to as a "hopper" or "nozzle" in the industry (Fig. 4). Its primary purpose is to extrude the viscose into the coagulating bath and control film thickness. Since the final thickness of the sheet must be controlled by nozzle adjustments within a few percent, that is, a few hundred thousandths of an inch, very fine and high precision extruders are required. There are many designs of extruders in the cellophane industry [48]. Generally, they are composed of a body with a hollow interior designed with divergent paths to give uniform pressure behind the opening, which is fitted with adjustable die lips. Adjustments are achieved by several mechanical means, one of which is a series of differential screws mounted 4 to 6 in. apart across the entire length of the nozzle. By varying the adjustments of these screws the nozzle lip is deflected to control the finished sheet. The nozzle is made of or coated with corrosion-resistant materials [49–51], for during the process of coagulation in continuous operation the buildup of material eventually deteriorates the lip of the nozzle so that it requires periodic rehoning to a very fine finish. This buildup also occurs at other places in the machine and is one of the primary factors for shutting it down periodically for cleanup. Along with film breaks, this is one of the factors limiting continuity of operation of a cellophane-casting machine. Great care is taken with the design, construction, and choice of materials of construction of the viscose extruder, since it is one of the

Fig. 4   Casting machine hopper. The viscose issues from the die at bottom in a thin stream, the thickness of which can be adjusted from point to point by the closely spaced bolts shown.

most critical mechanical elements of the process for producing regenerated cellulose film of salable quality.

When viscose is formed into a continuous gel sheet, it is said to have been coagulated.   Regeneration refers to the decomposition of cellulose xanthate by acid to form regenerated or pure cellulose.   These processes occur simultaneously in a casting process, for the viscose is extruded directly into a bath containing sulfate salts and sulfuric acid—the salt concentration being about 18%, the acid concentration about 12%. Bath temperatures are generally held between 40 and 55°C.   Many other bath systems also have been described for the manufacture of cellophane.   One of these is a two-bath system that uses aqueous ammonium sulfate followed by an acid sulfate regeneration bath [52]. Another system uses sodium phosphate and phosphoric acid as a regeneration medium [53].   These two systems accomplish the steps of coagulation and regeneration consecutively, and have been particularly effective in producing clear films of regenerated cellulose from viscoses with a higher than normal degree of substitution. Also, other methods of coagulation have been advanced; for example, coagulation by heat [54]. Further details are given in Volume 1, Chapter 4B and C.

Over the years, many attempts have been made to translate coagulation-regeneration technology from the rayon process to that of cellophane. This has not been very successful; for example, baths containing zinc and other viscose additives or modifiers which have been most effective in the production of high tenacity rayons produce cellophanes with unacceptable transparency.

Several mechanical techniques for the extrusion of viscose or for handling a regenerated cellulose sheet during the coagulation-regeneration phase have been presented. Among them are the so-called belt-casting and flume-casting processes and a process for the extrusion of viscose onto a continuous rotating drum [21, 55–59].

It is also possible to produce two sheets simultaneously on the same machine. This is referred to as double casting. The methods used by the various manufacturers are a function of their individual technology and the economics of the section of the world in which they are operating.

E. Further Wet Processing

Breakdown of the xanthate during coagulation and regeneration liberates carbon disulfide and hydrogen sulfide as by-products of the process. Thus the sheet emerging from a coagulating-regenerating bath is a white, opaque gel sheet of cellophane that contains minute particles or bubbles of carbon disulfide and entrapped hydrogen sulfide. This sheet then travels on to the purification section of the casting machine in which these and any other impurities are removed. In a series of 11 to 20 tanks the wet film is further treated to produce a clear film. These treatments bleach, purify, and plasticize the film. Anchorage-promoting and coloring treatments may also be carried out.

Wet-treatment tanks are rectangular in shape and are approximately 2 ft wider than the widest film (4 to 8 ft) processed and up to 8 ft long. Tank depth is 3 to 4 ft. Temperature control of the bath fluids is provided by internal or external (circulating steam) heaters or by submerged steam spargers [26]. Those tanks that hold sulfuric acid solutions are usually made of lead. Those that contain caustic are usually of iron.   Bleaching-solution tanks may be rubber lined.   Other materials may be used for the tanks or for linings, depending on the corrosion resistances required, but there has been little information published on this subject in recent years.

Each tank has a number of driven rollers mounted across the top and a similar number of freely turning or fixed rollers mounted inside near the bottom.   The latter are submerged in the bath liquid and may be adjustable to center film travel [34].   With the trend to wider machines, it has been suggested that a footbridge be installed along the length of the machine to give operator access to the center [60].

In operation the film is festooned from top to bottom rolls throughout the length of the tank the number of times required to obtain the desired contact times with the solutions at the speed of film travel being used (Fig. 5).   When two sheets are being processed simultaneously, this roll arrangement permits occasional separation of the sheets to release gas trapped between them or to adjust their relative lateral positions [61, 62].

Various materials, including hard rubber, porcelain and glass [3, 34, 63, 64], have been used for these rollers.   If one tank provides insufficient contact time, the next one or two tanks, as required, may contain the same solution flowing from tank to tank [26, 65].   Many modifications of these arrangements have been suggested, including internal partitioning of the tanks [2].   As the film emerges from a bath, it carries with it layers of bath liquid on each surface.   To prevent carryover of this liquid and contamination of the following bath, wipers or squeeze rolls are used between tanks [66–68].

The first of 1 to 5 tanks following the regeneration treatment contains water up to 70°C [37].   The remaining tanks are generally not heated [3, 26].   This treatment serves to drive off the carbon disulfide and hydrogen sulfide, to wash off the sulfuric acid on the film surface, and to wash out sulfuric acid and sodium sulfate dissolved in the film [2, 8].

Following this washing, the film may enter a desulfurizing bath which has been described variously as "a hot (approximately 90°C) dilute sodium sulfide-sodium hydroxide solution," [2] "segregated steeping caustic," [8] "NaOH 60 g/liter . . . 50°C," [26] "a sodium hydroxide bath (3–4%) . . . 70 to 80°C," [37] "NaOH, Na_2SO_3," [11] and "boiling soda," [6] where the remaining sulfur compounds are removed. As this happens, the cellophane loses its opacity and attains its familiar transparency.   The film then passes through two or three wash-water

Fig. 5 Close view of cellophane base sheet passing through the wet end of the casting machine. Shown are several hard rubber rolls in the washing section and a septum dividing two tanks. The film makes several passes vertically through the tank. Note liquid drag on the film at this high speed.

tanks, the first one or two of which may be heated to 70 to 80°C [8, 26, 37]. In some localities it is necessary to soften the wash water before use.

All tanks at the hopper end of the machine, up to and including at least the desulfurizing bath, are hooded to collect and exhaust the liberated carbon disulfide and hydrogen sulfide fumes [8]. These fumes are generally diluted to 25 ppm or less in the process. The hoods and exhaust system are so instrumented and interlocked that the rate of exhaust is increased sufficiently whenever any hood is raised to ensure that operating personnel are not exposed to any injurious concentration of gases. Here, as in the viscose preparation areas, air samples are regularly taken and analyzed to make certain that any errant fumes are being adequately exhausted.

For final decolorizing and purification the film is next given a one-bath bleaching treatment with an unheated 0.9 to 1.2% solution of sodium hypochlorite or peroxide [3, 37, 69]. If the solution contains sufficient iron to give a yellowish cast to the film, 1% oxalic acid may be added to remove and precipitate the iron as basic iron oxalate. The bleaching treatment is followed by water washing in four or five tanks to remove all traces of the bleach and products formed in the bleaching process [2, 8, 37].

Removal of the bleach residual may be speeded up and made more complete by sequential treatment after the first bath with a dilute solution of an antichlor material such as thiosulfate, $Na_2S_2O_3$, dilute hydrochloric acid to remove the antichlor, a water wash, a dilute soda solution to neutralize residual hydrochloric acid, and further washing [26]. Although this technique appears to be rather involved, it has the purpose of replacing chlorine (which is quite difficult to remove and could cause film embrittlement during drying) in a stepwise fashion with more readily removable and less injurious materials.

Although techniques for coloring cellophane by the addition of dyes to the coagulating bath or the mixing of dyes or pigments in the viscose have been proposed [70], it is probable that coloring is most often carried out after (or in place of) bleaching and before the plasticizing step [2]. The quantities of colored cellophane produced are relatively small and the number of different colors made is rather large, so that the many materials and conditions used could not be conveniently covered here even if they had all been adequately described in the literature. The most complete published descriptions are for a Feldmühle process in Germany [26] and the Celosul process in Brazil [37]. Generally, these processes involve an antichlor treatment to remove residual bleach completely and quickly, treatment with a dye solution, a chemical treatment to fix the color in the cellophane, and one or more washing steps.

The purified wet gel cellophane is now ready for what is customarily the final treatment before drying: plasticization. Although a great many materials have been suggested as plasticizers [3, 71–88] the commonly used materials in the United States appear to be glycerol and propylene glycol, either separately or in combination. Glycerol, propylene glycol, ethylene glycol, triethylene glycol, and urea are common in Europe and Japan, but the last three materials named have not been used in the United States since 1961.

The concentration of plasticizer in the softener bath is regulated to give the desired final concentration in the film. It has been shown that movement of glycerol and other glycol plasticizers into the film is a concentration equalization process between the bath solution and the water in the film [89, 90]. Therefore, to obtain 100 lb of "dry" uncoated film with a plasticizer content of 15% and a moisture content of 6%, 15 lb of plasticizer must migrate into a minimum of 237 lb of water contained in 79 lb of regenerated cellulose at this point. The plasticizer solution concentration required then is slightly greater than 6.3%, that is, $(15/237) \times 100$. If the gel cellophane at this point contains more or less than this amount of water, or a different final concentration of plasticizer is desired in the film, the plasticizer bath concentration must be adjusted accordingly.

Other treatments may be given to the cellophane simultaneously in the plasticizer baths, immediately before drying, or in a separate step after drying [2, 26, 91–97]. Such treatments are for the promotion of surface slip properties, flameproofing, and improvement of the anchorage of subsequent coatings. For slip promotion finely divided or dispersed inorganic materials, such as silicon dioxide and talcum, are most frequently used, although many others have been proposed and may be in use [3, 98]. Flameproofing is accomplished by the addition of amine salts, derivatives of phosphoric acid, or ammonium salts to the films by inclusion in the plasticizer bath [2, 8, 99–101]. Ammonium sulfamate appears to be the most commonly recommended flameproofing material.

A large number of materials has been suggested for use in improving the anchorage of subsequent coatings to the "raw" cellophane or "base film." (Both terms are commonly used to describe the uncoated cellophane as it leaves the end of the casting/drying machine.) Indeed, the magnitude of work being done in this area suggests strongly that significant changes and improvements in coatings adhesion will soon be apparent [2, 96, 97, 102–125]. At present, it appears that urea-formaldehyde, melamine-formaldehyde, and polyethyleneimine are most commonly used for anchorage promotion and that these treatments are made in the plasticizing baths.

F. DRYING OF UNCOATED CELLOPHANE

The wet gel cellophane leaving the last wet processing tank contains at least 3 lb of water per pound of regenerated cellulose in addition to its contents of plasticizers, anchoring agents, colorants, and slip agents [2]. To reduce the moisture content to 6 to 7% in approximately one minute the film is passed over a series of steam- or hot-water-heated driven cylinders with hot air blowing onto or across the surfaces of the film. The critical effects of drying conditions on the final properties of the film have been stressed by Kirk and Othmer [2], Brandenberger [126–128], Morgan [100], Alles and Edwards [129, 130], Herndon [131], Charch [132], Eberlin [133], Stevens [134], and Von Hoessle and Shade [135]. Also, a variety of equipment arrangements and process control schemes has been proposed with the purposes of improving the efficiency of the drying process and improving the uniformity and dimensional stability of the dried cellophane when it is subsequently exposed to varying humidity and temperature conditions. From this variety of proposals it must be assumed that the various manufacturers do not use identical processes or equipment.

The wet film may be dried initially at a rapid rate by exposure to gas-jet flames or radiant heating to reduce or eliminate transfer of materials, primarily anchoring and slip agents, to the rolls [2, 132, 133].

Because the film is primarily water at this point, rapid evaporation maintains the film temperature low enough to prevent significant plasticizer evaporation. During the remainder of the drying cycle roll and air temperatures are approximately 70°C [26] but may be lowered as the end of the drying process is approached [2]. It has also been suggested that the film be carried through a drying-rewetting cycle before leaving the dryer to obtain improved control of the final moisture content, improved appearance, and reduced shrinkage [128, 136, 137].

The dryer usually consists of two horizontal series of hollow drums, 1 to $1\frac{1}{2}$ ft in diameter, staggered to nestle together [2, 126, 127, 129], although a vertical arrangement has been proposed [135] with the objective of improving drying efficiency. The total number of drums has been reported to be 50 to 60 for film speeds up to 200 fpm [3, 26]. In the arrangement proposed by Brandenberger [126–129] each roll in the upper series rests between two rolls in the lower series and is driven by them. (Only one of the series is driven.) With this arrangement the cellophane is completely supported throughout the drying period to give maximum shrinkage restraint. This arrangement, however, presents definite hazards to both operators and equipment and some dryers in use have a separation between the series, with all rolls being driven.

For further control of the drying process a number of dryer-roll constructions [34, 132, 138–141] and air supply duct arrangements [142–145] have been proposed, and it is customary with United States manufacturers to enclose the dryer totally, including the air distribution system [8, 24].

As water is removed from the film, large shrinkage forces develop. These forces are resisted in the direction parallel to the machine (machine direction) by the tension used to pull the film along. Restraint in the across-machine (transverse) direction is effected by friction between the film and the dryer drums. Control of the shrinkage is obtained by appropriate choices of surface finishes on the dryer drums [131, 146] and by dividing the dryer drive into sections so that the speed in each section can be varied independently to stretch or relax the film slightly between sections. Various stretch-relax patterns, with accompanying specific drying patterns, have been proposed to improve appearance, shrinkage uniformity, flatness, and dimensional stability [130, 134, 136, 137, 147–153]. A technique has been described for some adjustable control of transverse shrinkage by raising and lowering selected dryer drums [154]. Also it has been stated that some separation of sheets is required during the drying of two film webs simultaneously to prevent lamination [155]. In general, cellophane is characterized by having a greater tensile strength and lower tear strength in the machine direction with greater elongation and greater dimensional change with moisture-content change in the transverse direction. Also properties of the film at the edges are somewhat different than in the rest of the film. Most of the techniques proposed for film handling during wet processing and drying therefore have been designed to minimize these differences to obtain a closer approach to a balance of properties in both directions (cf. Volume I, Chapter 4-B).

The dried film leaving the dryer is wound onto metal cores into rolls up to 5 ft wide, weighing as much as 1000 lb, and containing more than 5 miles of film (Fig. 6). These rolls of film are then transported to an area in which humidity and temperature are closely controlled for storage before coating or slitting (Fig. 7).

1. Controls

Three characteristics of the spinning process make extensive automatic controls necessary.

Fig. 6   Controls at the windup of the casting machine. Note the three-station mandrel which permits changing of rolls without stopping the machine.

1. Highly reactive chemical solutions and high temperatures are used which require complete enclosure of certain sections of the machine such as the coagulation and regeneration tanks and the dryer.

2. Large quantities of chemicals are in the machine in-process inventory at all times—a single wet process tank may contain 300 to 500 gal of a solution whose concentration must be closely controlled.

3. Reaction times for each treatment are short. With the film travelling at speeds of 165 to 650 fpm [21] along a path approximately three times the length of the machine [6] (150 ft to "nearly as long as a football field" [28]), the time from hopper to windup during which the many chemical reactions and physical processes described must be carried out accurately is no more than 3 min.

The most extensive descriptions of an automatic control system for cellophane casting and drying to have been published describe the

Fig. 7   Mill rolls of coated cellophane in lag storage.   From here they go to the finishing room where they are slit to customer specifications, inspected, and rewound for shipment.

control center installed in the 50-million-lb/yr plant built by American Viscose (now a Division of FMC Corporation) in 1957–1958. This installation (Fig. 8) was reported at the time to be the "largest and most precise control center of its type in any industrial plant outside the atomic energy field," with necessary control actions being made automatically for more than 95% of the variables monitored [28, 156].

Fig. 8   Control panel of the fully automated cellophane plant of American Viscose Division of FMC Corp. at Marcus Hook.

In addition, several control tests, such as composition analyses on the coagulation, regeneration, desulfurizing, bleaching, and plasticizing baths, bacterial count in the plasticizer baths, and both the composition and physical properties of the raw cellophane, are carried out on an around-the-clock basis.   Tests routinely run on raw cellophane include pH, thickness uniformity, weight per unit area, clarity, moisture content, and plasticizer content.

G. Coating Operations

Over the years cellophane has been an exceptionally versatile product because it combines a tough flexible, transparent substrate (not appreciably affected by heat or solvents) with numerous coatings, each tailored to do a specific packaging job. The base sheet contributes grease resistance and a low permeability to gases and vapors. The coatings contribute a wide variety of other properties to the finished product, including moistureproofness, oxygen permeability control, waterproofness, and machinability and receptiveness to printing inks. The four most important characteristics are moistureproofness, sealability, machinability, and printing receptivity.

Since moistureproof coating compositions for cellophane first appeared, the formulation of coatings has been an applied Edisonian type of development or research work. From such activities highly skilled formulators emerged, who, working more by art than science, developed the final intricate and specific properties desired in the various coatings. In recent years statistical techniques and computers have been added to the arsenal of tools useful in this area of technology.

Coatings for cellophane are applied from solvent solutions or emulsions, or by extrusion. Typical examples are nitrocellulose and vinylidene chloride coatings which are applied from solvent solutions, vinylidene chloride and vinyl-type coatings applied from emulsions, and polyethylene, polypropylene, or polyvinyl chloride coatings applied by extrusion.

1. Solvent Coating

Generally speaking, coatings applied to cellophane from solvent solutions are made up of a film former, a blending resin, a moistureproofing agent, and plasticizers. Added to these basic materials are trace amounts of slip and antiblocking agents such as clay and silica, heat sealer release agents, antistats, and other materials that improve machinability for specific applications. These materials are dissolved in solvent systems which are generally mixtures of low- and high-boiling organic solvents to form what is referred to as lacquer. By adjusting the relative amounts of the film former and the various other formulation ingredients it is possible to obtain different levels of moisture impermeability, heat sealability, and antisticking desired in various final product applications [157–162]. These lacquers are polyphase, heterogeneous systems that require specific handling conditions before they are applied to the substrate. The materials are usually blended, dissolved in the solvent under exact temperature conditions, and transported to the coating machine at constant temperature.

The over-all process of coating cellophane can be defined as coating application, doctoring (smoothing of the applied lacquer), drying (solvent evaporation), and rehumidification. A subsequent surface treatment may also be applied to obtain specific (e.g. antistatic) machinability factors.

The individual elements of a typical coating machine are an unwind stand, a prehumidifier, a coating solution applicator, doctoring rolls, a drying chamber, a rehumidification chamber, and a windup stand. The unwind and the windup stands are usually of the turret variety that permit tagging on of new rolls of film entering the coater and removal of coated rolls without stopping the machine (Fig. 9).

Several types of mechanical arrangement are used to apply lacquers. Most widely employed is a dip trough through which the film passes before passing between a pair of squeeze rolls which removes excess coating. Several variations of this specific setup are used, depending on the type of lacquer being applied to the substrate, but basically the same process and same equipment is required, whether it be a vertical or horizontal configuration [163]. In order to ensure uniformity of coating distribution, the lacquer solids, solvent compositions, lacquer temperature and viscosity, and squeeze roll setting have to be very critically controlled.

After the application of the lacquer it is doctored just before the film passes into the dryer, where very accurate control of temperature, humidity, solvent vapor concentrations, and air-flow rates are required to dry the material satisfactorily and remove the solvent. On leaving the drying chamber the film proceeds into a rehumidification zone in which it is exposed to solvent-free air highly loaded with moisture. After rehumidification the sheet is cooled by passing over a series of chilled rolls and then wound into a coated mill roll.

Solvent vaporized from the lacquer coating in the drying zone is diluted by a large excess of air. Enough of this solvent-air mixture is continually drawn off to maintain the solvent-air mixture considerably below the lower explosive limit and piped to a solvent recovery area in which it is passed through carbon bed absorbers to remove and retain the solvent vapors. When sufficiently loaded, the adsorbers are steamed out and the mixture of steam and released solvent vapors is fed to a recovery system in which separation and recovery of the solvents are accomplished by using conventional distillation and extraction processes. These recovered solvents are subsequently reused in fresh lacquer

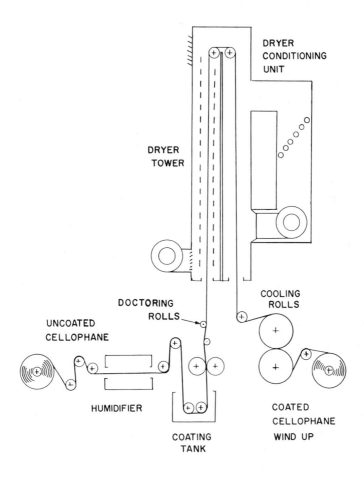

DRYER
CONDITIONING
UNIT

DRYER
TOWER

COOLING
ROLLS

DOCTORING
ROLLS

UNCOATED
CELLOPHANE

HUMIDIFIER

COATING
TANK

COATED
CELLOPHANE
WIND UP

COATER FOR CELLOPHANE

Fig. 9   Coater for cellophane, schematic, not to scale.   The dryer tower
may be 40 ft or more in height.

makeup. Since solvent recovery efficiency is a critical economic factor in the successful coating of cellophane, it is necessary that the entire coating process be highly instrumented.

2. Emulsion Coating

Since many polymers useful for coating cellophane can be prepared by emulsion polymerization, the emulsion coating process presents some potential advantages in not requiring purification of the polymer and subsequent dissolving in a solvent, thereby eliminating the costs of solvent recovery. There are, however, several process and product complications associated with emulsion-coating techniques. Heat requirements for removal of the water from emulsions are considerably greater and therefore more costly than for solvent evaporation from lacquer coatings. When high solids contents are used to reduce this difference, the coating may set up too rapidly to permit satisfactory smoothing or doctoring. Also the rapid penetration of moisture into the base sheet from the emulsion can cause distortion of the sheet during the coating process. In order to prepare an effective moisture barrier coating from an emulsion, it is necessary that the individual particles of the emulsion coalesce into a monolithic coating after removal of the water. Various means of minimizing these difficulties are described in detail in the literature [164, 165]. Elaborate instrumentation and controls are required, however, to effect a satisfactory emulsion coating of cellophane. Minor variations in residence time of the base sheet in the various unit processes of the coating operation can drastically affect the finished properties and the finished product performance of the individual materials coated.

Emulsion coatings can be applied by one of several coating techniques such as reverse-roll coating, air-knife coating, and gravure coating. In each of these processes the material is metered onto the substrate and the substrate is immediately introduced into a drying chamber where drying is effected by hot air or a combination of hot air and infrared heat. In some processes the initial set-up of coating and surface finish is achieved by coating against a highly polished, steam-heated, chromium-plated drum. After drying the film can be rehumidified, if necessary, cooled, and wound in the conventional fashion.

Emulsion coating for cellophane has not found wide use in the United States, although some specific products with unique characteristics that are highly desirable for certain market applications, particularly when antifogging characteristics are desired, have been manufactured. Emulsion-coating processes are used somewhat more extensively in European operations and the exceptionally high barrier coatings introduced in recent years have found wide application, particularly for products packaged or distributed in the tropics.

3. Extrusion Coating

Polyolefins are currently the most widely used extrusion coating materials. These coatings are generally moistureproof, but have considerably more permeability to gases, particularly oxygen, than other polymer coatings. Because of these properties, cellophane coated on one side with polyethylene is particularly useful for packaging fresh red meat. Also, in recent years converter-produced cellophanes coated on either one or two sides with a polyolefin have found application in packages requiring high durability, particularly under low-temperature conditions. This is an example of the combination of the properties of cellophane with those of plastic materials to produce materials of outstanding packaging characteristics. The high temperatures needed for extrusion, however, greatly limit the choice of formulation additives that can be used to tailor extrusion coatings for other specific end-use requirements.

The heart of the extrusion process is the melt extruder. The compound that is to be used as coating is introduced into the hopper (Fig. 10) and subsequently heated while it is being transported down the barrel of the extruder and into the die by an internal screw. Material is melted not only by the heat from the outside of the heated barrel but also by heat generated by the mechanical work of the screw on the polymer. The die is essentially a very accurately controlled narrow slit which is connected to the extruder by an adapter. The die is heated in sections which permits careful regulation of temperatures across the entire width of the opening. From the die the material is extruded onto the substrate in a nip between a chill roll and a back-up roll. The low temperature of the chill roll solidifies and cools the material which then passes to a surface treating unit for application of sizing and other materials for slip or antistatic control. The final product is then wound into mill rolls on conventional winding equipment.

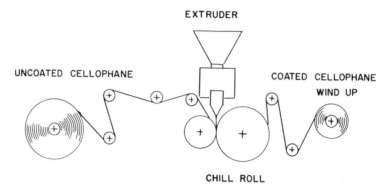

EXTRUSION COATING

Fig. 10   Typical arrangement for extrusion coating of cellophane.

IV. Final Handling

After coating, the mill rolls are transported to the final handling or finishing operation. For distribution purposes it is frequently convenient to retain a coated mill roll inventory. When this is done, the film is usually stored under controlled atmospheric conditions (normally at 75°F, 75% relative humidity). The finishing operation includes slitting or sheeting, final inspection, packaging, and shipping.

Over the years slitting equipment with elaborate controlled instrumentation has been developed for cellophane slitting (Figure 11). Several companies, both in the United States and abroad, have specialized in this area of technology and the film producers themselves have contributed heavily. Equipment is commercially available which operates in excess of 1000 fpm and provides a means for slitting the mill rolls into several widths and configurations simultaneously. In the United States films are sold in roll form from 8 to 30 in. outside diameter and in widths of $\frac{1}{4}$ to 83 in. in increments of $\frac{1}{16}$ in. Generally the 8- to 18-in. diameter rolls are wound on 3 in. diameter cores, and the 18- to 30-in. rolls on 6-in. diameter cores. During the slitting operation, the rolls are usually given a final inspection to meet exacting quality standards.

Several markets exist for cellophane sheets. In the sheeting operation the film is cut to size, inspected and packaged. Sheets are offered

Fig. 11 Slitting a mill roll of coated cellophane to customer requirements. The mill roll is at the rear of the machine; windup and controls are shown. (Photograph, courtesy of John Dusenbery Co., and American Viscose Division.)

in many sizes and shapes from $1\frac{1}{4} \times 1\frac{1}{4}$ in. to 40×50 in. in $\frac{1}{16}$ in. increments. Sheeting is usually done with guillotine or rotating knife cutters. These machines are standard equipment and are commercially available.

The packaging of cellophane is done with great care to ensure that the product will come out of the package in good condition for use by the customer. A myriad of packaging designs has evolved to meet specific requirements of the roll or sheet size, product type and customer-handling or shipping requirements. Some of the packaging is still done by hand, but more and more practical automatic or semiautomatic equipment has evolved.

Whatever the package design, shape, or size, cellophane producers the world over have learned that careful attention to packaging extends their quality control to the customer's door.

H. H. SINEATH AND WALTER R. PAVELCHEK

References

1. "Cellophane," *Encyclopedia Britannica*, **5**, 141 (1963).
2. L. L. Leach, "Cellophane," in *Encyclopedia of Chemical Technology*, R. E. Kirk and D. F. Othmer, Eds., **Vol. 3**, 280–292 (1949).
3. M. F. Monbiot et al., "Investigation of the German Regenerated Cellulose Film Industry," B.I.O.S. Rept. 858.
4. C. F. Cross, E. J. Bevan, and C. Beadle, British Patent 8700 (1892) and 4,713 (1896).
5. E. Weston, U.S. Patent 264,987 (1882).
6. J. E. Brandenberger "Notes on Cellophane," *J. Franklin Inst.*, **226**, 797 (1938).
7. C. H. Stearn, British Patent 1020 (Jan. 13, 1898); 1022 (Dec. 3, 1898).
8. G. C. Inskeep and P. Van Horn, *Ind. Eng. Chem.*, **44**, 2511 (1952).
9. J. E. Brandenberger, British Patent 15,190 (June 29, 1909).
10. J. E. Brandenberger, German Patent 231,265 (Nov. 24, 1908).
11. Williams L. Hyden, *Ind. Eng. Chem.*, **21**, 405 (1929).
12. J. E. Brandenberger, French Patent 414,518 (June 22, 1910).
13. J. E. Brandenberger, U.S. Patent 991,267 (May 2, 1911).
14. J. E. Brandenberger, British Patent 20,119 (Sept. 11, 1911).
15. J. E. Brandenberger, French Patent 434,104 (Nov. 16, 1911).
16. J. E. Brandenberger, French Patent 436,188 (Jan. 17, 1912).
17. J. E. Brandenberger, French Patent 438,774 (March 23, 1912).
18. J. E. Brandenberger, French Patent 438,775 (March 23, 1912).
19. J. E. Brandenberger, French Patent 438,776 (March 23, 1912).
20. C. M. Rosser, *Colloid Chemistry*, Reinhold, New York, **1**, 641, 1950.
21. J. Voss, *Svensk Papperstid.*, **64**, 863 (1961).
22. W. H. Charch and K. E. Prindle, U.S. Patent 1,737,187 (Nov. 26, 1929).
23. W. H. Charch and K. E. Prindle, U.S. Patent 1,826,696 (Oct. 6, 1931).
24. A. B. Clunan, *Good Packaging*, 96 (July **1960**).
25. R. N. Shreve, *Selected Process Industries*, McGraw-Hill, New York, 1950, 656–663.
26. B. G. Milov, *Bumazhn. Prom.*, No. 1, 41, (**1948**).
27. Williams Haynes, *American Chemical Industry*, Van Nostrand, New York, Vol. IV, 1948,351.
28. J. G. Mohlman, *Paper, Film & Foil Converter* (April, May, 1960).
29. Ref. 27, Vol. VI, 417.
30. Anon., "DuPont of Canada," *Chem. Ind.*, 1055 (**1958**).
31. Anon., "British Cellophane Ltd.," *Chem. Ind.*, 784 (**1938**).
32. *Pac. Pulp and Paper Ind.*, **12**, 63, Rev. No. 5 (May 1938).
33. Lawrence P. Lessing, "The World of Du Pont," *Fortune*, 93–94, 160–166 (October, 1940).
34. L. H. Smith, *Synthetic Fiber Developments in Germany*, Textile Research Institute, 1946, pp. 828–839.
35. W. C. McIndoe, "Cellophane Industry in the West," *The Paper Industry*, **33**, 780 (October 1951).
36. "Du Pont Cellophane Plant for West (near Eugene, Oregon)," *Pulp & Paper* **21** (6), 32 (May 1947).

37. A. Secondi and G. Calvino, "Celosul," PL., 40 (1949).
38. Anon., "Cellophane Meets Competition," *Chem. Eng. News*, **37**, 28, (Jan. 19, 1959).
39. R. A. Labine, "Controls Wrap Up Cellophane," *Chem. Eng.*, **67**, 104 (March 7, 1960).
40. Anon., "At Avisco the Emphasis is on Empathy," *Paper, Film & Foil Converter* (March **1964**).
41. H. L. Jacobs, *Sewage Ind. Wastes*, **25**, 296 (1953).
42. R. L. Mitchell, *Ind. Eng. Chem.*, **47**, 2370 (1955).
43. J. P. Beaudry, *Paper Trade J.*, **140**, No. 35, 24 (1956).
44. S. E. Seaman and R. P. King, U.S. Patent 2,490,097 (Dec. 6, 1949).
45. J. O. Smith and W. R. Weigham, U.S. Patent 2,805,924 (Sept. 10, 1957).
46. T. Dokkum, U.S. Patent 2,122,519 (July 5, 1938).
47. F. G. C. Klein, French Patent 696,411 (1930).
48. C. E. Coleman, U.S. Patent 2,387,718 (Oct. 30, 1945).
49. O. S. Petrescu, U.S. Patent 2,056,982 (Oct. 13, 1936).
50. American Viscose Corporation, British Patent 818,445 (Aug. 19, 1959).
51. American Viscose Corporation, British Patent 904,269 (Aug. 29, 1962).
52. H. G. Ingersoll, U.S. Patent 2,991,510 (July 11, 1961).
53. Private communication.
54. C. R. Price and V. C. Haskell, *J. Appl. Polymer Sci.*, **5**, 635 (1961).
55. E. Czapek, U.S. Patent 2,313,520 (March 9, 1943).
56. E. Czapek, U.S. Patent 2,344,603 (March 21, 1944).
57. B. L. Hinkle and F. C. Stults, U.S. Patent 2,962,766 (Dec. 6, 1960).
58. W. Bender, U.S. Patent 2,254,203 (Sept. 2, 1941).
59. J. A. Mitchell, U.S. Patent 3,073,733 (Jan. 15, 1963).
60. E. Bleibler, U.S. Patent 2,264,357 (Dec. 2, 1941).
61. A. E. Craver and W. H. Lamason, U.S. Patent 3,044,115 (July 17, 1962).
62. F. H. Reichel and A. E. Craver, Canadian Patent 550,873 (Dec. 31, 1957).
63. H. E. Roscoe, U.S. Patent 2,925,088 (Feb. 16, 1960).
64. W. H. Stevens, U.S. Patent 2,839,785 (June 24, 1958).
65. F. H. Reichel and A. O. Russell, U.S. Patent 2,286,645 (June 16, 1942).
66. H. McAlpine and H. F. Hoerig, U.S. Patent 2,634,221 (April 7, 1953).
67. H. E. Roscoe, U.S. Patent 2,807,891 (Oct. 1, 1957).
68. R. J. Iverson, U.S. Patent 2,936,468 (May 17, 1960).
69. H. O. Kauffman, U.S. Patent 2,189,378 (Feb. 6, 1940).
70. P. F. C. Sowter and R. Betteridge, U.S. Patent 2,216,793 (Oct. 8, 1940).
71. C. F. Cross, E. J. Bevan, and C. Beadle, U.S. Patent 604,206 (May 17, 1898).
72. O. Herrmann, U.S. Patent 1,979,936 (Nov. 6, 1934).
73. W. F. Underwood, U.S. Patent 2,074,349 (March 23, 1937).
74. W. F. Underwood and H. S. Rothrock, U.S. Patent 2,170,827 (Aug. 29, 1939).
75. F. M. Meigs, U.S. Patents 2,170,828–9 (Aug. 29, 1939).
76. H. S. Rothrock, U.S. Patents 2,170,839–40 (Aug. 29, 1939).
77. J. T. Power and E. G. Almy, U.S. Patent 2,172,406 (Sept. 12, 1939).
78. W. Konig, U.S. Patent 2,268,832 (Jan. 6, 1942).
79. W. Gellendien and J. Eggert, U.S. Patent 2,273,636 (Feb. 17, 1942).
80. W. L. Morgan, U.S. Patent 2,288,413 (June 30, 1942).

81. W. Gellendien and J. Eggert, U.S. Patent 2,312,708 (March 2, 1943).
82. H. S. Rothrock, U.S. Patent 2,328,679 (Sept. 7, 1943).
83. R. T. K. Cornwell, C. M. Rosser, and J. A. Yourtee, U.S. Patent 2,346,417 (April 11, 1944).
84. W. A. Hoffman and R. S. Schreiber, U.S. Patent 2,378,479 (June 19, 1945).
85. W. F. Gresham, U.S. Patent 2,438,909 (Apr. 6, 1948).
86. A. Akune, Japanese Patent 1390 (April 2, 1953); *Chem. Abstr.*, **48**, 2373e (1954).
87. B. H. Kress, *Am. Dyestuff Reptr.*, **48**, 33 (1959).
88. B. H. Kress, U.S. Patent 2,895,923 (July 21, 1959).
89. O. J. Sweeting, R. Mykolajewycz, E. Wellisch, and R. N. Lewis, *J. Appl. Polymer Sci.*, **1**, 356 (1959).
90. R. Mykolajewycz, E. Wellisch, R. N. Lewis, and O. J. Sweeting, *J. Appl. Polymer Sci.*, **2**, 236 (1959).
91. J. D. Pollard, U.S. Patent 2,206,046 (July 2, 1940).
92. D. E. Drew, U.S. Patent 2,095,129 (Oct. 5, 1937).
93. W. O. Brillhart, U.S. Patent 2,658,843 (Nov. 10, 1953).
94. R. P. Wymbs, U.S. Patent 2,658,835 (Nov. 10, 1953).
95. C. Schlatter, U.S. Patent 2,690,427 (Sept. 28, 1954).
96. W. H. Charch and D. E. Bateman, U.S. Patent 2,159,007 (May 23, 1939).
97. J. D. Pollard, U.S. Patent 2,394,009 (Feb. 5, 1946).
98. R. T. K. Cornwell, U.S. Patent 2,235,516 (March 18, 1941).
99. W. L. Morgan, U.S. Patents 2,071,558-9 (Feb. 23, 1937).
100. W. L. Morgan, U.S. Patents 2,071,353, 354 and 358 (Feb. 23, 1937).
101. M. E. Cupery, U.S. Patent 2,142,116 (Jan. 3, 1939).
102. W. J. Jebens, U.S. Patent 2,280,829 (Apr. 28, 1942).
103. G. Pitzl, U.S. Patent 2,432,542 (Dec. 16, 1947).
104. L. M. Ellis, U.S. Patent 2,523,868 (Sept. 26, 1950).
105. A. F. Chapman, U.S. Patent 2,533,557 (Dec. 12, 1950).
106. W. M. Wooding, U.S. Patent 2,546, 575 (March 27, 1951).
107. R. T. K. Cornwell, U.S. Patent 2,575,443 (Nov. 20, 1951).
108. W. M. Wooding and T. J. Suen, U.S. Patent 2,646,368 (July 21, 1953).
109. Zellstoffwerke, A.-G., German Patent 752,810 (1953).
110. W. Berry, C. R. Oswin, and J. Boyd, U.S. Patent 2,684,919 (July 27, 1954).
111. W. M. Wooding, U.S. Patent 2,688,570 (Sept. 7, 1954).
112. W. M. Wooding, U.S. Patent 2,688,571 (Sept. 7, 1954).
113. M. Morf, U.S. Patent 2,726,171 (Dec. 6, 1955).
114. E. Wellisch, U.S. Patent 2,728,688 (Dec. 27, 1955).
115. G. I. Keim, U.S. Patent 2,864,724 (Dec. 16, 1958).
116. W. M. Wooding, U.S. Patent 2,834,688 (May 13, 1958).
117. W. M. Wooding, Y. Jen, and E. H. Sheers, Canadian Patent 583,443 (Sept. 15, 1959).
118. W. M. Wooding, U.S. Patent 2,987,418 (June 6, 1961).
119. J. L. Justice and C. M. Rosser, U.S. Patent 2,999,782 (Sept. 12, 1961).
120. H. Wilfinger, U.S. Patent 3,009,831 (Nov. 21, 1961).
121. L. Hagan and V. D. Celentano, U.S. Patent 3,011,910 (Dec. 5, 1961).
122. G. H. Lacy and R. R. Chervenak, U.S. Patent 3,033,707 (May 8, 1962).
123. G. I. Klein, U.S. Patent 3,039,889 (June 19, 1962).

124. V. D. Celentano and L. Hagan, U.S. Patent 3,044,897 (July 17, 1962).
125. W. P. Kane, U.S. Patent 3,065,104 (Nov. 20, 1962).
126. J. E. Brandenberger, U.S. Patent 1,002,634 (Sept. 5, 1911).
127. J. E. Brandenberger, U.S. Patent 1,606,824 (Nov. 16, 1926).
128. J. E. Brandenberger, U.S. Patent 1,983,529 (Dec. 11, 1934).
129. D. H. Edwards, U.S. Patent 2,046,553 (July 7, 1936).
130. F. P. Alles and D. H. Edwards, U.S. Patent 2,115,132 (April 26, 1938).
131. L. R. Herndon, U.S. Patent 2,000,079 (May 7, 1935).
132. W. H. Charch, U.S. Patent 2,099,160 (Nov. 16, 1937).
133. W. C. Eberlin, U.S. Patent 2,099,162 (Nov. 16, 1937).
134. W. H. Stevens, German Patent 1,159,630 (Dec. 8, 1955).
135. C. H. Von Hoessle and O. M. Shade, U.S. Patent 1,890,832-3 (Dec. 13, 1932).
136. E. Bleibler, U.S. Patent 1,911,878 (May 30, 1933).
137. F. H. Reichel and A. O. Russell, U.S. Patent 2,698,967 (Jan. 11, 1955).
138. D. H. Edwards, U.S. Patent 2,046,553 (July 7, 1936).
139. H. P. Fry, U.S. Patent 2,792,642 (May 21, 1957).
140. W. H. Guy, U.S. Patent 2,792,643 (May 21, 1957).
141. L. B. Cundiff, U.S. Patent 2,919,904 (Jan. 5, 1960).
142. E. Bleibler, German Patent 569,597 (1933).
143. H. K. Steinfeld, U.S. Patent 2,616,188 (Nov. 4, 1952).
144. W. Buschmann and W. Bühren, U.S. Patent 2,869,246 (Jan. 20, 1959).
145. H. P. Fry, Jr., and L. B. Cundiff, U.S. Patent 2,929,153 (March 22, 1960).
146. S. Chylinski, U.S. Patent, 2,141,377 (Dec. 27, 1938).
147. H. Hampel, U.S. Patent 1,787,520 (Jan. 6, 1931).
148. F. P. Alles and D. H. Edwards, U.S. Patent 2,115,132 (April 26, 1938).
149. W. H. Charch and F. P. Alles, U.S. Patent 2,275,347 (March 3, 1942).
150. W. H. Charch and F. P. Alles, U.S. Patent 2,275,348 (March 3, 1942).
151. W. H. Stevens, U.S. Patent 2,746,166 (May 22, 1956).
152. C. P. Britton, J. T. Eiker, 3d., E. T. Ellis, and R. M. Lester, Jr., U.S. Patent 2,746,167 (May 22, 1956).
153. W. P. Kane, U.S. Patent 3,068,529 (Dec. 18, 1962).
154. A. Sunnen, U.S. Patent 2,798,303 (July 9, 1957).
155. L. Veyret, U.S. Patent 2,623,244 (Dec. 30, 1952).
156. Anon., *Paper Industry*, 324 (August 1959).
157. W. H. Charch, U.S. Patents 2,147,628-9 (Feb. 21, 1939).
158. M. V. Hitt, U.S. Patent 1,997,583 (April 16, 1935).
159. R. T. K. Cornwell, U.S. Patents 1,997,105-6 (April 9, 1935).
160. R. T. K. Cornwell, U.S. Patent 1,989,683 (Feb. 5, 1935).
161. J. A. Mitchell, U.S. Patent 2,079,379 (May 4, 1937).
162. J. A. Mitchell, U.S. Patent 2,213,252 (Sept. 3, 1940).
163. H. G. Rappolt and L. B. Case, U.S. Patent 1,979,346 (Nov. 6, 1934).
164. W. Berry, R. A. Rose, C. H. Phillips, and C. R. Oswin, U.S. Patent 2,910,385 (Oct. 27, 1959).
165. S. J. Wommack, H. A. Kahn, and J. F. E. Keating, U.S. Patent 2,918,393 (Dec. 22, 1959).
166. 1966 *Modern Packaging Encyclopedia*, p. 26.
167. 1968 *Modern Packaging Encyclopedia*, p. 19.

CHAPTER 8

CELLULOSE ESTERS

HUGH W. RICHARDS

Polymer Technology Division, Eastman Kodak Company,
Rochester, New York

I. Introduction

Unlike the many polymeric compounds that are produced by the polymerization of relatively simple low-molecular-weight components, the cellulose esters owe their polymeric nature to that of the basic component, cellulose itself. Cellulose is composed of a variable but extremely large number of anhydroglucose units chemically linked in a hemiacetal configuration. Since each unit has three available hydroxyl groups, the cellulose can assume the alcohol role in the conventional esterification and etherification reactions.

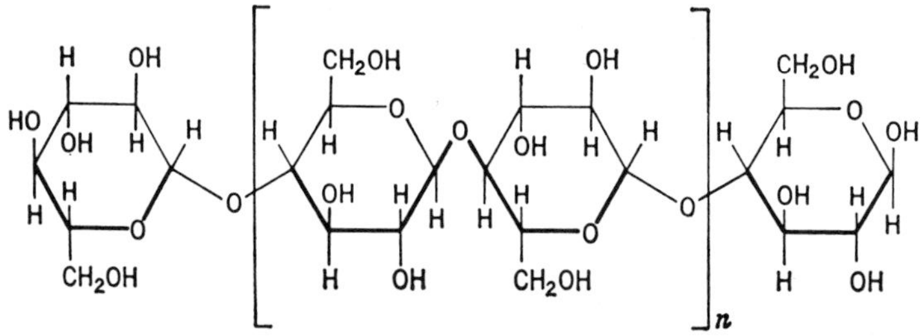

Fig. 1   Conventional representation of cellulose, showing the cellobiose repeat unit.

Reaction with acids produces the polymeric cellulose esters. Ethylcellulose, which is the subject of Chapter 9, is an example of the product of the etherification of cellulose.

By appropriate choice of esterifying agent organic esters, inorganic esters, or a combination thereof can be produced. Mixed esters, those having more than one type of acid group, are common. These possible variations plus the amount of unesterified hydroxyl which may exist in an ester give some suggestion of the many different cellulose esters that can be produced. This potentially large variety can be further augmented by the additional variations related to the degree of polymerization and the nature of the starting cellulose.

The starting cellulose may be a wood pulp or cotton linters. The degree of polymerization is a function of the degree of polymerization of the starting cellulose and the amount of degradation, as chain splitting, which occurs during the manufacture of the ester (cf. Volume I, Chapter 4-A, p. 99 ff.).

These several factors, together with the nature of the substituent or substituents, determine the properties, hence the potential uses of the cellulose esters. Cellulose esters find use in such diverse applications as the preparation of molding compositions, the formulation of surface coatings, and the manufacture of adhesives, explosives, yarn, and structural materials as well as in sheeting and films. The present concern is the use in films, particularly thin films.

The thin-film designation is understood here to include material up to 0.003 in. in thickness (3 mils). In the interest of showing more clearly the place of thin films and to describe the closely related areas, however, this discussion has not been strictly confined to that thickness range.

II. Cellulose Esters in Thin Films

A. BASIC CONSIDERATIONS

Several basic considerations are involved in determining the suitability of a cellulose ester for thin-film manufacture. The ester must be compatible with additives such as plasticizers and stabilizers which may be demanded in a particular end-use. In combination with any required additives it must be soluble in readily available, reasonably priced solvents in order that it may be cast from solution under feasible manufacturing conditions with subsequent removal and recovery of the solvent; otherwise it must have such thermal properties that together with any additives it can be extruded under conditions that are practical from both cost and operational standpoints. The cost of manufacture and of processing also exert a large influence on the position of the ester in relation to competing materials.

The physical and chemical properties inherent in a cellulose ester may be enhanced or modified by the incorporation of specific additives before film manufacture or by surface treatment or surface coating either during or after.

The practical application of any cellulose ester as a thin-film material is dependent on its adaptability to a feasible manufacturing operation and on the properties of the product. These factors are reviewed as they apply to the various cellulose esters that are finding use as thin films.

B. TYPES OF CELLULOSE ESTERS USED AS THIN FILMS

The cellulose acetate esters are those most widely used as raw materials in the thin film field. The closely related mixed ester, cellulose acetate butyrate, often referred to as cellulose butyrate, is also a suitable raw material for thin-film manufacture. Cellulose acetate propionate, usually referred to as cellulose propionate, is suitable for thin-film manufacture but is not generally available in gauges of less than 0.005 in. and for this reason cannot be considered a major thin-film material.

Cellulose nitrate, as a thin-film material, has been replaced by the less combustible aliphatic acid esters and by other polymers.

The alkali metal salts of cellulose esters containing both acyl and sulfate groups may be cast as thin film; however, commercial use of such esters is not significant. A key property encouraging their use is the fact that their salt form induces water solubility in addition to the organic solubility related to their organic nature. They may be cast from organic solvents and used in areas in which water solubility is required.

III. The Organic Acid Esters

A. INTRODUCTION

Here we are considering only the cellulose esters in which the substitution is accomplished by the use of acetyl, propionyl, and butyryl groups. The higher esters, however, have been prepared and characterized. Those esters through the stearate have been so described by Malm, Mench, Kendall, and Hiatt [1]. Cost is a major deterrent to exploitation of the higher esters. Investigation to date is limited in extent and does not involve commercial use as self-supporting films. This does not preclude the possibility of future manufacture to take advantage of their unique properties. Dibasic acid esters are also of potential commercial value in this field. To date their use has been restricted to coating applications rather than to self-supporting films.

Two basic types of cellulose acetate are used in the manufacture of thin films. The ester corresponding to an average acetyl content of approximately 39 % is often referred to as the "diacetate." It is actually intermediate between the diacetate and triacetate. Its use was spurred because it is readily soluble in acetone and therefore conveniently lends itself to casting from solution. The second type, commonly described as the triacetate and closely approaching the true triacetate acetyl level of 44.8 %, has an acetyl level of approximately 43 %. Its more restricted solubility retarded its growth as a thin-film material, but with the development of suitable solvent systems it now offers significant advantages over the lower acetyl material for certain uses.

The manufacture of both types of acetate is substantially alike in the early stages. In both cases the cellulose is fully esterified and then hydrolyzed to the desired acetyl level.

An example of a commercial mixed ester used in the manufacture of thin film is the acetate butyrate ester of 26% butyryl and 21% acetyl content. In the acetate propionate field a typical ester has a 47% propionyl and 6% acetyl content; as noted above the acetate propionate esters are not primarily thin-film materials. The balance in the acyl groups is by no means restricted to the examples described here and wide variations are possible. The choice in acyl balance will be made on the basis of the properties demanded in the end product and compatibility with manufacturing operations.

B. Manufacture of the Ester

Originally cotton linters was the prime source of cellulose for esterification. Specially refined wood pulp has now been established as a suitable alternate or replacement for cotton linters [2]. Sulfuric acid is the one major catalyst used in commercial production. Other catalysts have their places in historical and development work; for example, zinc chloride, perchloric acid, and sulfoacetic acid, among others [3].

Steps of a commercial manufacturing operation for the production of cellulose acetate are activation, esterification, hydrolysis, precipitation, washing, and drying. These steps are described briefly below.

1. Activation

The cellulose, cotton linters or wood pulp, is supplied by the manufacturer in roll or bulk form at approximately 5% moisture content. Sheeted cellulose is dry shredded; in the case of baled linters, the compressed cellulose is broken up into small clumps. It is then treated in the reaction vessel with acetic acid. A portion of the catalyst may be included with the acid. The breakdown of the cellulose into the shorter polymer units required in the final ester is initiated in this activation step. It may be preferable to omit any catalyst in the activation and achieve the viscosity reduction during the esterification, in which case a higher catalyst concentration or more severe esterification conditions may be required.

2. Esterification

Acetic anhydride is added in excess of that sufficient to allow full esterification of the cellulose and also to react with the water present. The mixture is cooled to facilitate temperature control. The major portion, or, as suggested above, the entire amount of sulfuric acid catalyst, is then added. During this step the sulfuric acid rapidly combines with the cellulose and acetyl groups are introduced. This exothermic reaction is allowed to continue with necessary temperature control until the cellulose is fully esterified. At this point the triester is a cellulose acetate sulfate, soluble in the reaction mixture. The solution is then held until it has become completely clear and the desired viscosity is obtained.

3. Hydrolysis

The addition of sufficient water in acetic acid to react with the excess anhydride and to provide excess water marks the end of the esterification and the beginning of the hydrolysis.

A major portion of the combined sulfate is removed at the water addition. The removal of acetyl groups then proceeds as the solution is held at constant temperature. The proper time to terminate the hydrolysis and assure the desired acetyl level is determined by periodic sampling.

4. Precipitation

The reaction solution is then precipitated into dilute acetic acid to obtain a solid suitable for washing. The acetic acid concentration is kept as high as possible to facilitate the economic recovery of the acetic acid for subsequent reuse.

5. Washing and Drying

The product is thoroughly washed to produce a product that is stable in storage. After washing it is dewatered and dried.

The manufacture of the propionate and butyrate esters parallels that of the acetate. The esterification is accomplished by an acid-anhydride mixture that introduces both the acetyl groups and the higher acyl groups. In the interest of economic operation it is preferable to restrict the use of the higher acid anhydride as much as possible, compatible with the requirements of the end product.

C. Recovery

It should be noted that a highly important phase of cellulose acetate manufacture is the acetic-acid recovery operation. The recovery of this acid and its return into the process is an essential factor in achieving economic operation.

The manufacture of the mixed esters results in the accumulation of a combination of recoverable acids. Here the more complex operation of separation and purification must be performed to return these components to the appropriate systems.

D. Key Properties Related to the Nature of Cellulose Esters

A compilation of properties of the three major thin-film esters—"diacetate," triacetate, and acetate butyrate—appears in Tables 1 to 4.

The cellulose triacetate by virtue of its high degree of esterification has considerably greater moisture resistance in films than the "diacetate" ester. Moisture permeability is also lower. The films from the cellulose acetates have good combustion resistance and are classified as slow burning or self-extinguishing. The nature and amount of plasticizer employed are determining factors in respect of combustion resistance.

The mixed acetate butyrate ester displays even greater water resistance than the triacetate and its films are less permeable to water vapor. They do, however, show considerably greater gas permeability than films from either of the acetate esters. The combustion characteristics of acetate butyrate are similar to those of the two acetates

The mixed esters such as acetate propionate and acetate butyrate have physical properties that relate to the chain length of the higher aliphatic acid group as well as to the acyl balance [4]. A comparison of the acetate ester with the acetate priopionate and acetate butyrate shows that as the chain length of a substituent is increased the ester displays reduced melting point, increased flexibility, increased elongation, decreased density, and improved water resistance.

Plasticizer compatibility is greater in the mixed esters. This permits the use of higher molecular-weight, less volatile plasticizers. In addition, the inherent properties of mixed esters often allow the use of a low plasticizer level or even the omission of plasticizer while achieving the necessary thermal properties for extrusion and the required properties in the finished film.

TABLE 1

General Properties of Commercial Cellulose Ester Films[a]
(1-mil material)

	ASTM test method	Diacetate	Triacetate	Acetate butyrate
Transmittance of visible light, %	D–1003–59T(B)	93	92	92
Refractive index, n_D	D–542–50	1.4855	1.4875	1.4828
Specific gravity	D–1505–57T	1.29	1.29	1.19
Moisture absorption (24 hr)	D–970–59aT			
Total water absorbed, %		4–6	2–4	1–2
Soluble matter extracted, %		3–5	0–0.10	0
Moisture content:70°F,50% rh,%[b]		1.8	1.2	0.5
Water vapor permeability				
(100°F to 90% rh), g/m²-24 hr	E–96–53T(E)	1200	790	790
Discoloration under UV (200 hr)		none	none	slight
Gas permeability	D–1434–58			
CO_2, ml/100 in.²-24 hr		860	880	6000
O_2		150	150	950
N_2		30	30	250

[a] Abstracted from data published by Eastman Kodak Company
[b] Karl Fischer method.

TABLE 2

Mechanical Properties of Commercial Cellulose Ester Films[a]
(1-mil material)

	ASTM test method	Diacetate	Triacetate	Acetate butyrate
Tensile strength (Instron), 10³ psi	D–882–56T(A)	13–15	12–15	7–9
Total elongation, %	D–882–56T(A)	25–35	25–35	50–60
Modulus of elasticity in tension,				
× 10⁻⁵ psi	D–882–56T(A)	4–6	5–6	2.0–2.5
Tear resistance (Elmendorf), g	D–689–44	5–10	5–10	5–10
Folding endurance (MIT),				
number of folds	D–643–43(B)	500–600	1000–2000	800–1200
Bursting strength (Mullen) psi	D–774–46	40–60	50–70	40–70

[a]Abstracted from data published by Eastman Kodak Company.

TABLE 3

Thermal Properties of Commercial Cellulose Ester Films[a]
(1-mil material)

	ASTM test method	Diacetate	Triacetate	Acetate butyrate
Flammability	D–568–56T			
Area burned (total: 12 in.^{2}), in.^{2}		12	1–2	12
Burning time, sec		4–5	Self-extin- guishing	10–17
Burning rate, in.^{2}/min		180–240		40–70
Thermal coefficient of linear expansion (70 to 120°F), $\times 10^5$ in./in.-°F		3–4	2.5–3.5	4–6
Tensile heat distortion temperature (2% elongation at 50 psi), °F	D–1637–59T	300–310	300–310	230–240

[a] Abstracted from data published by Eastman Kodak Company.

TABLE 4

Electrical Properties of Commercial Cellulose Ester Films[a]
(1-mil material)

	ASTM test method	Diacetate	Triacetate	Acetate butyrate
Dielectric strength (short time) V/mil	D–149–55T	3200	3700	3100
Dielectric constant	D–149–55T			
200 cycles		3.7	3.3	2.9
10^3 cycles		3.6	3.2	2.9
10^6 cycles		3.2	3.3	2.5
Power factor, %	D–149–55T			
200 cycles		0.92	1.06	0.81
10^3 cycles		1.3	1.6	1.3
10^6 cycles		3.8	3.3	4.4
Volume resistivity, Ω-cm	D–257–58	7.0×10^{13}	1.4×10^{15}	1.4×10^{15}

[a] Abstracted from data published by Eastman Kodak Company.

IV. Conversion of Cellulose Esters

A. GENERAL

The dried cellulose ester becomes a raw material in subsequent manufacturing operations. Certain esters are compounded with pigments and plasticizers and become raw material for molding compositions. Others become components of lacquers. Still others are doped for subsequent spinning into filaments.

Those destined for film or sheeting use are formulated for either casting or extrusion operations. In the case of thin-film materials the casting operation is that predominantly used. Both processes are reviewed here briefly.

The manufacturers of cellulose esters are in many cases also producers of films from these esters. At the same time the esters themselves or esters compounded with plasticizers and any other required additives are purchased under specification by converters for the extrusion of films. Converters that use the casting process incorporate any required plasticizers and additives during the solution step of their process. The incorporation of coloring agents and the introduction of matte surfaces are discussed under those specific headings.

B. CASTING

Basically the casting operation involves the flowing of a solution of the ester onto a moving belt or rotating wheel with sufficient removal of the solvent during the travel of the belt or wheel so that the sheet can be stripped and then transported through further drying sections for additional solvent removal. The cured film is then wound on take-up rolls. In this operation the sheet may also be surface-treated or colored as discussed under the pertinent headings. Economical operation demands that a suitable solvent recovery system be integrated in the casting operation.

The composition of the casting solution is dictated by the solubility of the ester and also by the end-use insofar as the additives are concerned. The "diacetate" can be readily dissolved in acetone, methyl acetate, and other common solvents or solvent mixtures [5]. These, then, are suitable coating solvents. The triacetate with its more restricted solubility requires a more powerful solvent such as mixed methylene chloride and methanol as the coating solvent [5]. The acetate butyrate is conveniently soluble in many solvents; it can be coated from methylene chloride, ethylene dichloride, and propylene dichloride and mixtures with methanol [5]. Solvents such as these may be combined with critical amounts of nonsolvents to provide a means of increasing coating speed by inducing gelation in the coated dope. This allows the sheet to be more quickly stripped from the coating surface and further allows solvent removal from both surfaces of the sheet [6].

The addition of plasticizers to the coating solution is primarily designed to impart toughness and flexibility to the final product. The phthalate esters—diethyl, dibutyl and dioctyl—may be added in amounts up to 30 to 40% of the cellulose ester. Triphenyl phosphate may be included to help minimize the burning rate [7]. The choice of plasticizer is limited by its compatibility with the system. In those end-uses in which contact with food or food products is involved the choice of plasticizer is also restricted by the regulations of the Food and Drug Administration (cf. Volume I, Chapter 16). The acceptability of film offered for food use is specifically described by the manufacturer.

The casting solution must be thoroughly filtered before being fed to the casting hopper. The filtering steps remove any gels, skins, or other insoluble materials that have been introduced during the ester manufacture or in the handling of the ester during the solution steps. Poor filtration could result in imperfections in the cast sheet and even disruption of the casting operation by causing a break in the sheet during curing.

Solvent casting has the advantage of inducing especially good surface to the sheeting. This surface is directly related to the highly polished surface of the casting belt or wheel. This method is less practical for medium- or heavy-gauge material because of the difficulty involved in removing the relatively large amount of solvent from a heavier sheet.

The lighter sheets in the medium-gauge range, however, are suited to the casting operation.

C. Extrusion

For the heavier medium-gauge material and the heavy-gauge material, extrusion offers the advantage of more economical operation but at some sacrifice of surface quality because of die lines and possible occlusions. Although the extrusion of thin-gauge material is accomplished commercially, it requires highly-refined techniques to obtain a high-quality product. Extruded thin-gauge material is entirely satisfactory for those end-uses for which extreme optical clarity is not required or in which certain subsequent operations such as metallizing would mask any surface imperfections. Improved surface, hence greater optical clarity, can be induced by the press-polishing operation described below. As noted there, matte surfaces, that is, nonglossy, granular surfaces, as well as highly polished surfaces can be achieved by pressing.

Esters intended for extrusion are filtered in solution during manufacture, since there is no convenient method of filtering during the extrusion operation. Extrusion is adaptable to the "diacetate," propionate, butyrate, and the mixed esters. The triacetate ester, because of its high softening point, is not ordinarily extruded.

Since no solvents are involved, there is, of course, no attendant solvent recovery system. Compatibility with and solubility in any required plasticizers is an important factor in formulation. It is also necessary that the extrusion charge be dry and free of volatile material to minimize the risk of sheet defects arising from trapped vapor.

In addition to the ester, the composition of the extrusion charge includes suitable plasticizers and any dye or pigment required to give the desired color or opacity (or both) to the sheet. Stabilizers against heat and ultraviolet light may also be incorporated in the composition. In addition to enhancing the strength and flexibility of the product, the plasticizer can also improve the flow characteristics in the extrusion operation. The amount of plasticizer used is dependent on the relative costs of the ester and the plasticizer as well as the quantity of plasticizer that will provide the desired handling properties and finished quality.

D. ESTER RECOVERY

Solvent recovery has been alluded to as an important economic factor in the film-casting operation. Scrap-film recovery is an important aspect of both the extrusion and casting operations. Any trim, scrap, or substandard material can be finely chopped and returned to the extrusion blend or the casting dope make-up in a proportion commensurate with product requirements and manufacturing performance.

The use of scrap must be controlled so that no incompatibility of esters is encountered to cause physical defects in the film. Plasticizer content of the scrap must also be considered in relation to the desired plasticizer level of the product being manufactured. When use of the film may involve food contact, any additives in the scrap must be carefully considered from the standpoint of Food and Drug Administration approval.

V. Operations Related to Film Manufacture and End-Use

A. PRESS POLISHING

Extruded film or sheeting does not have the superior surface smoothness of solvent-cast material. The die lines may be objectionable for certain optical uses. A highly polished surface can be induced by press polishing. In this operation the cut sheets are placed between highly polished metal plates and stacked in a press that is gradually heated by steam to the softening point of the polymer at a pressure of several hundred pounds per square inch. The sheets are held for a prescheduled period and the press is then cooled by injection of water. The sheets produced have a surface that approaches that of cast material. A cycle time of approximately 1 hr is required to load such a press and then accomplish the heating, pressing, cooling, and unloading.

The extruded product can be made available in continuous rolls unless press polishing is required to improve the surface characteristics. The pressing operation, however, limits sheet size to the practical dimensions of the press.

It is worthy of note that advances in techniques and equipment have resulted in such quality improvements in extruded material that there is now considerably less need for polishing. The cost factor enters the picture also in that there is little difference in the polishing cost as a function of gauge; therefore polished light-gauge material would bear approximately the same premium as heavy-gauge material on a square foot basis and a considerably greater premium on a weight basis. Polishing of the lighter gauge extruded material can therefore be justified only for very special uses.

B. MATTE FINISHING

In certain applications of cellulose ester film a high degree of transparency is not required; in fact, a matte surface is sometimes needed for decorative purposes; for instance, a matte film placed over a colored picture in a greeting card achieves an attractive effect. A matte surface is also useful in applications in which the end-use is a sheet on which one can write. Although films of both the acetate esters and the acetate butyrate esters may be matte-finished, in practice the acetate films are those usually found in uses that require this type of finish. This results in part from the lower cost of the acetate sheet and from the fact that the special properties of the mixed esters are not usually required in these cases.

A matte surface is usually applied to one side only, although material with two matte surfaces can be produced. In addition, some suppliers offer a choice in degree of matte. In cast material the matte surface can be induced during the casting operation. The cured sheet is passed over an embossed roll which is wetted with solvent as it rotates in a solvent supply pan. The solvent picked up by the rotating embossed rolls softens the surface of the sheet and the embossed nature is imparted to the film. After solvent removal, the matte surface material is wound in the conventional manner. A procedure such as this can be incorporated as an in-line part of the over-all casting operation, thus providing good economy. The nature of the operation allows cast matte material to be produced in continuous rolls.

Extruded material with a matte finish may be made by a pressing operation similar to that discussed under press polishing. Here the plates in the press have an embossed surface that induces the matte surface. Either one or both sides of the material may be matte finished by the appropriate choice of embossed and polished plates. Sheet size is limited by the practical dimensions of a press.

C. Colored Film

Special uses for colored film are based on the selective light transmission that can be achieved by the addition of color for decorative purposes and by its value for color coding.

Color is obtained by applying a surface coating of the dye in a suitable solvent. The method is difficult to control in respect to uniformity and it is difficult to obtain colors of any depth. Adding the coloring agent to a batch of cellulose ester dope overcomes both deficiencies. An effective variation based on this method involves metering a concentrated colored solution into the normal dope supply at a point close to the point of casting. This not only results in a product of more uniform color but it also allows a convenient means to effect a color change on a given casting machine. Only a small part of the system requires flushing in order to make a color change or to return to an uncolored ester solution.

Colored film finds particular use in pressure-sensitive tape, signs, window streamers, window shields, wrappings, laminates, shims and gaskets, and file tabs. In the last three uses the color factor serves as identification or coding. Colored films generally have an acetate base; the mixed ester, however, is equally suitable to the coloring operation.

D. PRINTING

Films produced from cellulose acetates or from cellulose acetate butyrate are suitable for printing by conventional printing methods. The choice of inks must be based on the printing method used with due regard for the low absorption of the sheeting compared with paper.

Frequently reverse printing is used so that the image is seen through the film. This gives a bright appearance and at the same time protects the image from abrasion. Reverse printing in combination with surface printing can be used to achieve a three-dimensional effect, especially when the double-printed film is used in a paper laminate such as a prestige brochure or a periodical cover. Printed film also finds use in laminates with metal foil, where again the brilliance of the lettering is achieved by the reverse printing. Here the film acts as a support for the more delicate metal foil that contributes the air-tight nature to the packets fabricated from it. A three-layer laminate of cellulose acetate, metal foil, and polyethylene provides a heat-sealable packet material which is airtight, attractive, and can contain the necessary decorative printing.

E. LAMINATION

There are two common methods of laminating cellulose ester films to other materials. Designated wet lamination or dry lamination, these methods differ in that the wet involves mating the components directly after application of the adhesive and the dry involves drying the adhesive after application and subsequently mating the components under heat and pressure.

Wet lamination is applicable to the manufacture of paper or cloth laminates when there is no problem in solvent removal. The sandwich lamination of screen or scrim between two layers of ester film also employs wet lamination. Dry lamination is required in metal-foil lamination for which solvent removal would otherwise be difficult to accomplish. It may also be used in paper lamination.

F. Sealing

In those applications in which cellulose ester film is sealed to itself, sealing may be accomplished by solvent-type adhesives or by certain resin-type adhesives. Resin-type adhesives may be used to bond the ester films to other materials, such as cardboard or paper, to produce packaging materials and other laminates.

VI. Uses of Cellulose Acetate Thin Films

End-uses related to the suitability of acetate films as a base for printing and to the incorporation of a matte finish or of color have already been alluded to. Other end-uses are discussed below.

A. Magnetic Sound-Recording Tape

Cellulose acetate film finds wide use as the base for magnetic recording tape. Both the high acetyl material and the lower acetyl materials are used. In some tape applications polyester base is preferred for the high strength it affords, particularly when low caliper such as 0.5 mil is a special requirement. The cellulose acetate base, however, is used extensively in the 1- and 1.5-mil tapes. The actual application of the oxide coating may be a separate step with stock rolls of film or it may be incorporated as an in-line step following the casting operation.

B. Release Sheet

Triacetate film finds varied uses in the release sheet field. Its moisture resistance is used to advantage in the release sheet for adhesive-backed bandages, for it is able to withstand the wet sterilization given the final product. It can be also printed in this use as required. On a larger scale triacetate film finds use as a release sheet in the molding of large plastic objects such as fiberglass reinforced boats. Here it may be used in a paper laminate or as the film itself.

C. Electrical Insulation

The superior thermal stability of cellulose triacetate compared with the lower acetyl ester makes it preferred for use in electrical applications in which high temperatures are encountered; for example, as layer insulation in coil windings or barrier layers in electrical cables.

D. Overlays

Transparent acetate sheeting finds wide use in overlay applications, as animated cartoon manufacture in which brilliant transparency is essential. They may also include applications in which a matte surface is required for the overlay to accept pencil or conventional writing inks. The triacetate with its superior dimensional stability is preferred in multiple overlay applications such as those used in reference or textbooks to show, successively, the internal parts of a complex subject.

E. Envelope Windows

The brilliant transparency of cellulose acetate sheeting makes it especially attractive for envelope windows. If not only serves to show clearly the name of the addressee, it can also show a message written on the inside back of the envelope after removal of the contents. In this application it may be used in equipment designed to handle glassine envelope windows. In fact, it may be desirable to be able to change from acetate film to glassine paper in a given installation. For this use the acetate sheeting is treated so that it can be glued in place with the same water-base adhesives used for the glassine material. For this application the sheeting is coated with a water-compatible layer such as gelatin. A parallel use related to the transparency factor is the suitability of acetate film for windows in boxes. The contents of the box are thus clearly visible, though protected. Display and packaging are thus coordinated by using a package with a clear window that allows the contents to be viewed.

F. Metallized Film

Cellulose triacetate compounded with nonvolatile plasticizers is a base for metallized film. High-vacuum deposition of metal on this base produces low-cost material with a metallic appearance. Metallized film is widely used in such diverse applications as the fabrication of components for electrical circuits, the manufacture of sequins, and the decoration of greeting cards and ornaments.

Another interesting application is that used in the manufacture of a dry-cell battery wrap, which also illustrates the printability and lamination aspects of metallized film. Cellulose triacetate sheeting is first reverse-printed and metallized to give a metallic background to the printed matter. It is then laminated to a paper backing. The laminate which forms the dry-cell wrap presents a brilliant, metallic appearance; the printed matter and the metal surface are both protected from abrasion by the acetate.

VII. Uses of Acetate Butyrate Thin Films

It has already been noted that there are end-uses in which colored film or matte-finished film could involve the use of cellulose acetate butyrate material. Uses more specific for this material are reviewed below.

A. METALLIC YARN

Among the cellulose ester films acetate butyrate is the most suitable for the manufacture of metallic yarn. The properties especially required in this application are good stretch to allow the film component to withstand the rigors of weaving the metallic yarns and good water resistance to prevent it from blushing during the dyeing of the fabrics in which it is incorporated.

Metallic yarn is produced by laminating a sheet of aluminum foil between two sheets of cellulose acetate butyrate. The laminate is then slit into the required thickness, twisted, and spooled. The metallic appearance of copper, silver, and gold as well as color effects may be produced by the use of the appropriate hue in the laminating adhesive. Metallic yarn finds use in upholstery material, draperies, clothing, and women's accessories. It is frequently used in combination with conventional yarns.

B. WINDOW-GLASS SUBSTITUTE

Cellulose acetate butyrate is also preferred to glass for windows in farm buildings and coldframes. This sheeting, laminated to both sides of cotton scrim or wire screening, offers the weather protection of glass with minimum risk of breakage. It allows greater ultraviolet light transmission than glass and is preferred to cellulose acetate because of its greater resistance to ultraviolet-light degradation and moisture.

C. ELECTRICAL INSULATION

The relatively high moisture resistance of the acetate butyrate films makes it useful also as primary insulation on electrical wiring. A narrow tape of film is wound around the wire in an overlapping spiral configuration. This end-use was exploited in particular during World War II when the rubber normally used for this purpose was not available. Electrical wire is now often plastic-coated by extrusion rather than by wrapping with foil; however, this application of acetate butyrate does continue to some degree.

VIII. Other Esters

A. Cellulose Nitrate

Cellulose nitrate has use in lacquers, plastics, and explosives and once commanded a major position in the photographic film field until replaced by the less flammable materials such as cellulose acetate. Advantages of cellulose nitrate include its low water absorption, good dimensional stability, and toughness. The flammability factor and its relatively low resistance to light and elevated temperatures are disadvantages. It has no significant place in self-supported thin-film manufacture, but in the coating of moistureproof cellophane it still plays a major role.

Although sheets as thin as 0.003 in. are produced, they are used primarily in lamination processes for producing heavier stocks of vari-colored material to impart a decorative effect that can be incorporated in subsequently manufactured articles. The heavier gauge material, which is itself suitable for molding operations, represents greater volume.

B. Sulfate Esters

The alkali metal salts of acylate-sulfate esters of cellulose can be cast from organic solvents to form water-soluble films. Such films are applicable to unit packaging in which the contents are dissolved in water in its end-use.

A method of producing such esters has been described by Malm and Rowley [8]. It involves the reaction of the activated cellulose in the lower fatty acid anhydride with an alkali metal salt of the acyl sulfuric acid. A catalyst such as methanesulfonic acid, phosphoric acid, or sulfoacetic acid is used.

To date the sulfate esters have not developed commercially to any appreciable degree.

References

1. C. J. Malm, J. W. Mench, D. L. Kendall, and G. D. Hiatt, *Ind. Eng. Chem.*, **43,** 684 (1951).
2. C. J. Malm, *Svensk Papperstid.*, **64,** 740 (1961).
3. A. G. Lipscomb, *Cellulose Acetate—Its Manufacture and Applications*, Ernest Benn, London, 1933.
4. C. J. Malm and G. D. Hiatt, in *High Polymers*, E. Ott, H. M. Spurlin and M. W. Grafflin, Eds., **Vol. V,** 2nd ed., Part II, Wiley-Interscience, New York, 1955, p. 797 ff.
5. E. Ott, H. M. Spurlin, and M. W. Grafflin, *High Polymers*, **Vol. V,** 2nd ed., Part II, Wiley-Interscience, New York, 1955, p. 1454 (from *Eastman Cellulose Esters*, Tennessee Eastman Co., Kingsport, Tenn., 1949).
6. C. R. Fordyce and W. F. Hunter, Jr. (to Eastman Kodak Company), U.S. Patent 2,319,052 (May 11, 1943).
7. C. J. Malm, N. G. Baumer, and G. D. Hiatt, *Ind. Eng. Chem.*, **47,** 2521 (1955).
8. C. J. Malm and M. E. Rowley (to Eastman Kodak Company), U.S. Patent 3,075,964 (Jan. 29, 1963).

CHAPTER 9

ETHYLCELLULOSE

GARTH H. BEAVER

Designed Polymers Research
The Dow Chemical Company
Midland, Michigan

I. The Polymer

A. HISTORICAL

Early research on commercial ethylcellulose was conducted by both The Dow Chemical Company and Hercules Powder Company. Ethylcellulose has been manufactured in the United States since 1935.

433

Lacquer coatings were the target of early development effort.   It soon became apparent that ethylcellulose had certain properties which made it useful as a thermoplastic molding powder.   Today ethylcellulose is sold for both coating and molding powder applications.

B. Current Manufacturing Processes

Ethylcellulose flake is manufactured by treating cellulose, which has been soaked in strong sodium hydroxide, with ethyl chloride under the appropriate conditions.   After ethyl chloride has reacted with the alkali cellulose and the crude ethylcellulose has been formed the general sequence of purification is isolation, washing with water, and drying. An excellent and complete review on ethylcellulose manufacturing is presented by A. B. Savage [1].   Ethylcellulose flake is a white, odorless free-flowing powder with a density up to 0.5 g/cm^3.   Products of the two commercial ethylcellulose processes operated in the United States, although they have apparently similar specifications, can vary widely in regard to specific properties for a designated application.

C. Ethylcellulose Grades

A majority of the ethylcellulose manufactured today falls into two grade classifications.

1. Ethylcellulose with an ethoxyl content between 45.0 and 46.5 wt%. This grade is used in certain industrial coating applications that require high impact resistance, in molding powder, and in plastic sheeting.

2. Ethylcellulose with an ethoxyl content between 47.5 and 49.5 wt%. This grade has a much wider compatibility with solvents and plasticizers and finds utility in industrial lacquers.   It also has limited pharmaceutical applications.

Both grades of ethylcellulose are available in a range of molecular weights, generally represented in terms of their viscosity as 5% by weight solutions in suitable toluene-alcohol mixtures [2].   Viscosity grades ranging from 7 to 200 cp are available.

D. Formulation

1. Plasticizers

Small amounts of plasticizer may be incorporated with most grades of ethylcellulose.   In contrast with other cellulose derivatives, ethylcellulose is rarely plasticized to more than the 15 wt % level. Commonly used plasticizers for ethylcellulose are Dow Plasticizer 1099 [3], which is bis(p-1,1,3,3,-tetramethylbutyl)phenyl ether, Dow Resin 276V9, which is an α-methylstyrene polymer, Dow Plasticizer No. 5 (2-biphenylyldiphenyl phosphate), Baker's No. 15 castor oil, dioctyl phthalate, and n-butyl stearate.

2. Stabilizers

Depending on the end-use of the ethylcellulose, a number of stabilizers may be used. For heat stabilization a sterically hindered phenol has been found effective. Butylated hydroxytoluene can be used, as can related compounds such as Santonox R, or Ionex 220. When the ethylcellulose is exposed to light for an extended period, ultraviolet light absorbers such as Dow salol, Uvinol 400, or Dow TBS are commonly used. For some molding applications it is necessary to add acid acceptors, such as epoxidized soya alkyds. Admex 710 or Paraplex G-60 may be used. Occasionally a neutral organic phosphite may be added as a color stabilizer.

E. ECONOMICS

Some factors that tend to influence ethylcellulose flake economics are the following:

1. The relative high cost of purified chemical cellulose, which is used as the polymer backbone.
2. An inefficient yield from ethyl chloride reagent. This is caused, in part, by the partial hydrolysis of ethyl chloride to yield alcohol and its tendency to form diethyl ether. The ethyl chloride yield problem has been partly solved by The Dow Chemical Company's "semicontinuous process" [1] which is aimed at improved reaction efficiency.
3. Purification costs for ethylcellulose are fairly high, for when the chemical manufacturing reaction is completed the polymer concentration is low in the unspent caustic, ethyl chloride, and reaction by-products. Costs of removing the large quantities of inorganic salt (such as sodium chloride and unreacted caustic) by existing manufacturing techniques are significant.
4. Quality control costs for the polymer are above the average for the plastics industry. This is largely the result of the large number of specific unit processes in a complete manufacturing operation.

Within the broad category of thermoplastic resins and solvent-soluble resins, ethylcellulose is an engineering plastic. It is relatively expensive, currently selling for approximately 70 cents/lb. Although the polymer market has been subject to considerable price attrition in recent years, ethylcellulose prices have remained firm.

II. Sheeting

A. SOLUTION-CAST SHEETING

1. Choice of Ethylcellulose Grade

Although both grades of commercially available ethylcellulose are solvent-soluble and thermoplastic, it has been found desirable to use a higher melting grade for sheeting. Figure 1 presents melting-point data on ethylcellulose. Ethylcellulose, with an ethoxyl content of less than 46.5%, is commonly used for cast sheeting. Figure 2 demonstrates that ethylcellulose with an ethoxyl content between 45.0 and 46.5% possesses greater strength than ethylcellulose with an ethoxyl content of 49.0% for a similar viscosity type. It should also be noted in Figure 2 that there is a relation between the physical properties of the film and polymer intrinsic viscosity. For practical purposes, however, ethylcellulose with viscosity above 20 cp can be satisfactorily used in commercial ethylcellulose sheeting.

2. Choice of Vehicle Solvent

The decision to use ethylcellulose with an ethoxyl content of 45.0 to 46.5% in cast sheet manufacture has some limitations. One is the fact that ethylcellulose with the lower ethoxyl content is not so widely soluble in organic solvents as the more highly substituted polymer. In general, only a few solvents can be used. Chlorinated solvents, such as chloroform, methylene chloride, and methyl chloroform can be used, although the latter is best utilized in combination with small amounts of a lower alcohol such as ethyl alcohol. Mixtures of an aromatic solvent and a lower alcohol are the most effective solvents. Solvents commonly used for the lower ethoxyl content ethylcelluloses are 70/30 benzene-methanol or 60/40 toluene-ethanol. Table 1 presents viscosity data as a function of polymer concentration for certain solvent systems containing ethylcellulose of the 50-cp type with an ethoxyl content between 45.0 and 46.5%. Three important considerations in the final

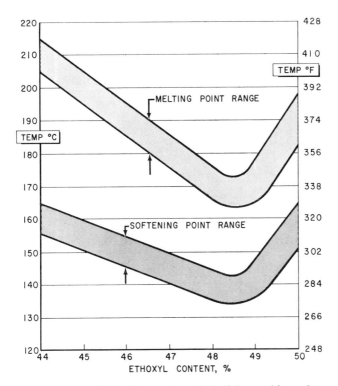

Fig. 1 Effect of ethoxyl content on ethylcellulose melting point.

selection of a solvent for use in cast ethylcellulose sheeting are (a) the quantity of sheeting solids being deposited per pound of solvent employed, (b) the ease of recoverability of the solvent, and (c) the toxicity and flammability of the solvent system.

The ease of recoverability is determined primarily by the type of equipment available and the efficiency of stills to recover an alcohol-containing aqueous phase, should an alcohol be used. Toxicity and flammability considerations favor the use of the chlorinated hydrocarbon solvents over the aromatic solvent systems.

Unplasticized sheet cast from any of the solvent systems listed in Table 1 will, at rupture, have a tensile strength of approximately 8000 psi and an elongation of about 25%.

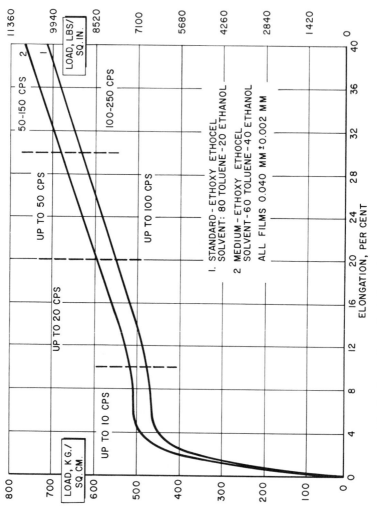

Fig. 2 Approximate load-elongation curves for Ethocel viscosity types.

438

TABLE 1

Viscosity of Various Concentrations of
Ethylcellulose in Various Solvent Systems

Solvent System[a]	Ethylcellulose[b] (wt %)	Viscosity[c] (poises at 90°F)
70/30 benzene-methanol	29	795
90/10 methyl chloroform-ethanol (190 proof)	14	226
90/10 methyl chloroform-ethanol (190 proof)	16	551
90/10 methyl chloroform-ethanol (190 proof)	18	1390
90/10 methyl chloroform-ethanol (190 proof)	20	4030
90/10 methyl chloroform-methanol	14	128
90/10 methyl chloroform-methanol	16	288
90/10 methyl chloroform-methanol	18	711
90/10 methyl chloroform-methanol	20	1500
Methylene chloride	14	164
Methylene chloride	16	426
Methylene chloride	18	1040
Methylene chloride	20	2300

[a] All ratios are by weight.
[b] 46% Ethoxyl, 50-cp type.
[c] ASTM D1343–56.

3. Formulation

Ethylcellulose cast sheeting is formulated with two concepts in mind: (a) the properties required in the end-use application, and (b) consideration of the ease of practicability of its manufacture in commercial equipment.

Considerations in formulation can best be put in the following categories:

1. Plasticizer considerations.
2. Stabilizer considerations.
3. Manufacturing considerations, which are twofold:
   (a) strippability from the casting surface;
   (b) the corrosive tendencies of the casting dope if it contains a polar solvent such as an alcohol.

Since it is desired to keep the final product as hard as possible yet impart machinability properties for further fabrication, only small amounts of plasticizer are normally used. Table 2 presents data on the effect of two plasticizers on the physical properties of a common type of ethylcellulose used in cast-sheeting manufacture.

Since ethylcellulose sheeting has excellent shelf-life stability, it is necessary to protect the sheeting only from the possibility of its constant exposure to ultraviolet light. An ultraviolet light absorber is almost always present in sheeting.

TABLE 2

Effect of Plasticizer on the Physical Properties of Ethylcellulose[a]

Plasticizer	Weight percent	Yield point, (kg/cm^2)	Tensile strength[b] (kg/cm^2)	Elongation[b] (%)
None	0.0	530	658	25
Dow Plasticizer P–1099	2	536	674	24
	10	466	522	20
Dow Plasticizer P–5	2	510	640	24
	10	388	506	24

[a] 46% Ethoxyl, 50-cp type.
[b] At rupture.

If continuous metal belts are used in casting the ethylcellulose solutions for sheeting, it will be necessary to ensure that the sheeting be easily released from the metal substrate. A small amount of release agent is commonly added to the casting dope for this purpose. A purified stearic acid or the lower esters of stearic acid are commonly used.

Depending on the precise composition of the casting solvent feasible for commercial use, it may be necessary to add a corrosion inhibitor to the ethylcellulose casting dope. This inhibitor is usually a salt of a weak organic acid and a strong alkali.

4. Manufacturing Techniques

Ethylcellulose cast sheeting can be manufactured by conventional film-casting techniques. The master batch of casting dope is usually stored for several days at a carefully controlled (elevated) temperature to ensure its complete de-airing. The dope, which has been filtered, is then metered onto the casting surface and the solvents are evaporated in a carefully controlled manner.   Solvents can be recovered by conventional techniques.

The correctly dried sheet is then stripped from its support, drying is completed, and it is edge-trimmed and wound onto master rolls.   The exact techniques employed in the drying operation are a highly skilled art and vary considerably, depending on the particular casting solvent employed and on the gauge of the sheet being manufactured.   The techniques of imparting slip as hereinafter discussed are considered trade secrets.

5. Grades

The principal criteria in determining grades of ethylcellulose sheet are (a) the slip, the ease at which the sheet slides over other similar sheets when undergoing further processing, and (b) the gauge of the sheet. Common gauges of sheet are between 2 and 12 mils.

Three grades of slip are usually recognized: standard product, medium slip (MS), and high slip (HS).

6. Properties

The properties of ethylcellulose cast sheeting are presented in Table 3. Ethylcellulose products are best noted for their high impact resistance and excellent low-temperature flexibility.   Cast sheeting is the hardest of the ethylcellulose sheet products.   Optical properties are excellent.

7. Applications

Because of the highly specialized nature of the packaging industry, the applications for an engineering sheet such as ethylcellulose cast sheet are rather closely guarded.   The sheeting can be used in greeting card overlays and as transparent boxes or box tops.   It is also used as sheet protectors in ring binders.

8. Economics

There are two basic considerations in the economics of ethylcellulose cast sheet:

1. The relatively high cost of the primary ethylcellulose flake, which constitutes at least 90 wt% of the final product.

TABLE 3

Typical Properties of Ethylcellulose Sheeting

Property	Extruded sheeting		Solution-cast sheeting		
	Test method ASTM	Value	Test method ASTM	Value	Thickness (in.)
Mechanical Properties					
Tensile strength 25°C (77°F), psi	D638–61T	5000		8000	
Elongation in tension, 25°C, %	D638–61T	10–30		20–35	
Modulus of elasticity in tension, 25°C, %	D638–61T	2×10^5		3×10^5	
Folding endurance, MIT, double folds, 23.9°C (75°F), 50% rh					
Bursting strength, psi				2750	0.001
Tear strength, Elmendorf, g/0.001 in. width				85	0.001
				97	0.005
Thermal Properties					
Softening temperature, °C, bar		149		154	
Melting temperature, °C		150		191	
Specific heat, Btu, lb.		0.35		0.348	
Weight loss on heating, maximum, %	D787–61T[a]	0.4			
Electrical Properties					
Dielectric constant, 10^3 cps	D150–59T	3.1	D130–56[a]	3.1	
10^6 cps		3.0		3.0	
Power factor, 10^3 cps	D150–59T	1.3	D130–56	0.4	
10^6 cps		1.6		2.0	

442

Property					
Dielectric strength, V/0.001 in.			D130–56	3500	0.002
Fabrication					
Specific gravity	D792–48	1.10	D71–27	1.15	
Extrusion temperature, °C		210			
Color possibilities		Transparent to opaque		Transparent to opaque	
Optical Properties					
Refractive index, n_D^{20}				1.47	
Transmission of white light, %				88–92	
Ultraviolet cutoff, Å				2200	
Transmission of infrared (except a narrow absorption band at 10^{-3} cm), %				90	
Fadeometer, 200 hr, yellowing				Almost none	
embrittlement				None	
Chemical Properties					
Water sorption, gain in 24 hr, 26.6°C (80°F), % 100% rh, 37.8°C (100°F), %	D570–42	1.2	D570–42	7.5 3.3	
Water solubility, maximum, %	D570–42	0.2			
Normal moisture content, 23.9°C (75°F), 50% rh, %				1.4–1.7	
Water vapor transmission, 95% rh, 37.8°C (100°F), g/100 in.^{2}/24 hr.[b]				3.5	0.002

Chemical resistance: weak acids, no effect; strong acids, severe attack; weak alkalies, no effect; strong alkalies, slight attack; organic solvents, widely soluble

[a] Approved as American Standard by the American Standards Association.
[b] Thwing-Albert Vaporometer in General Foods Cabinet: cabinet and procedure modified by The Dow Chemical Company.

2. The ability economically to recover casting solvents used in the conversion of the raw polymer flake into the cast sheet.

It will be fairly obvious from studying Table 1 that it is desirable to incorporate a certain amount of methyl or ethyl alcohol into the casting dope. Loss of alcohol containing weak water in the solvent recovery system can be high. This loss, of course, adds to the over-all cost of the ethylcellulose sheeting.

Another factor that determines the economics of the cast sheet, as contrasted with melt extruded sheet, is the relatively high cost of the casting operation, in terms of its production rate. Production rate is limited by the maximum speed at which solvents can be removed with retention of optical properties.

B. Melt Extruded Sheeting

1. Formulation

Commercially available ethylcellulose molding powders can be extruded into sheeting. Desirable thermoplasticity of ethylcellulose flake for melt extrusion can be obtained by adding considerably more plasticizer than is commonly used in the casting operation. A typical composition given by Savage [1] is as follows:

Ethylcellulose with degree of substitution of 2.3 (46 % ethoxyl content) of a 70-cp viscosity—84 parts

Bis(p-1,1,3,3-Tetramethylbutyl)phenyl ether (Dow Plasticizer 1099)—16 parts

Other plasticizers that may also be used with or in place of the Dow Plasticizer 1099 are white mineral oil from a highly naphthenic crude oil, dioctyl phthalate, or n-butyl stearate. Usually the ultraviolet light absorber used in a cast sheeting will be included in molding-powder formulations intended for melt-extruded sheeting.

2. Manufacturing Techniques

The extruded sheeting can be manufactured by commonly practiced techniques. Usually a flat die up to 5 ft wide is used for ethylcellulose. It is not practical to lower the gauge of the extruded sheeting much below 7 mils and still easily maintain gauge uniformity across the sheet. Sheeting $\frac{3}{8}$-in. thick has been manufactured. A three-heat stage extruder with a Dulmage torpedo is generally used. Auxiliary polishing rolls are commonly used to impart desirable optical properties. The use of polishing rolls is largely an art. To a great degree final quality of the sheet is dependent on the capability, skill, and the over-all techniques of the operators of the extrusion equipment.

3. Properties

The properties of ethylcellulose extruded sheeting are also presented in Table 3. In general, this sheeting is softer than the cast sheeting and is restricted to the heavier gauges that are rather difficult to manufacture by the casting technique. Optical properties of the extruded sheeting are generally inferior to those of the cast sheeting.

4. Applications

At the present time ethylcellulose extruded sheeting is believed to be unavailable commercially except on special order. Very little is known about its applications except as one might envision uses for sheeting up to $\frac{1}{4}$- or $\frac{3}{8}$-in. thick. Some sheeting is used in the blister packaging applications. Such uses as replacements for safety glass in laboratory safety shields have been important for this product. From time to time military demand for extruded sheeting has been significant.

5. Economics

The economics of extruded ethylcellulose sheeting are those of the plastics industry and are not discussed in any great detail. Extruded sheeting offers some formulation cost savings over cast sheeting, since more of the lower cost plasticizers are used in the extruded sheeting. In a thermoplastic operation the by-product scrap can be easily reused. The quality desired in the extruded sheeting product largely determines the economics of a particular operation. Although, theoretically, extruded sheeting should cost less than cast sheeting, customer specifications and demand usually dictate that each product may have an approximate cost of slightly more than one dollar a pound.

References

1. A. B. Savage, in *Encyclopedia of Polymer Science and Technology*, Vol. 3, Herman Mark and N. Gaylord, Eds., Wiley-Interscience, New York, 1965.
2. Standard Methods of Testing Ethylcellulose, ASTM Designation: D914–50, American Society for Testing and Materials, 1916 Race St., Philadelphia 3, Pa.
3. E. R. Kropscott and P. H. Lipke (to The Dow Chemical Company), U.S. Patent 2,524,812 (Oct. 10, 1950).

CHAPTER 10

POLYVINYL ALCOHOL

HERBERT K. LIVINGSTON

* *Department of Chemistry, Wayne State University*
Detroit, Michigan

In 1924 Willy O. Herrmann and Wolfram Haehnel, working in the laboratories of the Consortium für Elektrochemische Industrie in Munich, discovered that alcoholic potassium hydroxide saponified powdered polyvinyl acetate to form a water-soluble powder from which extremely strong films could be made [1]. They called this new polymer "polymerized vinyl alcohol," even though vinyl alcohol is not a known compound. Reactions that might be expected to form vinyl alcohol always produce instead its tautomer, acetaldehyde.

With time and the changes in usage in naming polymers the original name has been shorted to polyvinyl alcohol. The repeating unit in the polymer as Hermann and Haehnel visualized it was ($-CH(OH)CH_2-$) and the name polyhydroxyethylene would have given a better chemical description of this product.

It was not until about 10 years after the original discovery that significant commercial interest was developed for polyvinyl alcohol in Germany. In the United States, manufacture on a commercial scale

* Formerly affiliated with Electrochemicals Department, E. I. duPont de Nemours and Company, Wilmington, Delaware.

was initiated by Du Pont in 1941.   Early in this development it became apparent that samples of polyvinyl acetate that were less that 100% saponified were of considerable commercial interest.   In fact, materials with saponification numbers of 10 to 150 (corresponding to 99 to 87% saponification) represent a large part of the commercial sales of "polyvinyl alcohol."   It is now accepted that the inclusive term for materials made by 50 to 100% saponification of polyvinyl acetate is *polyvinyl alcohol*.

I. Manufacture

Polyvinyl alcohol is still manufactured by the basic process discovered by Herrmann and Haehnel.   Schematically, the desired reaction is

$$CH_2=CHOCCH_3 \xrightarrow{\text{catalyst}} (-CH_2-CH-)_n$$
$$\underset{O}{\overset{\parallel}{}} \qquad\qquad \underset{OCOCH_3}{\overset{|}{}}$$

$$\xrightarrow{\text{acid or alkali}} (-CH_2-CH-)_n$$
$$\underset{OH}{\overset{|}{}}$$

Vinyl acetate polymerization is initiated by free-radical catalysis and is carried out in a homogeneous or heterogeneous system.   Solution polymerization is generally preferred.   Polyvinyl acetate dissolved in methanol or in a similar solvent is then saponified.   Methanol is a preferred solvent because the polyvinyl alcohol is insoluble in the methanol, even in the presence of the byproduct methyl acetate, and precipitates out of the system.   It is isolated by filtration and washing.

In actuality there are side reactions that compete with the desired reaction indicated above.   The most significant in its effect on the over-all chemistry of the process is

$$(-CH_2-CH-)_n + CH_2=CHOC-CH_3$$
$$\underset{OCOCH_3}{\overset{|}{}} \qquad\qquad \underset{O}{\overset{\parallel}{}}$$

$$\longrightarrow (-CH_2-CH-OCOCH_2-CH_2CH-)_n$$
$$\underset{OCOCH_3}{\overset{|}{}}$$

This sidechain branching reaction, which occurs as a consequence of the high susceptibility of the acetate groups in polyvinyl acetate to transfer reactions, causes a rapid increase in the molecular weight of polyvinyl acetate with time of polymerization in methanolic solution. All of these branches, however, are cleaved on saponification [2], and the molecular weight of the resultant polyvinyl alcohol does not increase with polymerization time.

The main parameters affecting the properties of polyvinyl alcohol are molecular weight and saponification number. Molecular weight depends on the molecular weight of the initial polyvinyl acetate. It is difficult to measure exactly, and therefore the specification of molecular weight for commercial products is based on the viscosity of 4% aqueous solutions at 20°C. The time required to dissolve a polyvinyl alcohol powder in water at constant conditions increases with molecular weight.

The extent of saponification of a polyvinyl acetate that has been used to make polyvinyl alcohol is usually expressed in units that will reach 100% for true polyvinyl alcohol; that is, we speak of percent alcoholysis rather than saponification number. Tensile strength, elongation at break, and tear resistance increase with molecular weight, whereas water solubility reaches a maximum at 88% alcoholysis. Above 88% solubility in cold water decreases but the polyvinyl alcohols continue to be soluble in hot water.

Commercial grades of polyvinyl alcohol may fall anywhere in the viscosity range of 2 to 150 cp (4% aqueous solution at 20°C), though special grades may run as high as 500,000 cp and will cover the hydrolysis range of 50 to 100%.

II. Properties

A. SOLUBILITY

Water solubility coupled with high strength is the most remarkable property of polyvinyl alcohol. No other water-soluble polymer forms such strong films. On the other hand, polyvinyl alcohol is quite insoluble in nonpolar organic solvents and some of its earliest uses were as diaphragms or the like in mechanical systems involving exposure to lubricating oil or gasoline. Because of the numerous grades and the fact that surface exposure and dissolving time are important factors in all polymer solution processes it is difficult to give general data. As previously mentioned, the grades of lower molecular weight can be dissolved readily and vice versa. It can be stated that all polyvinyl alcohols that are more than 85% hydrolyzed are soluble in hot water (85 to 95°C) and those that are 85 to 90% hydrolyzed are soluble in cold water as well.

Water sensitivity can be reduced by heating the polyvinyl alcohol film at 100°C, by reaction at 70 to 120°C with water-soluble formaldehyde derivatives (such as dimethylolurea) or formaldehyde itself, or by reaction at room temperature with certain metal derivatives such as organic esters of orthotitanic acid, Werner chromium complexes, or cuprammonium hydroxide.

Solubility data in solvents other than water are summarized in Table 1.

TABLE 1

Solubility of Completely Hydrolyzed Polyvinyl Alcohol of
Medium Molecular Weight in Various Organic Solvents

Solvent	Solubility
Hydrocarbons	Insoluble
Methanol	Insoluble
Ethyl alcohol	Insoluble
1-Butanol	Insoluble
Ethyl ether	Insoluble
Ethyl acetate	Insoluble
Amyl acetate	Insoluble
Acetone	Insoluble
Carbon tetrachloride	Insoluble
Trichloroethylene	Insoluble
Glycerol	Soluble hot (120 to 150°C)
Ethylene glycol	Soluble hot
Ethanolamine	Soluble hot
Formamide	Soluble hot
Diethylenetriamine	Soluble cold
Triethylenetetramine	Soluble cold
Dimethyl sulfoxide	Soluble cold

The permeability of polyvinyl alcohol films to gases is unusually low in most cases. Measurements on a dry film made from 88% hydrolyzed polyvinyl alcohol of low molecular weight (viscosity = 5 cp for a 4% aqueous solution at 20°C) and plasticized with 35% glycerol showed permeabilities below 0.02 g/m^2-hr for hydrogen, helium, oxygen, nitrogen, and carbon dioxide, in each case with the test gas at 25°C and zero humidity. On the other hand, permeabilities to water and ammonia vapors are high, particularly if the film is allowed to equilibrate at relative humidities of 70% or more.

All commercial grades of polyvinyl alcohol have excellent film-forming ability. Films are laid down from solution and are homogeneous and strong as soon as the solvent has been removed. The best results are obtained with fully hydrolyzed high-molecular weight polymer (Figs. 1 and 2). Further improvement in tensile strength can be obtained by drawing the film; for example, fivefold increase in tensile strength resulted when a film was stretched five times it original length. Tear resistance is also high and increases with molecular weight and degree of hydrolysis.

To obtain flexibilities in the practical level it is usually desirable to add plasticizers such as glycerol. The effect of glycerol on tensile strength and elongation is shown in Figs. 3 and 4. As would be expected, at constant plasticizer content elongation increases with molecular weight (Fig. 5).

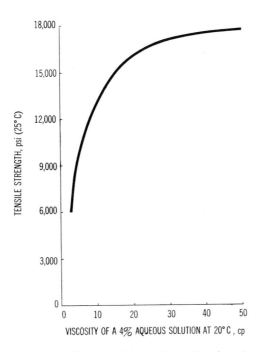

Fig. 1 Polyvinyl alcohol films: tensile strength as a function of molecular weight (as judged by viscosity). Data are for polyvinyl alcohol at 98 to 99% hydrolysis; films were conditioned at 35% relative humidity at 25°C.

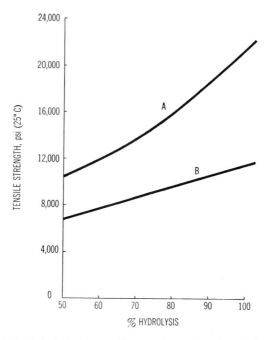

Fig. 2   Polyvinyl alcohol films: tensile strength as a function of degree of hydrolysis. Data are for samples of high molecular weight (viscosity = 50 cp for a, 4% aqueous solution at 20°C). A, film dried over sulfuric acid before testing; B film conditioned at 50% relative humidity before testing.

Plasticizers in commercial use are usually materials that are hygroscopic and therefore maintain sufficient water in the polymer to retain the plasticizing effect of water, even when the relative humidity of the surrounding air is low. Examples of good plasticizers are glycerol, ethylene glycol, diethylene glycol, 1,3- or 1,4-butanediol, 1,2,4-butanetriol, trimethylolpropane, or phosphoric acid.

B. Optical Properties

Since polyvinyl alcohols are used primarily for their water solubility and strength, their optical properties have not been studied in detail. To visible light they are colorless and transparent. Films have been exposed for two weeks to carbon arcs at a distance of 2 ft without

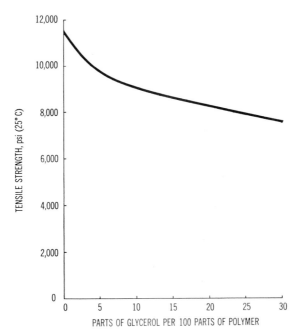

Fig. 3 Polyvinyl alcohol films: tensile strength as a function of plasticizer concentration. Data are for samples of high molecular weight (viscosity = 50 cp for a 4% aqueous solution at 20°C), 98% hydrolyzed, conditioned at 50% relative humidity before testing.

discoloring and have remained colorless after 16 years of indoor exposure [3]. Polyvinyl alcohol film is substantially transparent to ultraviolet light in the range of 254 to 400 mμ. The infrared absorption characteristics of polyvinyl alcohol have been summarized by Krimm [4].

When oriented by stretching, polyvinyl alcohol film polarizes light, and this has led to its use in photographic applications, sunglasses, lamps, and the like.

C. ELECTRICAL PROPERTIES

Polyvinyl alcohol films have no insulating properties. The hygroscopic nature of the films makes the electrical properties quite sensitive to humidity (Table 2).

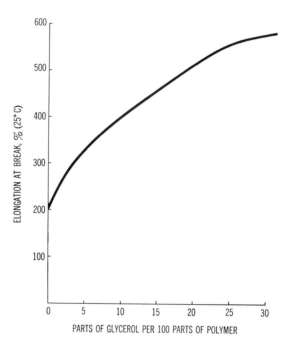

Fig. 4   Polyvinyl alcohol films: elongation at break as a function of plasticizer concentration. Data are for samples of high molecular weight (viscosity = 50 cp for a 4% aqueous solution at 20°C), 98% hydrolyzed, conditioned at 50% relative humidity before testing.

TABLE 2

Typical Electrical Properties of Polyvinyl Alcohol Film

	Dried Over P_2O_5		Conditioned at 50% Relative Humidity	
	60 Cycles	1.6 Megacycles	60 Cycles	1.6 Megacycles
Power factor	0.07–0.20	0.03–0.09	0.20–0.60	0.07–0.20
Dielectric constant	5–10	3–4	8–25	4–7

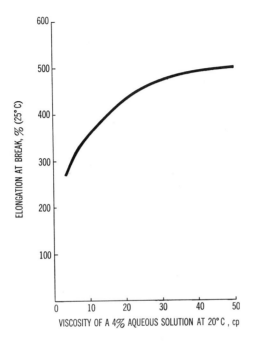

Fig. 5 Polyvinyl alcohol films: elongation at break as a function of molecular weight, as judged by viscosity. Data are for samples at 98 to 99% hydrolysis, plasticized with 15% glycerol, condititioned to 50% relative humidity at 25°C.

III. Commercial Status

Polyvinyl alcohol film has been available commercially in the United States since 1948 [5]. Sales were less than 1 million lb/yr in the United States in 1960 [6]. Acceptance in other countries has been at a somewhat slower rate [7]. The major commercial application, vacuum-bag molding, utilizes the excellent gas barrier and strength properties of polyvinyl alcohol film. Another growing use, unitized packaging of water-soluble particulate materials, also takes advantage of the fact that some grades of polyvinyl alcohol are soluble in both hot and cold water. Markets are growing steadily as the technical problems, particularly those involving film sealing, are being solved [8].

IV. Film Preparation

Polyvinyl alcohol film can be made by conventional solution casting from a 10 to 30% aqueous solution of any commercial grade of polyvinyl alcohol [3]. Selection of a grade will depend on the properties required. The casting solution will contain all desired ingredients, such as dyes, insolubilizers, and plasticizers. In some cases compatible hydrophilic polymers are included as permanent plasticizers. Surface-active agents may be incorporated either for their effect on the film properties or on the casting operation; for example, sodium alkyl sulfate improves the release of the film from the casting surface. Polished metal makes an acceptable casting surface. Drying temperatures below 210°F are chosen to avoid too rapid drying.

Post-treatments of the cast film will depend on the application visualized. Most industrial uses can be carried out with ordinary plasticized polyvinyl alcohol film that has not had any special coating or mechanical treatment.

V. Film Applications

Although polyvinyl alcohol films have good tensile and tear strength, most applications rely on the unique combination of water solubility, gas impermeability, and high strength possessed by polyvinyl alcohol film. Some of the most promising applications are listed in Table 3.

TABLE 3

Applications for Polyvinyl Alcohol Films

Applications Depending on		
Impermeability to simple gases	Resistance to organic solvents	Solubility in water
1. Airtight bags for molding	1. Bags and liners for oils, greases, paints or chemicals	1. Bagging unit quantities of bleach, bluing, detergent, or agricultural chemical
2. Oxygen tents	2. Oil-resistant diaphragms	2. Laundry bags
3. Wrapping polished metal sensitive to air or industrial fumes		3. Backing for lace made with Schiffli equipment
4. Protecting reactive materials such as tire camelback		4. Water-soluble thread

References

1. W. O. Herrmann and W. Haehnel, German Patent 450,286 (Oct. 5, 1927).
2. A. Beresniewicz, *J. Polymer Sci.*, **35,** 321 (1959); **39,** 63 (1959).
3. C. P. Argana, *Soap Chem. Specialties*, **35,** No. 8, 46 (1959).
4. S. Krimm, *Fortschr. Hochpolymer.-Forsch.*, **2,** 51 (1960).
5. T. Motoyama, *Kunststoffe*, **50,** No. 1, 33 (1960).
6. Anon., *Chem. Eng. News*, **38,** No. 51, 28 (1960).
7. Anon., *Brit. Plastics*, **35,** No. 2, 72 (1962).
8. T. S. Bianco, *Package Engineering*, **8,** No. 11, 71 (1963).

CHAPTER 11

POLYCARBONATES

LAWRENCE D. BURKINSHAW and DAVID W. CAIRD

General Electric Company
Pittsfield, Massachusetts

I. Historical Background

Polyaryl carbonate resins have been known as a class of polymers for many years, but commercially useful products have been available only in the last decade. In 1898 Einhorn allowed hydroquinone and resorcinol to interact with phosgene in pyridine to form low molecular weight polymers [1]. The same polymeric materials were prepared by Bischoff and von Hedenström in 1902, who used an ester exchange reaction [2]. No further references to polyaryl carbonate resins appear in the literature for more than 50 years, probably because of the unpromising properties of the original polymers investigated.

In the early 1930's W. H. Carothers and his co-workers included the aliphatic polycarbonate polymers [3] as part of their classical study of polymerization and ring formation. No commercially valuable polycarbonate resins were identified at that time.

Subsequent work on polyesters resulted in 1941 in the development of high-molecular-weight linear polyesters based on terephthalic acid and ethylene glycol. This, in turn, spurred the search for other high-melting polyester resins.

In the early 1950's chemists at the General Electric Company and Farbenfabriken Bayer A.G. undertook work on polyaryl carbonates by taking advantage of the more recent polyester technology. Both groups discovered that resins based on bisphenol-A (4,4'-dihydroxydiphenyl-2,2-propane) possessed mechanical properties that had substantial commercial promise.

Both companies started publishing information on the new polymer in 1956–1957. The basic patent, U.S. Patent No. 3,028,365, was subsequently issued to Bayer in 1962. Since the beginning, both companies have developed their separate technologies and manufacturing capabilities. Bayer produces Makrolon® resins at their plant at Uerdingen, West Germany. General Electric manufactures Lexan® polycarbonate resins at Mount Vernon, Ind. Mobay Chemical Company, initially a joint venture of Bayer and the Monsanto Company and now wholly owned by Bayer, produces Merlon® polycarbonate resins at New Martinsville, W. Va. Three companies, Teijin Limited, Mitsubishi Edogawa, and Idemitsu, produce polycarbonate resins in Japan. No basic patent has issued in that country.

By far the largest percentage of the polycarbonate resin produced has

® Registered trademarks.

been used in injection-molding applications. A solution-cast film was introduced in Germany by Agfa A.G. and in the United States by the Ansco Division of General Aniline and Film Corporation in 1959. Bayer has also produced a series of thin cast films for capacitor use, the most recent being a cast, oriented, crystallized film. Extruded films have been produced by custom extruders in limited volume ever since the resin was first commercially available.   In 1962 the General Electric Company started production of extruded polycarbonate films.

II. Chemistry

A. BISPHENOL-A

The chemistry of the polyaryl carbonates is such that an almost un-limited number of variations is possible. The present commercially available polycarbonate resins, however, are all based on bisphenol-A. The variations in the commercial polymers represent differences in molecular weight, additives and stabilizers, minor additions of co-reactants, and fillers.

Bisphenol-A is a constituent of epoxy resins and is produced in large volume for that use. It is produced by the acid catalyzed reaction between phenol and acetone.

$$2C_6H_5OH \ + \ (CH_3)_2CO$$

$$\downarrow \tag{1}$$

$$HO-\!\!\left\langle\!\!\bigcirc\!\!\right\rangle\!\!-C(CH_3)_2-\!\!\left\langle\!\!\bigcirc\!\!\right\rangle + H_2O$$

A much higher quality bisphenol-A is required for polycarbonate manufacture than for conventionally used epoxy resins. A subsequent purification step is usually required to obtain the necessary quality.

B. MANUFACTURE OF POLYCARBONATES

It was mentioned earlier that the original synthesis of polyaryl carbonates by Einhorn was by phosgenation and that the later synthesis by Bischoff and von Hedenström was by means of transesterification. These two methods still exist today as the basic commercial techniques for making bisphenol-A polycarbonate resins.

1. Transesterification

This process is carried out by reacting bisphenol-A and diphenyl carbonate in the presence of a basic catalyst in a stirred reactor.

An excess of diphenylcarbonate is used to stabilize the reaction, a two-step process. The bulk of the phenol is removed in the first step which is carried out at an elevated temperature at reduced pressure. The temperature is further raised (to the 280 to 300°C range) and the pressure reduced in the second step to strip off final traces of phenol and diphenyl carbonate to produce the high-molecular-weight viscous polymer. The polycarbonate resin is then forced out of the polymer kettle and pelletized. This process is similar to the standard reaction between dimethyl terephthalate and ethylene glycol to form polyethylene terephthalate, in which case methanol is removed from the reaction by heat and vacuum.

2. Phosgenation

The phosgenation of bisphenol-A to form polycarbonate resin is carried out by using two different techniques. In either case the reaction is the same.

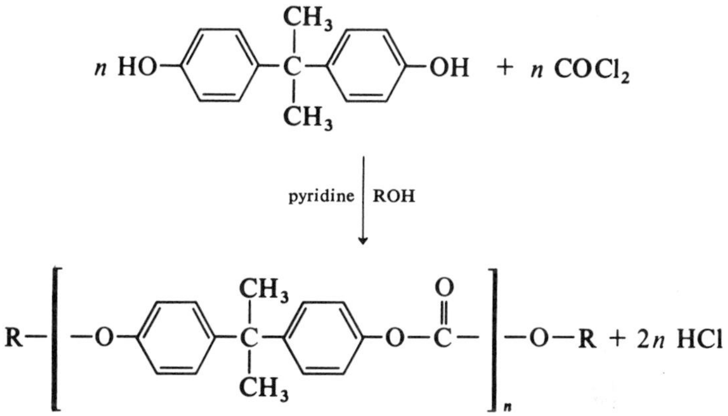

$$n \; HO-\!\!\!\!\bigcirc\!\!\!\!-\underset{CH_3}{\overset{CH_3}{C}}-\!\!\!\!\bigcirc\!\!\!\!-OH \;+\; n \; COCl_2$$

pyridine | ROH

(3)

$$R-\left[-O-\!\!\!\!\bigcirc\!\!\!\!-\underset{CH_3}{\overset{CH_3}{C}}-\!\!\!\!\bigcirc\!\!\!\!-O-\overset{O}{\overset{\|}{C}}-\right]_n-O-R \;+\; 2n \; HCl$$

Provision must be made for removing the hydrochloric acid from the reaction. This is usually accomplished with pyridine, caustic soda, or other alkaline hydroxides to remove the hydrogen chloride as it is liberated. Tertiary amines are used as catalysts, and a monofunctional aromatic hydroxy compound, such as phenol, is used as a chain stopper to control molecular weight. The basic processes are described as follows.

1. A solution of bisphenol-A in a mixture of pyridine or some other amine and an inert solvent (commonly methylene chloride) is phosgenated. The reaction is usually run at room temperature and atmospheric pressure. The polymer as formed stays in solution and is recovered from the reaction mixture by first washing to remove the pyridine hydrochloride and then precipitating.

2. The second technique involves a two-stage reaction in which bisphenol-A, water, sodium hydroxide, and methylene chloride are combined in an agitated reaction vessel. Phosgene is added in the first stage to produce a low-molecular-weight polymer. A tertiary amine is then added and the high-molecular-weight viscous polymer rapidly forms. The aqueous and organic phases are separated. The polymer solution is then washed and precipitated.

C. Molecular Weight

High purity grades of bisphenol-A, which are now available and normally used in commercial manufacture, can produce resins that span a wide range of average molecular weights. Resins in any desired range can be prepared readily by use of a suitable monofunctional chain stopper, such as phenol, in amounts predetermined by the process and by the molecular weight required.

Osmotic (number-average) molecular weights typical of the polycarbonate resins used in melt fabrication processes range from about 15,000 to roughly 25,000. Corresponding intrinsic viscosities, $[\eta]$†, range from about 0.50 and 0.65 dl/g. The number-average molecular weights of resins used in solution-casting processes range, typically, from about 35,000 to about 50,000, with corresponding intrinsic viscosities between 0.90 and 1.30 dl/g. Weight-average molecular weights, determined by light-scattering measurements on resins produced from high-purity bisphenol-A, are very close to twice the number-average values obtained from osmotic determinations. $\overline{M}_w/\overline{M}_n$ ratios derived by gel-permeation chromatography are between 2.0 and 3.3 for commercially available polymers. These relationships of weight-average to number-average molecular weight are characteristic of linear condensation polymers. Fractionation studies have shown a statistically most probable molecular-weight distribution. (See Volume I, pp. 34, 778, 790.)

III. Melt Extrusion

A. Extrusion Resins

Extruded polycarbonate films which have the characteristic clarity, physical strength, and dimensional stability of these resins can be produced with resins which have a spectrum of intrinsic viscosities ranging from 0.5 to 0.7 dl/g. If resins of lower molecular weight are used, certain film properties begin to suffer, notably those such as toughness that reflect ductility or ability to absorb energy by stretching. On the other hand, the melt viscosities of resins with intrinsic viscosities above 0.7 dl/g may become excessively high for fabrication unless undesirably high extrusion temperatures are used, with the danger of initiating degradation. Minor increases in ductility may be achieved

---

† Intrinsic viscosities referred to in this chapter were all determined in dioxane at 30°C.

with the higher molecular weight resins in this extrusion range. This effect is reflected primarily in greater retention of elongation on thermal aging.

Melt viscosities of polycarbonate resins increase rapidly with molecular weight and show about a tenfold increase over the intrinsic viscosity range from 0.5 to 0.7 dl/g. The correlation at 600°F and 8.2×10^5 dynes/cm^2 shear-stress is shown in Figure 1. This increase in melt viscosity as molecular weight increases can be compensated to a substantial extent by use of higher extrusion temperatures. Temperatures up to at least 580°F are feasible because of the good thermal stability inherent in clean, properly dried resins. The decrease in melt viscosity with increase in temperature of a typical extrusion resin is shown in Figure 2. At no temperature in the extrusion range do melt viscosities

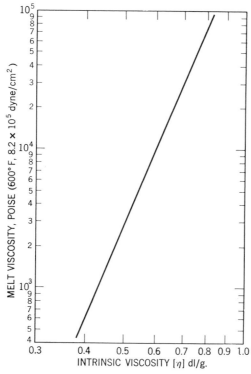

Fig. 1 Melt viscosity as a function of intrinsic viscosity of polycarbonate resins. Intrinsic viscosities were measured in dioxane at 25°C.

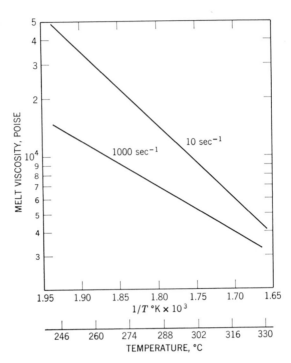

Fig. 2   Melt viscosity as a function of temperature and shear rate (sec^{-1}) for a poly-carbonate of $[\eta] = 0.56$ dl/g.

exhibit any abrupt changes or discontinuities versus temperature.   In conjunction with the high tenacity and noncorrosive nature of the resin these melt viscosity properties afford desirable characteristics for extrusion processing.

Extrusion resins are normally provided in pellet form by extrusion of carefully blended powders to ensure homogeneity.   Inhomogeneous mixing, or a too broad variation in molecular weight, can cause undesirable fisheye effects in extruded films.

B. Extrusion Processing

General theory and practice of extrusion is the subject of Chapter 8, Volume I.   Here only matters of explicit import are considered.

Temperatures feasible for extrusion of polycarbonate resins are limited at the low end to about 480°F by softening and melting properties and at the high end to about 580°F by initial thermal-oxidative-moisture degradation. Optimum extrusion temperatures are generally found between 500 and 530°F. These temperatures are high and good extrusion practice is essential, since thermal degradation is dependent on the residence time of the resin in the extruder and on the possible presence of trapped air and traces of moisture. Basic melting and stability properties of extrusion resins are shown in Table 1.

TABLE 1

Melt Properties of Polycarbonate Resins

	Mp, °F	Mp, °C
Heat distortion temperature (ASTM: 264 psi)	265–275	130–135
Softening range	420–50	215–231
Melting range (crystallized)	420–510	215–265
Extrusion range	480–580	249–304
Moisture degradation	280	138
Thermal decomposition range	600–650	315–340

Start-up temperature for extrusion is generally selected on the high side of the optimum range and the extruder and die are equilibrated to avoid excessive initial pressures. Final adjustments to temperatures optimum for the particular resin and die geometry are made while operating. A uniform temperature profile has been found effective. Water cooling adjusted at the throat to keep the cylinder somewhat cool in this region prevents premature softening and ensures a positive bite on the feed pellets. It also provides some degree of venting which reduces the tendency to trap air and moisture during compression and melt-down. Screw cooling is in general neither required nor used.

It is desirable to run die temperatures slightly lower than the extruder temperature, but a high differential is not recommended. Thermal homogeneity of the melt is required for producing film of high quality and uniform die temperatures are important. Thermal profiling of the die lips is useful in tailoring film gauge but should be used sparingly and only for fine adjustment.

If resins are improperly preheated or dried, the maximum operating temperatures can be severely limited. It is important to avoid dead spots in the extruder and die and to avoid introducing excess air into the barrel. It is particularly important to take precautions during extruding operations to ensure that the moisture content will be held below the

0.04% maximum level recommended by the manufacturers. Preheating and drying at 250°F in a suitable oven or drying hopper is essential.

Maintaining low moisture content in the pellets is possibly the one most important requirement for good-quality processing. Excessive moisture is indicated by bubbles or surface streaking which may appear without melt discoloration; such bubbles usually occur in readily detectable numbers. If oxygen is also present, the resin may become dark and less viscous, especially at elevated temperatures. Appearance of slugs of dark thermally degraded material is evidence of holdup and localized overheating.

C. Extruders

Polycarbonate resins are readily processed on modern commercial extruders capable of attaining thermostatically controlled temperatures to at least 600°F. It is desirable to have independent zone control of heaters on the cylinder, adapter, and die. Choice of cylinder liner and screw materials is flexible, since the polycarbonate melt is noncorrosive: Xaloy or nitrided stainless steel liners and nitrided stainless steel or chromium plated screws have performed satisfactorily.

Extruders with length to diameter ratios of 20 or higher are preferred, since they promote superior homogenization of the melt and uniform output. The thermal stability of the resin is such that considerable working can be tolerated.

D. Screws

Film gauge variations due to melt flow and pressure pulsations originating in the extruder are effectively minimized in polycarbonate films by proper selection and proper use of a long-metering type of screw design. It is desirable to use screws having metering sections with eight or more metering turns and compression ratios of about 2.5 to 1, or greater, to ensure uniformity in rate of extruder output. A long metering section achieves greatest constancy in output but of necessity tends to limit output rate because of the generation of frictional heat that may cause thermal over-ride conditions and possible thermal inhomogeneity in the melt at high screw speeds. Careful cooling of the extruder barrel may be required to maintain temperature in the high screw speed range, although in producing high-quality films it is generally unnecessary and undesirable to operate this type of screw at speeds exceeding about 40 rpm.

The flight pitch and flight length in screws for polycarbonate resins should be chosen to minimize the possibility of entrapping air in the feed and compression zones and to avoid any material holdup in the corners of the flights.

E. BACK-PRESSURE CONTROL

Extruders are normally designed with a breaker plate at the end of the barrel which provides support for a screen pack. The screen pack performs a dual function: it restricts melt flow, causes pressure buildup in the screw to promote good compression, melting and homogenizing of the resin, and provides a filter to remove particulate foreign contaminants. When a suitable metering screw is employed, as in film extrusions, a 40–80–100 mesh screen placed in that order against the breaker plate is generally satisfactory. Some extruders employ an adjustable valve in place of the breaker plate and screen pack for control of back pressure.

F. DIES

Melt viscosities of polycarbonate resins are quite high and in general tend to retain their high viscosity characteristics at the shear stresses encountered in film extrusion. Typical melt pressures in the die may vary from 500 to 3000 psi or higher, depending on processing conditions, including temperature, output rate, die geometry, and resin grade. It is important that die design be rugged and capable of withstanding these high internal pressures at elevated temperatures and over extended periods of operation, with minimal deflection at the die orifice. This is especially true in the case of wide film dies to prevent introducing critical gauge variations across the extrudate web.

Despite the characteristically high melt viscosities, polycarbonates of extrusion grade exhibit relatively low degrees of melt elasticity, or nerve, in the extrudate. Accordingly, it is unnecessary to employ excessively long die lands. In general, a 1-in. land length is adequate for producing good extrudate appearance and quality, although land marks and related optical irregularities can be further minimized by use of a somewhat longer land. Polished chromium-plated land surfaces, together with streamlining of the die that will avoid abrupt changes in the flow path to eliminate holdup and to promote uniform flow, are most effective in reducing normal extrusion defects. Furthermore, it is important to provide uniform heating to minimize thermal gradients in the die.

Flat film dies, designed for vertical or horizontal extrusions, are normally used, since flat extrudate web is most advantageously handled on present take-off equipment. Blown film dies have not been extensively used because of problems, due to wrinkling, in collapsing the tube. Such problems are made severer by the high glass-transition temperature and the low creep properties of polycarbonate films. The latter also tend to produce a detectable, permanent curvature imparted by the radius of the blown tube.

G. TAKE-OFF EQUIPMENT

Flat-web extrudates can be readily handled in conventional film take-off equipment with suitably heated chill rolls. The unusually high glass-transition temperature (284 to 293°F) at which these resins solidify and develop rigidity requires that temperatures approaching this magnitude be used in the chill rolls to prevent cooling strains which cause film distortion. Overcooling causes transverse ripples which are eliminated at proper chill-roll temperatures. Temperatures are generally in the range between 270 and 280°C. It is important that the chill-roll temperature during processing not exceed the glass-transition temperature of the resin, since the hot web may tend to tack to the roll surface. Oil heat readily provides the required temperatures and control and is generally used. Since machine-direction slippage on the take-off rolls can cause severe periodic gauge variations, it is important to provide for close chill-roll-temperature and web-tension control and to prevent translating any change in tension to the molten web at the die.

Molten polycarbonate resins have high tenacity, and the extrudate web can tolerate a high degree of drawdown. Die-gap settings of 20 mils or higher are frequently used, and drawdown ratios of initial to final web thickness, up to 20 to 1, may be used for producing thin films. Drawdown capability is unusually flexible and can be adjusted over this broad range to obtain the desired gauge. Excessive take-off speeds and drawdown ratios, however, may tend to cause ripples in the machine direction which are attributable to entrapment of air under the extrudate web as it travels over the chill-roll.

H. Precautions

If the extrusion run has to be interrupted for an appreciable period, the cylinder temperatures should be reduced to 320°F. This avoids degradation of the resin and also prevents damage to the cylinder wall by contraction of the adherent material when cooled below its glass-transition temperature.

On completion of the extrusion the extruder should be purged with crystal polystyrene or high-density polyethylene as the temperature is being lowered. The temperature should not be allowed to drop below 320°F until the polycarbonate resin has been cleared from the barrel of the extruder.

IV. Casting from Solution

A. General Considerations

Polycarbonate resins are advantageously adapted to solution-casting technologies: good casting solvents, such as methylene chloride and chloroform, are among the best polycarbonate solvents; resins in the high-molecular-weight range which cannot be readily fabricated by extrusion have optimum solution viscosities for casting; films from these resins have excellent toughness, physical strength, and dimensional stability characteristics. Furthermore, processing is inherently simple: casting dopes can be easily prepared and handled without difficulty on band or wheel machines to produce films in thicknesses up to 8 mils and widths up to at least 60 in. which have high clarity, excellent uniformity, and freedom from optical imperfections.

B. Casting Resins

Casting grades of polycarbonates are in a molecular-weight range which is defined by the resin solubility and the concentrated solution viscosity in the selected casting solvent. The basic relationships of these parameters are illustrated in Figure 3 for methylene chloride solutions. This solvent becomes saturated with resin at about 23% by weight, more or less independently of molecular weight. At lower concentrations solutions are stable and exhibit indefinitely long shelf-life. In this range the logarithm of solution viscosity η_s increases linearly with weight-percent concentration and is also an approximately

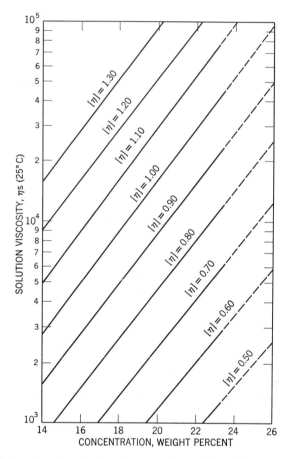

Fig. 3   Solution viscosity of polycarbonate resins of different intrinsic viscosities as
a function of concentration in methylene chloride.

linear function of intrinsic viscosity. The empirical correlation for
methylene chloride solutions is expressed by

$$\log \eta_s = 2.23[\eta] - 0.562 + \frac{[\eta]C}{6.62[\eta] + 1.25},$$ (1)

where concentration C is expressed in weight percent and the resin
intrinsic viscosity $[\eta]$ has been measured in dioxane at 30°C. At

concentrations exceeding 23%, solutions become supersaturated and the resin may precipitate as a crystallized phase. This solubility limit defines the maximum useful concentration for the casting dope, since solutions are unstable and shelf-life becomes uncertain.

Desired dope viscosity is dictated by the flow properties required by the casting process. Typically, viscosities of $20-25 \times 10^3$ cp, or somewhat higher, are used. Dope viscosities in this range, at resin concentrations approaching the maximum of 23 wt% in methylene chloride, are obtained with casting-grade resins which have intrinsic viscosities of approximately 0.95–1.15 dl/g. These intrinsic viscosities correspond to molecular weights $\overline{M}_w$ of 70,000 and higher. The optimum molecular-weight range for a casting resin varies if the resin is used with another solvent or solvent system, depending on the solubility limit and the solution viscosity properties in the new solvent. Solubility data are shown in Table 2 for solvents that might be used for casting.

TABLE 2

Solvent Behavior of Bisphenol-A Polycarbonate Resins[a] in Solvents Which Dissolve at least 1% of Resin at 25°C
(weight of resin to volume of solvent)

Solvent	Bp (°C)	Density[b] (g/ml)	Solubility at 25°C Resin/dl-g solvent		Weight percent, average
			Pellets	Powder	
Methylene chloride	40.1	1.335^{15}_4	36	35	22
sym-Tetrachloroethane	146.3	1.60	31	36	18.5
Chloroform	61.3	1.475^{25}_{25}	25	27	15
Methylene bromide	98.2	2.478^{25}_{25}	22		8
cis-1,2-Dichloroethylene	60.1	ca 1.28	21		14.5
Pyridine	115.5	0.978^{25}_4	16.5	18.5	16.5
1,1,2-Trichloroethane	113.5	1.4416^{20}_4	12		8
Thiophene	84.1	1.0617^{20}_4	10		9
1,2-Dichloroethane	83.7	1.252^{20}_4	7	11	7
Bromoform	149.5	2.902^{15}_{15}	7		2.4
sym-Tetrabromoethane	151/54 mm	2.964^{20}_4	5.5		1.8
Dioxane	101.5	1.030^{20}_4	5	6.2	5.1
Tetrahydrofuran	65.0	0.888^{20}_4	3.8	1.2	1.3–4.1

[a] $[\eta]$ from measurements in dioxane at 25°C = 0.6 dl/g.
[b] Values from Merck Index.

Other resin parameters which influence the definition of casting grade polycarbonates are those that promote best clarity, uniformity, and good physical properties. The molecular weight level, defined by viscosity and solubility considerations, ensures highest physical properties. Clarity and uniformity require absence of inhomogeneous contaminants, including gels, foreign inclusions, and agents which can promote crystallization. In addition to inhibited crystallizability resulting from the high molecular weight of casting resins, grades with further reduced tendency to crystallize have been developed by copolymerization with modifying amounts of components that introduce bulky substituents or asymmetry in the bisphenol-A polycarbonate chain.

C. CLEAR FILMS

Photographic and graphic-arts films require optical clarity. For these and related uses it is important in the casting process to minimize or eliminate crystallization which may produce hazing or clouding.

In general, casting-grade polycarbonate resin solutions in good casting solvents persist as clear supersaturated solutions for conveniently long periods, as concentration increases beyond the point at which crystallization can begin. Noncrystalline films are formed from these solutions as the concentration approaches 40 wt% of resin, unless solvent removal is retarded sufficiently to promote nucleation and crystallite growth. These solvent-swollen clear films remain subject to crystallization. Normal control of ambient air velocity, temperature, and vapor concentration effectively inhibit crystallization which can cause detectable hazing.

Present casting resins, for example, Lexan® casting grade, have a molecular-weight range and composition that permit casting clear films up to at least 20-mil thicknesses. Drying times, however, become uneconomical at thicknesses exceeding 8 to 9 mils.

In a typical operation for casting clear polycarbonate films of optical quality a casting resin powder is dissolved in the required amount of solvent by mixing in a suitable tank. The viscous raw dope is filtered through coarse and polishing filters. Dissolved air is removed by heating to boiling and the degassed solution is stored in a holding tank. From here it is piped to thermostatically heated casting tanks

---

® Registered trademark.

and brought to uniform temperature somewhat below the boiling point of the most volatile solvent component.

The conditioned casting dope is transferred continuously through insulated piping to the casting hopper in which constant hydrostatic head and carefully regulated thermal and vapor ambients are maintained to ensure uniform flow-out onto the polished casting wheel or belt. Highly polished stainless steel or a mirror-smooth surface of regenerated cellulose on a stainless steel or copper substrate is preferred for casting when optical quality is desired, since polycarbonate films faithfully replicate any imperfections of the casting surface.

The temperature and velocity of the circulating air and vapor ambient in the machine enclosure are programmed to facilitate uniform and rapid solvent diffusion, evaporation, and removal. Vapor-saturated air is continuously drawn off to a brine-cooled condensing system for solvent recovery, and fresh preheated air, at desired temperatures in the range of 60 to 100°C, is introduced at controlled rates in desired locations. Achievement of highest casting speeds and best film quality requires the optimization of these variables with regard to the film thickness being produced and the solvent volatility.

As previously mentioned, polycarbonate films are first generated from solutions at about 40% resin concentration. At this initial point the solvent-laden film is weak and requires support. At about 65 to 80% solids films can be detached from the casting surface under carefully controlled tension. Tension and amount of residual solvent affect adhesion and are optimized to eliminate stick-mark phenomena. The self-supporting free film is continuously finish-dried, with both surfaces exposed in an in-line oven enclosing conveyor rollers which are spaced and tensioned to prevent curling or stretching of the film. The oven environment is circulated heated air, with final drying temperatures at about 110 to 120°C.

Edges of the dried film are normally embossed to aid in windup and to protect the surface from scratching. The film is wound as it comes from the finish-drying oven enclosure.

D. Crystallized Films

At some sacrifice in optical clarity crystallized polycarbonate films can be prepared from casting grade resins. These films can be advantageously oriented and are useful as capacitor dielectrics and for other electrical applications.

Crystallization can be produced by retarding the rate of solvent removal, but it is more effectively induced by adding a volatile crystallization promoter to the casting dope. Promoters are swelling and plasticizing agents that increase mobility of the macromolecules and facilitate crystalline formation. They are nonsolvents for the crystalline polycarbonate phase. Agents in this class include toluene and aliphatic ketones, esters, and ethers of low molecular weight. It is desirable to choose a swelling agent with a boiling point sufficiently above that of the casting solvent to ensure its retention by the film during initial solvent removal. The plasticized film crystallizes rapidly during intermediate drying stages and the volatile promoter is removed in the final drying. Promoter concentrations of about 30 to 35 parts per 100 parts of resin are effective with bisphenol-A polycarbonate resins and process conditions are chosen to bring about maximum crystallization.

A second route for producing crystallized films is by treating the clear amorphous films with the swelling agents after casting. These films are readily dilated and plasticized by immersion in these liquids and crystallization is induced without destroying the integrity of the film. After drying the film is equivalent to one produced by the casting process.

In either process the molecular weight of the resin is an important factor in regulating the physical properties of the crystallized film. Casting-grade resins produce films that are strong and elastic when dilated. Polymers of lower molecular weight, such as extrusion grades, are adversely affected and become cheesy or embrittled on crystallization.

In the plasticized elastomeric condition the tough crystallized film can be readily stretch-oriented at rather low loadings and temperatures and subsequently dried to remove plasticizing solvent. Alternatively, the crystallized films can be predried and then stretch-oriented at elevated temperatures by usual methods. (See Volume I, Chapter 10.)

V. Film Characteristics

A. MECHANICAL PROPERTIES

It was mentioned earlier that a large difference [$\overline{M}_n$ = 15 to 25,000 compared with $\overline{M}_n$ = 35 to 50,000] exists in the molecular weights of extrusion-grade and casting-grade polycarbonate resins. In spite of this difference the mechanical properties of films made by solvent casting are only slightly better than those made by melt extrusion. The differences that do exist probably reflect the better gauge uniformity of the solvent-cast films as much as they reflect the higher molecular weight. However, the solvent-cast films do lend themselves to orientation and crystallization. Major changes can be made in the mechanical properties of cast films by these techniques. Table 3 gives selected property data on polycarbonate films prepared by various techniques.

It will be noted that the "as-extruded" and "as-cast" values are similar. The slightly lower average tensile modulus in the cast film probably is the result of traces of residual solvent in that film. Oriented polycarbonate films show moderate increases in tensile yields in the machine direction and much larger increases in ultimate tensile strengths. The transverse direction mechanical properties are largely unaffected by the monaxial orientation process. The crystallized films have a significant advantage in thermal properties when compared with noncrystallized forms. They retain excellent dimensional stability on heating to temperatures up to at least 165°C. This may significantly exceed the glass-transition temperature, about 145°C, of amorphous films.

Additional physical property data are given in Table 4 for 1-mil extruded polycarbonate film.

B. PERMANENCE AND CHEMICAL RESISTANCE

Polycarbonate films exhibit excellent resistance to oxidative embrittlement. After six months at 75°C extruded and solvent-cast films showed no measurable change in tensile yield and ultimate tensile. In one month at 125°C the tensile-yield point and the ultimate tensile strength of similar samples increased some 10%. Most of this change took place within the first 48 hr. The elongation of the same films dropped from an initial average of 135% to a final elongation of 120%.

No significant changes have been noted in tensile and elongation properties measured on films immersed in water at room temperature for several months. Immersion in boiling water causes an initial drop in elongation from 100 to 50% within a period of 4 hr. Thereafter there is a further gradual reduction in properties. The films still retain essentially all of their tensile-yield strength after immersion in boiling water for periods of as long as one week. Long-term immersion in water at elevated temperatures does result in hydrolysis and crystallization, however.

TABLE 3

Properties of Polycarbonate Films Prepared by Various Techniques

Property	Units	ASTM method	As extruded	As cast	Cast monaxially oriented		Cast monaxially oriented and crystallized	
					MD[a]	TD[a]	MD[a]	TD[a]
Tensile yield	psi	D 882–56T	8000–9500	7500–9500	10,000–13,500	7300–8300	9000–13,500	9000–10,200
Ultimate tensile	psi	D 882–56T	8500–12,000	8500–12,000	15,500–23,500	8500–12,000	15,000–35,000	8500–12,000
Elongation	%	D 882–56T	100–130	100–120	40–65	100–135	30–65	150–225
Tensile modulus		D 1530–55T	300,000–370,000	310,000–330,000	310,000–330,000	255,000–275,000	400,000–430,000	250,000–270,000
Tensile heat distortion [temperature at 2% deformation]	°C	D 1637–59T	150–155	153–156	150–155	150–155	165–170	160–195

[a] Machine direction and transverse direction, respectively, of the sheet as cast.

TABLE 4

Selected Properties of 1-Mil Extruded Polycarbonate Film

Property	Units	ASTM Test Method	Value
Area factor	in./lb		23,100
Burst strength	Mullen points	D–774	25–35 (on 4 mil film)
Tear strength	g/mil	Elmendorf	20–25
Tear strength	psi	D–1004	1150–1570
Folding endurance		D–643–43 [B]	250–400
Water absorption (24 hr)	%	D–570	0.35

The gas and water-vapor permeabilities of polycarbonate films are summarized in Table 5.

The optical properties of polycarbonate films are attractive for packaging and decorative applications. As shown in Figure 4, films are transparent at wavelengths above 275 mμ. Between 350 and 800 mμ transmission values are approximately 83 to 86% for 1.2 mil extruded film.

Polycarbonate film absorbs and is slowly degraded by radiation energies in the ultraviolet range. The natural grade of material is not

TABLE 5

Gas and Water-Vapor Permeability, 1-Mil Polycarbonate Film

Gas	ml/mil-100 in.^{2}-24 hr-atm(STP)
Air	70
Oxygen	200
Nitrogen	50
Carbon dioxide	800
Hydrogen	1800
Argon	95
Helium	1660
Sulfur hexafluoride	0.0011
Water[a]	

[a] 11 g/100 in.^{2}-24 hr, measured on 1-mil film at 100°F and 90% relative humidity (ASTM E96–53T).

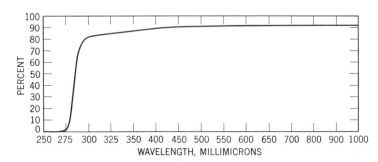

Fig. 4   Light transmission of Lexan polycarbonate resins.

recommended for continuous outdoor service. Ultraviolet-stabilized grades are commercially available. Samples of such ultraviolet-stabilized 3-mil film were exposed in Florida for six months. The films had an average elongation of 10% after this exposure and could be creased and folded without any cracking.

Polycarbonate films have good resistance to irradiation. Samples films have been exposed to electron-beam irradiation in dosages of 5.0×10^7 r. The films appeared stable to this dosage, although the elongation dropped from an average of 100 to 30%.

Extruded polycarbonate films have also been exposed to pile irradiation. At a dosage of 10^8 r the films had a residual elongation of about 40%. These samples turned slightly yellow but did not show any sign of embrittlement.

Extruded bisphenol-A polycarbonate films contain no plasticizers and have excellent outgassing properties. Extruded films have been tested at vacuums as high as 10^{-7} mm Hg at 100°C without detection of outgassing.

Bisphenol-A polycarbonate films have been sanctioned by the Food and Drug Administration for food packaging, handling, and dispensing. In addition, the material meets the requirements of the Meat Inspection Division of the U.S. Department of Agriculture for uses involving contact with meat and meat food products. Lexan® resists staining by common foodstuffs and neither contributes to nor picks up odors. It is also fungus-resistant when tested according to MIL-E 4970B, -5272C, -5422E, and MIL-F-8281A. (See also Volume I, Chapter 16.)

The chemical resistance of a film is usually important only as it relates to such items as food packaging, printing, laminating, and specific industrial uses.

Polycarbonate films are resistant to dilute organic and inorganic acids and aliphatic hydrocarbons. They are attacked by amines, alkalis, and ammonia. Polycarbonate films can be dissolved by chlorinated hydrocarbons. Methylene chloride is the solvent commonly used in resin manufacture and for bonding operations. Other good solvents are shown in Table 2.

It has been found that foodstuffs in general have no effect on polycarbonate films and the films impart no taste or flavor to food. Penetration of the films by moisture or oxygen, however, may affect the quality of the food. Their broad commercial usefulness is indicated by the listing in Table 6 of typical foodstuffs and other common materials to which polycarbonate films are fully resistant.

TABLE 6

Compatibility Chart. Typical Foodstuffs and Other Common Materials to Which BPA Polycarbonate Films are Fully Resistant

Acetic acid, 5%	Isopropyl alcohol, 70%	Pine oil
Bacon fat	Kerosene	Salt solution, 10%
Beer	Lemon juice	Sardine oil
Catsup	Light lube oil	Shortening (Spry)
Citric acid, 5%	Mayonnaise	Soap solution, 5%
Cocoa	Merthiolate (tincture)	Soya oil
Cod liver oil	Milk	Tea
Coffee	Mustard	Tomato juice
Detergent solution, 2%	Oleic acid	Vick's Vaporub
Grape juice	Oleo margarine (colored)	Wine
Hydrogen peroxide, 30%	Orange juice	Wine vinegar
Iodine (tincture)	Permanent ink	Whiskey

Polycarbonate films can be printed without surface preparation. Similarly, extruded films can be emulsion-coated with polyvinylidene chloride without using primers or prior surface treatment. No chemical attack or crazing has been noted in printed polycarbonate film, probably because the amount of solvent and contact time are so low. Solvent coating with rotogravure or reverse roll coaters has not been successful when the solvent was in a class that normally attacked the film.

Special care must be taken and tests made if the film is to be used in a chemical environment at an elevated temperature. Testing should be conducted at service conditions to determine whether the film is suitable for the application.

C. ELECTRICAL PROPERTIES

Polycarbonate films have properties that are particularly useful for electrical applications. These properties appear to be unaffected by molecular-weight differences.

The dielectric constant and power factor of the amorphous films are essentially unchanged over the temperature range of 0°C to the heat distortion temperature of the films (approximately 145°C) (Figure 5). A sharp break occurs in the power factor of films at that temperature. There is evidence that the oriented and crystallized films have lower dielectric losses over the entire temperature range and that the high temperature peak is moved up about 15 to 20°C.

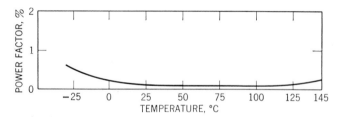

Fig. 5   Relation between power factor and temperature for Lexan polycarbonate resins at 60 cycles.

The values shown in Table 7 for the 4-mil cast amorphous films were determined at 23°C and 50% relative humidity. The values for the cast crystallized film are taken from published data.

Typical values of dielectric strength for polycarbonate films are as follows (s/s, V/mil in air at 23°C): 4-mil film, 1500; 1 mil film, 6000. Oriented and crystallized films generally give more consistent dielectric strength values than do extruded films, probably because of the better gauge uniformity of the former.

TABLE 7

Dielectric Constant and Power Factor Variation With Frequency

| Frequency | Cast amorphous film | | Cast crystallized film |
	Dielectric constant	Power factor (%)	Power factor (%)
60 cycles	2.99	0.10–0.23	0.25
1 Kc	2.99	0.13	0.20
10 Kc	2.99	0.19	
100 Kc	2.97	0.49	
1 Mc	2.93	1.10	0.55

VI. Applications

It has already been noted that the injection-molding applications of polycarbonate resin are much farther advanced than the film applications. This reflects the early development of molding resin compared with the relatively recent effort in film.

A. PHOTOGRAPHIC FILM

The use of cast polycarbonate resin for photographic applications was the first and is still one of the largest for polycarbonate film. It is a high performance product in which good optical properties are needed in combination with exceptional dimensional stability. In the United States the polycarbonates are competing with the polyester film base products for this market. The lack of a process for making high-quality extruded polycarbonate film necessitated the use of the less economical casting process. This resulted in polyester films of lower cost and severely limited market penetration by the polycarbonate films. Solvent-cast polycarbonate films are being offered by Bayer. There is currently no production of solvent-cast polycarbonate photographic film in this country.

B. ELECTRICAL APPLICATIONS

Polycarbonate films have found opportunities in two specialized electrical uses: (a) high-voltage power cable insulation and (b) capacitor film.

There is a substantial market available for film dielectrics in high-voltage power-cable insulation applications. Paper and oil combinations are now used in all cable constructions above 69 kV. At the higher voltages (345 kV and above) the electrical properties of the paper impose severe design limitations. The polycarbonate films hold much interest because of their low dielectric constant and power factor over the operating temperature range. The test periods involved in this application are very long because of the exceptional service life demanded (25 years or more). This is therefore a slowly developing but real market opportunity for polycarbonate films.

Polycarbonate capacitor films are used primarily in the higher temperature applications (up to 125°C ratings). Polyester film capacitors are generally derated when used at temperatures above 100°C because of their changing electrical properties with temperature. Both the insulation resistance and dissipation factor versus temperature curves for polycarbonates are much flatter than those of the available polyester films. As a result, polycarbonate film capacitors can more nearly carry the same ratings at elevated temperatures as at room temperature. The opportunities and demands in this market have been great enough to justify the development of a cast, oriented, and crystallized product specifically for the application.

C. OTHER USES

Considerable effort has been expended in the development of packaging applications for polycarbonate films. They have the advantages of clarity and excellent mechanical and good forming properties. The cost of the film demands that they be considered for high performance uses only, however.

A skin-packaging construction that consists of a lamination of 1 mil of extruded polycarbonate film and 2 mils of low-density polyethylene has been available for some time. The polycarbonate provides the mechanical protection. The polyethylene gives bulk to the laminate and acts as an adhesive in the construction. The laminate has excellent forming properties and is used to package sharp objects, such as hardware and electronic items.

Polycarbonate films are being used in a variety of industrial uses and evaluated in many more. The usefulness of the films derives equally from the excellent mechanical properties and their broad processing capabilities. Polycarbonate films have high sparkle and transparency. They can be readily printed by a variety of processes without prior surface treatment or special inks. These properties have led to their use in metallized and printed applications, particularly when a subsequent embossing operation is involved.

Colored films can be produced by direct extrusion. This permits greater flexibility and variety than is possible by the more common dyeing techniques for making colored metallized film.

Polycarbonate films can be readily vacuum-formed, hot- or cold-embossed, heat-sealed, and solvent-cemented. Thus the products have found use in such diverse applications as honeycombs, light diffusers, electrical spacers, solar stoves, and photographic assemblies. The development of these industrial-film uses is still in a preliminary stage. Their rapid maturity is certain because of the inherent processing and property advantages of the polycarbonate films.

References

1. A. Einhorn, *Ann.*, **300,** 135 (1898).
2. C. A. Bischoff and A. von Hedenström, *Ber.*, **35,** 3431 (1902).
3. W. H. Carothers and F. J. van Natta, *J. Am. Chem. Soc.*, **52,** 314 (1930).

CHAPTER 12

POLYURETHANES

CHARLES S. SCHOLLENBERGER and DONALD ESAROVE

The B. F. Goodrich Company Research Center   *Tremco Manufacturing Company*
Brecksville, Ohio   *Cleveland, Ohio*

487

I. Introduction

Polyurethanes constitute an interesting polymer class of unusually broad utility. Polymer chemists have taken advantage of the myriad opportunities this system offers to modify and tailor polymer structures to specific property levels, processing techniques, and end applications. Thus polyurethanes based on a host of monomers have been formed in polymerizers or *in situ*, in random or regular configuration, to provide soft-to-hard, elastic and rigid, cellular and solid, soluble and insoluble, and thermoplastic and thermosetting products. It is not surprising that such versatility has resulted in a large variety of polyurethane uses, which include film and sheet applications.

The purpose of this chapter is to discuss the current status of polyurethanes in film and sheet applications and to suggest possible future trends. An attempt has been made to describe the urethane systems used in, or adaptable to, sheet and film manufacture, the properties of polyurethanes which are relevant to such application (using values obtained from sheet and film whenever possible), and the technology involved in polyurethane sheet and film manufacture.

In preparing this chapter we have canvassed materials suppliers and fabricators known to, or expected to, have interest and experience in, or information regarding, polyurethane sheet and film. Use of such information here with appropriate acknowledgment was solicited. Information available to us from files of the B. F. Goodrich Company is also included. This procedure helped to provide the information appearing in the chapter. It has served to emphasize the fact that film and sheet applications of polyurethanes, although under way, appear to be in the primary stages of development.

II. Pertinent Polyurethane Types.   Nature and Chemistry

Polyurethane materials of several distinct general types adaptable to film and sheet production are commercially available.   They include (a) thermoplastic resins, (b) thermoplastic-thermosetting resins, (c) vulcanizable millable gums, and (d) vulcanizing liquid polymers.

The nature and chemistry of these various types are reviewed briefly and commercial examples of each type are cited.

A. THERMOPLASTIC POLYMERS

The members of this class of polyurethane resins are essentially linear polymers whose preparation follows the chemistry of

$$HO-R-OH + OCN-R'-NCO$$

$$\text{diol} \qquad \text{diisocyanate}$$

$$\downarrow$$

$$HO-R\left[O-CO-NH-R'-NH-CO-O-R\right]_n O-CO-NH-R'-NCO$$

$$\text{polyurethane}$$

or

$$Cl-CO-O-R-O-CO-Cl + NH_2-R'-NH_2$$

$$\text{dichlorocarbonate} \qquad 1° \text{or } 2° \text{ diamine}$$

$$\downarrow$$

$$ClCO_2-R\left[O-CO-NH-R'-NH-CO-O-R\right]_n O-CO$$

$$\text{polyurethane} \qquad\qquad\qquad\qquad\qquad\qquad |$$

$$NH\,R'\,NH_2$$

$$+ HCl$$

The urethane groupings are italicized in these formulas to identify them. Since approximately equivalent amounts of the diisocyanate and dihydroxy alcohol components are charged in the polymerization represented in (1), the average molecular weight of these polymers is usually high enough to eliminate significant reactive terminal group effects.

When R in these equations is small, such as the tetramethylene group, $-CH_2-CH_2-CH_2-CH_2-$, and R' is aliphatic (e.g., hexamethylene, $-CH_2CH_2CH_2CH_2CH_2CH_2-$) or aromatic (e.g., 4,4'-

diphenylenemethane, ⟨○⟩$-CH_2-$⟨○⟩$-$), the polymeric pro-

duct resulting is a true polyurethane. As usually prepared, such polymers contain a high concentration of closely and regularly spaced urethane groups in each chain. Consequently these polyurethanes are rigid plastics similar in nature to the polyamides (hard, tough, high melting, linear, thermoplastic). *Durethan U** [1] is a commercially available representative of this polyurethane type which is in use in Europe. It has a molecular weight in the range 10,000 to 40,000. True

* A glossary of trade names is given on page 521.

polyurethanes are not extensively used in this country, where the nylons are well established and apparently meeting all needs.

When R in (1) and (2) is itself polymeric in nature (molecular weight about 1000 to 3000) and polyester in structure such as tetramethylene adipate $[-(CH_2)_4-O-CO(CH_2)_4CO-O-]_n$ or polyether in structure such as 1,2-oxypropylene $[-O-CH_2CH-O-]_n$ or oxytetramethy-
$$\overset{|}{CH_3}$$
lene $[-O-(CH_2)_4-]_n$, the reactants containing it are liquid or low melting and yield flexible or elastic polyurethane products.

Certain advantageous properties result when combinations of diols, separately containing small and polymeric R groups, are used in preparing such polyurethanes. Depending on the point at which the small diols are added during polymerization, their incorporation in the polymer chains can be regulated to produce random or block polymer structures.

*Estane** thermoplastic polyurethane elastomers represent a family of commercial resins whose chemistry falls within the foregoing types [2]. Typical processing (calendering, extrusion, solution casting, and injection molding) and pigmentation characteristics mark these stable polymers as true thermoplastics [3]. They have molecular weights in the range 35,000 to 40,000 [2]. Other representatives of the thermoplastic polyurethane elastomer class include some Roylar* elastoplastics, certain Pellethane* resins, and certain Rucothane* resins. Tuftane* films for the packaging industry and specialty applications are commercially available products based on thermoplastic polyurethanes.

B. THERMOPLASTIC-THERMOSETTING POLYMERS

Application of the chemistry described in Section A, in which the diisocyanate component may be charged in some excess during polymerization to allow a degree of subsequent crosslinking by postcure of formed articles, apparently is practiced in the case of thermoplastic-thermosetting polyurethane elastomers. One might expect that the crosslinking reaction would involve allophanate link formation between the internal urethane groups of some polyurethane chains and the unusual number of terminal isocyanate groups of other polyurethane chains (resulting from the use of excess diisocyanate) as in (3):

* A glossary of trade names is given on page 521.

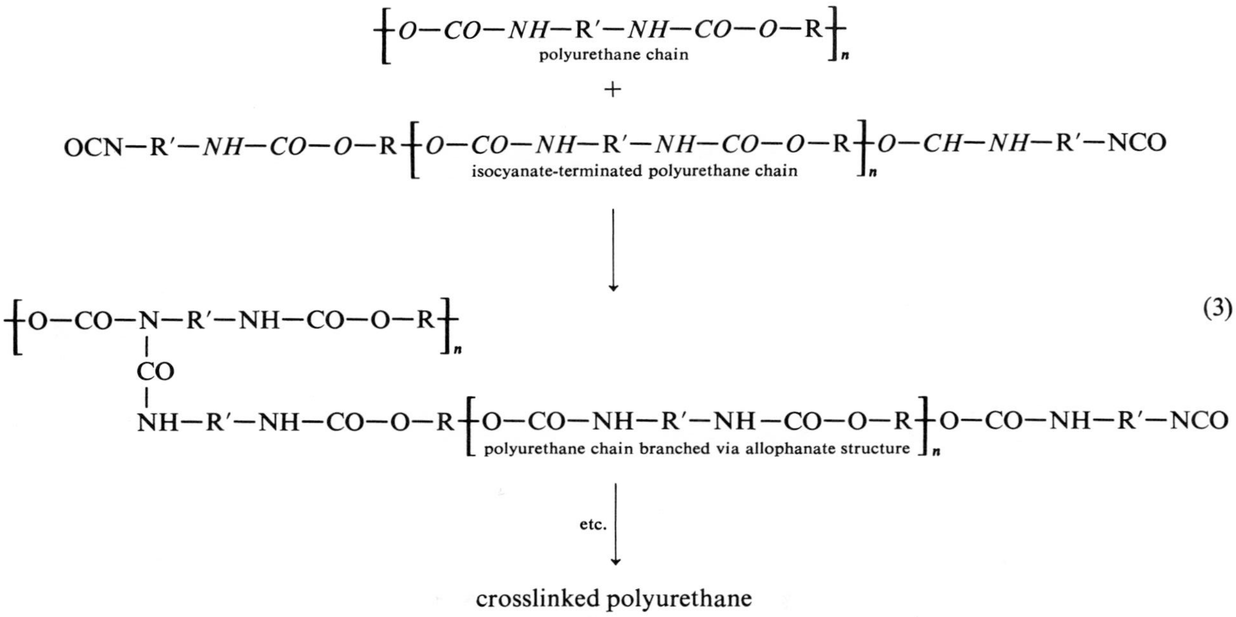

$$\Big[O\!-\!CO\!-\!NH\!-\!R'\!-\!NH\!-\!CO\!-\!O\!-\!R \Big]_n$$

polyurethane chain

$+$

$$OCN\!-\!R'\!-\!NH\!-\!CO\!-\!O\!-\!R\Big[O\!-\!CO\!-\!NH\!-\!R'\!-\!NH\!-\!CO\!-\!O\!-\!R \Big]_n\Big[O\!-\!CH\!-\!NH\!-\!R'\!-\!NCO \Big]_n$$

isocyanate-terminated polyurethane chain

$\longrightarrow$

$$\Big[O\!-\!CO\!-\!N\!-\!R'\!-\!NH\!-\!CO\!-\!O\!-\!R \Big]_n$$
$$\qquad\qquad\quad \overset{\displaystyle |}{\underset{\displaystyle |}{CO}}$$
$$NH\!-\!R'\!-\!NH\!-\!CO\!-\!O\!-\!R\Big[O\!-\!CO\!-\!NH\!-\!R'\!-\!NH\!-\!CO\!-\!O\!-\!R \Big]_n\Big[O\!-\!CO\!-\!NH\!-\!R'\!-\!NCO \Big]_n$$

polyurethane chain branched via allophanate structure

$\xrightarrow{\text{etc.}}$

crosslinked polyurethane

(3)

Texin is a commercially marketed poly(ester-urethane) resin of the thermoplastic-thermosetting elastomer type [4]. In its use the supplier recommends thermoplastic processing methods with ultimate property development in products realized on two weeks of aging at room temperature or several hours of cure at higher temperatures [5].

Uniroyal markets *Roylar* resins which are also described as thermoplastic-thermosetting polyurethanes [6], and some *Pellethane* resins are described as thermoplastic-thermosetting polyurethane elastomers. The reactive nature of such polyurethane systems becomes important in their use when moisture sensitivity, the reuse of scrap, and substantial pigmentation are considerations.

C. Vulcanizable Millable Gums

Since millable polyurethane gums are linear polymers of high molecular weight, the chemistry of Section A also applies in their formation. Millable poly(ester-urethane) gums, as in (1) (R is polyester), which can be vulcanized, are commercially available. They include *Genthanes S* and *SR, Urepan E,* and *Vibrathane 5003.* The dicumyl peroxide cure of these polymers is recommended [7–10] as is the case with *Vibrathane 5004* millable polyurethane gum [11].

In the case of the Genthanes one structure involved in such vulcanization has been suggested as the methylene bridge separating the benzene rings in the diisocyanate component

$$-O-CO-NH-\!\!\left\langle\bigcirc\right\rangle\!\!-CH_2-\!\!\left\langle\bigcirc\right\rangle\!\!-NH-CO-O-$$

[12]. In the case of Vibrathane 5003 the α-methylene position of the polyester component adipyl moiety is suggested [13].

Molecular similarities enable some thermoplastic poly(ester-urethanes) such as *Estane 5702* to be processed as a peroxide-vulcanizable gum resin if desired [3].

Genthanes S and SR may also be cured with polyisocyanates, according to the supplier [7, 8]. The curing mechanism may involve allophanate formation [see (3)].

Adiprene C, a polyether glycol-based millable polyurethane gum, may be cured with sulfur or peroxide curing systems [14].

Daltoflex 1 and *Daltoflex 1S* are also commercially available polyurethane raw rubbers. In use, the recommended cure of these gum polymers is accomplished with polyisocyanates [15, 16]. The limited shelf-life of the uncured polymer-isocyanate mixtures must be considered in their use, according to the supplier.

The millable polyurethane gum systems have the ability to accept high pigment loadings and to blend with other polymers. These features are also characteristic of the thermoplastic polyurethane elastomers and, to a lesser degree, the vulcanizing liquid urethane polymers.

This section has listed examples of commercially available, millable polyurethane gums but by no means has included all such systems available to the fabricator.

D. VULCANIZING LIQUID POLYURETHANES

The foregoing discussions have dealt with urethane polymers whose chains are essentially linear and have considerably high molecular weight as supplied. As a consequence the polymers are gum-like or tough, elastomeric solids. When the length of the polyurethane chains is deliberately limited during preparation by charging an excess of the diisocyanate to the polymeric glycol, liquid or low-melting products (prepolymers) result which bear reactive terminal isocyanate groups. The well-known liquid urethane casting prepolymers are prepared this way. Their structure may be represented by that of the isocyanate-terminated polyurethane chain in (3) in which n frequently averages 0 to 2, R is a polyester or polyether, and R′ is aromatic.

In usual practice the liquid or melted polyurethane prepolymer is rapidly mixed with a second component (chain extender and crosslinker) by the user, the mixture is promptly arranged to desired shape (poured into a mold, spread as a film, etc.) and then usually heated for a period of time to vulcanize it.

The chain extender and crosslinker may be one of several classes of di- and polyfunctional materials, which include small diols, aromatic diamines, and aminoalcohols. Its nature and level help determine the nature of the product, for example, hardness, and can be controlled to provide specific property levels. The practice of controlling product properties in this way has been termed "chemical compounding" and provides considerable versatility.

The first action of the chain extender and crosslinker is to couple rapidly the terminally reactive polyurethane prepolymer chains to produce much longer chains. Since this agent is usually charged in subequivalent amount, however, some prepolymer and extended pre-polymer chains with terminal isocyanate groups remain in the system. These then serve to crosslink other coupled chains during continued heating and effect gradual vulcanization of the polymer.   Alternatively, a small triol or other polyol chain extender and crosslinker may be used. The specific chemistry involved in the use of the various chain extender and crosslinker types is described in the following paragraphs.

When the chain extender and crosslinker is a small diol such as 1,4-butanediol, chain extension occurs through urethane link formation. Subsequent crosslinking of these chains involves allophanate link forma-tion as in (3).

If a diamine chain extender and crosslinker such as 4,4'-methylene-bis(2-chloroaniline) is used, polyurethane prepolymer coupling occurs through urea link formation.   In this case crosslinking of the resulting poly(urethane-urea) chains occurs predominantly through biuret link formation.   Equation 4 illustrates these reactions.

$$\sim NCO \quad + NH_2-R-NH_2 + \quad OCN\sim \quad \longrightarrow$$
prepolymer chain end   diamine   prepolymer chain end

$$\sim NH-CO-NH-R-NH-CO-NH\sim \quad \xrightarrow{OCN\sim}$$
prepolymer chains coupled through urea links

$$\sim NH-CO-N-R-NH-CO-NH\sim \quad \xrightarrow{etc.} \quad \text{crosslinked poly(urethane-urea)} \quad (4)$$
$$\underset{\text{diamine-coupled prepolymer branched through biuret link}}{\overset{|}{CO-NH\sim}}$$

Amino alcohol chain extenders and crosslinkers of structure $HO-R-NH_2$ have been used in the foregoing manner.   They undergo reactions characteristic of both the diols and diamines.

The use of polyol chain extenders and crosslinkers such as trimethylol propane, $C_2H_5-C(CH_2OH)_3$, with polyurethane prepolymers results in rapid and extensive polymer network development through simple urethane link formation.   For this reason they may be used in equiva-lent proportions.   Frequently they are used in admixture with diols.

Water may also serve as chain extender and crosslinker with isocya-nate-terminated polyurethane prepolymer.   Water couples the pre-polymer molecules through urea linkage formation with carbon dioxide evolution (5).

$$\sim NCO + HOH + OCN\sim \quad \longrightarrow \quad \sim NH-CO-NH\sim + CO_2 \quad (5)$$
prepolymer chain end   prepolymer chain end   prepolymer chains coupled through urea link

These coupled chains then spontaneously crosslink, likely through biuret link formation [see (4)]. Thus it is clear that the effects of moisture must be considered in the processing and use of polyurethane systems containing unreacted and partially reacted isocyanate components. Water can be used to advantage [17], but for the most part it must be regulated if not excluded in the interest of product uniformity.

Adiprene L [18], a poly(tetramethylene oxide)glycol-based liquid urethane elastomer [19] and *Vibrathane 6001, 6004, 6005, 6006* liquid polymers [20–23] represent commercially available liquid urethane polymer systems suitable for sheet and film manufacture, according to their suppliers. Castable poly(ether-urethane-urea) elastomers based on polyoxypropylene-1,2-glycol [24, 25] are also claimed to be adaptable to sheet and film manufacture by a supplier (Wyandotte Chemical Corp.) who has developed formulations.

The present section has dealt with an exemplary list of commercially available liquid polyurethane systems but by no means has it included all that are available.

III. Pertinent Polyurethane Properties

The physical properties that can be developed in polyurethanes are characteristically high level and usually reflected in high performance products. Pertinent properties of commercial polyurethanes are listed in this section. The materials are grouped according to the classification established in the foregoing section and deal with the same examples: thermoplastic polyurethanes, *Durethan U, Estane (5701, 5702, 5707, 5710)*; thermoplastic-thermosetting polyurethane resins *Texin E-224, Roylar*; vulcanizable, millable gums, *Genthane (S and SR), Adiprene C, Vibrathane (5003, 5004, 5005), Daltoflex (1 and 1S), Estane 5702*; vulcanizing liquid polymers, *Adiprene L, Vibrathane* liquid polymers (*6001, 6004, 6005, 6006*).

A. THERMOPLASTIC POLYMERS

The only commercial example of the true polyurethane (nylonlike) class of rigid thermoplastic polyurethanes that has come to our attention is *Durethan U*. It is considered representative and is the basis of discussion of this class.

Durethan U is supplied as dry pellets in three grades U_0, U_{20}, U_{50}; all have specific gravities of 1.21 but different hardnesses and property ranges [1]. Durethan U has a sharp, high melting point and high melt fluidity. Although these characteristics have been a deterrent to the use of some common plastics processing and fabrication methods, such as milling, calendering, kneading and pressing, they ideally suit the material to injection molding and sometimes extrusion techniques.

Relative to the polyamides, lower water absorption, resulting in better dimensional stability and improved electrical properties, is obtained. Even so, as in the case of polyamides, virgin material and scrap for processing must be thoroughly dried before use to avoid defects in finished products. Special mention is made of the resistance of Durethan U to esters, aliphatic and aromatic hydrocarbons, and to most chlorinated hydrocarbons. Outstanding resistance to vegetable, animal, and mineral oils and fats and neutral salt solutions is also a reported characteristic of Durethan U. Resistance to lower aliphatic alcohols, water, and alkaline solutions, especially soap and ammonia solutions, is found. Phenols, substituted phenols and formic acid act as solvents.

Agents reported to deteriorate Durethan U include strong oxidizing agents (chlorine, hydrogen peroxide, etc.) and oxygen (at highly elevated temperatures). High mineral acid concentrations rapidly decompose Durethan U, but dilute solutions (less than 5%) of both mineral and organic acids are generally adverse only at elevated temperatures. Since the above are generalizations, exposure tests are recommended to assess resistance to specific chemicals and conditions.

The tensile strength of Durethan U is good and may be increased several-fold by orientation. Toughness and abrasion resistance are reported to be outstanding.

The sharp high melting point of Durethan U permits repeated steam sterilization of products at 120 to 130°C. Molten Durethan U is susceptible to oxygen and during incorrect processing its surface may darken and gas may be evolved.

Some properties of the various Durethan U grades are listed in Table 1.

The elastic or flexible class of thermoplastic polyurethanes is represented here by one commercial material, *Estane* [3]. Consequently our discussion is again confined to a single family of products.

Estane is a completed, essentially linear, storage-stable polymer which exhibits true thermoplastic processability, including the repeated reuse of scrap. Certain polar solvents such as tetrahydrofuran, and even methyl ethyl ketone in the case of some grades, dissolve the Estanes. This

TABLE 1

Some Properties of Durethan U

	Durethan grade		
	U_0	U_{20}	U_{50}
Hardness,[a] Rockwell	104	95	70
Melting point,[b] °C	180–5	170–5	150–60
Yield strength,[c] lb/in.^{2}	7823	6543	4267
Elongation at yield point,[c] %	15	25	35
Heat distortion temperature,[d] method (a), °C	80–90	60–70	45–55
Heat distortion temperature,[d] method (b), °C	130–140	95–105	75–85
Weight change on immersion,[e] %			
water	+1.1	+1.6	+2.7
gasoline	< +0.1	< +0.1	< +0.1
mineral oil	< +0.1	< +0.1	+0.1
linseed oil	+0.2	+0.2	+0.1
ethyl acetate	+0.2	+0.6	+2.1
benzene	+0.3	+0.5	+1.3
alcohol, denatured	+1.7	+2.7	+8.2
trichloroethylene	+1.4	+2.1	+4.7

[a] Method A, scale L, ASTM D 785–51, 0.5 in. × 0.5 in. × 1.0 in. sheet.
[b] Kofler method.
[c] DIN 53455, dumbbell No. 2, tension.
[d] ASTM D 648–56, 0.5 in. × 0.5 in. × 5 in. bar.
[e] 20 days at 25°C.

permits important solution applications [26], but, in addition, Estane displays many of the useful properties of a tough, abrasion-resistant, rubbery vulcanizate, including high elongation with elastic recovery, excellent resistance to oils and hydrocarbons, and a practical degree of resistance to many other solvents and chemicals. For these reasons it has been described as a "virtually crosslinked" elastomer.

As with many amorphous and low crystallinity polymers, when heated at recommended processing temperatures the Estanes retain high viscosity and do not yield the fluid melts characteristic of some more crystalline polymers. This rheology adapts Estane particularly well to conventional thermoplastic mixing and fabrication.

Estane is supplied in cube or granule form in moistureproof bags. The latter practice guards against the mechanical absorption of water which may cause porosity in fabricated products (the equilibrium moisture content of Estane 5701 granules at 25°C is 0.6% at 50% relative humidity and 1.6% at 100% relative humidity). Five raw resins, Estane 5701, 5702, 5707, 5710, and 5714, all having specific gravity 1.20 but varying in hardness and other properties, are representative of polymers now available. In addition, several compounds are supplied. Some properties of raw Estanes 5701 and 5702 are listed in Tables 2 and 3. Properties of a one-mil film of raw Estanes 5707 and 5710 are listed in Table 4.

The water-vapor transmission of a 3-mil Estane 5701 film is moderate, showing a value of 26.3 g/100 in.^{2}-24 hr. Gasoline diffusion is 0.067 fl oz/ft^2-24 hr.

The gas permeability of Estane proves to be low. Values determined for Estane 5701 at 25°C and expressed in ft^3/mil-ft^2-day-psi are air, 2.46×10^{-4}; oxygen, 6.58×10^{-4}; hydrogen, 2.99×10^{-3}; nitrogen, 1.78×10^{-4}.

After exposure to 100 Mr of gamma radiation in air Estane 5701 still retained 61% of its original tensile strength and 96% of its original elongation, demonstrating relatively good resistance to adverse change.

B. THERMOPLASTIC-THERMOSETTING POLYMERS

Texin [4,5], a commercial thermoplastic-thermosetting poly(ester-urethane) elastomer, specific gravity approximately 1.23, is available in grade E-224 as a raw material for the manufacture of calendered film. The resin supplier specifies processing characteristics which render it ideally suited to this application, as well as typical outstanding polyurethane elastomer properties (high hardness with high elasticity, toughness, abrasion resistance, resistance to oils and common household solvents, and flexibility at temperatures approaching −100°F). Exposure of Texin E-224 to actinic light, as in prolonged outdoor exposure, degrades the polymer, but it should be noted that this is true of unprotected poly(ester-urethanes) in general. Consequently, stabilization by compounding is recommended where such exposure is encountered.

Table 5 lists some properties of Texin E-224 film which are representative.

Roylar is described by the supplier as a family of urethane elastomers that processes like thermoplastics and elsewhere [6] as a thermoplastic-thermosetting polyurethane. Adaptability to sheet and film manufacture by calendering, extrusion, solution casting and compression molding is claimed. Roylar properties are reported to recommend their use

TABLE 2

Some Mechanical Properties of Estane

	Estane grade	
	5701	5702
Hardness,[a] Shore	88A, 60C	70A
Tensile strength,[b] psi	5800	5000
Elongation,[b] %	500	700
300% modulus,[b] psi	1300	<500
Tear strength,[c] lb/in.	350	<100
Abrasion resistance,[d] mg loss	5	150
Extension set,[e] %	47	>100
Compression set,[f] %		
22 hr/25°C	25	43
22 hr/70°C	87	79
Low-temperature brittleness,[g] °C	<−80	<−100
Gehman freezing point,[h] °F	−24	−16
Flexlife,[i] flexures to indexes 1:2	639,000; 1 million	
Processing temperature,[j] °F	340	260

[a] ASTM D 676.
[b] ASTM D 412, Die D dumbbell, 20 in./min.
[c] Graves angle tear, ASTM D 624.
[d] Taber abrader, CS17 wheel, 1000 g weight, 5000 cycles, ASTM D 1044–49T.
[e] Extend 300% at 2 in./min, hold one minute, relax five minutes, measure,
[f] Method B, ASTM D 395–55.
[g] ASTM D 746.
[h] ASTM D 1053.
[i] Demattia, ASTM D 813.
[j] Stock temperature.

TABLE 3

Weight Change on Immersion of Estane 5701 in Several Media

Medium	Days' immersion at 25°C	Weight change (%)
Water	28	+ 1.05
Gasoline, boron	28	+11.92
Gasoline, white	28	+ 2.62
Type I Fuel A	28	+ 0.58
Type II Fuel B	28	+ 9.51
Carbon Tetrachloride	28	+44.93
Methanol	28	+14.6
ASTM Oil No. 1	7	+ 1.47
ASTM Oil No. 3	7	+ 1.9
Wesson Oil	7	− 0.12

TABLE 4

Estane Polyurethane Film Properties

| | | Estane | |
		5707	5710
Film thickness, mils		1.0	1.0
Specific gravity		1.20	1.20
Hardness, Shore[a]		50D	78A
Tensile strength, psi	MD	11,000	9000
	TD	8000	8000
Elongation at break, %	MD	600	700
	TD	500	800
Stress at 10% elongation, psi	MD	550	160
	TD	500	160
Impact strength, in.-lb/mil[b]		90	50
Tear strength, g/mil[c]	MD	100	200
	TD	300	200
Abrasion resistance,[a] g loss (cf. Table 2[d])		0.002	0.004
% return after 150% stretch	MD	97	97
	TD	96	97
Film haze, %		17	50
45° gloss		60	25
Permeability to oxygen, ml/100 in.^{2}/24 hr-mil		50	100
Permeability to water vapor, g/100 in.^{2}/24 hr-mil		27	40
Gehman low-temperature modulus[a] °C			
T_2		+2	−9
T_5		−9	−14
T_{10}		−17	−16
T_{100}			−27
freezing point		−23	−23
Brittle point[a] °F D 746		No breaks at −100°F	
Chemical resistance, 28 days' immersion at room temperature: medium used, % change in tensile strength			
Wesson Oil		0	0
ASTM Fuel A		0	0
ASTM Fuel B		−2	−7
Transmission oil		−1	0

[a] These properties were obtained with a 75-mil sheet.
[b] Falling ball method.
[c] Determined with Elmendorf Tear Tester.

500

TABLE 5

Representative Properties of Texin E-224 Film

	Value	Test method
Shore A hardness	86–9	ASTM D 676
Tensile strength, psi	5500–8000	ASTM D 412
Elongation at break, %	525–625	ASTM D 412
300% modulus, psi	1800–2200	ASTM D 412
Tear strength, lb/in.	400–600	FTMS 601–M 4221

in applications requiring excellent resistance to oxidation, high load-bearing capacity, toughness and resistance to tearing, excellent resistance to abrasion, excellent resistance to fuels, oils and many solvents including chlorinated solvents, good heat resistance. In addition, specific formulations are claimed to provide stocks showing good flexing resistance, high or low resilience, low heat buildup, good dampening or shock absorbing quality, low compression set. Some representative properties claimed by the supplier for Roylar include: 70A–65D Shore hardness; 3000 to 8000 psi tensile strength; 250 to 700% elongation; 1500 to 5000 psi 300% modulus; 100–400 lb/in. tear strength (ASTM D 470).

C. VULCANIZABLE MILLABLE GUMS

Several suppliers of millable polyurethane gum elastomers claim adaptability of their products to vulcanized film and sheet applications. A summary of the properties of such polymers appears in the following section.

Genthane S, density 1.19, 50 ±10 Mooney viscosity (ML-4/212°F). is available as a rubberlike gum in 1-in. slabs. It is soluble in ketones, esters and chlorinated solvents and in use is vulcanized with isocyanates or peroxides or mixtures of both. Cured Genthane S will not crystallize, has good creep properties and will not stain. Vulcanizates show excellent resistance to hydrocarbons up to 212°F, dry-cleaning solvents, ozone, and oxygen. They exhibit an excellent balance of physical properties, including good performance at low temperatures and up to 300°F, as well as outstanding abrasion resistance. Stress-strain data for Genthane stocks, determined on 75-mil-thick dumbbells unless indicated otherwise, follow. A representative 60 min at 300°F dicumyl peroxide gum cure of Genthane S shows 4315 psi tensile strength, 830% elongation and 223 psi modulus at 300% elongation. A white compound cured with dicumyl peroxide and toluene diisocyanate dimer at 30 min at 320°F has about 86 Shore A hardness. A sheet of this compound 15 mils thick shows 6865 psi tensile strength, 480% elongation, and 1600 psi 300% modulus. A black compound cured with dicumyl peroxide for 45 min at 310°F shows 75 Shore A hardness, 4675 psi tensile strength, 475% elongation, 2940 psi 300% modulus, 330 lb/in. crescent tear strength, and 16% compression set (ASTM B, 22 hr, 158°F).

Genthane SR is a grade recommended where continuous immersion in hot oil or hot water is required. A black Genthane SR stock vulcanized with dicumyl peroxide yields the following properties on 25 min at 320°F cure and 60 min at 300°F postcure: 62 Shore A hardness, 4900 psi tensile strength, 570% elongation, 1775 psi 300% modulus.

Vibrathane 5003, 50 Mooney (ML-4/212°F) is available as a stable gum elastomer which must be vulcanized to achieve physical properties and can be processed on conventional rubber equipment, including mills, calenders, and by solvation. Stocks containing dicumyl peroxide curing agent are nonscorchy and have good storage life. The cured stock is a high-quality product with excellent abrasion, tear and heat resistance qualities, coupled with good stress-strain properties at elevated temperatures and excellent resistance to oxygen, ozone, chlorinated solvents, fuels, and oil. Vibrathane 5004 (density 1.15, 50 Mooney viscosity) and Vibrathane 5005 are also available. The Vibrathane gums may be compounded with various pigments and their peroxide cures effected by heating in the range of 250 to 305°F for 30 to 60 min. Postcure is claimed to be unnecessary. The 5003 and 5004 polymers may also be cured rapidly in 3 to 4 min at temperatures of 350 to 400°F. The supplier lists the following property range for Vibrathane millable gum elastomer vulcanizates: 50-90 Shore A hardness; 2500 to 5000 psi tensile strength; 100 to 550% elongation; 700 to 4200 psi 300% modulus; 250 to 770 lb/in. tear strength (die C). The suggested maximum service temperature for Vibrathane millable gum vulcanizates in extended use is 250 to 300°F.

Estane 5702, specific gravity 1.19, although strong enough for many useful applications as a thermoplastic resin in the unvulcanized state and without reinforcement, may also be pigmented and vulcanized with peroxides to enhance solvent, temperature, and compression set resistance. A pure gum cure at 45 min at 310°F with dicumyl peroxide produces the following vulcanizate properties: 6% compression set (ASTM B, 22 hr/158°F); 6000 psi tensile strength (microdumbbell); 440% elongation (same); 500 psi 300% modulus (same); 210 lb/in. tear strength (Graves tear test); insolubility. More rapid peroxide cures have been achieved in 3 to 4 min at higher temperatures.

Adiprene C, specific gravity 1.07, 55 ±5 Mooney viscosity (MS-10/212°F) is supplied as soluble (in tetrahydrofuran, methyl ethyl ketone, of dimethyl formamide), amber, and transparent sticks which have been stabilized with a staining-type of antioxidant. Excellent storage stability up to 212°F is claimed. It can be processed satisfactorily on conventional rubber-processing equipment and can be converted to sheet or film, as by calendering. In use Adiprene C is vulcanized with sulfur-accelerator or peroxide cure systems and can be pigmented. Vulcanizates are noncrystallizing. Reinforcing fillers are necessary to develop the good strength and abrasion resistance of Adiprene C. Supplier claims for Adiprene C vulcanizates also include outstanding low temperature characteristics, good resistance to heat deterioration, ozone cracking, weathering, radiation, oil and solvent swell, and good resilience. Black Adiprene C stock cured 45 min at 310°F with sulfur and accelerator affords the following vulcanizate properties: 64 Shore A hardness, 5150 psi tensile strength, 540% elongation, 2475 psi 300% modulus, 20% compression set (ASTM B, 22hr/158°F), 380 lb/in. tear strength (Graves).

Daltoflex 1 and 1S, specific gravity 1.2, are supplied as irregularly shaped, light-colored 3- to 5-in. squares which have good stability when stored in cool, dry, closed containers. Daltoflex 1S is soluble in organic solvents. Both polymers can be processed by conventional rubber-processing techniques and machinery. Thin sheet can be obtained by calendering and thicker sheet, by molding. The polymers may be pigmented and are considerably reinforced by fillers. In use both polymers are vulcanized with isocyanates. A high-boiling liquid diisocyanate (Suprasec M) is recommended or, alternatively in the case of the 1S grade, a mixture of polyisocyanates (Suprasec G and K). These curing agents are active enough to necessitate efficient cooling of the stock during mill compounding and give rise to compounded stock of short shelf-life. According to supplier claims, Daltoflex 1 vulcanizates are superior to natural and synthetic rubber compounds in tensile strength and resistance to oxidation, abrasion, and solvents. In the case of the latter two resistance is rated excellent. In addition, Daltoflex vulcanizates exhibit high tear resistance and the ozone resistance associated with polyurethane rubbers. Daltoflex 1 gum cured with Suprasec M for 20 min at 150°C yields the following properties: 64 B.S. hardness; 4352 psi tensile strength; 700% elongation; 398 psi 300% modulus; 512 psi tear strength; 17% compression set (25% compression, calculated on compression, 24 hr/70°C).

D. VULCANIZING LIQUID POLYURETHANES

Adiprene L is commercially available as a honey-colored liquid, specific gravity 1.06, and a viscosity of 14,000 to 19,000 cp at 86°F (500 to 600 cp at 212°F). It is soluble in most common solvents and has excellent storage stability at room temperature in the absence of moisture. Adiprene L is a liquid urethane rubber which can be cured to a strong rubbery solid, including film and sheet forms, with high tensile strength and resilience and excellent resistance to abrasion, compression set, oils, solvents, oxidation, ozone, and low temperatures. Heat and radiation resistance are also important attributes. Diamines are the preferred chain extender-crosslinker class for the general purpose cure of Adiprene L. Use of these agents results in relatively short mix pot life and rapid cure as well as high vulcanizate strength, hardness, and abrasion resistance and low compression set. One diamine, 4,4'-methylene-bis (2-chloroaniline), or MOCA, as the supplier has termed it, is particularly suitable. Adiprene L can be compounded with dry, inert pigments, thickening in the process. A typical MOCA cure of Adiprene L involving a 25-min set-time, a 180-min cure at 212°F, and 14 days' after-cure at room temperature affords the following set of properties: 88–90 Shore A Hardness; 4000+ psi tensile strength; 450% elongation; 2100 psi 300% modulus, 500 lb/in. tear strength (Graves, ASTM B, 22 hr/158°F). MOCA level may be varied and other chain extender-crosslinkers, including polyhydroxy compounds and mixtures, may be used to develop specific property variations. This method of regulating properties has been termed "chemical compounding" as mentioned earlier.

Vibrathane polyurethane liquid polymers are available in several grades: 6001, 6004, 6005, 6006. Vibrathane 6001 is a viscous liquid of 450,000 cp at 80°F which thins out to 7500 cp at 195°F. It is soluble in several organic solvents. Vibrathane 6006 has 600 cp at 212°F. Use of the different Vibrathane reactive liquid polymers with various levels of the usual diamine and polyol chain extender-crosslinker agents provides formulations with the following spread of properties: 25A-70D Shore hardness; 300 to 9000 psi tensile strength; 300 to 800% elongation; 500 to 6000 psi 300% modulus; 60 to 1000+ lb/in. (Die C) or 7 to 250 lb/in. (ASTM D 470) tear strength. The supplier makes the following claims for such polymers, which are general for the group and not necessarily all characteristic of each composition: excellent oxidation resistance, high load-bearing capacity, toughness and resistance to tearing, excellent abrasion resistance, excellent resistance to fuels, oils, and many solvents, including chlorinated ones, and good heat resistance (maximum temperature for extended use is 210 to 250°F). In addition, specific formulations are available to provide good flexing, high or low resilience, low heat buildup, good dampening or shock absorbing quality, and low compression set. Formulations based on Vibrathane liquid polymers may be used as 100% solids to cast, spray, or compression-mold sheet or film or in solvated form by casting, spraying, or dipping. Vibrathane 6001 formulations are mixed at 195 to 275°F with pot lives of about 6 and 1.5 min, respectively. Cure is effected in 30 min at 305°F. Optimum product properties are realized after seven days' postcure at room temperature in free circulating air of at least 50% relative humidity. Other Vibrathane liquid polymer formulations require similar conditions but recommended curing temperatures range from 73 to 305°F, whereas optional postcures range from 73 to 212°F.

IV. Physical Compounding

A. Thermoplastic Compositions [27, 28]

The thermoplastic polyurethanes are quite stable and as a consequence can be compounded widely. Yet there are some limitations. In general, highly acidic and highly basic materials are to be avoided. Transition metals such as iron, lead, and tin tend to induce degradation at elevated temperature. Although these restrictions also pertain to other polyurethane types, they are specifically mentioned here, since, as a group, the thermoplastic polyurethanes have a high tolerance for additives.

As a matter of practice fillers are added to lower the cost and increase the hardness, tear strength, and modulus. Some improvement in processing and a decrease in tensile strength may also be noted. Although carbon blacks and silicates tend to be preferred, for they reduce the tensile properties least, the effect of most fillers is evident beyond the 10 part-per-hundred of polymer level. At a 25-part level the tensile properties are lowered to the extent of 20 to 30%.

There are few compatible plasticizers for the thermoplastic polyurethanes. Those that can be used seem to soften these materials and rapidly degrade the polymer's properties.

Many other polymers have been compounded with the thermoplastic polyurethanes to alter properties. Included in this group are the vinyls, vinyl copolymers, copolymer nylons, ABS, polycarbonates, limited amounts of polyolefins, and a wide range of elastomers. The effect on properties is highly dependent on the blending polymer. In general, the hardness, moduli, and elongation move toward that of the blending polymer. The effect on tensile strength is dependent on the degree of compatibility.

Lubricants are normally used to aid in processing. Powdered polyolefins, synthetic and natural waxes, and stearates are commonly used. These materials are required for proper release in calendering and molding and may be used in extrusion to increase the emission rate and dampen blocking.

Stabilizing agents to improve resistance to hydrolysis, ultraviolet, and fungus are available. The hydrolytic stabilizers, based on polycarbodiimides, are added at levels up to 5 parts per hundred. The ultraviolet and fungus resistance additives are characteristic of those in current use. They are normally used at levels up to about 2 parts per hundred.

B. THERMOPLASTIC-THERMOSETTING COMPOSITIONS [5, 29, 30]

The generally reactive nature of the thermoplastic-thermosetting materials tend to limit most types of compounding. Polyolefin and synthetic waxes, lubricants, certain color master batches, and silicate fillers can be used. The expected results are obtained. The lubricant levels are normally under 2 parts per hundred. Fillers are used up to about 20 parts.

C. Millable Gums

1. Polyesters [7–11, 30, 31]

Most poly(ester-urethanes) are cured with peroxides (normally 2 to 4 parts of dicumyl peroxide or 4 to 11 parts of a 40 % dispersion of dicumyl peroxide). Most rubber chemicals containing sulfur tend to interfere with this cure and cause depolymerization. Therefore, when rubber equipment is used, proper cleanup is essential.

Both silicates and carbon-black fillers can be used. The normal effect is an increase in hardness and moduli and a decrease in elongation.

Highly styrenated resins and phenolics can be compounded with the poly(ester-urethanes) to increase hardness.

A limited number of plasticizers is reportedly compatible with the millable gums; phthalate, phosphate, and dibenzoate esters are cited as the most suitable. A drop in hardness, moduli, and tear strength is the primary change up to the inclusion of 20 parts per hundred.

Waxes and stearates at a 0.10 to 1 % level are recommended to aid in processing.

2. Polyethers [14, 32, 33]

The poly(ether-urethanes) can be cured with peroxides or conventional sulfur systems. About one part of dicumyl peroxide is the suggested peroxide level. A typical sulfur system includes 3/4 part of sulfur, 4 parts of MBTS, and 1 to 2 parts of MBT. The addition of zinc-containing materials is recommended to accelerate the sulfur cure. A zinc chloride/MBTS compound is generally used. Although the zimates also accelerate the cure, they are reported to be appreciably slower than the specific combination recommended.

The poly(ether-urethanes) require fillers to develop optimum properties; for reinforcing, 25–30 parts are considered the minimum needed. Either silicates or reinforcing carbon blacks can be used. As could be anticipated, the addition of filler increases hardness, moduli, and tear strength and decreases elongation.

Dibenzoates, phosphates, pentaerythritols, and phthalates are reported to be suitable plasticizers for the poly(ether-urethanes). A decrease in hardness, modulus, and abrasion resistance is usually observed. Plasticizers with unsaturation tend to interfere with peroxide cures and are therefore often avoided.

Lubricants based on low-molecular-weight polyethylene and micro-crystalline waxes are used for calendering. A 3 to 5-part level is recommended. In addition, stearates (0.5 to 1 part) are often added to improve processing.

D. Cast Systems [19, 20, 22, 23, 25, 34–38]

Since cast systems are handled in a liquid form, compounding materials are selected primarily for their effect on properties and curing rather than on processing. In general, the properties of the resulting polymer depend on the amount and type of chain extender-crosslinker charged and the resulting frequency of urethane or urea groups: the greater the frequency, the higher the hardness. In addition, diamine extender-crosslinkers tend to produce greater hardness than triols. Since the area of polymer formation (chemical compounding) has been discussed earlier, the following discussion of additives for the liquid cast system will be confined to what is normally conceived of as physical compounding.

The properties of the basic cast systems can be altered by the addition of fillers and plasticizers. It is essential, however, that the compounding additives be (a) free of moisture, (b) noncatalytic, and (c) unreactive with the curing reactants under the condition used.

Most additives increase the required amount of curative and curing time. The additives are normally blended with the nonisocyanate-containing part of the reacting mixture or metered separately as the third or fourth ingredient into the mixing head (see Section V). The latter approach has the advantage of versatility, particularly when short runs are involved.

Subject to the above restriction, a wide range of carbon blacks, silicates, clays, and whitings can be added. The amount of solid pigment that can be incorporated is limited by its thickening action on the system, which must remain fluid enough to be metered, to release trace amounts of entrapped moisture and air, and be discharged. The usual result is a decrease in tensile strength and an increase in hardness. The effect on elongation and modulus is dependent on the filler. Some fillers increase these values and others decrease them.

Plasticizers generally soften the polyurethanes but usually markedly reduce the tensile and moduli values. The phosphates, phthalates, and a formal polyether are reportedly compatible up to about 40 parts.

Liquid epoxies are compatible with many of the cast bulk systems. A full range of combinations characterized by the properties of the unmodified polyurethane at one end and the epoxy at the other is possible.

V. Methods of Polyurethane Sheet and Film Production

Most of the methods currently in use for preparing polyurethane sheet and film were developed early in this century and have persisted with some modification to this date. These methods fall into four general types: casting, calendering, extrusion, and compression molding.

Included in the casting types are the solution, latex, and liquid bulk systems which are cast as liquid and then solidified as the result of a chemical reaction or a physical change in phase caused by heat and the evaporation of the conveying medium, if any.

Calendering is generally done on equipment that is an adaptation of the similar steel mill machines. The stock is heated and repeatedly squeezed into a thinner sheet by passing it between rotating steel rolls.

The third type, extrusion, encompasses a continuous masticating and forming technique under pressure. A modification of this technique is film blowing which provides for the extrusion of a tube and its subsequent in-line expansion into a thin-walled tube which is then slit to produce film (cf. Volume I, Chapter 8).

Compression molding is an intermittent forming of material under conditions of high temperature and pressure into the configuration of a mold.

A. CASTING TECHNIQUES

The area of casting techniques readily subdivides into three general material categories: bulk, solution, and latex casting.

1. Liquid

Liquid casting techniques, as described earlier, refer to the preparation of crosslinked polyurethane polymers through the reaction of a liquid or liquefied diisocyanate with a polyester glycol or polyether glycol and the subsequent or simultaneous crosslinking and chain extension with polyols or diamines. The reaction is rapid and extremely sensitive to contaminants such as moisture and metallic compounds. As a consequence equipment systems have been developed to provide the necessary control. Although the basic techniques used were designed primarily for molding, they have been adapted to the continuous or intermittent production of sheet and film.

Two raw materials systems are in use at present. The most widely used system requires the preparation or the purchase of a reactive prepolymer before the forming step. The prepolymer is heated (110°C), combined with carefully metered amounts of heated (110°C) polyol (triol and diol) or diamine (and occasionally a catalyst), agitated, and deposited onto a moving belt or into molds on a continuous or intermittent basis. Degassing to remove entrapped air and moisture is used both before metering the streams and during the mixing.

The alternate approach (one-shot method) involves the blending of a polyester or polyether with the appropriate triols, diamines, diols, and catalyst and the mixing of this blend with a diisocyanate such as tolylene diisocyanate or diphenylmethane diisocyanate. This reacting mixture is then deposited in a manner similar to the procedure used in the prepolymer system.

Pigments, plasticizers, and fillers can be used in limited quantities, As already indicated, however, a great deal of care must be exercised in their choice. The need to exclude moisture and metal contaminants and materials which will readily react with diisocyanates is essential. These additives are metered into the mixing head separately or blended with the nondiisocyanate reactants.

Figure 1 is a general schematic diagram with alternative gauging and take-up methods for producing liquid cast polyurethane sheet and film. Rather obviously, this does not cover all contingencies or systems, but it does serve to describe the preponderance of existing operations.

The following discussion relates to some of the more important processing variables at work in the liquid cast systems.

1. The reactants are normally heated to provide lower viscosities which, in turn, facilitate better metering and mixing action. This heat also produces a higher rate of polymerization. There are some systems in which the reactants are sufficiently low in viscosity to be accurately pumped and mixed at ambient temperatures. These systems normally require catalysts and have a much longer pot life.

2. The degassing step is used to pull off any remaining moisture and accumulated gases. Some manufacturers suggest successive or continuous degassing throughout the processing cycle.

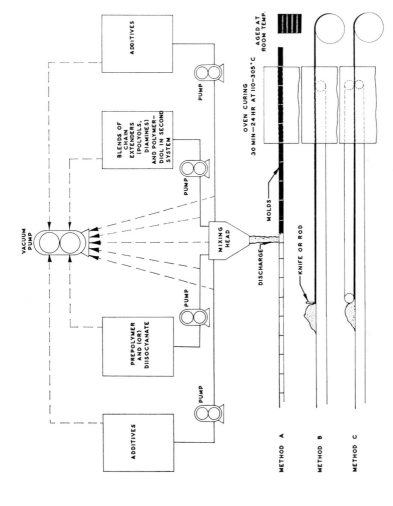

Fig. 1 Schematic diagram of liquid casting lines. Three methods of sheet formation. All four reservoirs are held at 100 to 110°C.

3. Complete intermixing and discharge are quite important. Failure to accomplish either will produce a highly variable material and in time may result in plugging the mixer.

4. Gauge control of the finished sheet or film has been obtained in a number of ways. The most direct methods (Method A) are those involving the intermittent or continuous filling of molds and a subsequent light surface-pressing of the solidifying mass. A somewhat more sophisticated approach (Method B) involves continuously casting the reacting mixture onto a moving belt and passing the belt beneath a knife or bar to obtain the proper thickness. Problems associated with buildup on the knife or bar have limited the use of this technique, however. More recently, several firms have begun experimenting with a fixed-space continuous top belt (Method C) which could be described as a continuous molding operation. This technique seems to circumvent the buildup problem satisfactorily. Heavily catalyzed formulations are employed to achieve rapid solidification [39].

5. The belt or mold surfaces which touch the polyurethane are usually coated with a releasing agent such as a silicone or Teflon. Although the mixing head and discharge nozzle are designed to be self-flushing, they are cleaned periodically to remove any buildup.

6. Curing is usually accomplished in two steps: an accelerated $\frac{1}{2}$- to 24-hour step at 110°C and a 3- to 7-day step at room temperature. Final properties are not attained for about one week and with an all-polyol curing system, two weeks.

Those systems that reportedly do not require heat anywhere in processing depend solely on catalysts to accelerate the crosslinking reaction. Longer aging periods are usually required for the development of final properties in these systems. During this period humidity control is important to ensure uniformity [24].

In practice a portion of the heat aging is provided by the heat retained in the rolled material. This is particularly true for the thinner gauge materials which attain final properties in 12 to 24 hours.

7. In all instances in which the materials are heated some shrinkage occurs. When molds are involved, slightly oversize dimensions are normally used (cf. Section V-D).

Materials ranging from a thickness of about 5 mils to 1 in. have been produced commercially. The thicker sections are normally produced by molding techniques and the thinner by belt-casting methods.

2. Solution

Solution techniques can be applied to any of the systems noted, with perhaps one exception, the thermoplastic-thermosetting type. In general the components or component blends for the vulcanizing systems are dissolved separately in suitable solvent systems (toluene, xylene, cyclic ethers, methylene chloride, and ketones), combined, and spread onto a releasing surface more or less continuously. Knife blades, roll coaters, and bars are used to control the thickness deposited (see Volume I, Chapter 11). The solvent is then evaporated and the film remaining is either taken up and stored or cured, depending on the types, as noted in the applicable sections.

Solution-cast systems involving liquid casting reactants have several notable advantages over 100% solids systems. They attain pumpable viscosities at ambient temperatures, reduce the likelihood of entrapping moisture and air, and retard the rate of reaction. The obvious limitations are thickness per pass, solvent costs, and the need for explosion-proof motors in the evaporating area.

The thermoplastic polyurethanes are prepared by agitating the appropriate polymer with solvent systems which range from simple ketones such as methyl ethyl ketone to dimethylformamide. The resulting solutions have good pot life and can be stored indefinitely in contrast to the reacting systems. A full range of coating procedures for depositing a continuous film has been used: knife, bar, roller, reverse rolls, and air-knife. Stainless-steel belts coated with a silicone or Teflon are normally used to provide the needed release. The solvent in the deposited film is evaporated at room temperature, the film is force-dried with hot air, and taken up. An aging or curing step is not required.

In any of the solution systems proper choice of solvents is required to provide a balance of viscosity, total solids, and evaporation rate.

3. Latex [40]

Latex systems for polyurethanes reportedly based on polyether diols, and probably tolylene diisocyanate, have appeared in the field. The chemistry of these systems has not been described in detail.

Knife or roll casting is the normal method of application of the latexes. Total solids are in the 35 to 50% range. Either air or hot-air drying can be used.

Most of the applications noted involve adhesives, coatings, or saturants.

B. CALENDERING

Calendering techniques can be applied to three of the four general classes of polyurethanes.   The liquid casting systems do not lend themselves to this form of processing.

1. Thermoplastic Polymers [41]

The thermoplastic polyurethanes may be handled in a manner closely related to other thermoplastics.   In general, they can be prepared for calendering with an internal mixer or a mill at stock temperatures of 290 to 340°F.   Since these are "high working" materials, the required stock temperatures are attained with production equipment set in the range of 220 to 250°F.   When lubricant levels above 1 % are used, somewhat higher equipment temperatures are required.   The calender roll temperatures are usually in the 250 to 300°F range, depending on the thickness and compounding additives.

Since these materials do not require a curing step, the sheet or film issuing from the calender is ready for fabrication.

Calendered sheet ranging from about 3 to 60 mils has been produced. Close control of the roll and stock temperatures is essential for the thinner gauges.   Feed and stock temperature maintenance is the limiting factor at heavy gauges.

2. Thermoplastic-Thermosetting Polymers [5, 29, 30, 42]

The thermoplastic-thermosetting polyurethanes can be calendered with thermoplastic processing conditions.   The reactive nature of these materials tends to make careful control essential.

In general, they are processed in a manner similar to the thermoplastic polyurethanes.   Somewhat higher equipment temperatures (260 to 330°F) and stock temperatures (up to 375°F) are required.   Only limited reworking of the trim and scrap stock is possible.   A postcuring step (12 to 48 hr at 150 to 200°F) is recommended by the suppliers.   In practice, a 7- to 14-day storage at ambient temperature is used by the converter.

Sheet and film varying from about 5 to 40 mils have been produced. Problems similar to those noted in the thermoplastic section are applicable.

3. Millable Gums [7, 8–11, 14, 30–33]

The millable gum polyurethanes are handled in a manner similar to conventional elastomers (e.g., SBR and nitrile). They are supplied as a relatively stable gum and therefore must first be masticated in an internal mixer or a mill. A working stock temperature of 230 to 250°F is recommended for most of the poly(ester-urethanes). As indicated in the preceding section, the equipment temperature will be appreciably lower (140 to 200°F). The stock is then cooled to about 200°F and a peroxide (normally dicumyl peroxide) is added. The compounded material is transferred to a calender and sheet is produced in a normal manner. The roll temperatures are generally in the 150 to 210°F range. Subsequent curing in an inert atmosphere (to avoid oxygen poisoning) under pressure (to prevent bubble formation) is required; for example, 30 min at 300 to 310°F.

Poly(ether-urethanes) are handled similarly, but in addition to peroxide curatives some can be crosslinked with sulfur. When a peroxide cure is chosen, conditions similar to those noted for the polyester may be used, except that somewhat higher initial stock temperatures are usually required (250 to 270°F). When a sulfur system is employed, the curing additives may be combined with the stock during initial mastication or directly afterward without appreciable cooling. Processing is normally accomplished at somewhat higher temperatures when a sulfur system is used: 270 to 300°F.

Calendering of the poly(ether-urethane), according to available literature, suggests a roll temperature of 140 to 150°F without a lubricant and 180 to 200°F with lubrication. The peroxide-containing stocks are cured as noted in the polyester section and the sulfur system is cured from 30 min to 2 hr at 290 to 310°F.

In addition to the curing agents mentioned above, most of the millable gums can be crosslinked with diisocyanates and polyisocyanates. In general, however, the resulting blend of an isocyanate-containing material with a millable gum produces a "scorchy" mix, particularly at elevated temperatures. As a consequence only a very limited amount of work involving this curing system is performed. Most of that which does exist involves the preparation of stock for molding sheet and film.

Although almost all elastomer and plastic equipment designed for calendering may be used, cross-contamination, particularly from sulfur and sulfur compounds, should be avoided when peroxide curatives are employed. These additives interfere with the crosslinking mechanism. Calendering is normally used to prepare sheet stock about 20 to 50 mils thick. Heavy sections that need peroxide curatives require high pressures during curing to avoid bubble formation.

A schematic diagram of a typical calendering line is shown in Figure 2.

C. Extrusion and Film Blowing [41]

Although millable gum, thermoplastic-thermosetting, and thermoplastic materials all have the potential to be extruded into sheet and film, only the thermoplastic materials are being commercially converted by this method. This situation seems to stem from the need for a highly stable material for sheet and film extrusion.

The thermoplastic polyurethanes have been extruded and film blown into sheet and film $\frac{1}{2}$ mil to 250 mils thick (0.0005 to 0.250 in.). Those gauges of less than 2 mils are prepared by film-blowing techniques and of more than 2 mils by conventional extrusion. A full discussion of both methods of film manufacture is included in Volume I, Chapter 8.

A wide range of equipment is currently in use. The highly stable nature and extremely low working viscosities of this class of polyurethanes apparently are responsible for this diversity. Extruders with length to diameter ratios of 17:1 to 24:1 and screws varying from those with a constant pitch-decrease depth to those with sudden compression metered sections (nylon and olefin type) have been used. Both end- and center-fed dies have been employed successfully. The best unsupported sheet has been reportedly obtained with coat-hanger die configurations.

Most of the materials used are at least minimally compounded with lubricants and fillers to provide a dry, tack-free surface. Very successful sheet extrusions based on the unmodified polymer have been reported, however.

Blown polyurethane film is produced with conventional equipment designed for this purpose. The main difficulty arises from tackiness during takeup. The magnitude of this problem is closely associated with a given installation, the amount and type of compounding, and equipment modification. To reduce tack lubricants and fillers are used and the amount of cooling and equipment dwell-time is increased somewhat over normal practice. Rates in a commercial range have generally been experienced in these operations.

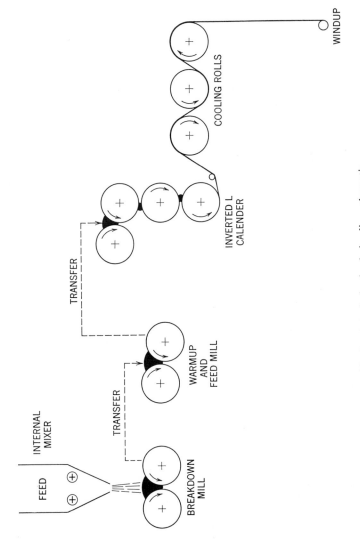

Fig. 2   A typical calendering line schematic.

517

D. Molding

Sheet and occasionally film of specific and somewhat limited dimensions are produced by compression-molding techniques. All of the polyurethanes are adaptable to this technique. The compounding and general processing are similar to that described earlier.

In the case of the liquid casting systems, the reacting mixture is discharged into a mold and a minimum amount of pressure is applied. The preparation of heavy bubble-free sheets, however, requires somewhat increased pressures. Curing is accomplished as noted in previous discussions.

Relatively high pressures (500 psi and up) and temperatures (290 to 360°F) are required for the other systems. The exact time varies widely with the sheet thickness. Slabs of milled sheet or granules are commonly used. A preheat of several minutes before the application of pressure is usual [3].

The mold surfaces require a releasing character for all systems, and therefore they are normally coated or lined with silicones or Teflon. This surface needs to be renewed frequently, in many instances with each cycle.

Some mold shrinkage can and does occur. In part, this is compensated for by the pressures involved; however, it should be anticipated in the mold design and the compound; 1 to 1.5% shrinkage is not uncommon.

Sheet produced by molding tends to be in the 20-mil and thicker range. When peroxide curatives are used, bubble formation is a problem. When several thicknesses of material are used, air inclusion is frequent. As a result the heavier sections ($\frac{1}{4}$ in. and thicker) are normally produced with bulk cast systems. Occasionally, laminating techniques are used to prepare heavier sections from the other polyurethane systems.

The thermoplastic systems, and to some degree the thermoplastic-thermosetting systems, need to be cooled under pressure to prevent distortion during removal from the mold. As a consequence presses that can be heated *and* cooled are necessary.

VI. Fabrication

Fabricating techniques are generally dependent on the type of polyurethane involved.

The thermoplastic polyurethanes are fabricated according to the techniques developed for flexible thermoplastic materials. A full range of methods is practiced: heat sealing (both dielectric and bar type), cement welding, vacuum forming, cement and heat lamination, and direct solution or melt casting onto various substrates. Standard equipment is normally used.

In general the thermoplastic-thermosetting materials are fabricated by the in-line or adhesive lamination to substrates. Some heat sealing and vacuum forming may be performed.

As a class the millable gums are fabricated with standard rubber techniques: direct molding, solution casting, curing on substrates, adhesive lamination, and mechanical shaping.

The liquid (bulk) systems are fabricated by direct coating, adhesive lamination, and mechanical shaping.

VII. Applications

The applications for polyurethane sheet and film are scattered over a wide range of uses. In general the high-performance properties of these materials tend to direct their application to particularly demanding uses and into composite constructions with other base materials, The most satisfactory and enduring applications are considered to be those in which the polyurethanes are designed into the construction to take full advantage of their properties.

Some of the earliest applications of polyurethanes were in areas that required high abrasion or fuel- and oil-resistance. Since that time, however, the scope of applications has expanded rapidly and now includes uses in which high film strength, a soft elastic hand, breathability, low air and oxygen permeability, good low-temperature flexibility and nonmigration are required. Listed below is a partial compilation of known applications of polyurethane sheet and film and a brief note on the primary motivations for their use.

Application	Property Exploited
Fuel cells	Fuel and oil resistance, flexibility
Oil containers	Fuel and oil resistance, flexibility
Petroleum-storage liners	Fuel and oil resistance, flexibility
Leather laminants	Flexibility, abrasion resistance, high water-vapor transmission rate

Application	Property Exploited
Conveyor-belt facings	Abrasion resistance, flexibility
Large hose liners and jackets	Abrasion resistance, flexibility
Ink pouches	Solvent resistance, flexibility
Tear seals	Adhesion strength
Magnetic computer tapes	Abrasion resistance, high loading capacity
Laminants for rainwear	Flexibility, high water-vapor transmission rate
Fabricated surgical and household gloves	Flexibility, strength
Laminated panel facings	Scuff and abrasion resistance
Transmission-belt facings	Flexibility, abrasion resistance
Diaphragms and gaskets	Flexibility, fuel and oil resistance
Chute liners	Abrasion resistance
Packaging	Toughness, fuel and oil resistance
Floor coverings and runners	Abrasion resistance, fuel and oil resistance
Upholstery	Abrasion resistance, fuel and oil resistance
Pad protectors	Abrasion resistance
Liners (ship hulls, trains, trucks)	Abrasion resistance
Step protectors	Abrasion resistance
Shoe uppers	Toughness, abrasion resistance
Ankle supports	High modulus
Outerwear laminants	Abrasion resistance, ability to be cleaned
Hygienic bed sheeting	Stain resistance, flexibility, water repellence
Inflatables	Film strength, low gas permeability

VIII. Future Trends

There seems to be little doubt that applications for polyurethane sheet and film will continue to grow. At the moment most of this growth seems to be in the areas of thin films for packaging and lamination to other materials, belts, hoses, abrasion liners, upholstery, and synthetic leather. When better latex polymers and more environmentally resistant polymers are available, however, the list of applications should further expand. In addition, it is likely that in time the cost of the polyurethanes will decrease to a level which, at present, is generally characterized by the specialty elastomers. As this occurs it is probable that polyurethane sheet and film will find many more large-volume applications.

Based on today's trends in usage, an increasing awareness of the polyurethane properties, some improvement in environmental resistance, and a projected price decline, we would expect the primary areas of future growth to include general upholstery, packaging, inflatables, "breathable" fabrics, fuel- and oil-storage linings, abrasion-resistant facings, and industrial and commercial wall coverings.

Glossary of Trade Names

Adiprene	E. I. du Pont de Nemours and Co., Elastomer Chemicals Department
Daltoflex	Imperial Chemical Industries, Ltd., Dyestuffs Division
Durethan	Farbenfabriken Bayer
Estane	B. F. Goodrich Chemical Co.
Genthane	General Tire and Rubber Co.
Pellethane	CPR Division, Upjohn Co.
Roylar	Uniroyal
Rucothane	Ruco Division, Hooker Chemical Corp.
Texin	Mobay Chemical Co.
Tuftane	B. F. Goodrich Chemical Co.
Urepan	Farbenfabriken Bayer
Vibrathane	Uniroyal

References

1. Farbenfabriken Bayer, "Durethan," Technical Bulletin Leverkusen (April 1, 1960).
2. C. S. Schollenberger, H. Scott, and G. R. Moore, *Rubber World*, **137** (4), 549 (January 1958).
3. B. F. Goodrich Chemical Company, "Estane Polyurethane Materials," Service Bulletin G18, Cleveland, Ohio (Rev. July 1960).
4. K. A. Pigott, J. W. Britain, William Archer, B. F. Frye, R. J. Cote, and J. H. Saunders, *Ind. Eng. Chem., Prod. Res. Develop.*, **1**, 28 (1962).
5. Mobay Chemical Company, Technical Bulletin, "Processing Methods for Texin Urethane Elastomer Resins," Pittsburgh (1964).
6. *Modern Plastics*, **41**, No. 10, 162 (June 1964).
7. The General Tire and Rubber Co., "Genthane S," Technical Bulletin GT–S3, Akron, Ohio.
8. The General Tire and Rubber Co., "Genthane SR," Technical Bulletin GT–SR 1, Akron, Ohio.
9. Farbenfabriken Bayer, "Urepan E, An Ester-Based Urethane Rubber for Cross-linking with Peroxides," Technical Bulletin, Leverkusen (July 1, 1961).
10. Naugatuck Chemical Co., Division of United States Rubber Co. (Uniroyal), "Vibrathane 5003," Technical Bulletin, Naugatuck, Connecticut.
11. Naugatuck Chemical Co., Division of United States Rubber Co. (Uniroyal), "Vibrathane 5004," Technical Bulletin, Naugatuck, Connecticut.

12. The General Tire & Rubber Company, Research Division.
13. L. B. Weisfeld, J. R. Little, and W. E. Wolstenholme, *J. Polymer Sci.*, **56**, 455 (1962).
14. E. I. du Pont de Nemours and Co., Elastomer Chemicals Department, "Adiprene C, a Urethane Rubber," Development Products Report No. 4, Wilmington, Delaware (July 15, 1957).
15. Imperial Chemical Industries, Ltd., Dyestuffs Division, "Hard Polyurethane Rubbers from Daltoflex 1 and Suprasec M," Urethane Chemicals Literature No. 6, Manchester, England (June 1961).
16. Imperial Chemical Industries, Ltd., Dyestuffs Division, "Surface Coatings and Adhesives from Daltoflex 1S and Suprasecs M, G, and K, Urethane Chemicals Literature No. 7, Manchester, England (January 1962).
17. C. E. Brockway, U.S. Patent 2,741,800 (April 17, 1956).
18. R. J. Athey, J. G. dePinto, and J. S. Rugg, "Adiprene L, a Liquid Urethane Elastomer," Development Products Report No. 10, Elastomer Chemicals Department, E. I. du Pont de Nemours & Co., Wilmington, Delaware, (March 15, 1958).
19. R. J. Athey, "Liquid Urethane Elastomers," Contribution No. 138, Division of Rubber Chemistry, American Chemical Society National Meeting, Cincinnati, Ohio, May 14, 1958.
20. Naugatuck Chemical Division, United States Rubber Company (Uniroyal), "Vibrathane 6001 Liquid Polymer," Technical Bulletin No. 2.
21. Naugatuck Chemical Division, United States Rubber Company (Uniroyal), "Vibrathane 6004 Liquid Polymer."
22. Naugatuck Chemical Division, United States Rubber Company (Uniroyal), "Vibrathane 6005 Formulated with Vibrathane 3005—Low Hardness Urethane Elastomers"; "Vibrathane 6005—Low Hardness Liquid Polymer."
23. Naugatuck Chemical Division, United States Rubber Company (Uniroyal), "Vibrathane 6006—High Hardness Liquid Polymer."
24. S. L. Axelrood, L. C. Smith, and K. C. Frisch, "Machine-Cast Urethane-Urea Elastomers Produced by a One-Shot Method," Polymer Research Department, Wyandotte Chemical Corp., Wyandotte, Michigan.
25. S. L. Axelrood, C. W. Hamilton, and K. C. Frisch, "A One-Shot Method for Urethane-Urea Elastomers," *Ind. Eng. Chem.*, **53**, 889 (1961).
26. T. T. Stetz and James F. Smith, "Thermoplastic Polyurethane Solutions," Paper No. 46–13, 20th Annual Technical Conference, Society of Plastic Engineers, Inc. (Jan. 27–30, 1964).
27. B. F. Goodrich Chemical Company, Technical Service Reports TSR–63–26 and 64–18 and Technical Newsletter No. 2 (June 1961).
28. D. Esarove and J. F. Smith, "The Versatility of Thermoplastic Urethanes," paper presented to the Rubber Chemistry Division of the American Chemical Society National Meeting, Cleveland, Ohio, October 1962.
29. Mobay Chemical Company, "Texin Urethane Elastomers for Film Applications," a technical report.
30. U.S. Rubber Company, Naugatuck Chemical Division (Uniroyal), private communication.
31. The General Tire and Rubber Company, private communication.

32. Thiokol Chemical Company, "Elastothane 455 Polymer," a technical report.
33. O. C. Elmer and A. E. Schmucker, *Rubber Age*, **94** (3), 438 (1963).
34. S. L. Axelrood and K. C. Frisch, *Rubber Age*, **88** (3), 465 (1960).
35. L. H. Dickenson, *Rubber Age*, **82** (1), 96 (1957).
36. K. A. Pigott, *Rubber Age*, **91** (2), 629 (1962).
37. C. H. Smith and C. A. Peterson, *Modern Plastics*, **38** (11), 125 (1961).
38. Thiokol Chemical Company, "Solithane 291 Resin," Bulletin UR–8.
39. Automatic Process Control Company, private communication.
40. National Starch and Chemical Company and Wyandotte Chemical Company, private communications.
41. B. F. Goodrich Chemical Company, "Estane Sheet and Film," a bulletin, November 1964.
42. Correspondence with the Seiberling Rubber Company.

CHAPTER 13

FLUOROCARBON POLYMERS

HARVEY A. BROWN and GEORGE H. CRAWFORD

3M Company
Saint Paul, Minnesota

I. Introduction to the Fluorocarbon Polymers

This chapter is devoted to polymers derived from monomers containing at least two fluorine atoms per monomer unit. Polymers based on vinyl fluoride, the first member of the series of fluorine-containing ethylene derivatives, are thoroughly dealt with in Chapter 15 and are not included here.

Fluorocarbon films are still in the class of *specialty* or *high-performance* products whose areas of practical utility are delineated by their unusual properties and often limited by their cost. The per-pound cost of the base polymers is in the range of $3 to $15; hence these films are not likely in the near future to invade the mass markets of general packaging, photographic film backings, and tape backings or in similar applications that are the province of the polyolefins, vinyls, cellulosics, and polyesters.

Some of the features possessed by individual fluorocarbon films which qualify them for speciality applications are stability at elevated temperatures, low permeability to moisture and oxygen, resistance to solvent and chemical attack, low coefficient of friction, low surface energy, good optical properties, and good outdoor weathering characteristics. The latter is possessed in particular by polyvinyl fluoride and by polyvinylidene fluoride and promises to provide these fluorocarbon films with a comparatively large volume outlet.

In this discussion we have chosen to go beyond simply presenting the properties, fabrication techniques, and uses of fluorocarbon films. The fluorocarbons and the polymers derived from them constitute a unique class of materials of growing significance. We therefore believe it is also useful to cover some of the basic fluorine chemistry pertinent to the preparation and polymerization of the commercially important monomers and to relate the properties of the films to the fundamental nature of the fluorocarbon class of compounds.

A large number of fluorine-containing polymers have been prepared and studied but only a few have been produced on an industrial scale. Excluding polyvinyl fluoride, four fluorocarbon polymer types are used at present in film manufacture. All fall in the so-called *vinyl* or *addition polymer* class. Although a number of interesting polycondensates have been synthesized, based, for example, on the fluorocarbon dicarboxylic acids, none is commercially available today.

The four classes of fluorocarbon film-formers are listed below in approximately the chronological order of their development.

1. Polytetrafluoroethylene, $(-CF_2-CF_2-)_n$.
2. Polychlorotrifluoroethylene, $(-CF_2-CFCl-)_n$.
3. Polyperfluoroethylenepropylene, $(-CF_2-CF-CF_2-CF_2-)_n$.
$$\underset{\displaystyle CF_3}{\big|}$$
4. Polyvinylidene fluoride, $(-CH_2-CF_2-)_n$.

These resins, along with the currently accepted abbreviations and the trade names used by the respective resin manufacturer, are listed in the accompanying table.

The abbreviation for the film or resin under consideration (with ASTM designation of specific type when established and applicable) is used hereafter.

Fluorocarbon Film-Forming Polymers

Chemical Name	Abbreviation	Trade Name	Manufacturer
Polytetrafluoroethylene	TFE	Teflon[a]	E. I. du Pont de Nemours & Co.
		Halon[a]	Allied Chemical Corporation
		Tetran[a]	Pennsalt Chemicals Corporation
Poly(tetrafluoroethylene/ perfluoropropylene)	FEP	Teflon FEP[a]	E. I. du Pont de Nemours & Co.
Polychlorotrifluoroethylene	CTFE	Kel–F[a]	3M Company
		Plaskon[a]	Allied Chemical Corporation
		(Aclar)[b]	Allied Chemical Corporation
Polyvinylidene fluoride	PVF$_2$	Kynar[a]	Pennsalt Chemicals Corporation

[a] Registered trademark.
[b] Applied to films only.

Since the four principal fluoroolefins copolymerize with one another readily, copolymerization is often used to adjust the properties of a given base resin; FEP is a particularly striking example (see below). Also available are grades of CTFE which are copolymers containing vinylidene fluoride, which confers improved processability.

II. Properties of the Fluorocarbons as a Chemical Class

A. THERMAL PROPERTIES

The fluorocarbons are *prima facie* the reaction products of the elements fluorine and carbon. Fluorine, the most electronegative element, is also among the most reactive. It combines, often violently, with nearly all the elements of the periodic table, including certain of the "inert" gases. Compared with their hydrocarbon analogs, the aliphatic fluorocarbons have extreme resistance to thermal decomposition.

The most stable fluorocarbon is perfluoromethane, since it lacks a C—C bond, which even in the fluorocarbons is weaker than the C—F bond; that is, the C—C bond dissociation energy in perfluoroethane is 83 kcal/mole, whereas in perfluoromethane the C—F bond energy is ca. 120 kcal/mole [1].

Direct fluorination reactions and thermal decompositions of higher fluorocarbons tend to yield CF_4 as the most stable end product. Carbon tetrafluoride can be decomposed, however, by passing it through a carbon arc, thus giving radical fragments that coalesce to mixed products [2, 3]. The next higher member, hexafluoroethane, is somewhat less stable but is not decomposed appreciably up to 840°C [4].

It will be seen that polytetrafluoroethylene, the high-polymer homolog of the simple aliphatic fluorocarbons, although extremely stable in its own right, is "not so stable as it should be," since at temperatures above 500°F it begins to decompose via chain unzippering or random scission, depending on the experimental conditions. This is probably because of weak points, unstable end groups, branching, unsaturation, or incorporated impurities such as trifluoroethylene segments. The replacement by another element of any of the fluorine in a fluorocarbon polymer invariably reduces thermal stability. On the other hand, the introduction of backbone hydrogen is advantageously employed to provide the chain flexibility necessary for elastomeric behavior in the fluorocarbon rubbers such as Kel-F† elastomer, Fluorel,† and Viton.‡ The hydrogen sites here serve a dual purpose, since they also provide sites for crosslinking by diamines or peroxides.

B. Chemical Characteristics

When free of hydrogen, chlorine, or other substituents, the aliphatic fluorocarbons are remarkably resistant to chemical (particularly oxidative) attack. Aside from being chemically inert, they are also physically incompatible with most chemical reagents in contradistinction to the hydrocarbons which are readily oxidized and miscible with many reagents. Fluorocarbons are also suprisingly resistant to reducing agents, being attacked only by hot alkali metals, metal alkyls, and the like. In fluorocarbon films, the latter type of reaction, which causes surface defluorination, is used to render the film bondable with ordinary adhesives.

† Registered trademark of 3M Company.
‡ Registered trademark of E. I. du Pont de Nemours & Co.

The high C—F bond energy accounts in great measure for the chemical inertness of fluorocarbons, but this is not the entire story. The size of the fluorine atom is ideally suited to provide shielding of the carbon backbone of the compound from attack. The fluorine atom is larger than hydrogen but smaller than chlorine. Higher perchloro compounds are unstable because of the strain induced in attempting to pack these large atoms along a carbon-carbon chain. The approximate relative size of the hydrogen, fluorine, and chlorine atoms can be seen in photographs of molecular models elsewhere in this chapter.

Although the simple aliphatic fluorocarbons are unreactive, fluorocarbon derivatives having functional sites are often much more reactive than their hydrocarbon analogs, even to the point of not existing under ordinary conditions. In many cases it is necessary to insert a methylene group between a perfluoroalkyl group and a functional site to avoid spontaneous decomposition (often via HF elimination). Thus CF_3CH_2OH, $CF_3CH_2NH_2$, and CF_3CH_2SH are stable, whereas CF_3CF_2OH, $CF_3CF_2NH_2$, and CF_3CF_2SH are not. Similarly, in the fluorine-containing siloxane elastomers the CF_3 groups must be insulated from the silicon atom, as, for example, in

$$
\begin{array}{c}
CH_3 \\
| \\
[-Si-O-]_n \\
| \\
CH_2-CH_2-CF_3
\end{array}
$$

Although trifluoroacetic acid, CF_3CO_2H, is a stable compound, the powerful electron withdrawing effect of the CF_3 group causes it to resemble the inorganic halogen acids rather than acetic acid in the extent of dissociation in ionizing solvents.

C. Physical Properties

The fluorocarbons are dense materials with liquid densities in the region of 2. They have low bulk and solution viscosities in relation to their molecular weights. As mentioned before, fluorocarbons have an extreme aversion to mixing with other chemicals. They have the lowest surface free energies of any class of compounds. These properties carry over into the polymers, so that the fluorocarbon films have excellent nonsticking properties, low coefficients of friction, and good optical properties (when amorphous).

Since fluorocarbon chains slide past one another very easily, the completely fluorinated polymers tend to have low tensile and compressive strengths, and in some cases poor creep resistance under load.

Fluorocarbons also have the lowest indexes of refraction of any class of compounds: for example heptane, $n_D^{20} = 1.375$; perfluoroheptane $n_D^{20} = 1.2618$. They are also very transmissive to light in the visible and near-infrared regions.

We have briefly examined some of the characteristic features of the fluorocarbon family of compounds and have given a few examples to illustrate in a general way the manner in which these characteristics are manifested in the fluorocarbon polymers and films.

We now move to the specific fluorocarbon monomers, their preparation and polymerization. We have abstracted much of the data on film properties and other information in the text and tables that follow from product bulletins and technical brochures provided by Allied Chemical Corporation, E. I. du Pont de Nemours & Co., 3M Company, and Pennsalt Chemicals Corporation.

III. Synthesis of Fluorocarbon Monomers

Since the important film-forming fluoropolymers are all based on wholly or partially fluorinated olefins, these compounds will be emphasized here.

The fluoroolefins are derived from the simple chlorofluorocarbons which were originally developed as refrigerants. The chlorofluorocarbons are prepared by the fluorination of simple chlorine-containing hydrocarbons:

$$CHCl_3 + HF \xrightarrow{\ SbCl_3\ } CHF_2Cl$$

These chlorofluorocarbons, known commercially as Freons,† Genetrons,‡ and Isotrons,§ are attributed to the pioneering research of Midgley and Henne [6], and still constitute the principal basic raw materials of the fluorocarbon industry. They are also widely used as aerosol propellants and solvents.

† Registered trademark of E. I. du Pont de Nemours & Co.
‡ Registered trademark of Pennsalt Chemicals Corp.
§ Registered trademark of Allied Chemical Corp.

Fluorocarbons, in contrast to chlorinated organic compounds which may be prepared by direct reaction of hydrocarbons with chlorine, are not made commercially by elemental fluorination. Although it is true that commendable progress has been made in controlling the reactions of elemental fluorine with organic materials, the tendency is to fragment the carbon skeleton and to go all the way to CF_4. The methods normally employed to prepare fluorocarbons, for example, electrochemical fluorination [5] or reaction of chlorinated organics with certain metal fluorides [6] cannot be applied to the direct synthesis of fluorinated olefins because these methods normally add fluorine across double bonds. For this reason fluorocarbon olefins are prepared by processes that involve the indirect fluorination of hydrocarbons or chlorinated hydrocarbons and subsequent disproportionation and elimination reactions. Some of these processes are illustrated in the following sections on monomer synthesis.

A. Tetrafluoroethylene

O. Ruff [7], an early investigator in fluorine chemistry, first prepared tetrafluoroethylene, a gas, bp $-76.3°C$, by pyrolysis of carbon tetrafluoride in a carbon arc.

$$CF_4 \xrightarrow[\text{arc}]{\text{electric}} CF_2{=}CF_2 + CF_3{-}CF_3 + \cdots$$

The resulting products were separated from unreacted CF_4 by low-temperature fractional distillation. The tetrafluoroethylene was separated from the other reaction products by adding Br_2 to give $CF_2Br{-}CF_2Br$ which was easily isolated. Treating the dibromo derivative with zinc dust in alcohol regenerated the desired olefin.

Park and his associates [8] later found that the pyrolysis of $CHClF_2$ under controlled conditions gave excellent yields (90 to 95%) of tetrafluoroethylene.

$$2CF_2ClH \xrightarrow{\Delta} CF_2{=}CF_2 + 2HCl$$

Commercial routes to tetrafluoroethylene employ pyrolytic dehydrochlorination reactions of this type [9,10] with conditions set to optimize the yield of the C_2 olefin.

It should be noted here that tetrafluoroethylene can be highly hazardous, particularly if stored in the liquid state under pressure. If oxygen is present as a contaminant, a runaway exothermic polymerization reaction which proceeds with explosive violence can be initiated.

B. Chlorotrifluoroethylene

The monomer, $CF_2=CFCl$, bp $-27.9°C$, was first prepared [11] by dechlorinating 1,2,2-trichloro-1,1,2-trifluoroethylene obtained by the fluorination of hexachloroethane using $SbF_3 + SbCl_5$ [12].

$$CCl_3-CCl_3 + SbF_3/SbCl_5 \longrightarrow CF_2Cl-CFCl_2$$

$$CF_2Cl-CFCl_2 \xrightarrow[\text{alcohol}]{Zn} CF_2=CFCl + ZnCl_2$$

Trifluorotrichloroethane was also prepared by reaction of SbF_3 with hexachloroethane under pressure in a steel autoclave [13, 14]. The commercial preparation of chlorotrifluoroethylene is based on the dehalogenation of $CF_2Cl-CFCl_2$ [15]. It is claimed [16] that chlorotrifluoroethylene can be prepared in high yield by pyrolyzing $CF_2Cl-CFCl_2$ in the presence of hydrogen.

$$CF_2Cl-CFCl_2 + H_2 \xrightarrow[\text{iron tube}]{550-600°C} CF_2=CFCl$$
$$90\%$$

C. Hexafluoropropylene

Hexafluoropropylene, bp $-33°C$, was isolated from products of the pyrolysis of polytetrafluoroethylene [17]. This monomer is conveniently obtained in the laboratory in good yields by the thermal decarboxylation of sodium perfluorobutyrate [18].

$$C_3F_7CO_2Na \xrightarrow{\Delta} C_3F_6 + CO_2 + NaF$$

Several other methods have been developed which involve the pyrolysis of relatively cheap starting materials, such as CF_2HCl, tetrafluoroethylene, polytetrafluoroethylene, and tetrafluoroethylene-fluorocarbon mixtures, the yield being determined by reaction conditions [19–21]. By pyrolyzing CF_2HCl and introducing $CF_2=CF_2$ near the exit end of the pyrolysis tube the yield of $CF_3CF=CF_2$ is increased considerably [22]. Such processes may involve the thermal generation of difluorocarbene as a transient intermediate, that is,

$$C_2F_4 \xrightarrow{\Delta} 2:CF_2$$

$$:CF_2 + CF_2=CF_2 \longrightarrow CF_3CF=CF_2$$

A recent patent [23] describes the preparation of $CF_3CF=CF_2$ by the disproportionation reaction:

$$2CF_3CF=CFCl \xrightarrow{325-425°C} CF_3CF=CF_2 + CF_3CF=CCl_2$$

D. VINYLIDENE FLUORIDE

In his classical research on the synthesis of fluorine-containing organic compounds, Swarts [24] found that replacement of the halogens, chlorine, bromine, and iodine, with fluorine could be effected by reaction of halogen-containing organic compounds with antimony(III) fluoride in the presence of catalytic amounts of pentavalent antimony. By this method 1,1-difluoro-2-bromoethane was prepared.

$$3CH_2BrCHBr_2 + 2SbF_3/Br_2 \xrightarrow{100°C} CH_2BrCHF_2$$

By treating the product with sodium amylate Swarts obtained vinylidene fluoride, bp $-84°C$.

$$CH_2BrCHF_2 + NaOC_5H_{11} \longrightarrow CH_2=CF_2$$

1,1-Difluoro-2-bromoethane can also be prepared [25] by reaction of the same starting material, $CH_2BrCHBr_2$, with mercuric fluoride formed *in situ* by reaction of mercuric oxide with hydrogen fluoride.

Recently more economical routes for preparing vinylidene fluoride have been devised; for example, the thermal dehydrohalogenation of chlorodifluoroethane has been thoroughly studied [26–28] and may be used for preparing this monomer on an industrial scale.

$$CH_3CClF_2 \xrightarrow{500-1700°C} CH_2=CF_2 + HCl$$

E. OTHER FLUOROCARBON MONOMERS

Endeavors in the realm of fluorocarbon research have yielded many unsaturated compounds theoretically capable of being polymerized.

Indeed, many have been homopolymerized or copolymerized with hydrocarbons or other fluorocarbons. In some cases, however, fluorocarbon monomers have stubbornly refused to homopolymerize except under drastic conditions and then only low-molecular-weight polymers were obtained. A case in point familiar to us and undoubtedly to many other workers is hexafluoropropylene.

With the exception of vinyl fluoride, only the four monomers presented above have reached industrial significance as far as film-forming fluorocarbons are concerned. The reader who desires further information concerning typical synthetic procedures used to prepare fluorocarbon olefins should consult [8, 26, 29–31].

IV. Polymerization of Fluorocarbon Monomers

The powerful electron-withdrawing effect of fluorine profoundly affects the polymerization characteristics of the fluoroolefins. This is reflected in the e-values assigned to the fluoroolefins in the Alfrey-Price [32] Q-e scheme for relating copolymerization behavior to resonance (Q) and polar (e) factors in the monomer structure. Fluoroolefins have high positive e values; for example, for TFE and CTFE $e = +1.22$ and $+1.48$, respectively, whereas for ethylene and vinyl chloride $e = -0.20$ and $+0.20$, respectively. This polarization about the fluorine substituents render the π-bond in highly fluorinated olefins extremely "electron poor." One would thus expect the fluoroolefins to be susceptible to anionic but not to cationic polymerizations.

Although anionic polymerizations of some fluoroolefins have been carried out in the laboratory, industrial processes employ free-radical techniques in aqueous suspensions or emulsion systems. In the emulsion homo- and copolymerization of fluoroolefins specialized recipes which employ fluorocarbon emulsifiers must often be used if adequate solubilization of the fluoroolefin is to be achieved and undesirable side reactions, avoided.

A. TETRAFLUOROETHYLENE

The inadvertent discovery of polytetrafluoroethylene by Plunkett in 1938 is a classic of early fluorocarbon lore [33].

The monomer had been stored in a steel cylinder, and when less than the known quantity of monomer was recovered the cylinder was cut open and a white, waxy solid, which was shown to be the polymer, was found. The polymer could neither be dissolved nor molded and might well have been given up as a useless laboratory curiosity. The early investigators, however, were quick to recognize the potential value of the material. Commercially workable polymerization techniques [34, 35] were developed which today allow the manufacturer to produce many grades of TFE suited to a broad spectrum of end uses and fabricating methods.

The polymerization of tetrafluoroethylene is usually conducted in water with free-radical initiators such as ammonium persulfate at elevated pressures. Redox initiator systems such as persulfate-bisulfite-copper sulfate can be used [36, 37]. Organic peroxides, for example, acetyl peroxide and benzoyl peroxide, are also used as initiators [38]. Dibasic acid peroxides such as disuccinic acid peroxide are used in aqueous recipes to give stable TFE suspensoids [39]. Recently the stereospecific polymerization of tetrafluoroethylene has been reported to give slow conversion to crystalline polymers [40]. The Ziegler-Natta catalysts used in these experiments were based on titanium tetraalkoxides and aluminum alkyl derivatives.

B. CHLOROTRIFLUOROETHYLENE

The original commercial preparation of CTFE resin was by polymerization in bulk in the presence of a free-radical initiator at low temperatures [41]. Other processes have also been patented [42, 43], including bulk and solution polymerization using the Ziegler type of catalysts [44]. Various other processes utilize aqueous suspension and emulsion techniques [45–47]. The latter are readily adapted to the manufacture of CTFE copolymers, such as copolymers of CTFE and VF_2.

C. HEXAFLUOROPROPYLENE

Hexafluoropropylene ($CF_3CF{=}CF_2$), as well as the higher perfluorinated 1-olefins, has not been *homo*polymerized to high polymer in aqueous emulsion systems such as those that have been successfully applied to the fluorinated ethylenes. Various free-radical initiators, ultraviolet light, and nuclear radiation have been tried without success [48–50]. Eleuterio has reported the preparation of amorphous, glassy polyhexafluoropropylene by using special free-radical catalysts such as mercury bis(trifluoromethyl mercaptide) at 2000 to 10,000 atm pressures [51, 52]. Sianesi [40] describes the stereospecific polymerization of HFP in the presence of certain Ziegler-Natta catalysts, such as triisobutyl aluminum/titanium tetraisopropoxide in methylene chloride, but low polymerization rates and yields were obtained.

The commercial utility of hexafluoropropylene lies in its ability to copolymerize with the fluorinated ethylenes in aqueous emulsion systems [53]. In such copolymerizations the reactivity ratio of hexafluoropropylene with respect to other olefins approaches zero and therefore it resists incorporation to an extent greater than 50 mole percent. It is, in fact, difficult to incorporate more than a few percent of hexafluoropropylene into a copolymer with TFE. This, however, is enough to provide the necessary crystallinity disruption for melt processing in the commercial FEP grade of TFE.

The most notable comonomer for hexafluoropropylene (aside from TFE monomer) is vinylidene fluoride ($CH_2=CF_2$). This copolymerization reaction can produce polymers with an extremely broad spectrum of properties, ranging from thermoplastics to elastomers, depending on combined ratio. (The thermoplastic film former derived from the homopolymerization of vinylidene fluoride is treated elsewhere in this chapter, Section VIII.) The elastomeric copolymers of vinylidene fluoride and hexafluoropropylene are currently marketed under the trade names *Fluorel* and *Viton* and are the result of complete or near-complete disruption of the crystallinity of the

$$[-CH_2-CF_2-CH_2-CF_2-]_n$$

chain [54–56]. These materials constitute a highly useful class of high-temperature-serviceable, chemically resistant rubbers. They do not come within the scope of this chapter but illustrate the crystallinity disrupting power of the pendant CF_3 group. When the energy barrier to internal rotation about the C—C bonds in the backbone is low enough that the inherent glass transition temperature T_g of that backbone is below use temperature (the T_g of polyvinylidene fluoride is $-38°C$ [57]), disruption of crystallinity results in an elastomeric material.

D. VINYLIDENE FLUORIDE

Vinylidene fluoride (CH_2=CF_2) found early commercial utilization in copolymerization reactions with CTFE monomer [58]. This reaction proceeds readily in emulsion systems of the aforementioned type by employing anionic fluorocarbon or chlorofluorocarbon emulsifiers. Combined in small quantities, vinylidene fluoride serves to modify the properties of the rigid crystalline CTFE homopolymer, and thus produces more readily melt-processable materials at 5 to 10% VF_2 and ketone-ester soluble resins at 20 to 30% VF_2. When larger quantities of VF_2 are introduced, elastomers are obtained. Here the effect is one of mutual disruption of crystallizing tendencies in the two combined monomers. As in the copolymerization with hexafluoropropylene, introduction of the hydrogen-containing monomer provides chain flexibility for elastomeric behavior. Vinylidene fluoride has also been copolymerized with ethylene and vinyl fluoride [59].

In an early study of the polymerization of vinylidene fluoride McBee [60] found that by conducting the reaction in a nickel autoclave under autogenous pressure at 100°C for 48 hr a polymer with a softening point of 269°F could be obtained. Acetyl peroxide was used as a polymerization catalyst.

$$nCH_2\text{=}CF_2 \quad \xrightarrow[\text{pressure, heat}]{\text{peroxide}} \quad [-CH_2-CF_2-]_n$$

Vinylidene fluoride can also be polymerized in water-suspension systems that employ peroxides and persulfates as catalysts at moderately elevated temperatures and pressures up to 1000 atm [61]. The polymer so obtained softens at 293 to 320°F.

A recent patent [62] describes a process for polymerizing vinylidene fluoride in the presence of alkylene oxides, which utilizes organic peroxides as catalysts. High yields of PVF_2 are achieved.

Continuing research on monomer synthesis and polymerization techniques has made possible the commercial development and production of the crystalline thermoplastic homopolymer of vinylidene fluoride.

E. OTHER FLUOROCARBON ADDITION POLYMERIZATIONS

It is interesting to note that unsaturated heteroatomic functional groups attached to fluoroalkyl radicals often readily enter into addition polymerization reactions, thus giving rise to stable polymers of high molecular weight. Polymerizations through the double bonds in carbonyl, thiocarbonyl, nitrile and nitroso groups have been achieved:

a. 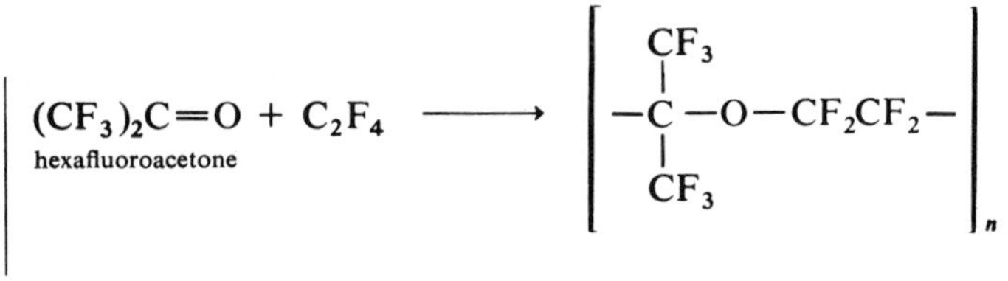
$$(CF_3)_2C{=}O + C_2F_4 \longrightarrow \left[\begin{array}{c} CF_3 \\ | \\ {-}C{-}O{-}CF_2CF_2{-} \\ | \\ CF_3 \end{array} \right]_n \qquad [63]$$

hexafluoroacetone

b. 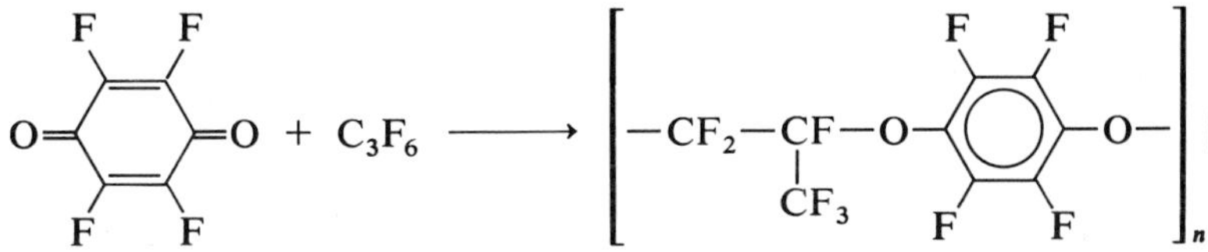
$$O{=}\!\!\!\bigcirc\!\!\!{=}O + C_3F_6 \longrightarrow \left[{-}CF_2{-}\underset{\underset{CF_3}{|}}{CF}{-}O{-}\!\!\!\bigcirc\!\!\!{-}O{-} \right]_n \qquad [64]$$

c. 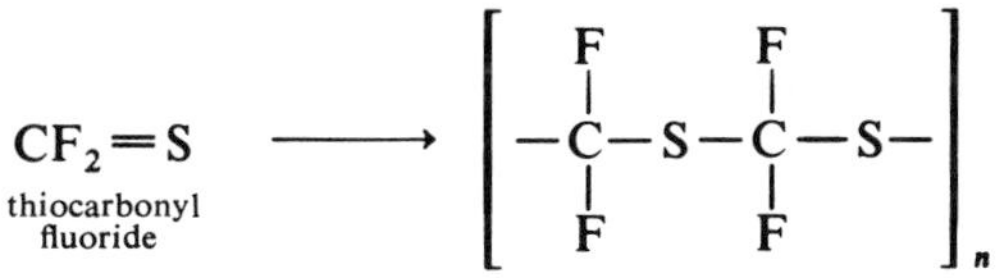
$$CF_2{=}S \longrightarrow \left[\begin{array}{cc} F & F \\ | & | \\ {-}C{-}S{-}C{-}S{-} \\ | & | \\ F & F \end{array} \right]_n \qquad [65]$$

thiocarbonyl fluoride

d. 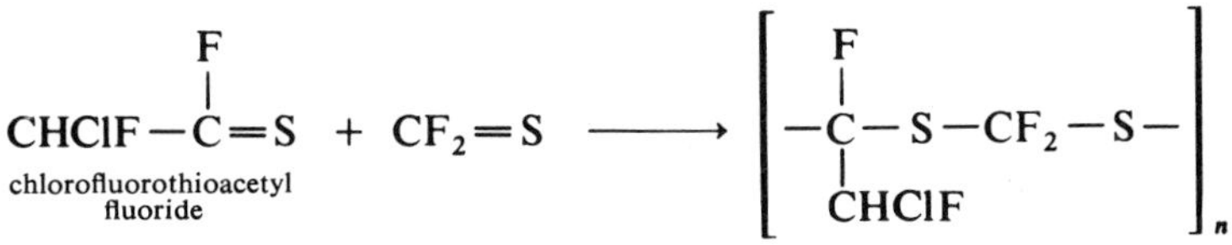
$$\underset{\text{chlorofluorothioacetyl fluoride}}{CHClF{-}\overset{\overset{F}{|}}{C}{=}S} + CF_2{=}S \longrightarrow \left[\begin{array}{c} F \\ | \\ {-}C{-}S{-}CF_2{-}S{-} \\ | \\ CHClF \end{array} \right]_n \qquad [65]$$

e. 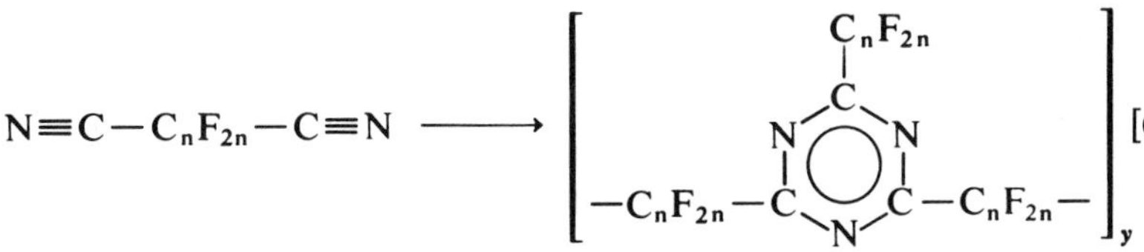
$$N{\equiv}C{-}C_nF_{2n}{-}C{\equiv}N \longrightarrow \left[\begin{array}{c} C_nF_{2n} \\ triazine \\ {-}C_nF_{2n}{-} \end{array} \right]_y \qquad [66]$$

f. 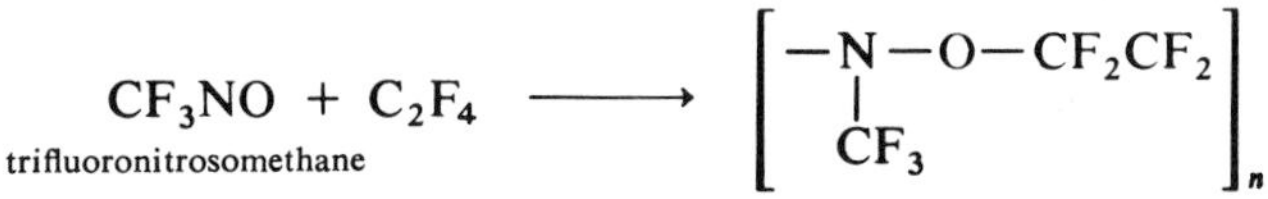
$$\underset{\text{trifluoronitrosomethane}}{CF_3NO + C_2F_4} \longrightarrow \left[\begin{array}{c} {-}N{-}O{-}CF_2CF_2 \\ | \\ CF_3 \end{array} \right]_n \qquad [67]$$

540   HARVEY A. BROWN AND GEORGE H. CRAWFORD

The last reaction is unusual because it is initiated and propagates spontaneously by a free-radical mechanism at temperatures as low as $-100°C$ when CF_3NO and TFE are mixed [68].

The high-molecular-weight product known as "nitroso rubber" is rendered curable by carboxyl groups introduced by terpolymerization with $ON(CF_2)_nCO_2H$ [69].

Also worthy of mention here are the polyfluorocarbon ethers prepared by the ring opening of fluorocarbon epoxides. The low polymers are useful as stable lubricants and hydraulic fluids.

$$CF_3CF\overset{O}{\overbrace{-}}CF_2 \xrightarrow[\text{on carbon black}]{CsF} \left[CF\underset{CF_3}{-}CF_2-O\right]_n$$

In the foregoing sections we have reviewed the chemistry of the monomers and polymerization reactions related to the commercial film-forming resins and have also reviewed some of the newer polymerization reactions. The fluorocarbons have a fascinating chemistry. There are several excellent books and series to which the reader is referred for additional information [25, 29–31]. We now consider the individual fluorocarbon film formers.

V. Polytetrafluoroethylene (TFE)

Polytetrafluoroethylene was the first fluorocarbon polymer to be commercialized (1941) and today enjoys the largest sales volume, being greater than all of the others combined. Sales of TFE for 1967 are estimated at 80 to 90 million dollars, with a large portion of recent growth being due to TFE-coated cookware and other hardware.

A. General Characteristics

TFE is a white, opaque, highly crystalline material of unusually high molecular weight. Values by end-group analysis indicate molecular weights ranging from about 0.4×10^6 to 8.9×10^6 [71].

Its structure, in essence, constitutes very long chains of CF_2 units. Bunn concluded from x-ray diffraction analysis that these units are equally spaced along the chain, suggesting a regular helix [72]. A fully extended chain of singly linked carbon atoms is usually represented as a planar zigzag. In reality the chain must adopt helical configuration in

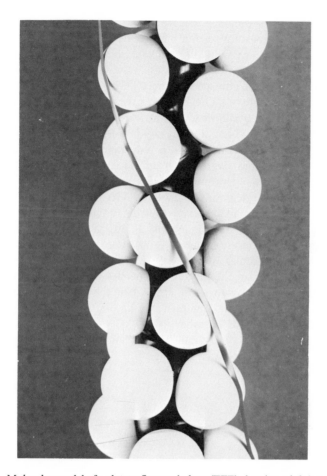

Fig. 1   Molecular model of polytetrafluoroethylene (TFE) showing a left-hand twist
of 180° every 13 $-CF_2-$ units.

order to accommodate the fluorine atoms.   A 360° twist of the chain
occurs in every 26 $-CF_2-$ units.   Figure 1 represents a molecular
model showing a left-hand twist of 180° every 13 $-CF_2-$ units [73].
At about 30°C a much larger angular displacement occurs, and at higher
temperatures changes in diffraction pattern suggest further uncoiling.
Random motion results until at the melting point (327°C) all crystalline

order is lost and the polymer goes from an opaque solid to a transparent rubbery gel. It does not form a melt in the conventional sense.

It has been shown by x-ray diffraction studies that although quenched TFE has reduced crystallinity it possesses the same unit cell and helical chain structure as the annealed polymer [74].

The thermal stability of TFE is impressive. It is suitable for applications that require continuous service at 500°F or intermittent service up to 600°F. At elevated temperatures (>600°F) degradation occurs and leads to a mixture of products that is indicative of random chain cleavage. However, rapid degradation of TFE does not occur below 735°F. At lower temperatures and very low pressures in the absence of oxygen the major degradation product (nearly 100%) is the monomer [75]. Madorsky and his associates concluded that a mechanism involving the unzippering of monomer units at free-radical ends of chains is important, so that once unzippering starts it continues in most cases to the end of the molecule.

Any presentation of the physical properties of TFE resin misses the point if it does not show the manner in which these properties are maintained over temperature extremes; for example, the room-temperature tensile strength of TFE is rather low (2000 to 3000 psi); however, it retains much of its strength at 500°F, at which many other polymers are undergoing thermal degradation, melt flow, or both. This is further illustrated in Table 1.

TABLE 1

Typical Mechanical Properties of Polytetrafluoroethylene Resin as a Function of Temperature

Temperature (°F)	Tensile Strength lb/in. D 638–61T[a]	Tensile Yield (psi) D 638–52T[a]	Elongation, (%) D 638–61T[a]	Flexural Modulus (psi) D 747–50[a]	Compressive Strength (psi)
−400	20,000	19,000	2	730,000	30,500
−200	9,000	11,500	90	560,000	10,000
−100	6,000	8,000	160	375,000	5,500
+ 32	3,000	2,000	250	150,000	2,500
+ 77	2,700	1,200	275	90,000	1,700
+250	1,300	500		20,000	500
+500	500				

[a] American Society for Testing and Materials test methods [86].

The very low coefficient of friction (0.04) is independent of temperature up to the melting point. The electrical properties all remain constant over the entire service range.

For additional information relating to the properties of TFE the reader is referred to published work [76].

TFE resin typifies fluorocarbon behavior in its resistance to attack by chemicals and solvents. It is attacked only by reagents such as molten sodium, sodium-ammonia, and elemental fluorine at elevated temperatures. It is completely nonflammable and is highly resistant to outdoor weathering.

Gas and liquid permeabilities are very low but can be affected by voids arising from the processing techniques (*vide infra*). Permeability data appear under "Film Properties."

TFE maintains a low dissipation factor and dielectric constant over a wide range of operating temperatures [77]. Its volume resistivity, greater than 10^{15} Ω-cm, is almost entirely unaffected by immersion in water for long periods. Arc resistance is good, since the arc vaporizes the polymer at the point of contact rather than leaving a conductive carbonized trail. The corona resistance of TFE is poor.

B. Processing into Sheet and Film

The processing methods of chief importance to the film maker include casting from dispersions, calendering, and extruding with an extrusion aid. Compression molding and extrusion are used in forming the billets used in making skived films.

1. Hot Forming of TFE

Since TFE does not form a true melt, it was necessary to develop an entire technology to form the polymer into useful shapes. This, in turn, has led to the development of granular, paste, and dispersion grades of TFE tailored to specific processing techniques and properties requirements in the finished article.

ASTM designation D 1457–62T groups TFE powders into four general types which are shown in Table 2 along with property differences in molded specimens as reflected in the ASTM minimum requirements specification [78].

Although TFE is classified as a thermoplastic resin, it does not, as

TABLE 2

Polytetrafluoroethylene Molding Powder Types
(ASTM D 1457–62T)[a]

Type	Description	Particle size average diameter (μ)	Specific gravity		Tensile strength (psi)	Elongation (%)
			Minimum	Maximum		
I	Granular powder for general-purpose molding or extrusion	575 ± 150	2.13	2.18	2000	125
II	Powder is of finer particle size than I	325 ± 75	2.13	2.18	2000	140
III	"Extrudable" TFE coagulated from dispersion used with volatile extrusion aid					
	Class 1	500 ± 150	2.19	2.24	2300	400
	Class 2	500 ± 150	2.15	2.20	3000	200
IV	Granular powder (extremely fine particles)		2.13	2.19	2700	170

[a] [78].

mentioned above, have the melt and flow characteristics of other thermo-plastics. The crystalline properties of TFE are retained at temperatures up to 620°F. Beyond this it becomes an amorphous transparent gel but with severely limited flow properties. At gel temperatures, however, the molding powder particles become quite tacky and can be welded to themselves under pressure. This phenomenon, which occurs in all TFE "melt" processing operations, coupled with the relationship between percent crystallinity and thermal history, leads to a significant dependence of mechanical properties on processing conditions.

All of the molding powder processing techniques involve essentially the same three basic steps: (a) cold forming, (b) sintering, and (c) cooling.

In Step 1 the powder is densified by pressing. This causes the powder particles to cold-flow and adhere together in varying degrees, depending on the pressure and molding powder.

In Step 2 the proximate particles are fused and caused to interpene-trate by heating above the melting point so that the aforementioned gel results. Here, pressure in varying degrees may be applied, depending on the process. The time and temperature of sintering are critical factors, since the polymer undergoes slow degradation at a rate that increases with increasing temperature when the fusion temperature is exceeded.

In Step 3 the percentage of crystallinity is controlled by the cooling rate. The film can thus be annealed to provide up to 90% crystallinity or quenched down to 45% crystallinity, depending on TFE molding-powder type. Thus the processor, by controlling Steps 1, 2, and 3, controls void content, molecular weight, and crystallinity, respectively.

Inherent polymer properties such as dielectric constant, power factor, chemical resistance, and coefficient of friction are not significantly affected by the processing method. Properties that *are* affected are flex life, stiffness, tensile properties, creep rate, dielectric strength, and permeability.

As one would expect, the void content, molecular weight, and crystal-linity affect these properties in an interrelated and complex way.

Specific gravity is commonly used as a quality index. This value decreases with increasing molecular weight, increases with crystallinity, and decreases, of course, with void content. Since the processor usually can predict one or more of these characteristics from the processing variables, the specific gravity can be used as a check on, for example, void content. In film applications void content should be essentially zero, and correspondingly a minimum density of 2.12 to 2.14 is usually required.

Elimination of voids in the cold compacting stage requires pressures of about 6000 psi or greater with Type 1 TFE because of the large particle size. However, if Type III material is used, the smaller primary particle size allows void-free material to be obtained at only 400 psi. Here it is necessary to distinguish between the primary particles that relate to the original latex particles and the size of the particles of coagulum in the molding powder.

The effect of sintering temperature and cooling rate on crystallinity is shown in Figure 2. Reduction in molecular weight increases the crystallizing tendency.

Increasing crystallinity to 85 to 90% can affect physical properties markedly and bring about a 30-fold reduction in carbon dioxide permeability, a 50% reduction in compressibility, a 70% reduction in recovery, a 50% reduction in tensile strength, and a 100% increase in ultimate elongation. Excessive crystallinity can seriously affect flex life as shown in Figure 3. The effect of crystallinity on flexural modulus of elasticity test at various temperatures is shown in Figure 4 and the effect on yield strength is shown in Figure 5.

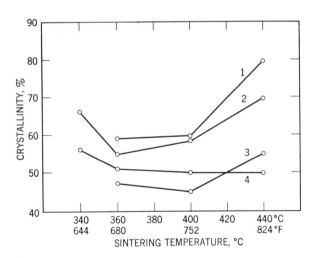

Fig. 2   Percentage of crystallinity as a function of sintering temperature for TFE: 1, sintered 3 hr, cooled 2°C/min; 2, sintered 0.5 hr, cooled 2°C/min; 3, sintered 3 hr, air-quenched; 4, sintered 0.5 hr, air-quenched.

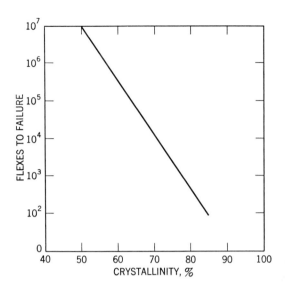

Fig. 3   Effect of crystallinity on flexural fatigue resistance, TFE.

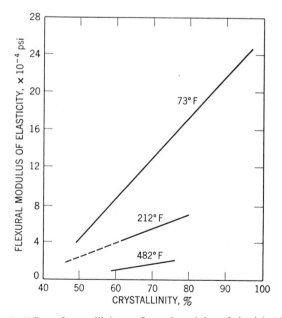

Fig. 4   Effect of crystallinity on flexural modulus of elasticity, TFE.

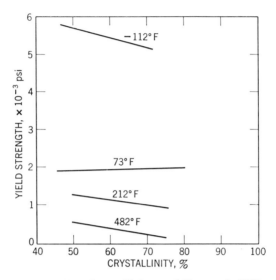

Fig. 5   Effect of crystallinity on yield strength, TFE.

Excessive void content (ca. 6%) generally has a deleterious effect on these physicals.   The reduction in molecular weight likely to be encountered in processing is not nearly so serious, for it has no effect in many cases but can cause a 25% reduction in tensile strength.

2. Calendering

The calendering process has been used [79] in forming TFE film or sheet and takes advantage of the high degree of cohesion between cold-pressed particles of TFE.   The molding powder (e.g., ASTM, Type II) is fed to the calender rolls, which compress the powder into sheet of the desired thickness and width.   The preform has sufficient cohesion to retain its shape until it is passed into a sintering oven or liquid bath. The sintering temperature is in the neighborhood of 790 to 850°F, somewhat higher than that required for compression molding.

3. Casting

The calendering techniques described above can be used to produce sheets or films of TFE but are normally limited to film thicknesses greater than approximately 1 to 2 mils. Much thinner supported and unsupported films can be formed by casting and dip-coating techniques. TFE aqueous dispersions, from which films may be cast, have been prepared by a process disclosed by Renfrew [39] for preparing suspensoids directly from tetrafluoroethylene. The suspensoid is obtained by polymerizing tetrafluoroethylene with a dibasic acid peroxide catalyst (e.g., disuccinic acid peroxide) in an aqueous system. Frequently anionic fluorocarbon emulsifiers are employed [80]. Such suspensoids are quite stable and are lyophobic sols consisting of negatively charged particles approximately 0.1 to 0.3 μ in diameter.

Commercially available TFE dispersions contain 59 to 61 % (by weight) of TFE as solids and are stabilized by nonionic wetting agents such as mentioned below.

Possibly the most important property of aqueous dispersions is the size of the TFE particle. By way of comparison, ASTM Types I and II molding powders have an average particle size of about 500 μ and are obtained by grinding. The extremely small particle size of the dispersion allows the casting of relatively thin films which are rapidly fusible.

Dispersions of TFE are available which are best applied to various surfaces by dipping or flowing onto the surface and subsequent drying to remove water. Drying may be accomplished in forced-air conversion ovens or with infrared heating lamps. Following the drying operation, the film is fused or sintered in air ovens maintained at 675 to 750°F.

During this period the nonfluorocarbon materials, particularly the wetting agents such as Triton X-100 (Rohm & Haas) used in preparing and stabilizing the dispersions, are volatilized and the particles of TFE are fused together. Sintering occurs almost instantaneously once the film reaches the required temperature and baking times as short as 3 min can be employed for thin sections (0.001 in. or less). The final step consists of stripping the film from the metal support, which is accomplished by rapid cooling from the sintering temperatures in cold air or water.

Films of various thicknesses can be prepared by adjusting the initial coating thickness or by applying additional coats over the sintered film and repeating the drying and sintering operations described above. TFE films prepared by the casting process are available in a typical range of thicknesses of 0.00025 to 0.004 in. and in various widths. Table 3 lists typical cast film properties.

TABLE 3

Typical Properties of Tetrafluoroethylene Cast Film

Property	Value	ASTM method [78]
Specific gravity	2.1–2.3	D 792–50
Tensile strength, lb/in.^{2}	2500–3500	D 412 Mod.
Elongation, %	200–300	D 412 Mod.
Tear strength, lb/in.	400–800	D 624
Stiffness 77°F, lb/in.^{2}	40,000–60,000	D 747–50
Water absorption, %	0.00	D 570–42
Water-vapor permeability, g/in.^{2}-day	0.2	
Specific heat, BTU/lb-°F	0.25	100–260°F
Coefficient of linear thermal expansion, °F	5.5×10^{-5}	D 696–44T (73–140°F)
Brittleness temperature, °F	< -100	
Maximum continuous service temperature, °F	500	
Solvent resistance	Not affected	
Chemical resistance	Excellent	
Dielectric constant at 1000 cps	2.0–2.2	D 150–54T
Power factor at 1000 cps	<0.0002	D 150–54T
Volume resistivity, Ω-cm	$>10^{15}$	D 257–52T
Dielectric strength, V/mil	3200	D 149–55T (short time 0.001 in.)
Threshold breakdown voltage, V/mil	150–200	

4. Skiving

TFE film can be manufactured by skiving. In this process a large molded billet of TFE, hot formed by techniques described above, is turned in a device similar to a lathe, and a thin section is cut or peeled off with a knife blade. Typical skived TFE film is available in thicknesses ranging from 0.0007 to 0.01 in. and in various widths.

5. Paste Extrusion

Although extrusion of compacted TFE molding granules through a hot die, followed by sintering and annealing, is commonly used for thick sections, thin films cannot ordinarily be produced in this way. In producing thin films it is generally recommended that an extrusion aid such as naphtha thickened with Vistanex (product of Enjay Company of New

Jersey) or white oil be employed in conjunction with ASTM Type III extrusion powder. The addition of extrusion aids greatly improves the extrusion characteristics and allows processing at fairly low pressures. In one procedure the extrusion mixture, which is a paste, is converted to a preform under light pressure (about 100 psi) and then loaded into a ram-type extruder. The mixture is then extruded through a die. The film, as it leaves the extruder, is flexible and fairly strong. The extrusion aid is then removed (volatilized) by heating at temperatures greater than 200°F. The extrusion aid must volatilize at the proper rate without leaving a carbon residue. Sintering is carried out at temperatures ranging from 720 to 750°F.

Among the foregoing processing methods leading to sheets or films paste extrusion using Type III or similar TFE extrusion powder in conjunction with a volatile hydrocarbon extrusion aid appears to be preferred, since it provides, in a continuous process, strong, uniform films free from pinholes or voids.

Paste extrusion of TFE thin films is a complex and sensitive process requiring careful selection and control of pressure, feed rate, temperature, resin type, and ancillary materials, Typical properties of paste-extruded film are shown in Table 4.

C. APPLICATIONS FOR TFE FILMS

Important applications of TFE film are in the electrical field in which they are utilized in the form of tapes of various kinds; for example, tapes that have been through the extrusion process but not fused are available in Military Standard 104 colors for use in wrapping and color coding lead wires. In this application the tape is wrapped around the cable and then fused to itself to form a continuous sheath of TFE.

Extruded film is available with one or both sides treated to allow bonding with conventional adhesives. This priming technique takes advantage of the aforementioned reaction of fluorocarbon film surfaces with alkali metals and related materials [81].

Bondable films are used for interlayer and interphase insulation in coils, transformers, relays, etc. Extruded bondable TFE films are used also as backings for pressure-sensitive adhesive tapes. Silicone adhesives are used for class "H" temperature conditions, whereas acrylic adhesives are used for less severe applications. These tapes are available in thicknesses ranging from 0.0035 to 0.0060 in. and are widely

TABLE 4

Typical Properties of Polytetrafluoroethylene Paste-Extruded Fused Film

Property	Value	ASTM method [78]
Specific gravity	2.1–2.4	D 792
Tensile strength, lb/in. width		
2-mil film	20	D 1000
5-mil film	45	D 1000
Tear strength, lb/in.	400–800	D 624
Stiffness, 77°F, lb/in.^{2}	40,000–60,000	D 747–50
Elongation, %	125–200	D 1000
Water absorption, %	0.0	D 570
Specific heat, Btu/lb/°F	0.25	
Coefficient of linear thermal expansion	5×10^{-5}	D 696–44
Maximum continuous service temperature, °F	500	
Solvent resistance	Excellent	
Chemical resistance	Excellent	
Dielectric constant, 1000 cycles	2.0–2.2	D 150–54T
Power factor, 1000 cycles	0.0002	D 150–54T
Surface resistance, Ω	5×10^{13}	D 25
Volume resistivity, Ω-cm	$>10^{15}$	D 257–52T
Dielectric strength, V		
2 mil	8500	D 1000
5 mil	1400	D 1000
10 mil	1800	D 1000
Water-vapor permeability, g/100 in.^{2}-day	0.2	

used for electrical harness wrapping and for coil, transformer, and relay insulation in which thermal resistance and/or imperviousness to moisture and chemicals are required.

Another extruded TFE tape has a bondable backing on one side and a thermosetting silicone adhesive on the other.

Outside the electrical field TFE films which have one side rendered bondable are widely employed as nonstick coverings for process rolls. Extruded, unfused TFE tapes take the place of pipe dope when a tight, noncontaminating seal is required, particularly at elevated temperatures. The unique lubricating properties of TFE make this application possible even with threaded joints that tend to gall, such as 316 stainless steel.

Other applications are chute and tray liners, cap liners for bottles containing corrosive chemicals, diaphragms, and moisture barriers.

TFE films have many applications in common with FEP fluorocarbon films discussed later. We have necessarily limited ourselves to a few film applications. The nonfilm applications for TFE resins are many and the list is growing rapidly. These applications invariably take advantage of one or a combination of the special properties of this polymer, that is, high temperature stability, chemical inertness, good electrical properties, and low surface friction.

The application that has produced the most recent rapid growth for TFE is low-surface-energy and thermally stable coatings for metals. TFE coated cookware has become a common household item.

The principal failing of these materials was their susceptibility to scratching by spatulas, which has been overcome in large degree by the use of fused rough granules in admixture with the TFE coating. The spatula "rides" on these and does not mar the underlying coating. Other "tough" TFE coatings employ TFE in combination with high-temperature-stable organic polymers, which are used for nonstick saw blades, etc.

VI. Polychlorotrifluoroethylene (CTFE)

A. FAMILY RELATIONSHIPS

CTFE stands in relation to TFE as polyvinyl chloride (PVC) does to polyethylene. This obvious chemical formula relationship also appears to have some validity in terms of physical properties. To the hand and eye polyethylene and TFE bear a close physical resemblance; for example, both TFE and polyethylene (particularly the low-pressure, linear variety) yield sheets and films that are white and translucent. The films are limp and generally rather weak in spite of a relatively high crystallinity level. Both TFE and polyethylene have low dielectric constants and dissipation factors. Both have low surface energies with a characteristic waxy, greasy feel.

When one of the hydrogens (or fluorines) in the monomer unit of each polymer is replaced by a chlorine atom both polymers become transparent. They both become stiffer and exhibit significantly increased tensile strengths. The degree and form of crystallinity is altered and the solubility is increased. In short, PVC and CTFE physically resemble "typical" rigid thermoplastics more closely than their chlorine-free parents.

B. Morphology

Clearly, chlorine produces marked effects when introduced into a fluorocarbon as well as into a hydrocarbon chain. In the case of CTFE the effect is a reduction of the powerful ordering tendency of the TFE structure, manifested in the crystal melting point and in the melt viscosity observed after melting.

CTFE homopolymer may exhibit varying degrees of crystallinity, depending on thermal history. Polymer that has been rapidly quenched after melt fabrication is considered "amorphous" and is transparent and softer than annealed polymer which tends toward opacity and is commonly referred to as "crystalline."

The detailed structure of CTFE has not been completely elucidated. Studies of x-ray fiber diagrams of crystalline CTFE by several investigators [82–84] show very long fiber repeat distances which indicate that the polymer chains in the crystals are in the helical form with complete rotation occurring every 14 to 16 monomer units. The high percentage of crystallinity and high melting point (ca. 213°C) further imply stereoregularity.

A syntactic left-hand helix model representation is shown in Fig. 6.

Tiers and Bovey [85] have presented evidence based on studies comparing nuclear magnetic resonance spectra of low-molecular-weight CTFE and model compounds which indicate that the polymer is highly irregular with both isotactic and syntactic regions, the latter probably predominating.

C. Modifications of CTFE

Since chlorotrifluoroethylene readily copolymerizes with monomers such as vinylidene fluoride (VF_2), it is common practice to modify the properties by copolymerization. The VF_2 probably acts by providing segments with lowered energy barriers to chain rotation. Other comonomers (as well as VF_2) introduce disorder and partially disrupt crystallinity. These modified CTFE resins are generally softer, lower melting, more transparent, and more easily molded than the parent homopolymer.

The 1968 ASTM revision of Tentative Specifications for CTFE Molding and Extrusion Materials (ASTM D 1430–65T) covers resins that consist of at least 90% chlorotrifluoroethylene [86]. The remainder may be comonomer but not polyblend or filler. The resins are

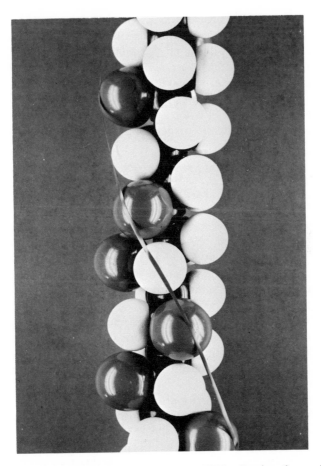

Fig. 6   Model of polychlorotrifluoroethylene (CTFE), showing the syndiotactic left-hand helix.

classified into types and classes according to composition and molecular weight, respectively. These are Type I-Homopolymer, Type II-Modified Copolymer, and Type III-Copolymer. The copolymer types differ in molecular weight: Type II, 150,000; III, 185,000. The commercially available films are based on these types.

By more radical changes in polymerization conditions as well as amount of comonomer introduced into the polymerization process a broad spectrum of physical characteristics is attainable in chlorotri-fluoroethylene-based polymers; for example, copolymerization of chlorotrifluoroethylene with about equimolar quantities of vinylidene fluoride results in a useful higher temperature serviceable rubber sold under the trade name "Kel-F 5500 Elastomer." Here, however, we are concerned only with the previously mentioned types I, II, and III polymers in which $CF_2=CFCl$ is the sole or principal monomer. CTFE is also available in plasticized grades containing fluorinated oils.

D. Film and Sheet Properties

CTFE resins are tough, hard thermoplastics. The mechanical properties of the homopolymer (Type I sheet) appears in Table 5, in which variations in properties as a function of thermal history and test temperature should be noted. The 0.110- and 0.064-in. specimens were obtained by hot pressing at 250°C and followed by quenching or annealing to obtain "amorphous" or "crystalline" material, respectively. Mechanical properties of films in the three grades are compared in Table 6.

The films from CTFE exhibit greater strength and rigidity than TFE or FEP films. Although CTFE films are slippery to the touch, this effect is less pronounced than with TFE or FEP. CTFE films can be monoaxially or biaxially oriented with a corresponding enhancement of tensile properties.

The gas- and moisture-barrier properties are unparalleled in organic polymer films. Typical values for water-vapor transmission are 0.020–0.060 g/100 in.^{2}-24 hr-mil at 25°C [ASTM E 96-66(E)].

CTFE films have outstanding electrical properties. The dielectric constant and dissipation factor remain reasonably low in spite of the permanent dipole due to the presence of the Cl atom (Table 7). CTFE retains useful properties over a wide temperature range (-400 to $+400$°F); at ca. 415°F it undergoes true crystal melting and can be fabricated as a conventional thermoplastic (see Processing Conditions). When subjected to excessive heating, CTFE undergoes degradation via a random scission process, as opposed to unzippering. The rate of this reaction becomes measurable above 500°F and results in degeneration of the physical properties of the polymer through progressive lowering of the average molecular weight.

Since nonvolatile degraded polymer rather than volatile fragments are initially formed by this process, thermogravimetric analysis does not

TABLE 5

Typical Mechanical Properties of Chlorotrifluoroethylene Homopolymer

Property	Temper-ature (°F)	Crystalline	Amorphous	Caliper (in.)	ASTM test method [86]
Tensile strength, psi	77	4630	4650	0.110	D 638–56T
	77	5200	5260	0.064	
	158	3550	2900	0.064	
	258	540	575	0.064	
Elongation, %	77	120	160	0.110	
	77	125	180	0.064	
	158	330	330	0.064	
	258	400	400	0.064	
Yield point, psi	77	5200	4800	0.110	
	77	5300	4700	0.064	
	158	2700	1600	0.064	
	258	560	340	0.064	
Yield strength, psi	77	2450	2000	0.110	
0.2% offset	77	3350	2600	0.064	
	158	1600	1100	0.064	
	258	350	180	0.064	
Modulus of elasticity, psi	77	190×10^3	157×10^3	0.110	
Tensile, psi	77	190×10^3	160×10^3	0.064	
	158	97×10^3	55×10^3	0.064	
	258	20×10^3	6×10^3	0.064	
Flexural strength, psi	77	9600	8600	0.110	D 790–59T
	158	5070	3150	0.110	
	258	1700	700	0.110	
Modulus of elasticity, psi	77	238×10^3	185×10^3	0.110	
Flexure, psi	158	133×10^3	79×10^3	0.110	
	258	37×10^3	15×10^3	0.110	
Shear strength, psi	77	5400	5600	0.064	D 732–46
Deformation under load, %	77	0.20	0.40		D 621–59
24 hr/1000 psi	158	0.40	7.12		
	258	4.00	25.00		
Heat deflection, °F, 66 psi		258			D 648–56
264 psi		167			
Impact strength, ft-lb/in.					
of notch, IZOD		3.01	5.10		D 256–56
Hardness					
Knoop, KNP		10.9	7.5		D 1474–57T
Durometer, Shore D		79	76		D 1706–59T
Rockwell, S Scale		84	79		D 785–51
R Scale		110	108		
Specific heat, cal/g-°C		0.2			
Specific gravity	77	2.124	2.107	0.064	D 792–50
Zero strength time (ZST), sec	482	370	375	0.062	D 1430–58T
Clarity		Translucent	Transparent		

TABLE 6

Typical Mechanical Properties of Polychlorotrifluoroethylene Films

Property	ASTM test method [86]	Test Temperature (°F)	Resin type I			Resin type II		Resin type III			
			Film thickness (mils)	Amorphous	Crystalline	Film thickness (mils)	Crystalline	Film thickness (mils)	Amorphous	Film thickness (mils)	Crystalline
Specific gravity		73.4/73.4		2.10–2.12			2.15		2.07–2.12		2.10
Tensile strength, psi MD[a]	D 882	77	4	8000		0.3 / 5	8000 / 6000	0.5 / 4	13,000 / 7000	1 / 4	8000 / 5000
Tensile strength, psi TD[b]	D 882	77						0.5 / 4	4000 / 5000	1 / 4	4000 / 5000
Elongation at break, %	D 882	73	4	200	180	0.3 / 5	15 / 200	0.5 / 4	25 / 200	1 / 4	25 / 200
Elongation at break, % TD	D 882	73	4	200	180	0.3 / 5	15 / 200	0.5 / 4	100 / 350	1 / 4	35 / 300
Modulus of elasticity, psi, MD	D 882	73	4	110,000	110,000			4	110,000	4	110,000
Bursting strength, mp	D 774		2	42				2	40–42	2	40–42
Tear strength, lb/in. (starting) MD	D 1004	73	1	330–900	330–900	1	800	1	930	1	950
Shrinkage, %	D 1204	300, 24 hr	1	0.00	0.00	1	−0.1	1	+10	1	−0.5
Water-vapor transmission, g/100 in.^{2}-24 hr-atm-mil				0.03	0.01		0.015		0.18		0.05

[a] Machine Direction.

[b] Transverse Direction.

TABLE 7

Electrical Properties of Polychlorotrifluoroethylene Films

Property	ASTM test method [86]	Conditions	Resin type		
			I	II	III
Dielectric strength V/mil	D 149	13–15 mil thickness	1100		1000
Dielectric constant	D 150	100 cps, 77°F	2.59	2.6	2.8
		10,000 cps, 77°F	2.41	2.5	2.6
Dissipation factor	D 150	100 cps, 77°F	0.0215	0.023	0.019
		10,000 cps, 77°F	0.0229	0.022	0.029
		100,000 cps, 77°F	0.0144	0.008	0.022
Volume resistivity, Ω-cm	D 257	12–15 mil thickness	4×10^{16}	10^{17}	10^{18}
Surface resistivity, Ω	D 257		10^{18}	10^{16}	10^{16}
Arc resistivity, sec	D 495		360		130–140

give a true picture of thermal degradation rate. A simple empirical test known as the Zero Strength Time (ZST) was developed to detect degradation during fabrication [86]. The test measures the time required for a loaded specimen to break at a fixed temperature (usually 482°F).

CTFE films have typical fluorocarbon inertness to chemical and solvent attack. These dense impermeable materials are unaffected by inorganic acids and bases and by most organic solvents and compounds. Table 8 presents the results of immersion testing of quenched hompolymer samples. Solvent resistance can be enhanced by annealing to give the "crystalline" form of the polymer. Room temperature swelling values for CTFE films, Types I, II, and III, appear in Table 9. It is seen that CTFE is swelled by certain halogenated solvents at elevated temperatures and pressures and is also attacked by ammoniacal sodium and molten alkali metals. CTFE has a liquid oxygen (LOX) impact test value at 162 ft-lb/in.^{2} (RMD method), making it one of the limited number of polymers suitable for service in LOX.

Gamma radiation simultaneously crosslinks and degrades the polymer. A 24-megarad dose produced the following results in test specimens.

	Crystalline Specimen	Amorphous Specimen
Tensile, retained, %	65	60
Elongation retained, %	3	80
Yield strength retained, %	80	85
Modulus of elasticity retained, %	105	110

Resistance to ultraviolet radiation is good

E. PROCESSING

1. Melt Extrusion

CTFE resins can be processed into films by standard extrusion techniques with conventional equipment. Their behavior in such equipment is comparable to that of rigid polyvinyl chloride but requires somewhat higher temperatures and pressures.

Uniform films with excellent properties are attainable, provided temperatures, residence times, and shear rates are accurately controlled and the extruder parts are designed to avoid hold-up pockets. This

TABLE 8

Chemical and Solvent Resistance of Quenched Polychlorotrifluoroethylene

Material	Temperature (°F)	Immersion (days)	Weight gain (%)
Acetone	77	7	0.1
Acetyl chloride	77	7	0.1
Arochlor 1242	77	7	0.0
Benzene	77	7	0.2
Benzene	275	7	107.0
Carbon disulfide	77	30	0.5
Carbon tetrachloride	77	7	0.4
Carbon tetrachloride	275	7	600.0
Chlorosulfonic acid	77	30	0.0
CTFE monomer	67–77	7	9.1
Chromic acid cleaning solution	77	7	0.0
Cresol	284	7	2.0
unsym-Dimethylhydrazine	77	7	0.1
Dimethylformamide	77	7	0.0
Ethyl acetate	171	1	5.9
Freon 12	77	7	3.0
Heptane	194	7	1.8
Hydrofluoric acid, anhydrous	122	60	0.0
Hydrogen peroxide, 30%	77	30	0.0
Methyl ethyl ketone	194	7	4.6
Nitric acid, fuming, 85–95%	77	7	0.0
Nitric acid, white fuming	194	7	0.3
Nitrogen tetroxide	41	7	9.9
Ozone, 5% in oxygen	302	2	0.0
Potassium hydroxide, 50%	bp	7	0.1
Sulfuric acid, 95%	347	7	0.0
Tetrahydrofuran	147	1	8.2
Toluene	230	7	5.0
Xylene	280	7	27.0

criticality stems from the high melt viscosity of CTFE resins near their melting point (Type I melts ca. 415°F); CTFE is thermally stable well above this point so that compression molded films and sheets can be prepared at 500°F with no danger of degradation. Here the high melt viscosity is no problem.

In extrusion operation, as in injection molding, much higher temperatures—up to 650°F—are usually required to obtain sufficiently low

TABLE 9

Comparative Chemical Resistance of CTFE Types

Chemical exposure	Weight change, %		
	Type I	Type II	Type III[a]
Acetone	0.1	0.5	1.5–5.2
Benzene	0.1	0.6	1.1–2.4
Ethanol	0.0	0.0	0.0
Methyl ethyl ketone	0.2	1.2	2.6–5.9
Trichloroethylene	1.9	7.8	7.1–10.9
Ethyl ether	4.6	5.2	5.6–6.7
Nitric acid, fuming	0.0	0.04	0.7
unsym-Dimethylhydrazine	0.1		4.1[b]
Nitrogen tetroxide	9.9		21.6

[a] Shows a spread of values for various grades.
[b] Severe chemical attack.

melt viscosity.   At such a temperature CTFE resins degrade at a measurable rate, particularly when the processes involve flow in thin sections. Under these conditions there is danger of overheating, accompanied by excessive degradation.   If in such processes the melt is underheated and the viscosity is too high, there is a danger of melt fracture; hence the necessity for proper equipment and control.

CTFE resins evolve toxic gases under processing conditions involving temperatures above 500°F and therefore adequate ventilation should be installed in processing areas, particularly when continuous operations such as film extrusions are being carried out.

The processing conditions of CTFE resins of Types II and III are somewhat less critical than Type I, since, as mentioned above, the structure modification has resulted in lowered melt viscosities.

2. Casting

CTFE resins are available as organic dispersions.   This makes it possible to apply film coatings of CTFE to a broad variety of surfaces. This involves priming, spray-coating, and baking operations.   Since melt extrusion is available to CTFE film processors, dispersion casting is not ordinarily used in making unsupported films.

It is not the purpose of this chapter to provide detailed information on processing conditions, since these vary with resin type, film gauge, and operating equipment. The reader is referred to technical brochures available from the resin manufacturers.

3. Heat Sealing and Bonding

All three types of CTFE film are capable of being heat-bonded to themselves. To meet diverse sealing requirements a considerable variety of techniques has been developed for applying heat to the seal zone. Irrespective of the technique employed, a good bond is obtained only if the interfaces of the surfaces to be sealed are brought up to the completely melted state (390 to 440°F, depending on type). It is necessary to maintain pressure between the sealing faces to cause inter-flow between the films at the joint. Excessive heat and pressure will thin out the film and result in weakness along the sealing line; insufficient heat will result in poor fusion and a weak bond.

Since CTFE resins become tacky like polyethylene when heated to sealing temperatures, they should be cooled below 150°F before removing the pressure after sealing; CTFE has slow heat-transfer characteristics. A good rule of thumb is to double the heating and cooling cycle time used for carrying out the same heat-sealing operation with polyethylene. The crystalline films require slightly higher sealing temperatures than the amorphous films.

The different types of sealing equipment now being used successfully are thermal impulse sealers, dielectric sealers, hot-bar sealers, and ultrasonic sealers.

CTFE resins can be rendered bondable by thermoset adhesives by preliminary treatments which involve surface dehalogenation with alkali metals and related compounds [82]. This feature of CTFE films greatly extends their utility in allowing them to be firmly bonded to an almost endless variety of substrates.

A practical example is the lining of tanks and reactor interiors with CTFE film to provide a non-contaminating, corrosion-resistant interior surface.

4. Applications

CTFE films with their unique combination of high tensile strength, transparency and especially high degree of impermeability to water vapor, oxygen, and carbon dioxide find many applications in the packaging field.   This is especially true when absolute protection of expensive or delicate items from atmospheric attack is required; for example, delicate electronic gear for aerospace applications, precision bearings, and other mechanical parts are dry packaged in CTFE films under an inert nitrogen atmosphere.   Packaged in this way, such items can be stored for long periods even under tropical conditions of humidity and temperature.

In the medical field CTFE films provide reliable transparent and protective packaging of serums, plasmas, pharmaceuticals, and other sensitive materials.   Suture materials are sealed in alcohol in envelopes of CTFE.   The presence of the alcohol around the seal area does not interfere with the sealing operation.   The suture material is thus kept sterile and pliable and ready for immediate use.   Since CTFE is stable at high temperatures, it is useful for packaging items that require steam sterilization.

CTFE resin is used as a transparent encapsulating overlay for electroluminescent phosphors, when complete exclusion of atmospheric moisture is required.

Printed circuits etched in copper, fused to a CTFE film backing, and then sandwiched with a second layer in a melt-sealing operation, are kept free of moisture and are serviceable over the entire 700°F temperature range of the film.

The films have been used as cementable linings for fuel tanks and as splash covers where corrosive materials such as red fuming nitric acid are being handled.   Their compatibility with LOX makes them broadly applicable in the rocket industry.   Many of the CTFE film applications involve laminations with other materials, either by fusion or adhesive bonding of the pretreated film.   A process for conversion of seawater by electrodialysis employs specially processed CTFE films as membranes. In such applications resistance to salt, alkalies, and acids in water is required.

To sum up, the aforementioned applications of CTFE films usually stem from their composite-properties picture rather than from any single property.   CTFE films are unique in combining the following.

1. Useful physicals between −400°F and +400°F.
2. Flexibility and toughness at 77°F.
3. Resistance to deformation at high temperatures.

4. Good abrasion resistance.
5. Radiation resistance.
6. Chemical inertness.
7. Good electrical properties.
8. Zero moisture absorption.
9. Excellent optical properties.

VII. Polyperfluoroethylenepropylene (FEP)

A. GENERAL DESCRIPTION

FEP resin represents an extension of the polytetrafluoroethylene (TFE) family of resins [87] and is a development of the Du Pont Company [88, 89]. This resin has much in common with TFE insofar as its chemical and physical characteristics are concerned and for this reason might have been included with TFE in this review. However, since it followed CTFE in time and represents another approach to rendering the polyperfluoromethylene chain processable by ordinary means, we have chosen to treat FEP separately.

FEP resin is obtainable by the copolymerization of perfluoropropylene with tetrafluoroethylene. Figure 7 is a photograph of a model of FEP. The structure of the resultant copolymer chain may be

It will be noted that the sequence of $-CF_2-$ units is occasionally interrupted by a $-CF(CF_3)-$ unit. This pendent CF_3 group, although it occurs with comparative infrequency along the backbone, has a surprisingly pronounced ability to interfere with the ordering and crystallizing tendency of the parent chain. This has a dramatic effect in lowering the melt viscosity of the polymer at molding and extrusion temperatures and pressures so that the FEP resin, in contrast to TFE resin can, like CTFE, be extruded and injection-molded in conventional equipment. (The specialized techniques necessary for the fabrication of the TFE homopolymers were discussed in Section V-B above.)

This structural modification does not seriously impair the unusual

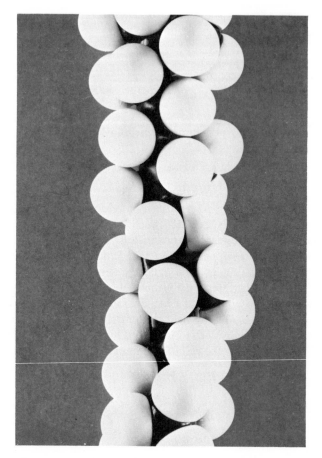

Fig. 7   Molecular model of polyfluoroethylenepropylene (FEP), showing occasional pendent $-CF_3$ groups.

thermochemical stability characteristics inherent in the parent polyper-fluoromethylene chain, although the tertiary F now present could be expected to be a labile site. As one might expect, there is some lowering of the maximum continuous-service temperature at which unfilled fabricated shapes will retain their dimensional integrity, that is, ca. 400°F versus ca. 530°F for TFE.

The effect imparted by many chlorine atoms in CTFE is achieved in

FEP by a few —CF_3 groups. The polymer is thus still a perflouro-carbon and the thermochemical stability loss due to the presence of chlorine is not experienced. The softness, limpness, slipperiness, and low tensile strengths characteristic of TFE are in large degree retained. The polymer becomes more transparent because of the reduction in crystallinity and freedom from the microscopic voids which often result from the processing methods necessary with TFE.

B. Properties of Polymer and Film

Several important physical properties of perfluoroethylenepropylene polymer are summarized in Table 10.

TABLE 10

Thermal and Mechanical Properties of FEP Polymer

Property	Value	Conditions	Test method, ASTM [78]
Melting point	$518 \pm 36°F$		Fisher-Johns apparatus
Melt flow rate	Minimum 4 g/10 min Maximum 6	701.6°F	D 1238
Tensile strength	2500 psi	Std. Lab. Atm. D 618	D 638
Cont. Serv. temperature	400°F		
Elongation	275%	Std. Lab. Atm. D 618	D 638
Tensile modulus	0.5×10^5 psi		D 638
Hardness, Rockwell	R25		D 785
Specific gravity	2.14–2.17 (73.4/73.4°F)		

1. Mechanical Properties

The mechanical properties of FEP films appear in Table 11. The films are waxy to the touch and highly flexible.

2. Electrical Properties

FEP films exhibit good electrical properties, particularly from the standpoint of invariance with temperature and frequency. The dielectric constant does not vary from a value of 2.1 over temperatures

TABLE 11

Mechanical Properties of FEP Film

Property	Typical measured value	Conditions (°F)	Test method ASTM [86]
Tensile strength	3000 psi	77	D 882–61T
Yield point	1700 psi at 3%	77	D 882–61T
Elongation	300%	77	D 882–61T
Tensile modulus	43,000 psi	77	D 882–61T
Tear strength, initial	270 g/mil	77	D 1004–59T
Tear strength, propagating	125 g/mil	77	D 689–44
Flex life	4000 cycles	77	D 643–43
Bursting strength, Mullen, 1 mil	11 psi	77	D 774–46
Coefficient of friction	0.09	77	Inclined plane against steel
Refractive index	1.34		D 452–50
Area factor	12,900 in.^{2}/lb-mil		

ranging from room to 400°F and frequencies from 100 cps to 300 mc. The dissipation factor also remains quite low and fairly constant, averaging about 0.0004 over this temperature-frequency range. The dielectric strength and surface volume resistivities are also high due in part to the low moisture absorption (<.01%).

Electrical properties of FEP films appear in Table 12.

TABLE 12

Electrical Properties of Perfluoroethylenepropylene Film

Property	Measured value	Conditions	Test method ASTM (78)
Dielectric strength	4000 V/mil (1-mil film)		D 149
	1700 V/mil (15-mil film)		
Volume resistivity	>10^{17} Ω-cm	−40 to 240°C	D 257
Surface resistivity	>10^{16} Ω-cm	−40 to 240°C	D 257
Dielectric constant	2.1 ± 0.1	73°F, 100 cps to 100 Mc	D 150–54T
	2.13–2.04	−40 to +437°F, 1000 cps	
Dissipation factor	0.002–0.007	73°F, 100 cps to 100 Mc	D 150–54T
	0.002	−40 to +437°F, 1000 cps	

3. Chemical Properties and Permeability

Since the trifluoromethyl group itself and its bond to the perfluoro-methylene backbone are both stable, FEP shares with TFE resin a high degree of imperviousness to most chemical reagents, being subject to attack only by hot alkali metals, elemental fluorine, and some of the halogen fluorides such as ClF_3. The fact that FEP resin (like the fluoropolymers discussed earlier) is attacked by alkali metals has been utilized to render FEP film surfaces bondable by organic adhesives (see below).

Solvent swelling by chlorofluorocarbons and some of the lower fluorocarbons occurs, particularly at elevated temperatures and pressures, but does not involve chemical attack.

FEP films consistently show good resistance to permeation by gases and liquids, as illustrated in Table 13.

TABLE 13

Permeability of Perfluoroethylenepropylene to Gases and Liquids at 77°F.

	Gases (g/100 m²-hr-mil)	Liquids (g/100 in.²-day-mil)
Water	30	0.02
Carbon dioxide	40	
Oxygen	30	
Nitrogen	12	
Ethanol	4	0.05
Benzene	36	0.15
Acetophenone		0.50
Carbon tetrachloride	20	0.11
Nitric acid, red fuming		7.5
Sulfuric acid, 98%		0.00001
Sodium hydroxide, 50%		<0.01

FEP films have excellent weatherability, resulting in part from lack of absorption bands between 300 mμ and 2μ. Neither "XW Weatherometer" exposure for 5000 hr nor exposure to high-intensity ultraviolet at 140°F and 50% relative humidity for 6000 hr produces any measurable change.

4. Thermal Properties

Like the other fluorocarbons, FEP films excel when service temperatures go up. The processability obtained by introducing the perfluoropropylene into the TFE chain is accompanied by some sacrifice in dimensional stability at elevated temperatures. Compare Tables 14 and 3.

TABLE 14

Thermal Properties of FEP Film

Property	Typical measured value
Melting point	545 to 563°F
Max. service temp., continuous	−425 to +400°F
Max. service temp., intermittent	−425 to +525°F
Tensile strength	ca 16,000 psi at −320°F
	ca. 5,700 psi at −100°F
	ca. 3,000 psi at +77°F
	ca. 650 psi at +390°F
	ca. 500 psi at 450°F
Coefficient of thermal expansion	4.61×10^{-5} in./in.-°F (-100°F)
	5.85×10^{-5} in./in.-°F ($+160$°F)
Specific heat	0.28 Btu/lb/°F
Thermal aging	No change, 10 months at 392°F

C. Processing and Fabrication

FEP resin is processed into film by melt extrusion in conventional equipment at 700 to 750°F. Its melt viscosity at extrusion temperatures is in the same range as other film-processing resins which are ordinarily processed by melt extrusion, considerably lower than CTFE, and far lower than that of TFE resin.

In common with other fluorocarbon resins FEP evolves gaseous products at a measurable rate when heated above a certain point (700°F). These are essentially the same as those liberated from TFE resin and can be hazardous to exposed personnel; therefore extruders and other melt-processing equipment should be provided with adequate ventilation. It is recommended that corrosion-resistant alloys be employed in extruder parts.

FEP sheets can be thermoformed by vacuum and other techniques at 480 to 550°F; FEP film is readily heat-sealed to itself by the techniques discussed under CTFE.

D. FILM APPLICATIONS

FEP films, having many properties in common with TFE films may be expected to find applications wherever similar properties are required. To a greater extent, however, FEP films are used when many of the properties of TFE are required, and when, in addition, transparency and ease of fabrication are necessary.

The antistick properties of FEP films make them valuable for refrigerator trays, release liners for molds, process roll coatings, and bearing surfaces.

Their chemical resistance, particularly over a broad temperature range, makes them useful for gaskets, diaphragms, jet and rocket fuel bladders, and liners and seals for corrosive chemical-process equipment.

Since FEP films are truly melt processed, they are pinhole free. This, in combination with the polymer's inherently good electrical properties over a wide temperature and frequency range, renders the films broadly applicable in the electric field. The films can be tightly wrapped around wire bundles and then fused to form solvent- and moisture-resistant sheaths.

FEP film is used in conjunction with a polyimide backing as a wire wrap for class H high-temperature service. Here the FEP is fused to bond the polyimide backing to the wire, thus taking advantage of the polyimide's high cut-through temperature. FEP films are utilized in capacitors, particularly those requiring high reliability at critical service temperatures. FEP film is finding application in printed circuitry because of its melt-bonding capacity. The metal can be bonded to the film, etched, and then sandwiched with another layer of film to provide a completely sealed circuit.

In applications such as those cited above, surface-modified film is often employed to aid in adhesive bonding.

FEP film is now available in the form of heat-shrinkable tubing which can be used to encase cylindrical objects ranging from process rolls to electrical connectors. Shrink-down is accomplished by heating the prestressed tubing to 300°F.

VIII. Polyvinylidene Fluoride (PVF$_2$)

A. General Description

Polyvinylidene fluoride (PVF$_2$) is a high-molecular-weight homopolymer of 1,1-difluoroethylene (CH$_2$=CF$_2$) developed commercially by the Pennsalt Chemicals Corporation. The abbreviation "PVF$_2$" is preferred by the manufacturer [90] and is used here.

PVF$_2$ is a predominantly linear high polymer of the indicated structure. The value of n is 5000 to 10,000.

$$\begin{array}{c} \text{F}_2 \\ \text{C} \end{array} \left[\begin{array}{c} \text{F}_2 \\ \text{C} \end{array} \right.\begin{array}{c} \\ \text{C} \\ \text{H}_2 \end{array} \left. \begin{array}{c} \text{F}_2 \\ \text{C} \end{array} \right]_n \begin{array}{c} \\ \text{C} \\ \text{H}_2 \end{array}$$

Although only half of the carbons in the chain are fluorinated, the polymer actually contains 59% fluorine by weight. Its properties and performance characteristics are also such that it is rightly included in a treatise on fluorocarbon polymers.

As should be expected, there is some loss in thermal stability and chemical resistance caused by replacing alternate CF$_2$ groups with CH$_2$ groups. The alternating arrangement, however, confers several unique and desirable properties which make PVF$_2$ more attractive than either the perfluorinated or conventional organic polymers for certain applications.

B. Morphology

Vinylidene fluoride, like similar vinyl monomers, polymerizes to give a predominantly head-to-tail structure, as represented above. Recent F^{19} nuclear magnetic resonance studies of PVF$_2$, prepared by use of diazo initiators, indicates the presence of head-to-tail sequences, that is, $-CF_2-CH_2-CH_2-CF_2-$ [91].

It has been reported [92] that the PVF$_2$ molecule, like TFE, has the form of a helix with a zigzag disposition of the links, shown in Fig. 8. The model was arbitrarily made to twist 360° for every 11 monomer units.

PVF$_2$ is a crystalline polymer, T_m, approximately 340°F. The degree of crystallinity can be affected by thermal history, that is, quenching or annealing from the melt. PVF$_2$ is highly susceptible to ordering and

Fig. 8   Polyvinylidene fluoride (PVF$_2$) molecular model showing a 360° twist every 11 monomer units.

concomitant enhancement of crystallinity by orientation, either by drawing while still in the semi-molten state or by cold orientation.   In the latter case it is preferable to quick-quench from the melt to inhibit crystallization in the unoriented material.   Orientation can greatly enhance the tensile strength of PVF$_2$.

The glass transition temperature of the polyvinylidene fluoride chain is comparatively low ($-38°$F) [57].   This means that, in contrast to the perhalogenated polymers, the PVF_2 chains are not prohibited by internal energy barriers from rotation about $C-C$ bonds at ordinary temperatures. This rotation, however, is severely restricted by the crystalline character of the polymer. Nonetheless, this inherent freedom of the dipolar $-CH_2-CF_2-$ segments to rotate and align themselves in response to energy stimuli is an important factor in determining the orientability, electrical properties, crystallizing tendency, and ultimate mechanical characteristics of PVF_2 films.

C. Properties of PVF_2 Polymer and Film

1. Chemical Properties

The resistance of PVF_2 to attack by most corrosive chemicals and solvents is good (Table 15). It is degraded by fuming sulfuric acid. Oxidizing agents, alkalis, and halogens (except elemental fluorine) do not generally affect the polymer; PVF_2 is swelled and partially dissolved by hydrogen bond acceptor solvents such as acetone, ethyl acetate, and methyl ethyl ketone. In highly polar organic solvents such as dimethylacetamide and dimethyl sulfoxide the polymer dissolves to give solutions with some colloidal character.

Although PVF_2 is only slightly affected by most organic and inorganic bases at ordinary temperatures, it is severely attacked at elevated temperatures by strong organic bases such as n-butylamine. This reaction is of practical interest because rubbers based on vinylidene fluoride are curable by aliphatic diamines. The mechanism by which PVF_2 is degraded by n-butylamine has not been established with complete certainty, but it is quite probable that the initial attack occurs by the splitting out of hydrogen fluoride [93, 94] by one of several routes:

1. $(-CH_2-CF_2-)_n + RNH_2$
    $$\longrightarrow \quad (-CH=CF-)_n + RNH_2 \cdot HF$$

2. $(-CH_2-CF_2-)_n + 2RNH_2$
    $$\longrightarrow \quad (-CH_2-CF-)_n + RNH_2 \cdot HF$$
    $$\qquad\qquad\qquad\quad | $$
    $$\qquad\qquad\quad NHR$$

Following the initial attack of the amine, a series of similar reactions leads to partial or complete degradation of the chain, the degree depending on reaction conditions; TFE is unaffected by n-butylamine [93] under similar conditions.

TABLE 15

Chemical Resistance of Polyvinylidene Fluoride Film

Compound	Tensile strength after exposure[a] (psi)	Compound	Tensile strength after exposure (psi)
Isooctane	6000	Sodium hydroxide (50%)	7200
Kerosene	7100	Triethylamine	6400
Toluene	5900	n-Butylamine	Degraded
Carbon tetrachloride	6700	Aniline	5400
Ethylene dibromide	6500	Hydrazine hydrate	7200
$CF_2ClCFCl_2$	6900	Hydrazine	5800
Ethanol	6000	unsym-Dimethylhydrazine	4700
Ethylene glycol	7100	Sulfuric acid, 95%	7100
Ethyl acetate	5000	Sulfuric acid, fuming	Degraded
Acetone	Partially dissolved	Hydrofluoric acid, 28%	6700
Acetophenone	5400	Chlorine (dry, 77°F)	7000
Acetaldehyde	5200	Phosphorus oxychloride	4800
Nitrobenzene	5000	Dinitrogen tetroxide	6500
Dimethyl sulfoxide	Partially dissolved	Hydrogen peroxide (90%, 77°F)	5500
Dimethylacetamide	Partially dissolved	Bromine (dry, 77°F)	5500

[a] Value before exposure: 7200 psi.

2. Thermal Stability

In spite of the presence of hydrogen in the molecule, PVF_2 has good thermal stability. The thermal degradation of PVF_2 has been studied in detail by Madorsky and his associates [75], who found that the substitution of one or more hydrogen atoms for fluorine on the TFE polymer chain changed the decomposition path. Pyrolysis of TFE produced nearly quantitative yields of the monomer, whereas PVF_2, under similar conditions, gave hydrogen fluoride and chain fragments of various sizes. The decreasing order of thermal stability for polyethylene and fluorine-containing polymers is $(-CF_2-CF_2-)_n > (-CF_2-CH_2-)_n > (-CF_2-CFH-)_n > (-CH_2-CH_2-)_n > (-CFH-CH_2-)_n$. Note that PVF_2 is more thermally stable than the more highly fluorinated polytrifluoroethylene.

Other thermal properties include the following:

Specific heat: 0.33 Btu/lb-°F.
Heat distortion temperature (ASTM D 648):
  66 psi, 300°F,
  264 psi, 195°F.
Coefficient of linear expansion (ASTM D 696):
  8.5×10^{-5} in./in.-°F.
Thermal conductivity, 77 to 325°F:
  0.11–0.14 Btu/hr-ft-°F.
Low temperature embrittlement: < -80°F.
Melt flow (ASTM D 1238-57T, condition J):
  1–2 g/10 min (depending on grade).
Flammability: self-extinguishing.

3. Electrical Properties

Compared with polyethylene and polytetrafluoroethylene, PVF_2 has an unusually high dielectric constant and a high loss factor over a wide temperature and frequency range. These results reflect the presence of permanent dipoles; that is, the

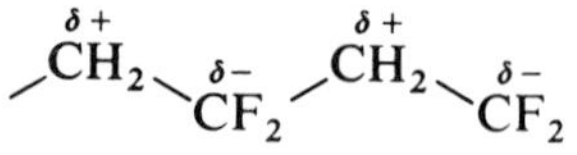

$$\overset{\delta+}{CH_2}\!\!\diagdown_{\overset{\delta-}{CF_2}}\diagup\!\!\overset{\delta+}{CH_2}\!\!\diagdown_{\overset{\delta-}{CF_2}}$$

units in zig-zag array along the chain and may limit the use of PVF_2 as an insulation material for high-frequency applications. The following is a comparison of these properties in PVF_2, TFE, and polyethylene:

Polymer	Dielectric Constant	Loss Factor
PVF_2 10^3 cps	7.72	0.019
10^{10} cps	2.29	0.11
TFE 10^3 cps	2.1	<0.0003
10^{10} cps	2.0	<0.0003
High-density polyethylene (ASTM Type III)		
10^3 cps	2.25–2.35	0.0002
10^6 cps	2.25–2.35	

The volume resistivity of PVF_2 is similar to that of other fluorocarbon plastics (ca. 10^{14}–10^{16} Ω-cm).

Further data are presented under film properties.

4. Mechanical Properties

If one were required to single out the one most outstanding characteristic of PVF_2 it would be "toughness." This is particularly important, of course, in many film applications. The measured properties that define the "toughness" and other characteristics of this material are set forth in Table 16.

TABLE 16

Mechanical Properties of PVF_2 Resin

Property	ASTM method [86]	Test temperature (°F)	Value
Tensile strength, psi	D 638	77	7,200
		212	5,000
Elongation, %	D 638	77	300
		212	400
Compressive strength, psi	D 695	77	10,000
Modulus of elasticity, psi			
tensile	D 638	77	1.2×10^5
compressive	D 695	77	1.2×10^5
Izod impact, ft-lb/in.			
notched	D 256	77	3.8
unnotched	D 256	77	30
Durometer hardness, Shore D-scale			80
Abrasion resistance, Tabor, mg/1000 cycles (CS–17, 1/2 kg load)			17.6

5. Properties of Polyvinylidene Fluoride Film

Typical physical properties of films cast from dispersion appear in Table 17.

TABLE 17

Properties of Polyvinylidene Fluoride Film

Transparency	Clear (1–3 mil) to slightly hazy (3–10 mil)
Specific gravity	1.76
Specific volume, in.^{3}/lb	15.7
Refractive index n_D^{25}	1.42
Gloss	High
Tensile strength, psi	7200
Elongation, %	150–500
Bursting strength, Mullen, psi/mil	15–20
Tear strength, Elmendorf, g/mil	40–60
Flex life, 3-mil film, cycles	75,000
Flammability	Self-extinguishing
Thermal stability	
weight loss	None, 1 year, 300°F
color change	None, 1 year, 300°F
Ultraviolet resistance	Excellent
Water absorption, %	<0.04
Water-vapor permeability, g/mil-24 hr	0.6
Gas permeability, ml/mm-sec-cm^2-cm Hg	
oxygen	24×10^{-12}
nitrogen	5.5×10^{-12}
carbon dioxide	9×10^{-12}
Dielectric strength, 8-mil film, V/mil	1280

D. Weathering and Radiation Resistance

Another outstanding feature of PVF_2 is its radiation resistance; for example, after a one-year exposure to a G.E. S-1 lamp (2000–4000 Å, ASTM D 795) a 2-mil film under 0.5-kg load only dropped from 5×10^5 cycles to failure to 4.5×10^5 cycles to failure (MIT Fold Endurance Test). PVF_2 transmits most outdoor radiation and will withstand 100 megarad Co^{60} gamma dosages without serious deterioration of properties. PVF_2 can be crosslinked by high-energy radiation. This is the basis for a "heat shrinkable" PVF_2 tubing.

E. Processing

PVF_2 is available in two grades (Kynar 18 and 21) which differ principally in molecular weight. The higher polymer ($\overline{M}_w = $ ca. 600,000) gives the optimum properties, that is, toughness and flex life, but the lower melt flow, 1 g/10 min, can be a limiting factor in melt processing. In such cases there is a lower polymer ($\overline{M}_w$ = ca. 300,000) with a lower melt viscosity (melt flow = 2 g/10 min).

PVF$_2$ is also available in powder and pellet form and as dispersions and solutions. The latter two forms are of principal interest here, since PVF$_2$ films are manufactured commercially by casting. PVF$_2$ solutions employ dimethylacetamide as a solvent (bp 330°F) and contain 20% solids. Their use is limited to films of 2 or 3 mils. The films are cast on aluminum foil or on paper carriers and fused at 400 to 460°F, then rapidly quenched by water spray or immersion to yield a continuous, high-strength film.

When thicker films (up to 10 mils) are required, dispersions are employed. The dispersions contain 45% solids by weight in dimethyl phthalate (44%) and diisobutyl ketone (11%). After casting, fusion is carried out at 460 to 480°F. In both cases rapid quenching produces films with the highest tensile, elongation, tear resistance, flexibility, and transparency.

F. Applications

PVF$_2$ resin is finding application in the electrical-insulation and chemical-process markets. Of particular interest are pigmented exterior coatings for metal siding. Film markets are less well defined.

Since PVF$_2$ films can be heat-sealed, they are adaptable to many packaging applications; for example they find use as sterilizable packages for medical instruments and pharmaceuticals, and chemicals, lubricants, and propellants can be hermetically sealed in PVF$_2$. Drum and tank liners of PVF$_2$ are durable and resistant to chemicals.

Tensilized, irradiated films should be useful in wrap-on and shrink-fit applications. Composite films in which the PVF$_2$ solutions are combined with glass cloth or similar substrates make possible many combinations of desirable properties. Films laminated to metals, wood, and other base materials provide protection from chemicals and weather.

The excellent weatherability of PVF$_2$ films makes them attractive for long-term protection of equipment and materials exposed to outdoor conditions.

IX. Conclusions

For quick reference we have included Table 18 in which the typical properties of the fluorocarbon polymer films are compared. In the

TABLE 18

Comparison of Typical Properties of Fluorocarbon Polymer Films

Property	Polyvinylidene fluoride (PVF$_2$)	Polychlorotrifluoroethylene (CTFE)	Polytetrafluoroethylene (TFE)	Polyperfluoroethylene-propylene (FEP)
Specific gravity	1.76	2.12 (77°F)	2.15	2.15
Refractive index, n_D^{25}	1.42	1.435 (77°F)	1.35	1.341–1.347
Tensile strength, psi	7000 (77°F)	4630 (77°F)	2500–3500	2500–3500 (73°F)
Elongation, %	300 (77°F)	120 (77°F)	200–300	300
Coefficient of linear thermal expansion, in./in.-°F	8.5×10^{-5}	4.8×10^{-5} (below T_g)	5.5×10^{-5} (73° to 140°F)	$4.6–5.9 \times 10^{-5}$ (−100° to +160°F)
Coefficient of friction, to steel	0.14/0.17		0.04	0.09
Maximum continuous service temperature, °F	300	390	500	400
Minimum service temperature, °F	−80	−320 to −400	−275 to −450	−120 to −425
Water absorption, %, 24 hr	0.04	0.00	0.00	0.01

Dielectric constant, 60 cycles	8.40	2.63 (100 cps)	2.0–2.2	2.0
10^6 cycles	6.43	2.40 (10^5 cps)	2.0–2.2	2.1
Dissipation factor, 60 cps	0.049	0.062 (100 cps)	0.0002	0.0002
10^6 cps	0.159	0.014 (10^5 cps)	0.0002	0.0007
Dielectric strength, V/mil	1280 (8 mils)	3700 (1 mil)	3200 (1 mil)	6500 (1 mil)
Volume resistivity, Ω-cm	2 × 10^{14}	2.5 × 10^{16}	10^{15}	10^{17}
Chemical resistance[a]				
Acids	C: Fum. H_2SO_4, $ClSO_3H$	E	E	E
Bases	D: n-$BuNH_2$ and other primary amines, otherwise E	E	E	E
Solvents	S: by certain strongly polar solvents, otherwise E	S: certain halogenated and aromatic solvents otherwise E	E	E
Other chemicals	All are attacked by ClF_3, F_2 at elevated temperatures and by molten alkali metals.			
Method of Processing Film	Casting	Extrusion	Extrusion, casting, skiving	Extrusion

[a] Symbols: C, chemical attack; D, degraded; E, excellent; S, solvent attack.

compilation of such a table there is considerable danger of falling into the error of comparing apples and oranges. Recognizing this, the authors have concentrated on data on fundamental resin properties, taken primarily from manufacturers' technical bulletins. When differences in type, grade, class, or molecular weight existed, the best properties representative of the basic polymer were used.

Certainly it is impossible to say that there is a *best* fluorocarbon resin or film. Each of the four listed has property combinations that can make it the preferred material for a particular end-use.

Since the fluorocarbon resins became available in the 1940's, annual consumption has increased steadily. Because of the lack of published production data by manufacturers, it is difficult to make an accurate statement of the total production and consumption of fluorocarbon resins, films, and elastomers. However, it has been estimated [95, 96] that in 1963 about 15 million lb were consumed and in 1968 between 16 and 17 million lb [97]. Most market forecasts place fluorocarbon resin consumption at 30 to 35 million lb in 1970 [95, 96]. The largest part of this has been TFE polymer, which probably accounts for some 80% of the total market.

This growth has occurred in spite of the comparatively high price of fluorocarbon resins. Prices have dropped steadily, however, since the introduction of the fluorocarbon polymers, and should continue to be lowered as methods of monomer synthesis and resin manufacture improve. In some cases another factor in the price picture is entry into the market by new domestic and foreign manufacturers.

Although initial costs may be high, consumers are finding that use of fluorocarbon resins and films can be economical over the long pull because of lowering maintenance and replacement costs.

It can be anticipated that new resins and films will be introduced. Many, particularly in the near future, will be modifications of existing polymers achieved by interpolymerization with other comparatively simple fluoroolefins or by novel polymerization methods.

The emergence of new fluorocarbon polymers along with the discovery of new uses for present fluorocarbon resins and films combine to provide a picture of continuing growth for these unique materials.

References

1. L. A. Errede, *J. Org. Chem.*, **27**, 3425 (1962).
2. O. Ruff and O. Bretschneider, *Z. anorg. allgem. Chem.*, **210**, 173 (1933).

3. J. H. Simons, R. L. Bond, and R. E. McArthur, *J. Am. Chem. Soc.*, **62**, 3477 (1940).
4. L. White, Jr., and O. K. Rice, *J. Am. Chem. Soc.*, **69**, 267 (1947).
5. J. H. Simons, U.S. Patent 2,519,983 (Aug. 22, 1950).
6. T. Midgley, Jr., and A. L. Henne, *Ind. Eng. Chem.*, **22**, 542 (1930).
7. O. Ruff and O. Bretschneider, *Z. anorg. allgem. Chem.*, **210**, 177 (1933).
8. J. D. Park, A. F. Benning, F. B. Downing, J. F. Laucius, and R. C. McHarness, *Ind. Eng. Chem.*, **39**, 354 (1947).
9. Brit. Patent 581,045 (1946).
10. A. F. Benning, F. B. Downing, and R. C. McHarness, U.S. Patent 2,384,821 (Sept. 18, 1945).
11. H. S. Booth, P. E. Burchfield, E. H. Bixbey, and J. B. McKelvey, *J. Am. Chem. Soc.*, **55**, 2231 (1933).
12. H. S. Booth, W. L. Mong, and P. E. Birchfield, *Ind. Eng. Chem.*, **24**, 328 (1932).
13. E. G. Locke, W. R. Brode, and A. L. Henne, *J. Am. Chem. Soc.*, **56**, 1726 (1934); A. L. Henne, U.S. Patent 2,007,198 (July 9, 1935).
14. H. W. Daudt and M. A. Youker, U.S. Patent 2,062,743 (Dec. 1, 1936).
15. O. A. Blum, U.S. Patent 2,590,433 (Mar. 25, 1952).
16. J. T. Rucker and D. B. Stormon, U.S. Patent 2,760,997 (Aug. 27, 1956).
17. A. F. Benning, F. B. Downing, and J. D. Park, U.S. Patent 2,394,581 (Feb. 12, 1946).
18. L. J. Hals, T. S. Reid, and G. H. Smith, Jr., *J. Am. Chem. Soc.*, **73**, 4054 (1951); U.S. Patent 2,668,864 (Feb. 9, 1954).
19. D. A. Nelson, U.S. Patent 2,758,138 (Aug. 7, 1956).
20. J. S. Waddell, U.S. Patent 2,759,983 (Aug. 21, 1956).
21. E. H. Ten Eyck and G. P. Larson, U.S. Patent 2,970,176 (Jan. 31, 1961).
22. L. A. Errede and W. R. Peterson, U.S. Patent 2,979,539 (Apr. 11, 1961).
23. H. Agahigian and C. Woolf, U.S. Patent 3,081,358 (Mar. 12, 1963).
24. F. Swarts, *Bull. Acad. Roy. Belg.*, **39**, 383 (1901).
25. A. L. Henne, *Organic Reactions*, **Vol. II**, John Wiley & Sons, New York, 1944, p. 49.
26. F. B. Downing, A. F. Benning, and R. C. McHarness, U.S. Patent 2,551,573 (May 8, 1951).
27. C. B. Miller, U.S. Patent 2,628,989 (Feb. 17, 1953).
28. C. F. Feasley and W. A. Stover, U.S. Patent 2,627,529 (Feb. 3, 1953).
29. J. H. Simons, Ed., *Fluorine Chemistry*, **Vol. I–V**, Academic Press, New York, 1964.
30. M. Stacey, J. C. Tatlow, H. G. Sharpe, Ed., *Advances in Fluorine Chemistry*, **Vol. 1, 2**, Butterworths, London, 1960.
31. A. M. Lovelace, D. A. Rausch, and W. Postelnek, *Aliphatic Fluorine Compounds*, Reinhold, New York, 1958.
32. G. E. Ham, *Copolymerization*, Interscience Publishers, New York, 1964, Chap. II, Sec. III.
33. R. J. Plunkett, U.S. Patent 2,230,654 (Feb. 4, 1961).
34. M. M. Brubaker, U.S. Patent 2,393,967 (Feb. 5, 1946).
35. R. M. Joyce, Jr., U.S. Patent 2,394,243 (Feb. 5, 1946).

36. M. I. Bro, and R. C. Schreyer, U.S. Patent 3,032,543 (May 1, 1962).
37. R. H. Halliwell, U.S. Patent 3,110,704 (Nov. 12, 1963).
38. W. E. Hanford and R. M. Joyce, Jr., *J. Am. Chem. Soc.*, **68**, 2082 (1946).
39. M. M. Renfrew, U.S. Patent 2,534,058 (Dec. 12, 1950).
40. D. Sianesi and G. Caporiccio, *Makromol. Chem.*, **60**, 213 (1963).
41. W. T. Miller, U.S. Patent 2,564,024 (Aug. 14, 1951).
42. W. T. Miller, U.S. Patent 2,579,437 (Dec. 18, 1951).
43. W. T. Miller, A. L. Dittman, and S. K. Reed, U.S. Patent 2,586,550 (Feb. 19, 1952).
44. G. H. Crawford, Jr., U.S. Patent 3,089,866 (May 14, 1963).
45. H. J. Passino, A. L. Dittman, and J. M. Wrightson, U.S. Patent 2,783,219 (Feb. 26, 1957).
46. H. J. Passino, A. L. Dittman, and J. M. Wrightson, U.S. Patent 2,820,026 (Jan. 14, 1958).
47. A. N. Bolstad and R. L. Herbst, Jr., U.S. Patent 2,874,152 (Feb. 17, 1959).
48. R. M. Adams and F. A. Bovey, *J. Polymer Sci.*, **9**, 481 (1952).
49. D. S. Ballantine and B. Manowitz, Brookhaven Natl. Laboratory, 294 (1954).
50. R. N. Haszeldine, *J. Chem. Soc.*, 3559 (**1953**).
51. H. S. Eleuterio, U.S. Patent 2,958,685 (Nov. 1, 1960).
52. H. S. Eleuterio and E. P. Moore, Paper presented at the Second International Symposium on Fluorine Chemistry, Estes Park, Colorado, 1962.
53. S. Dixon, D. R. Rexford, and J. S. Rugg, *Ind. Eng. Chem.*, **49**, 1687 (1957).
54. J. C. Montermoso, C. B. Griffis, A. Wilson, and J. F. Osterling, *Proc. Inst. Rubber Ind.*, **5**, 97 (1958).
55. D. R. Rexford, U.S. Patent 3,051,677 (Aug. 28, 1962).
56. Brit. Patent 823,974 (1959).
57. L. Mandelkern, G. M. Martin, and F. A. Quinn, Jr., *J. Res. Nat. Bur. Std.*, **58**, 137 (1957).
58. A. L. Dittman, H. J. Passino, and J. M. Wrightson, U.S. Patent 2,689,241 (Sept. 14, 1954).
59. T. A. Ford, U.S. Patent 2,468,054 (Apr. 26, 1949).
60. E. T. McBee, H. Hill, and G. B. Backman, *Ind. Eng. Chem.*, **41**, 70 (1949).
61. T. A. Ford and W. E. Hanford, U.S. Patent 2,435,537 (Feb. 3, 1948).
62. M. Hauptschein, U.S. Patent 3,012,021 (Dec. 5, 1961).
63. Brit. Patent 1,020,678 (1965).
64. W. J. Brehm and A. S. Milian, U.S. Patent 3,053,823 (Sept. 11, 1962).
65. W. J. Middleton, H. W. Jacobson, R. E. Putnam, H. C. Walter, D. G. Rye, and W. H. Sharkey, *J. Polymer Sci.*, A-3, 4115 (1965).
66. C. G. Fritz and J. L. Warnell, U.S. Patent 3,317,484 (May 2, 1967).
67. G. H. Crawford, U.S. Patent 3,399,180 (Aug. 27, 1968).
68. L. H. Piette and G. H. Crawford, Abstracts, Am. Chem. Soc. 142nd Meeting, Div. Inorg. Chem., Atlantic City, September 1962, 10 N.
69. G. H. Crawford and D. E. Rice, U.S. Patent 3,321,454 (May 23, 1967).
70. E. P. Moore, A. S. Milian, Jr., and H. S. Eleuterio, U.S. Patent 3,250,808 (May 10, 1966).
71. R. C. Doban, A. C. Knight, J. H. Peterson, and C. A. Sperati, Abstracts, Am. Chem. Soc. 130th Meeting, Atlantic City, September, 1956, 9 S.

72. C. W. Bunn and E. Howells, *Nature*, **174,** 549 (1954).
73. E. S. Clark, *Z. Krist.*, **117,** 119 (1962).
74. E. S. Clark and H. W. Starkweather, Jr., *J. Appl. Polymer Sci.*, **6,** S41 (1962).
75. S. Madorsky, V. Hart, S. Straus, and V. Sedlak, *J. Res. Nat. Bur. Std.*, **51,** 327 (1953).
76. R. C. Doban, C. A. Sperati, and B. W. Sandt, *Soc. Plastics Eng. J.*, **11,** No. 9, 17, (1955).
77. P. Ehrlich, *J. Res. Nat. Bur. Std.*, **51,** 185 (1953).
78. *1968 ASTM Standards*, Part 26, American Society for Testing and Materials, Philadelphia.
79. "Teflon" Product Brochure, E. I. du Pont de Nemours & Co., Inc., 1954, p. 36.
80. A. E. Kroll, U.S. Patent 2,750,350 (June 12, 1956).
81. R. J. Purvis and W. R. Beck, U.S. Patent 2,789,063 (Apr. 16, 1957).
82. H. S. Kaufman, *J. Am. Chem. Soc.*, **75,** 1477 (1953).
83. A. V. Ermolina, G. S. Markova, V. A. Kaugin, *Kristallografiya*, **2,** 623 (1957).
84. C. Y. Liany and S. Kremin, *J. Chem. Phys.*, **25,** 563 (1956).
85. G. V. D. Tiers and F. A. Bovey, *J. Polymer Sci.*, A–1, 833 (1963).
86. *1968 ASTM Standards*, Part 26, American Society for Testing and Materials, Philadelphia.
87. W. T. Miller, U.S. Patent 2,598,283 (May 27, 1952).
88. M. I. Bro and B. W. Sandt, U.S. Patent 2,946,763 (July 26, 1960).
89. R. A. Darby and E. K. Ellingboe, U.S. Patent 3,069,404 (Dec. 18, 1962).
90. Private Communication, L. E. Robb, Pennsalt Chemicals Corporation.
91. R. E. Naylor, Jr., and S. W. Lasoski, Jr., *J. Polymer Sci.*, **44,** 1 (1960).
92. S. S. Leshchenko, V. L. Karpov, and V. A. Kargin, *Vysokomolekulyarnye Soedineniya*, **1,** 1538 (1959); *Chem. Abstr.* **54,** 15998 (1960).
93. M. I. Bro, *J. Appl. Polymer Sci.*, **1,** 310 (1959).
94. K. L. Paciorek, L. C. Mitchell, and C. T. Lenk, *J. Polymer Sci.*, **45,** 405 (1960).
95. *Chem. Eng. News*, **42,** No. 29, 32 (Mar. 9, 1964).
96. *Chemical Economics Handbook*, Stanford Research Institute, Menlo Park, Calif., Sec. 580.0720 (June 1967).
97. Plastics review, *Modern Plastics*, **46,** No. 1, 49 (January 1969).

CHAPTER 14

POLYESTER FILMS

CARL J. HEFFELFINGER and KENNETH L. KNOX

Film Department
E. I. du Pont de Nemours and Co.
Circleville, Ohio

I. Introduction

Polyesters belong to the broad class of organic high-molecular-weight compounds called condensation polymers [1]. Condensation reactions eliminate a by-product molecule from the reaction mixture; so a typical polyester molecule lacks some atoms that were present initially in the

587

monomers or the hypothetical monomers from which it is formed. Several subcategories of condensation polymers have been described [2]. Extensive industrial and academic research has developed many types of polyesters and copolyesters which find use as films, fibers, and adhesives. As a result new techniques of molecular and structural analysis have led to a substantial understanding of the chemistry, physical behavior, and usefulness of polyesters. Although many polyesters are known, few have been offered for sale as films. Criteria have been established to permit comparisons among the many possible polyester film candidates for potential utility. The criteria relating chemical structure and physical properties involve considerations of polyester melting points, tendency to crystallize, cohesion energies, molecular flexibility and shape factors, structural regularity and types of interchain bonding [2–5]. (See Volume I, Chapters 1–3.)

At the present time films made from polyethylene terephthalate (PET), an aromatic polyester, dominate the field of industrial polyester films. This polymer has a regular linear structure with the connecting ester carbon attached directly to the aromatic ring in the backbone of the molecular chain (Table 3). This contrasts with those polyesters, such as the cellulose esters, in which the ester groups are lateral to the main chain or to the polycarbonates in which the ester carbonyl is separated from the aromatic ring.

Films from polyethylene terephthalate are strong, tough, flexible, and have low water absorption and low gas permeabilities coupled with excellent resistance to chemical attack. Industrial uses also capitalize on their excellent dielectric properties. PET films are offered for sale in thicknesses of 0.15 to 14.0 mils. The versatility in properties and in thickness broadens the use spectrum of these films into such diverse applications as packaging, decorative yarns, capacitor dielectrics, industrial belting, magnetic tapes, and satellite balloons.

The emphasis of this chapter is directed to an explanation of the PET film system. This has the advantage of permitting a detailed discussion of the science and technology of a typical aromatic polyester film. Moreover, within this framework are found the principles applicable to the synthesis and treatment of a variety of crystalline polyester films.

II. Discovery and Growth of Polyester Films

The commercial importance of polyester films is relatively recent, but polyesters generally have had a significant role in the technology and use of natural products and in the development of synthetic organic chemistry [6–8]. An oriented fiber from an aliphatic polyester was reported first in 1932 by W. H. Carothers and J. W. Hill [9]. This aliphatic polyester was low melting (about 75°C) and had poor hydrolytic stability, a common characteristic of aliphatic polyesters, which precluded their use as textile fibers. Attention was directed therefore to the higher melting and more hydrolytically stable polyamides for use as textile fibers. Consequently until 1940 research on polyesters was largely neglected.

Following the guidelines presented in the work of Carothers [10], J. R. Whinfield of the Calico Printers Association in England restudied the polyesters, in particular, the symmetrical aromatic systems. The result of this work was the discovery (with J. T. Dickson) of the crystalline, high-melting but tractable polyester, polyethylene terephthalate [11, 12]. This polymer has found broad applicability in both fiber and film form.

The commercial significance of PET films is attested to by the large number of producers throughout the world (Table 1). Polyester films are useful in many applications because of the broad range of mechanical, optical, and electrical properties available. As a result, the growth rate has been rapid (Figure 1) [13]. During 1966 the consumption of polyester films in the United States reached 55 million lb, a 5-million-lb increase over 1965. General polyester film applications are illustrated in Figure 2 [14], and a more detailed list is provided in Table 2.

The magnetic-tape industry is one of the largest consumers of polyester film in the United States. At the present time about 90% of the magnetic tape made is on a polyester base and is divided between audio and computer tapes (80%) and video and instrumentation tapes (20%) [13a, b]. Polyester film is preferred for magnetic tapes because of its higher strength and durability and low dimensional change with humidity, compared with other polymeric tape bases. The availability of particular types of polyester film in several thicknesses coupled with its relative insensitivity to tape-processing conditions are distinct assets to the tape manufacturer.

The excellent dielectric properties of polyester film combined with its high strength, flexibility, and hydrolytic stability provide superior electrical insulating systems. Certified as a class "B" (130°C) insulating material, polyester film is used extensively in electric motors, transformers, capacitors, printed circuits, wire and cable wrap, and in many

TABLE 1

Polyethylene Terephthalate Film Manufacturers

Country	Manufacturer	Product trademark
Belgium	Gevaert Photo-Production N.V. (photographic)	Gevar
France	Kodak-Pathé (photographic)	Estar
France	La Cellophane S.A.	Terphane
Germany	Kalle AG	Hostaphan
Great Britain	Imperial Chemical Industries Ltd.	Melinex
Great Britain	Kodak Ltd. (photographic)	Estar
Japan	Fuji Photo Film Co. (photographic)	Fuji Film
Japan	Mitsubishi Plastics Industries Ltd.	Diafoil
Japan	Teijin-Konishiroku Film Co. (photographic)	Sakura Film
		Koni Film
	(other)	Teteron Film
Japan	Toyo Rayon Co.	Lumirror
Luxembourg	Du Pont de Nemours (Luxembourg) S.A.	Mylar
Netherlands	I.C.I. (Holland) N.V.	Melinex
Switzerland	Celfa AG (photographic)	Folex
United States	Celanese Plastics Co.	Celanar
United States	E. I. du Pont de Nemours & Co.	Mylar
United States	E. I. du Pont de Nemours & Co. (photographic)	Cronar
United States	Eastman Kodak Co. (photographic)	Estar
United States	FMC Corporation (American Viscose Division)	Avistar
United States	Goodyear Tire & Rubber Co.	Videne
United States	Minnesota Mining & Mfg. Co.	Scotchpar

other electrical applications. Its strength is an asset when automated manufacturing equipment is used to fabricate electrically insulated items.

Polyester films are also utilized in packaging applications. Here they are used principally as a base for a heat-sealable laminate or coated structure with polyethylene or polyethylene-polyvinylidene chloride. The high strength, transparency, and flex durability of the polyester film, plus the excellent gas and moisture barrier properties of the coating compositions yield packaging films that are tough, transparent, and heat-sealable. Moreover, the versatility in polyester film properties permits the production of specific types with mechanical and optical characteristics tailormade for packaging uses [15].

High strength, clarity, and dimensional stability over a wide range of

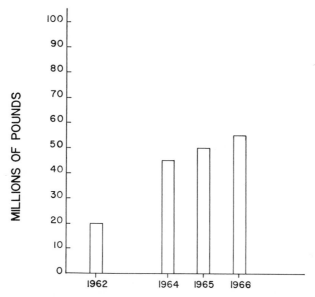

Fig. 1 Annual pounds of polyester film used in the United States [13].

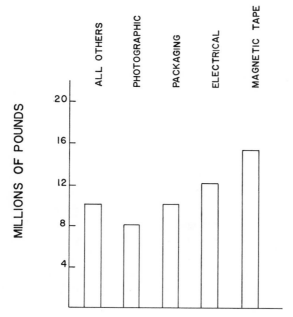

Fig. 2 Major markets for polyester films in the United States [14].

TABLE 2

Uses of Polyethylene Terephthalate Films

Electric motors, slot liners	Packaging
Wire and cable insulation	(a) boil-in-bag applications
Transformer insulation	(b) heat-shrinkable wrappings
Capacitors	(c) window cartons
Printed and laminated circuits	(d) formable packages
Magnetic recording tapes	Engineering reproduction
Magnetic memory cards	Graphic art films
Gaskets, spacers	Decals
Microphone diaphragms and loud-	Industrial belting
speakers	Book jackets
Typewriter ribbons	Stationery supplies
Punched computer tapes	Roll leaf applications
Pressure sensitive tapes	Apparel stays
Metallic yarn	Acoustical tile
Decorative laminates in automobile	Spring roll shelving
panels	Ticker tape
Echo satellite balloons	Pipe wrap
Missile diaphragms	Photographic film base
Glazing applications	

temperature and humidity conditions are the assets offered by polyester films as a base for photographic and x-ray films. Polyester films are rapidly displacing cellulose acetate in this technically sophisticated application.

III. Polyethylene Terephthalate Film Properties, Types and Uses

By means of suitable chemical, structural, and other modifications polyester films are made with a broad range of properties that serve many diverse types of application; for example, E. I. du Pont de Nemours offers eight basic types of Mylar† polyester film in more than 30 varieties. Each type is designed to meet the needs of specific industries as illustrated in Table 3. Other polyester film manufacturers have broadened this list to include heat-shrinkable tubing and specially coated products of many kinds. The properties of PET films must be discussed in terms of the particular type and variety under consideration.

† Du Pont Registered Trademark

TABLE 3

Types and Thickness of Mylar Polyester Film,
E. I. du Pont de Nemours and Company

1. Type A

A strong, tough polyester film, biaxially stretched and heat-set, with low electrical fault count. Transparent in 0.5 and 1.0 mil thickness and translucent in thicker films.

THICKNESS (mils). 0.5, 1.0, 1.5, 2.0, 3.0, 5.0, 7.5, 10.0, 14.0

USES *a. Electrical.* Slot liners, wedges, and phase insulation for motor and field coils; insulation for magnet wire and as a barrier and insulation tape in cable construction; insulation between turns in transformer coils; backing for mica.
b. Nonelectrical. Base for magnetic recording tape and other specialty tapes; surfacing material for acoustical tile; laminations, release sheets for reinforced plastics, stationery supplies, engineering reproduction materials, and apparel stays.

2. Type D

A highly transparent, strong, tough, dimensionally stable polyester film with superior surface characteristics.

THICKNESS (mils). 3.0, 5.0, 7.5, 10.0

USES. Engineering reproduction applications and stationery supplies.

3. Type S

A highly transparent, strong, tough, dimensionally stable film with superior surface characteristics.

THICKNESS (mils). 0.25, 0.35, 0.50, 0.75, 1.0, 1.5, 2.0

USES. Stationery supplies, graphic arts, metallic yarn; base for metallizing; surfacing material for paneling; decorative laminations; labels; roll leaf.

4. Type HS

A heat-shrinkable polyester film (ca. 50% at 100°C).

THICKNESS (mils). 0.65

USES. Packaging.

5. Type M

Polyester film polymer coated for improved gas and moisture barrier properties.

THICKNESS (mils). 0.50, 1.00

USES. Packaging.

TABLE 3 *(continued)*

6. Type W

A film resistant to outdoor exposure (several varieties).

THICKNESS (mils).   5.0, 10.0

USES.   Greenhouse windows, storm windows, etc.

7. Type T

A strong, tough polyester film with exceptional tensile strength in the longitudinal direction.

THICKNESS (mils).   0.50, 1.0, 1.5

USES.   Similar to those listed for type A but when very high unidirectional strength is required.

8. Type C

A film with low electrical fault count and good high temperature insulation resistance, primarily used as a dielectric in high-temperature capacitors.

THICKNESS (mils).   0.15, 0.25, 0.35, 0.50

USES.   Electronic capacitors.

Basically the commercial advantages of biaxially oriented and heat-stabilized PET base films can be summarized as follows.

1. Highly resistant to most organic solvents and mineral acids.

2. No plasticizers and very low moisture retention (less than 0.5% at room temperature and 50% relative humidity).   These properties make PET films particularly desirable for vacuum metallization.

3. High strength, toughness, and durability; excellent flex durability.

4. Large number of specific modifications available.

5. Excellent electrical properties.

To illustrate the scope of the properties of a typical PET film the example of a biaxially stretched, heat-set film with the polymer characteristics given in Table 4 is used.   A comparison of this typical polyester with other polymer films is given in Figures 3 to 6 and covers the four basic property areas of interest for an industrial film.   In Table 5 two types of PET film are compared, one with a balanced orientation

TABLE 4

Analysis of a Typical PET[a] Melt Polymer

Intrinsic viscosity (dl/g)[b] (limiting viscosity number)	0.54–0.58
Carboxyl end groups, g equivalents/10^6 g polymer	15–50
Minor constituents, wt %	1.3–1.8
Density (amorphous) g/cc at 25°C	1.333
Ratio of weight average molecular weight/number average molecular weight	1.5–1.8
Refractive index, n_D^{25} (amorphous film)	1.5760

[a] Polyethylene terephthalate:

$$HO \left(CH_2CH_2O_2C - \bigcirc - CO_2 \right)_n CH_2CH_2OH$$

[b] Solution viscosities measured in 60/40 wt % phenol/s-tetrachloroethane at 30°C.

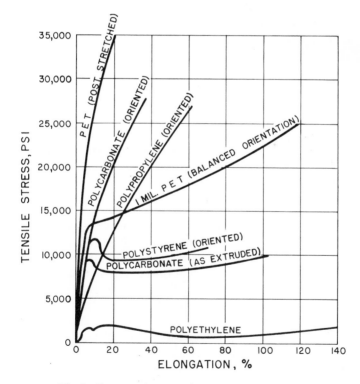

Fig. 3   Stress-strain curves for several polymer films.

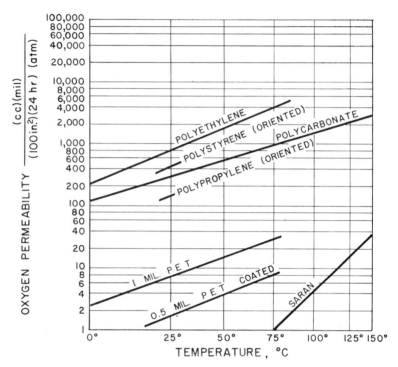

Fig. 4   Typical chemical property: oxygen permeability as a function of temperature for several polymer films.

and the other unbalanced.   The effect of temperature on several poly-ester film properties is presented in Figures 7 to 10.   These data typify the breadth of properties available commercially in PET films.

One of the most valuable characteristics of PET films is their electrical properties.   It is not known whether direct current conduction is ionic [16] or electronic [17, 18], but thin PET films are such good insulators that substantial improvements in the design of electrical components have resulted.   Reddish [19, 20] has studied comprehensively the temperature and frequency dependence of the dielectric constant and loss of polyester films.   In some electrical markets polyester films are replacing other materials (e.g., paper in foil-wound capacitors and electric motors), but in others they fulfil a need that had not been satisfied.

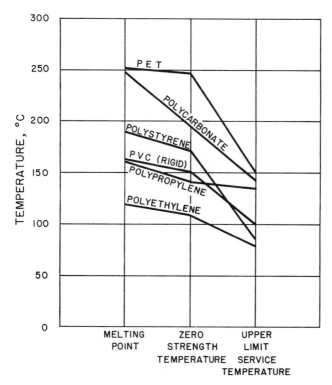

Fig. 5 Thermal properties of some polymer films.

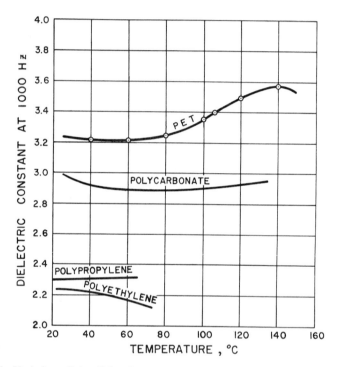

Fig. 6 Variation of the dielectric constant with temperature for several polymer films.

TABLE 5
Properties of Mylar Polyester Film[a]

Property	Typical value 1 Mil "Balanced"	1 Mil "Tensilized"	Condition	Method[b]
PHYSICAL PROPERTIES				
Ultimate tensile strength (MD),[c] psi	25,000	40,000	23°C	D 882–64T
Yield point (MD), psi at 4%	12,000	Indeterminate	23°C	Method A–100% elongation/ min
Stress to produce 5% elongation, psi	14–15,000	21–25,000	23°C	D 882–64T
Ultimate elongation (MD), %	120	50	23°C	D 882–64T
Tensile modulus (MD), psi	550,000	800,000	23°C	D 882–64T
Impact strength, kg-cm	6.0	6.0	23°C	T. M. Long pneumatic impact
Folding endurance (MIT), cycles	300,000		23°C	D 2176–63T (1 kg loading)
Tear strength, propagating (Elmendorf), g	15	12	23°C	D 1922–61T
Tear strength, initial (Graves), g	600	450	23°C	D 1004–61
Bursting strength (Mullen), psi	66	55	23°C	D 774–63T
Density, g/cc	1.395	1.377	23°C	D 1505–63T
Coefficient of friction (kinetic, film-to-film)	0.45	0.38	-	D 1894–63
Refractive index (Abbe), n_D^{23}	1.64		23°C	D 542–50
THERMAL PROPERTIES				
Melting point, °C	250–265			
Service temperature, °C	−60 to 150		20 to 50°C	
Coefficient of thermal expansion, in./in.-°C	1.7×10^{-5}			
Coefficient of thermal conductivity, cal-cm/cm²-sec-°C (1 mil balanced)	8.96×10^{-5}		100°C	
Specific heat, cal/g-°C	0.315			

TABLE 5 *continued*

Properties of Mylar Polyester Film[a]

Property	Typical value	Elonga-tion retained (%)	Tear strength retained (%)	Condition	Method[b]
Shrinkage, %	2–3			30 min at 150°C	
ELECTRICAL PROPERTIES					
Dielectric strength (1 mil), V/mil	7500			23°C, 60 Hz	D 149–64
	5000			150°C, 60 Hz	D 149–64
Dielectric constant	3.30			23°C, 60 Hz	D 150–65T
	3.25			23°C, 1 kHz	D 150–65T
	3.0			23°C, 1 mHz	D 150–65T
	2.8			23°C, 1000 mHz	D 150–65T
	3.7			150°C, 60 Hz	D 150–65T
Dissipation factor	0.0025			23°C, 60 Hz	D 150–65T
	0.0050			23°C, 1 kHz	D 150–65T
	0.016			23°C, 1 mHz	D 150–65T
	0.003			23°C, 1000 mHz	D 150–65T
	0.0040			150°C, 60 Hz	D 150–65T
Volume resistivity, Ω-cm	10^{18}			23°C	D 257–66
	10^{14}			150°C	D 257–66
Surface resistivity, $\Omega/in.^2$	10^{16}			23°C, 30% rh	D 257–66
	10^{12}			23°C, 80% rh	D 257–66
	Tensile strength retained (%)	Elonga-tion retained (%)	Tear strength retained (%)	Days immersed at room temperature	

CHEMICAL PROPERTIES

Chemical resistance

					ASTM method[b]
Sodium hydroxide, 2%	100	100	70	31	
Ammonium hydroxide, conc.	0	0	0	3	
Trichloroethylene	100	100	100	31	
Hydrocarbon oil	92	88	87	500 hr at 100°C	
				Baked 168 hr at 150°C	
Water absorption, %	<0.8			Immersion for 24 hr at 23°C	D 570–63
Hygroscopic coefficient of expansion, in./in.-% rh	1.1×10^{-5}			20–92% rh	
Permeability, 1 mil film					
Gas, cc/100 in.^{2}-24 hr-atm					
Carbon dioxide	16			23°C	D 1434–66
Hydrogen	100			23°C	D 1434–66
Nitrogen	1			23°C	D 1434–66
Oxygen	6			23°C	D 1434–66
Vapor,[d] g/100 in.^{2}-24 hr					
Acetone	2.22			40°C	Modified E 96–66
Benzene	0.36			40°C	Modified E 96–66
Carbon tetrachloride	0.08			40°C	Modified E 96–66
Ethyl acetate	0.08			40°C	Modified E 96–66
Hexane	0.12			40°C	Modified E 96–66
Water	1.8			40°C	Modified E 96–66
Fungus resistance	Inert			12 months' soil burial	E 96–66T

[a] *Mylar* is a Du Pont Registered Trademark.

[b] American Society for Testing and Materials, Philadelphia, Pa., *1968 Standards*, except as noted.

[c] Machine (longitudinal) direction.

[d] Vapor permeabilities are determined at the partial pressure of the vapor at the temperature of the test.

601

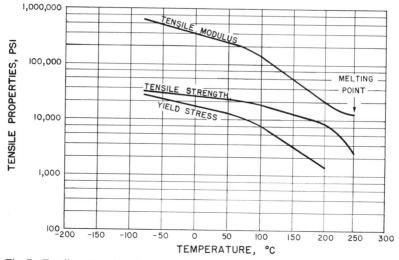

Fig. 7 Tensile properties of oriented heat-set polyethylene terephthalate film as a function of temperature.

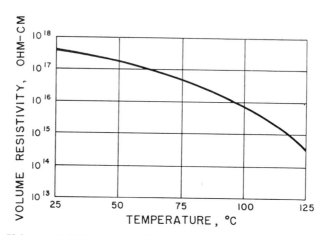

Fig. 8 Volume resistivity of oriented heat-set polyethylene terephthalate film as a function of temperature.

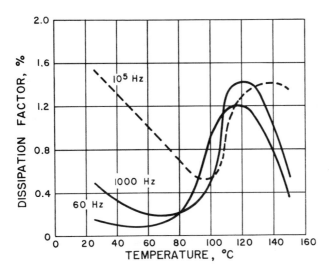

Fig. 9   Dissipation factor of oriented heat-set polyethylene terephthalate film as a
function of temperature.

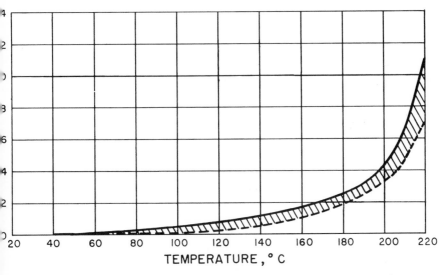

Fig. 10   Shrinkage of oriented heat-set polyethylene terephthalate film as a function
of temperature.

IV. Polyethylene Terephthalate (PET) Manufacture

A. CHEMISTRY

PET can be synthesized with several different starting materials [2, 5, 21, 22]. All synthesis routes can be considered to follow the procedure of first making bis(β-hydroxyethyl) terephthalate (DHET)

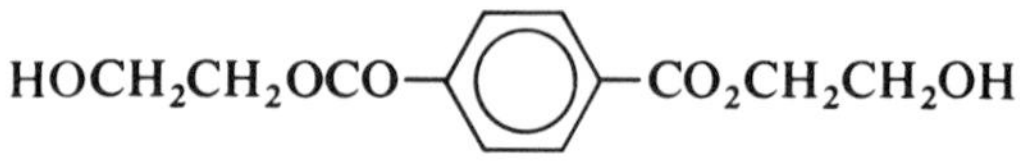

$$HOCH_2CH_2OCO-\langle\bigcirc\rangle-CO_2CH_2CH_2OH$$

which in a sense is the monomer for the polymerization of PET. Monomer formation can be carried out in several ways, as illustrated in the following four examples.

1. Transesterification of dimethyl terephthalate with ethylene glycol.

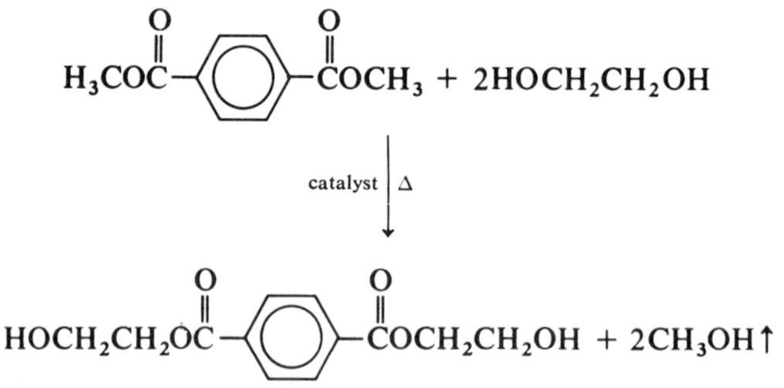

$$H_3CO\overset{\overset{\displaystyle O}{\|}}{C}-\langle\bigcirc\rangle-\overset{\overset{\displaystyle O}{\|}}{C}OCH_3 + 2HOCH_2CH_2OH$$

catalyst | Δ

$$HOCH_2CH_2O\overset{\overset{\displaystyle O}{\|}}{C}-\langle\bigcirc\rangle-\overset{\overset{\displaystyle O}{\|}}{C}OCH_2CH_2OH + 2CH_3OH\uparrow$$

2. Direct esterification of terephthalic acid and ethylene glycol.

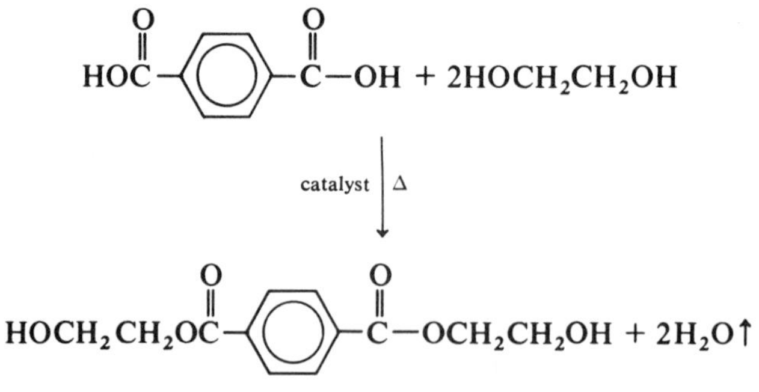

$$HO\overset{\overset{\displaystyle O}{\|}}{C}-\langle\bigcirc\rangle-\overset{\overset{\displaystyle O}{\|}}{C}-OH + 2HOCH_2CH_2OH$$

catalyst | Δ

$$HOCH_2CH_2O\overset{\overset{\displaystyle O}{\|}}{C}-\langle\bigcirc\rangle-\overset{\overset{\displaystyle O}{\|}}{C}-OCH_2CH_2OH + 2H_2O\uparrow$$

3. Reaction of terephthalic acid and the cyclic carbonate of ethylene glycol.

$$\text{HOCH}_2\text{CH}_2\text{OC} - \langle\bigcirc\rangle - \text{C} - \text{OCH}_2\text{CH}_2\text{OH} + 2\text{CO}_2\uparrow$$

4. Reaction of terephthalic acid with ethylene oxide.

PET is then formed from the monomer by an alcoholysis reaction with elimination of ethylene glycol; for example,

$(n\text{-}1)$ $\text{HOCH}_2\text{CH}_2\text{OH}\uparrow$ + $\text{HOCH}_2\text{CH}_2 \left(\text{OC} - \langle\bigcirc\rangle - \text{COCH}_2\text{CH}_2 \right)_n \text{OH}$

Most commercial PET is made by the transesterification of dimethyl terephthalate (DMT) with ethylene glycol. The choice of DMT over terephthalic acid (TPA) is based on ease of purification, solubility, and lower melting point.

Purity of ingredients and stoichiometric balance of the reactants are essential to the production of high-molecular-weight linear condensation polymers [6]. For polyesters made from diols and dibasic acids the stoichiometry is established initially by utilizing an excess of the diol to ensure complete reaction between the acid and alcohol. A monomer molecule is produced with a hydroxyl group on each end. For those polyesters made from monomers with a hydroxyl and carboxyl group on the same molecule, for example, HO—R—COOH, the system is self-stoichiometric and subsequent polymerization occurs by the reactions of hydroxyl and carboxyl groups with the elimination of water. In contrast, polymerization of PET involves the reaction between monomer molecules terminated at each end with the same functional group. Polymerization proceeds by the elimination of one molecule of ethylene glycol for each reaction step.

When DMT and ethylene glycol are used as the starting materials, a molar equivalent ratio of 2.1 to 2.3 ethylene glycol/DMT is normally required and ester interchange is accomplished with appropriate catalysts over the temperature range 150 to 220°C. All that remains, to form the polymer is to drive the reaction forward by removing the ethylene glycol (usually by heat and vacuum) as it is generated by the alcoholysis reaction. Ultimately a ratio of unity of terephthaloyl to glycol residues is approached in the polymer chain.

Not shown in these simplified equations are the side reactions that produce impurities. The influence of these side reactions is important and is mentioned as part of the discussion of a typical PET polymerization system. Several excellent sources describing the chemistry, synthesis, and analysis procedures for polyesters will be found in [2, 5, 21, and 22].

B. POLYMER PROCESSES

The three processes used to produce PET are batch autoclave, continuous melt, and solid phase. Each has commercial advantages and disadvantages that must be considered in terms of economics, polymer purity, and product quality [23]. All three processes require the formation of dihydroxyethyl terephthalate (DHET) from the starting materials DMT and ethylene glycol.   In practice the monomer formed is usually a mixture of DHET and higher oligomers and the chemical composition may vary with process conditions.   This monomer is then polymerized in the melt by batch or continuous processes or in some cases by solidification, grinding, and solid-phase polymerization.

Solid-phase polymerization [24] is carried out by heating the powdered monomer (and higher "mers") below the melting point and removing the by-product ethylene glycol with a vacuum or inert gas.   Fluidized bed systems are found to be particularly effective [25].   The solid-phase process has the advantage that polymer is produced with minimal thermal degradation but the disadvantage that the starting material should be reduced to a small particle size.   Even with small particle size, however, the molecular-weight distribution of the resulting polymer is broader than in polymer produced in a melt system [26].   The latter effect is understandable, since removal of the ethylene glycol from the particle surface is easier than from the interior.   Moreover, re-equilibration and degradation reactions are retarded, since the temperatures used for solid-phase polymerization are below the melting point of the polymer.   Consequently the molecular weight throughout the particle is usually nonuniform.   Solid-phase polymer is highly crystalline.

A consequence of the dynamic and reversible nature of the alcoholysis reaction is that subsequent melting of solid-phase polymer tends to narrow the molecular-weight distribution.   Moisture is deleterious and all PET polymers are dried thoroughly before melting to minimize degradation [27].   Hartley, Lord, and Morgan have discussed the importance of complete melting and have emphasized the influence of the premelt crystallization history on the kinetics of subsequent polymer crystallization [28].

Most PET is made by a continuous- or batch-melt process which utilizes DMT and ethylene glycol as the starting materials,   Continuous polymer processes consist of at least three stages [23, 29]:

1. Ester interchange (in which the methyl groups of the DMT are exchanged for a glycol residue).

2. A reduced pressure first-stage polymerization or prepolymerization (in which the major fraction of the by-product ethylene glycol is removed).

3. A second-stage polymerization or vacuum finishing, a continuation of the polycondensation to obtain the molecular weight desired.

This continuous polymer process is illustrated schematically in Figure 11.   There are many variations of primary and auxiliary equipment that satisfy the requirements for such a continuous polymerization process.

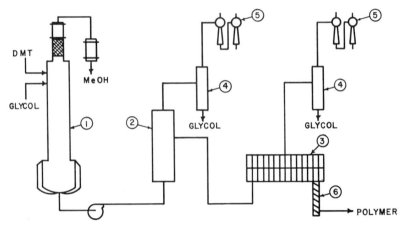

Fig. 11 Continuous melt polymerization process: 1, ester interchange column; 2, first-stage polymerizer; 3, second-stage polymerizer; 4, glycol condensers; 5 steam-jet vacuum pumps; 6, screw pump.

1. Ester Interchange

Ester interchange of the DMT and ethylene glycol can be accomplished in a continuous distillation column used as a reaction vessel, with methanol recovered as the by-product [30, 31].   This process is generally carried out at atmospheric pressure with appropriate catalysts (Table 6) and at temperatures sufficient to complete the reaction.   A temperature gradient exists from the top of the column to the bottom (generally 160 to 220°C) because of the changing plate composition that occurs as the reaction proceeds.   Process vessels are fabricated from stainless steel to prevent product contamination, and the monomer is removed continuously from the bottom of the column.

Prolonged heating of the reaction mass promotes the formation of undesirable diethylene glycol (DEG) caused from the dehydration of ethylene glycol:

$$2HOCH_2CH_2OH \xrightarrow[\Delta]{catalyst} HOCH_2CH_2OCH_2CH_2OH + H_2O\uparrow$$

This reaction generates a monomer product that contains a mixture of diols, and some of the DEG eventually becomes part of the polymer

TABLE 6

Examples of Catalysts Used in the
Formation of Polyethylene Terephthalate

Catalyst	
PbO	CaH_2
Sb_2O_3	NaOH
$(CH_3COO)_2Ca$	$LiAlO_2$
$(CH_3COO)_2Zn \cdot 2H_2O$	$NaAlO_2$
LiH	MgO

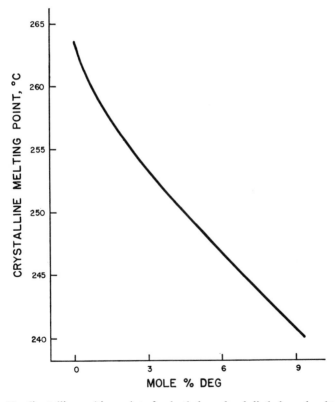

Fig. 12 Crystalline melting point of polyethylene glycol-diethylene glycol terephthalate copolyesters [36].

chain as 3-oxapentamethylene terephthalate units,

$$-p\text{-}O(CH_2)_2O(CH_2)_2OCO-C_6H_4-CO-$$

Such unwanted segments disrupt the geometric regularity of the polymer chain and influence the rate and level of polymer crystallization and the film properties. In general, the melting point of crystallizable polymers is decreased with increases in the amount of comonomer content [32]. The dependence of the melting point of a PET copolymer containing small amounts of 3-oxapentamethylene terephthalate units is shown in Figure 12. It has been suggested that the decrease in melting point is a consequence of a smaller average crystallite size [33, 34].

2. Prepolymerization

Monomer contains two glycol residues per terephthaloyl unit and polymerization occurs by elimination of one molecule of ethylene glycol for each reaction between monomer, dimer, trimer, etc.; for example, if 10 molecules of monomer react in sequence, a decamer is

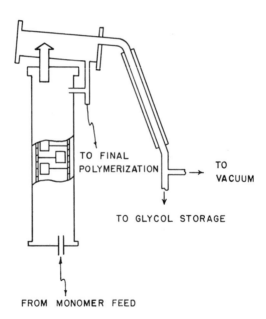

Fig. 13. Continuous polyester prepolymerizer [35].

formed with the elimination of nine molecules of ethylene glycol. The ethylene glycol is removed by raising the temperature and reducing the pressure in the reaction chamber.

A single vessel for the continuous processing of monomer to prepolymer is illustrated in Figure 13. Monomer enters the bottom and prepolymer is removed at the top. The temperature is increased and pressure is reduced from the bottom to top of the vessel [35]. Most of the ethylene glycol is removed at this stage of a continuous process. Even so, the prepolymerizer product may be of low melt viscosity with

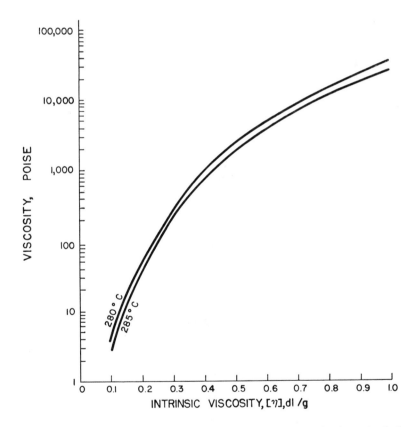

Fig. 14 Polyethylene terephthalate melt viscosity as a function of polymer intrinsic viscosity [36].

an intrinsic viscosity† of about 0.15 and number average molecular weight of perhaps only 2000. As more ethylene glycol is removed by further condensation, the melt viscosity increases rapidly as illustrated in Figure 14 [36].

3. Finishing

To complete the polymerization to the desired molecular weight, a vessel is used in which a large surface area is generated continuously [37]. The formation of new surface aids in the rate of removal of ethylene glycol and is accomplished either in a stirred autoclave as in a batch process or in a continuous finisher.

A continuous finisher described by Vodonik is a horizontal vacuum chamber containing a pool of molten polymer [37]. The pool is sampled continuously by rotating screens to expose a thin film of the polymer to an inert atmosphere (Figure 15). Pressures of about 1 torr

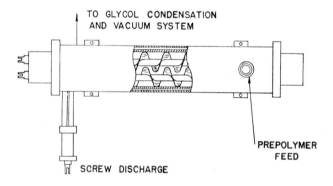

CROSS SECTION

Fig. 15   Continuous polyester finisher [37].

† All solution viscosities reported are based on measurements in the solvent s-tetrachloroethane/phenol (40/60 weight ratio) at 30°C.

and temperatures of 270 to 300°C are used. The polymer changes in viscosity from inlet to exit, ultimately attaining a value of 2000 to 3000 poises at 280°C. A continuous process for the production of polyethylene terephthalate is also described by Hippe [38] and by Scheller [39]. In each process well-designed mechanical agitation, lack of places for polymer to collect and degrade, and uniformity of gas flow are described. Although it is necessary to develop the required surface area at reasonable rates, overheating the polymer because of the input of heat from mechanical work must be avoided. The ability to remove the ethylene glycol rapidly from the polymer mass by diffusion can limit the rate of polymerization.

From the finisher the polymer is moved continuously into a transfer line by means of a screw pump where it is either extruded directly into film or processed into flake or chip for subsequent re-extrusion. As an alternate to continuous processing, polymer may be formed in batch equipment, as illustrated in Figure 16.

The statistical nature of the polycondensation reaction also causes the formation of several kinds of low-molecular-weight material constituting 1.3 to 1.8 wt% of the polymer mass. The major portion of these minor constituents is the cyclic trimer, but many other compounds have been found and analyzed [40–42]. Stockmayer [43] has discussed the cyclization process and many of the low-molecular-weight oligomers of PET have been synthesized and characterized [44–46].

Destructive thermal side reactions also occur which result in the formation of some polymer chains that end in one or two carboxyl end groups. The thermal degradation of PET has been studied extensively and has been attributed to random-chain scission with an energy of activation of 32 kcal/mole [47–49]. Traces of oxygen and water are particularly harmful to the molten polymer in producing chain degradation. The need to account quantitatively for a mixture of two different chain end groups forms a major part of the effort of molecular weight determination by end-group analysis [50]. A typical polymer analysis for PET made in a continuous melt system is given in Table 4.

V. Processing of Polyethylene Terephthalate Films

Commercial film-forming processes for PET films include melt extrusion of the polymer, quenching the melt, orientation, and thermal stabilization or relaxation. The process shown schematically in Figures 17 to 19 is a composite of the technology described in [51–106]. Molten

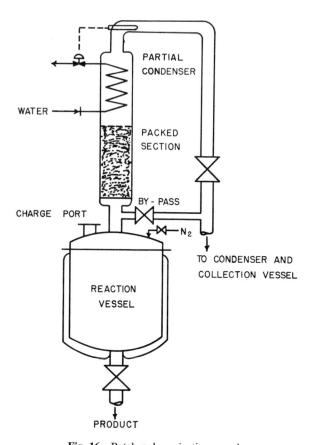

Fig. 16 Batch polymerization vessel.

polymer is normally extruded from a slit die. This technology has been discussed extensively in Volume I (Chapters 8, 10). Since the viscosity of the molten polymer is of the order of 2500 poises at 280°C, a self-supporting film can be maintained over the space from the die to the cold quenching roll. The film passes quickly from a fluid to a glass with a density of about 1.33 g/cm^3 and the description *cast amorphous* film is generally applied. This amorphous film can now be oriented by stretching or rolling.

Orientation by stretching may be accomplished in the direction of

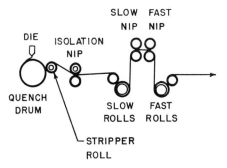

Fig. 17 Stretching in the longitudinal (machine) direction with nip rolls.

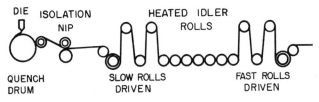

Fig. 18 Longitudinal direction stretching with idler rolls.

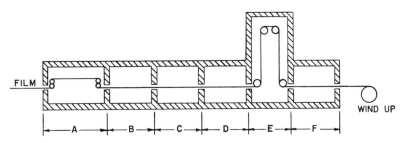

Fig. 19 Orientation and heat setting process: *A* longitudinal stretching zone; *B*, lateral stretching zone; *C*, buffer zone; *D*, heat set zone; *E*, heat relaxing zone; *F*, cooling zone.

film travel (machine direction), at right angles to the direction of film travel (transverse direction), or in both directions. Stretching in the machine and transverse directions may be carried out simultaneously, for example, expanding a tubular film, or sequentially (e.g., the flat sheet process illustrated in Figure 19). Film can be stretched under conditions in which the product will shrink as much as 50% of the stretched dimensions on being heated above the processing temperature. It is sold as a heat-shrinkable polyester film. Film can also be heat-set or crystallized while restrained and will exhibit much less residual shrinkage (ca. 1% at 105°C) on being heated. A dimensionally stabilized film with still less residual shrinkage (ca. 0.1%) can be prepared by heating the film under conditions of low stress (10 to 300 psi) to temperatures above the required-use temperature. Consequently many process variations which can produce different combinations of polyester film properties are possible. For illustrative purposes a general process for making a biaxially stretched film with balanced physical properties is described.

A. Extrusion and Quenching

Melt extrusion through a linear or circular die is used to form an unoriented web that may be in sheet or tubular form. The film extrusion process may be integrated with the polycondensation process or the chipped polymer can be re-extruded as a nonintegrated operation. In a nonintegrated extrusion operation the chipped polymer flake is dried to prevent polymer hydrolysis and the dried chip blanketed with an inert gas to prevent oxidative degradation.

The extruded polyester film, if cooled slowly, will crystallize spherulitically and be hazy, brittle, and difficult to stretch. When properly quenched, an amorphous, transparent, ductile web is obtained that is suitable for orientation. Quenching is normally accomplished by extruding the film onto one or more refrigerated rolls or into cold liquids. The maximum attainable thickness of amorphous film depends on the rate of polymer crystallization and on the heat transfer rates attainable in quenching. Amorphous film thicknesses in the range of 0.001 to 0.2 in. are normally attainable.

B. Orientation and Heat Setting

Amorphous film is preheated above the glass transition temperature and passed between two sets of heated nip rolls. Stretch or draw is developed by having the speed of the second set of nip rolls two to six times faster than the first set. Alternatively, stretching may be done over a series of heated idler rolls flanked on the entrance end by driven slow rolls and at the exit by driven fast rolls. The forces involved in machine direction stretching must satisfy a number of relationships simultaneously. Frictional forces in nips can be related by equations of the form $T_1 - T_2 = \mu N$, where $T_1 - T_2$ is the maximum increment permissible in web tension in passing through a nip, μ is the coefficient of friction, and N is the normal force in the nip. When roll wraps are used to restrain the film, web tensions are related by the expression $T_1/T_2 = \exp \mu\theta$, where θ is the angle of wrap. Roll and film surface characteristics and temperature are important in determining the coefficient of friction. Film that has been stretched usually has a lower coefficient of friction to metals than the amorphous sheet. The force required for stretching is a function of the film thickness, temperature, and amount and rate of deformation.

The uniaxially drawn and partially crystalline film then passes into a transverse direction stretcher called a tenter frame. This equipment is an adaptation of textile processing machinery and consists of film gripping devices attached to a moving endless chain. Film is transported by the chain through a series of independent zones which stretch and heat or cool the film. As the film leaves the longitudinal or machine-direction stretcher it is positioned automatically in the gripping devices and transported into a zone in which the film is again heated to increase the ductility. The chain path then diverges and the film is stretched in the transverse direction. Often the chain holding the film-gripping devices is attached to movable supports which permit variations to be made in the amount of transverse stretch. Typically the amount of transverse stretch is controlled to provide balanced mechanical properties in the film. At the end of the stretch zone the doubly drawn film sheet exhibits high mechanical properties in both directions; it is tough and flexible but will shrink if heated above the stretching temperature. To decrease the tendency of the film to relax dimensionally it is crystallised under restraint in one or more zones by utilizing temperatures in the range of 150 to 230°C. This process, called heat setting, increases the film density to about 1.38 g/cc. Typically, a heat-set film may still shrink slightly (fractions of a per cent) when heated above 80°C because of residual strains. This shrinkage can be reduced further by heating the film to a temperature lower than the heat-set temperature but with little or no restraint [63].   Before winding the film onto a roll it is cooled to ambient temperature and the film edges that were in the gripping devices are cut off.

A tensilized film (i.e., one more highly oriented in one direction than the other) can be made by additional processing of the biaxially oriented film.   This is done by stretching again, usually in the machine direction [68, 81].

Many variations of orientation and thermal treatments are described in the patent literature.   Included are devices to stretch the film in both directions simultaneously and a process in which the film is rolled.   The list of references given is of sufficient breadth to cover the essential characteristics of many polyester film processes.

VI. Auxiliary Film Treatments

Polyester films can be treated to make products with particular types of property for special uses.   Treatments such as coating, priming, metallizing, sizing, and weatherproofing are often applied.

Films with varying degrees of surface roughness are often required. Polyester films have surfaces that are normally smooth but may contain asperities ranging in height from 1 to 100 μin.   Smooth surfaces are desirable for transparency and brilliance and are required in some end-uses, such as metallized film.   Rougher surfaces are required for delustered films.   Several available methods for controlling the amount and kind of roughness include dispersing small particles [107, 108] of silica, titania, or bentonite in the polymer before extrusion, embossing the film, roughening the surface with a brush [109], sandblasting the surface with an abrasive [110] which has been accelerated either with a wheel or compressed air, coating, sizing [111], and chemical etching of the surface [112].

Oriented, crystalline polyester films do not heat-seal easily.   Heat-sealable, crystalline polyester films are made by melt coating the polyester base film with polyolefins, dispersion coating with polyvinylidene chloride copolymers [113, 114] or application of other heat-sealable coatings [115].

Bonding of untreated polyester films to other materials is sometimes difficult [116], but good adhesion of polyester film can be obtained in a number of ways.   Priming treatments that can improve adherability include flame treatment, electrostatic discharge [117], electrostatic discharge in the presence of monomers [118], chemical treatment [119], and solvent etching and coating the polyester with an adherable subcoat [120]. Ultraviolet light [121] can be used to improve the adhesion between a polyethylene and polyester film. The ultraviolet light is transmitted through the polyethylene but absorbed strongly by the polyester.

Polymer coatings can be applied to oriented PET films from aqueous dispersions or from solutions in organic solvents. Coating compositions based on vinylidene chloride copolymers impart water and oxygen impermeability as well as heat-sealing characteristics to the film [122, 123]. Inorganic dispersions for abrasive, ink receptive surfaces, or dispersions of magnetic oxides for magnetic tapes can be applied to PET films by coating processes [124, 125].

Weatherable polyester film may be prepared by impregnating either or both surfaces of the film with an ultraviolet light absorber. Coating equipment is used and the ultraviolet absorber is applied to the film in a glycol solution, the glycol being removed in a dryer [126].

Metallized polyester film slit to narrow widths is used as a decorative yarn and in capacitors. Metal is usually deposited by exposing the film to a metal vapor in a vacuum but metallic deposition can also be done by solution processes.

Polyester films have a high propensity for electrostatic charge. The charge on the film must be controlled to keep the film free of particulate contamination and to avoid electrical discharge or sparking when flammable solvents are present. This is particularly critical when the film is to be used as a photographic base.

VII. Polyethylene Terephthalate Film Structure and Analysis

The properties of PET films result from an ordered structure developed by means of stretching and thermal teratments. Ordinarily the terms *orientation* and *crystallinity* are applied to help explain this ordered structure without describing specifically what is meant; both terms are imprecise (cf. Volume 1, Chapter 7). In this section we illustrate conceptually the importance of these ideas as they are related to PET film structure and properties. We also describe some of the more commonly used methods of structural analysis of PET film.

Molten PET is highly viscous and can be extruded and quenched into a glassy state in film form. Glasses are formed typically by substances containing long chains, networks of linked atoms, or those that possess a complex molecular structure. Normally such materials have a high viscosity in the liquid state. When rapid cooling occurs to a temperature at which the crystalline state is expected to be the more stable, molecular movement is too sluggish or the geometry too awkward to take up a crystalline conformation. Therefore the random arrangement characteristic of the liquid persists down to temperatures at which the viscosity is so high that the material is considered to be a solid. The formation of an amorphous or glassy film is essential to the development of the desired ordered PET structure. An amorphous PET film is in a metastable state at room temperature.

Some polymers (e.g., polyethylene, isotactic polypropylene), even when quenched rapidly from the melt, crystallize into spherulites [127]. PET can also be crystallized spherulitically by cooling the melt slowly, by heating the amorphous polymer, or by treatment with appropriate liquids. Spherulitic PET films are translucent and brittle, and orient nonuniformly and with difficulty at normal processing temperature. Keller has discussed the structure of PET spherulites [128].

Amorphous PET film is of little commercial importance, since it has low strength and a tendency to be brittle. However, by appropriate combinations of deformation (stretching and/or rolling) and thermal processes exceptionally strong and tough films can be made. Temperature has a major role in the deformation processes of amorphous polyester films. At low temperatures molecular flexibility is suppressed, but at a sufficiently high temperature the material passes from a relatively rigid glass to a flexible, ductile viscous state. The temperature at which this transition occurs is commonly referred to as the glass transition temperature (T_g) [129]. For PET the T_g is about 70°C and the film is generally processed at temperatures in the range 70 to 90°C. At these temperatures the film can be drawn or rolled easily. During the deformation of the heated film the stresses developed cause the uncoiling and translational motion of the molecular chain segments. Concurrently, nucleation and strain-induced crystallization occur. Therefore the net result of the deformation process is the partial alignment of molecular segments in the direction of deformation, coupled with the formation of a new phase. This new phase is well enough ordered to diffract x-rays coherently according to the Bragg law.

PET films can be stretched in one direction or two directions, either simultaneously or sequentially. The general term *orientation* has often been applied to describe these processes without specifying the interplay

of other concomitant effects.   The stress-strain curves for the deformation processes have been studied extensively and are described by
Thompson [130] for fibers.   Curves of the general type illustrated in
Figure 20 are found to occur.   The viscoelastic equations necessary to
describe completely the deformation in polyester films would be extremely complicated, but qualitatively we can understand the process in
the following way.

Region A up to the yield point of the stress-strain curves can be
thought of as the work required to move the molecular segments from
their equilibrium positions, that is, simple elastic extension.   Region B
is indicative of massive viscoelastic deformation of the film in which the
molecular chain segments rotate, translate, and unfold and become
aligned in the direction of draw.   Once the system is set into motion
little additional force is required.   Ultimately, however, the ordering
process is complete enough to require a greater force for further deformation and a region (C) of reinforcement occurs.   The system then becomes more ordered, increasing with the position on the ascending

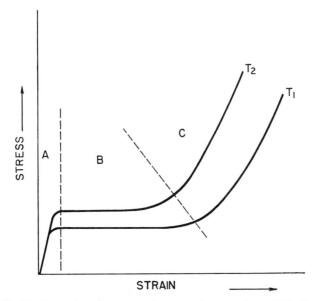

Fig. 20   Examples of stress-strain curves for amorphous PET film.
$T_1 > T_2$ (for $T > T_g$).   See text for explanation of A, B, C.

portion of the curve.   Finally, the stress will exceed the strength of the film and fracture occurs.   The shape and magnitude of the stress-strain curve depend on the temperatures used and on many other process parameters as well. Commercial deformation methods are designed to utilize these concepts to orient the structural elements in the film in one or more directions.

The terms commonly used to describe the initial glassy structure may be imprecise; for example, Kargin [131] postulated a structural pre-ordering in amorphous polymers.   Later G. L. Berestneva and colleagues [132] demonstrated that precrystalline order exists in unoriented PET film, and recently Yeh [133] has reported the existence of a ball-like structure of about 75 Å average diameter.   Consequently the assumption that a truly amorphous film exists in quenched PET may be in error. The inadequacy of the term *amorphous* has been recognized, and the same imprecision also applies to the polymer chemist's use of the terms *crystallinity* [134, 135] and *orientation* [136].   For discussion purposes we consider as crystalline that fraction of the structure that coherently diffracts CuK_α x-rays according to the Bragg law.   Crystalline polymers are far from perfect in their crystallinity, but it is convenient to utilize the concept of a crystalline/amorphous ratio or "crystalline" content.

Several hypotheses and models concerning the meaning of polymer crystallinity [4, 137] and film texture are covered in Volume I (Chapter 7).   The fringed-micelle model (i.e., an imperfect two-phase system of crystalline and amorphous domains) is valuable and useful as a descriptive tool to describe the structure of oriented PET films.

The structural factors required to describe the texture of oriented, crystalline films fully are the following:

1. The orientation distribution of crystallites.

2. The orientation and stress distribution of segments in the amorphous phase.

3. The size distribution of crystallites.

4. The morphology of the macrostructure.

5. Anisotropy in the thickness direction.

6. The temperature-dependence of chain segmental motion as influenced by orientation, crystallization, and strain relaxation.

Some progress has been made in describing quantitatively a few of these factors.

A. Orientation

In semicrystalline polymers both the crystalline and amorphous regions contribute to the properties and a range of order exists [138]. In the manufacture of PET films many process parameters contribute to the over-all orientation such as stretching temperature, stretching rate, amount of stretch, geometry, prior and post-thermal treatments, polymer purity, and molecular weight. The direct determination of the orientation distributions of both the crystalline and amorphous regions is important to establish the mechanisms that occur during the stretching process. Metallurgists have a similar problem concerning the orientation of the crystal-grain boundaries produced in metals by rolling and shaping. Many of the techniques used for polymer structural investigations (i.e., pole figure analysis and single crystal techniques) have been adapted from the voluminous investigations of the metal-processing industry. The application of these concepts to oriented polymer systems encompasses various degrees of theoretical and mathematical rigor [138–145], as applied to general and specific polymer systems.

A number of different concepts and terminology have been used in the literature to describe orientation, and communication has not always been precise. Therefore a pictorial representation of orientation was proposed [136] to clarify these ideas for films. Two requirements are needed:

1. A systematic geometrical system of reference of the several types of preferred orientation that can be conceived.
2. Experimental methods for the independent determination of the distributions of each kind of textural domain, for example, crystalline, amorphous, or other.

These definitions and concepts are not unique to any one type of structural organization, since they are equally applicable for describing the distribution of directions of (a) crystallite planes or axes, (b) the transition moments of functional groups within chain segments, that is, phenyl rings and amide or polyester linkages, or (c) the average direction of chain segments. To illustrate, this has been done for crystallite planes and axes in Figure 21.

Six types of orientation are thought to be possible, and the names of these orientation types are derived from the geometry of the reference

NAME	GEOMETRY	REFERENCE	OPERATOR	SCHEMATIC STRUCTURE	POLE FIGURE	
RANDOM	NO ORIENTATION	NONE	NONE			DOTS–CRYSTAL AXES
PLANAR	GIVEN CRYSTAL AXIS PARALLEL TO A REFERENCE PLANE	PLANE	AXIS			
UNIPLANAR	GIVEN CRYSTAL PLANE PARALLEL TO A REFERENCE PLANE	PLANE	PLANE			LINES–NORMALS TO CRYSTAL PLANE
AXIAL	GIVEN CRYSTAL AXIS PARALLEL TO A REFERENCE AXIS	AXIS	AXIS			
PLAN–AXIAL	GIVEN CRYSTAL PLANE PARALLEL TO A REFERENCE AXIS	AXIS	PLANE	NOT APPLICABLE TO FILM GEOMETRY BUT CAN EXIST IN FIBERS		
UNIPLANAR– AXIAL	GIVEN CRYSTAL AXIS PARALLEL TO A REFERENCE AXIS AND A GIVEN CRYSTAL PLANE PARALLEL TO A REFERENCE PLANE	PLANE AXIS	PLANE AXIS			

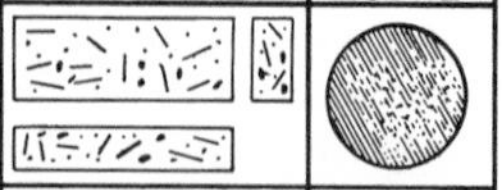

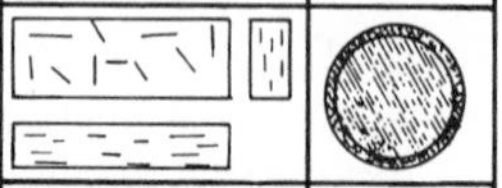

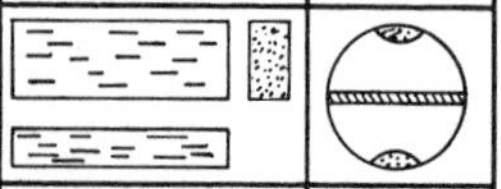

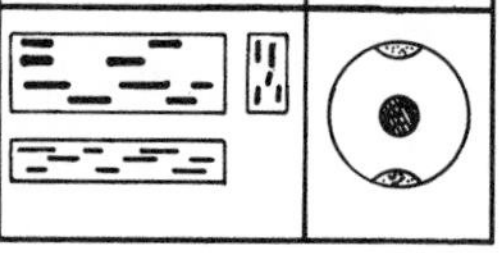

Fig. 21 Classification of orientation types [136].

system and not the geometry of the crystallite.   Orientation will not be perfect in films, and distributions of the directions of crystallite planes or axes will be observed.   It is also possible for several orientation types to coexist in films, as is sometimes found in metallurgical specimens. The definitions proposed are limiting cases designed to provide a simple and uniform terminology for describing orientation.

The structure of PET films has been studied with x-rays, infrared, nuclear magnetic resonance, optical diffraction, chemical and other classical techniques [132, 133, 136, 146–177].   Generally, much information of use is obtained by combining results from the several techniques outlined here.

1. Refractive Index and Birefringence

One of the easiest methods used to follow the progress of orientation in transparent films is the refractive index.   The only equipment needed is a suitable refractometer or a polarizing microscope with compensators. Amorphous PET film, stretched or rolled in one or both directions, becomes birefringent, thus indicating a preferred alignment of polarizable groups. The nomenclature proposed by Dulmage and Geddes [147] is used to illustrate the three principal refractive indexes measured (Figure 22). A distribution of refractive indexes exists between these

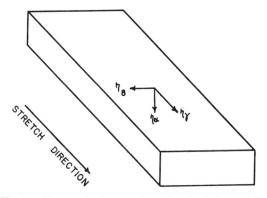

Fig. 22 Frame of references for refractive indexes [147].

reference directions, and the distribution depends on specific combinations of the process parameters used. For PET a positive birefringence always occurs in the direction of the highest orientation; for example, if the film is stretched more in the γ-direction than the β-direction, then $(\eta_\gamma - \eta_\beta) > 0$.

Typical values of refractive index or of the birefringence versus the amount of stretch in the γ-direction are shown in Figures 23 and 24 [147, 148]. Curves A and B of Figure 23 illustrate the effect of a post-stretching thermal crystallization on the birefringence of a one-way-drawn PET film and Curve C, the effect on the birefringence of a decrease in the stretching temperature. A minimum in η_β is found to occur at about 250% elongation. This minimum corresponds to the substantial enhancement of the alignment of the phenyl ring and the (100) crystallite planes parallel to the plane of the film [148]. One-way-stretched PET films approximate closely a uniplanar-axial type of orientation.

To observe the changes in the refractive index for various amounts of

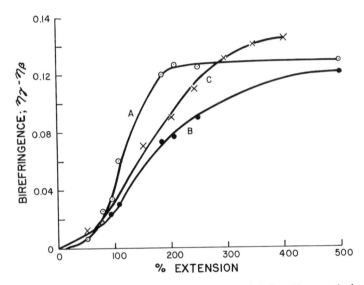

Fig. 23   Birefringence $(n_\gamma - n_\beta)$ of polyethylene terephthalate film stretched at 85 and 90°C: *A*, stretched at 90°C but thermally crystallized; *B*, stretched at 90°C but not thermally crystallized; *C*, stretched at 85°C but not thermally crystallized.

biaxial stretching requires sequential analysis. A particular point on the refractive index curve for a one-way stretched film is chosen (e.g., 200% elongation).   Then the changes in η_γ and η_β are determined as stretching is performed in the β- or transverse direction.   Refractive index versus elongation values like those illustrated in Figure 25 are produced.   Normally the film is considered to be balanced when the birefringence $(\eta_\gamma - \eta_\beta \simeq 0)$ approaches zero.   Birefringences for a biaxially oriented film are positive or negative, depending on the amount of stretch and the frame of reference chosen.   A frame of reference often used in commercial practice is to name the direction of film extrusion as the machine direction (MD) and the direction normal to the (MD) as the transverse direction (TD), regardless of the sequence by which the stretching has been done.

2. X-Ray Analyses

a. Crystallite Orientation Distributions.   The greater part of the orientation analysis of films uses x-ray diffraction techniques because

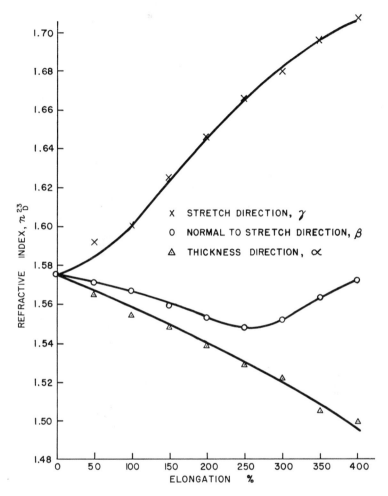

Fig. 24 Refractive indexes versus elongation of unidirectionally stretched poly-ethylene terephthalate films.

627

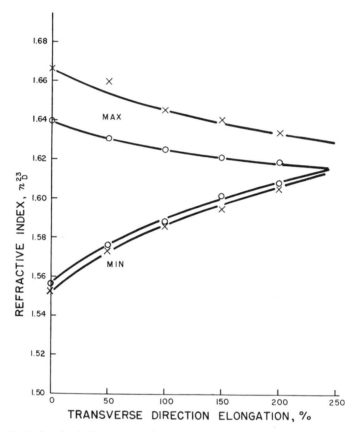

Fig. 25 Refractive indexes versus elongation in the transverse direction of biaxially drawn film: O, 200% stretch MD; × 250% stretch MD.

they are specific to the crystalline regions. Crystalline PET has a triclinic unit cell with the dimensions shown in Figure 26 [146]. Several crystallographic planes are of interest to describe the orientation distribution of the crystalline fraction in the film. Heffelfinger [136, 148] has used the (100), (010), and ($\bar{1}$05) planes; Dumbleton and Bowles [171], the ($\bar{1}$05); Krigbaum and Balta [152], principally the (100), (011), (003), and (010); Farrow and Bagley [172] and Kuriyama [173], the (010) and (100). A variety of orientation distributions is produced by stretching,

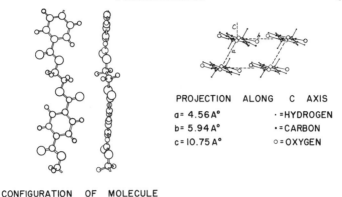

PROJECTION   ALONG   C   AXIS

a= 4.56 A°   · =HYDROGEN
b= 5.94 A°   • =CARBON
c= 10.75 A°   ○ =OXYGEN

CONFIGURATION   OF   MOLECULE

Fig. 26   Unit cell of polyethylene terephthalate [146].

and therefore it is essential to analyze reflections from several non-parallel crystallite planes.   Only in this way is it possible to eliminate the intensity contributions of over-lapping reflections to those intensity distributions of interest.   As a minimum, we prefer inspection of the combination (100), (010), and ($\bar{1}$05) for PET films, since under most conditions one or more of these three principal planes permit the sorting out of the information required.   Combinations of the distribution curves (x-ray intensity versus sample angle of rotation) yield an orientation surface.   Models can then be assembled from the distribution curves which illustrate visually in three dimensions the sample orientation types discussed earlier.   Regardless of how the orientation distribution is measured, it is necessary to describe the orientation surface quantitatively.   This can be done most rigorously by the inverse pole figure analysis of Krigbaum and Roe [144, 145], but in most cases approximation methods are used to describe the crystallite orientation distributions.

Perfect orientation is never achieved and the distributions are not always smooth or well behaved.   Sometimes the distribution curves e.g., c-axis distributions from MD to TD) contain undulations indicative of a nonuniform stress pattern developed in the film during the stretching process.   The stress patterns across the width and along the length of the film sheet depend on the equipment geometry and other process parameters.   Consequently, appropriate statistical sampling techniques are required.

b. Crystallite Length.   In PET the broadening of the ($\bar{1}$05) diffraction peak, observed at $2\Theta = 42.8°$, is used to measure the crystallite length. The ($\bar{1}05$) plane makes an angle of $88.0°$ with the c-axis and can, for the purposes of crystallite length measurement, be considered normal to it. Crystallite lengths of 30 to 80 Å have been reported and shown to change systematically with the stretching parameters [148].

3. Infrared Analysis—Structural Isomerism

The rotational conformation of the ethylene glycol portion of the PET repeat unit has been studied extensively. The conformation in the crystalline regions is *trans*, whereas the amorphous regions consist of a mixture of *trans* and *gauche* conformations (Fig. 27). Infrared studies have also provided information about the orientation of functional groups in PET films and are a valuable adjunct in helping to define structurally what occurs in the amorphous regions. Schmidt [153] has defined parameters characteristic of the molecular axial and uniplanar structure which correlate well with the x-ray orientation distributions.

No one instrumental method of film analysis provides all of the information needed. Interpretation of what occurs as a result of the stretching or rolling processes begins to become evident only by combining the results of several measurement techniques that see the film structure differently. Several major problems exist concerning the interpretation of results from any of these methods. The most troublesome is the precise meaning of the terms crystalline and amorphous in an oriented, highly stressed anisotropic structure.

B. Crystalline/Amorphous Ratio

Farrow and Ward [174–175] have concluded that crystallinity measurements by infrared, nuclear magnetic resonance, density, and x-ray diffraction on oriented fibers cannot be correlated. They indicate the cause to be a nonconstant amorphous density and that each of the experimental methods views the structure in a different way. Statton [176] agrees on the basis of x-ray and density studies of PET fibers and Bosley [177] has reviewed this problem from the standpoint of crystallite size and perfection. Several investigators have described measurement procedures which use x-rays to determine a relative crystallinity. We would caution that whatever method is used to define crystalline content

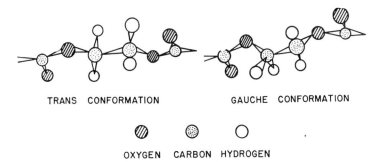

TRANS CONFORMATION   GAUCHE CONFORMATION

OXYGEN   CARBON   HYDROGEN

Fig. 27 *Trans* and *gauche* conformations of the glycol sections in polyethylene terephthalate.

in films it is essential, for clarity, to describe exactly how the measurements were made.

One way of defining the so-called crystalline content of PET is by use of the fractional change in density relative to the density difference between the PET crystal and amorphous polymer. This is accomplished in a density gradient tube [178] of carbon tetrachloride/*n*-heptane at 30°C.

C. Stress-Strain Curves as Indicators of Structural Change

Figure 28 illustrates the influence of film orientation and test direction on the shape of the room-temperature stress-strain curves measured on an Instron or similar device. A tensile test can be considered as a reorientation process. As a result the influence of the initial film structure to resist further deformation in a given direction is observable from the shape and magnitudes of the stress-strain curves (e.g., curves 1 and 2). Specifying the test direction is important. Rotation of the test direction by 90° (e.g., of the samples that generated curves 1 and 2) yields curves 1a and 2a which denote a paucity of orientation in that direction. Curve 1a has a shape much like that of an amorphous film (curve 4). At directions between the extremes of the maximum and minimum orientation directions the curve shapes follow an orderly progression. Consequently for films with unbalanced orientation, the tensile modulus, strength, elongation, and general shape provide a valuable insight into the average anisotropy of the structure.

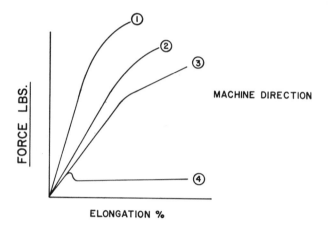

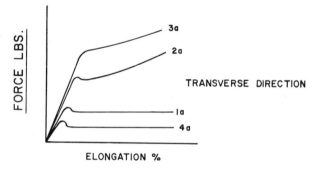

Fig. 28 Force-elongation curves of polyethylene terephthalate films having various levels of orientation and crystallinity: 1, uniaxially stretched only; 2 tensilized film; 3, biaxially stretched; 4, "amorphous" unoriented.

VIII. Summary

Semicrystalline aromatic polyester films offer unique and valuable combinations of physical, optical, and electrical properties. They are high melting, inert to many chemicals, and may be fabricated into films by means of melt extrusion or solvent casting. The most commercially useful of these materials are the polyethylene terephthalate films which are available in a wide range of thicknesses. The breadth of properties obtained from this polymer results primarily from the changes in physical structure produced by orientation and thermal processes. Great latitude is possible in the amount, direction, and permutations of the orientation and thermal treatments. Consequently a variety of film products and properties results.

Valuable additional properties are added to those of the basic polyester film by using auxiliary treatments such as coating, sizing, and weatherproofing.

References

1. W. H. Carothers, *J. Am. Chem. Soc.*, **51**, 2548 (1929).
2. V. V. Korshak and S. V. Vinogradova, *Polyesters*, Pergamon, New York, 1965.
3. R. E. Wilfong, *J. Polymer Sci.*, **54**, 385 (1961).
4. J. W. S. Hearle and R. H. Peters, *Fibre Structure*, The Textile Institute, Butterworths, London, 1963.
5. I. Goodman and J. A. Rhys, *Polyesters*, **Vol. I,** American Elsevier, New York, 1965.
6. P. J. Flory, *Principles of Polymer Chemistry*, Cornell University Press, Ithaca, N.Y., 1953.
7. C. Ellis, *The Chemistry of Synthetic Resins*, Reinhold, New York, 1935.
8. C. S. Marvel and E. C. Horning, *Organic Chemistry, An Advanced Treatise*, 2nd Ed., Wiley, New York, 1943.
9. W. H. Carothers and J. W. Hill, *J. Am. Chem. Soc.*, **54**, 1579 (1932).
10. H. Mark and G. S. Whitby, Eds., *Collected Papers of Wallace Hume Carothers on High Polymeric Substances*, Interscience, New York, 1940.
11. J. R. Whinfield and J. T. Dickson, Imperial Chemical Industries, British Patent 578,079, J. R. Whinfield and J. T. Dickson, E. I. du Pont de Nemours and Co., U.S. Patent 2,465,319 (Mar. 22, 1949).
12. E. F. Izard, "Scientific Success Story of Polyethylene Terephthalate," *Chem. Eng. News*, **32**, 3724 (1954).
13a. *Oil, Paint, Drug Reptr.*, **189**, 38, 57 (June 13, 1966).
13b. *Chem. Eng. News*, **47**, 18 (Jan. 27, 1969).
14. *Mod. Plastics*, **44**, No. 4, p. 110 (Dec. 1966).
15. *Chem. Eng.*, **45**, 86–88 (Jan. 16, 1967).
16. L. E. Amborski, *J. Polymer Sci.*, **62**, 331 (1962).
17. L. F. Fowler, *Proc. Roy. Soc. (London)*, **236**, 464 (1956).
18. F. S. Smith and C. Scott, *Brit. J. Appl. Phys.*, **17**, 1149 (1966).
19. W. Reddish, *Trans. Faraday Soc.*, **46**, 459 (1950).
20. W. Reddish, *Pure and Appl. Chem.*, **5**, 723 (1962).
21. C. S. Marvel, *An Introduction to the Organic Chemistry of High Polymers*, Wiley, New York, 1959.

22. W. R. Sorenson and T. W. Campbell, *Preparative Methods of Polymer Chemistry*, Interscience, New York, 1961.
23. P. Elwood, *Chem. Eng.*, **24**, 98 (1967).
24. E. F. Izard, E. I. du Pont de Nemours and Co., U.S. Patent 2,534,028 (Dec. 12, 1950).
25. J. S. Perlowski, R. D. Coffee, and R. B. Edwards, British Patent 1,041,853 (Sept. 17, 1966).
26. Chul-Yung Cha, *Polymer Preprints*, Am. Chem. Soc. Div. of Polymer Chem., **6**, No. 1, 84, Detroit Meeting, April 1965.
27. E. G. Edwards, Salford, and R. J. W. Reynolds, Imperial Chemical Industries, U.S. Patent 2,503,251 (Apr. 11, 1950).
28. F. D. Hartley, F. W. Lord, and L. B. Morgan, *Phil. Trans., Royal Soc. London*, **247**, 23 (1954).
29. B. V. Petukhov, *The Technology of Polyester Fibers*, Chapter IV, Macmillan, New York, 1963.
30. J. L. Vodonik, E. I. du Pont de Nemours and Co., U.S. Patent 2,829,153 (Apr. 1, 1958).
31. J. L. Vodonik, E. I. du Pont de Nemours and Co., U.S. Patent 2,681,360 (June 15, 1954).
32. T. Alfrey, Jr., J. J. Bohrer, and H. Mark, *Copolymerization*, (*High Polymers*, Vol. 8), Interscience, New York, 1952.
33. C. W. Smith and M. Dole, *J. Polymer Sci.*, **20**, 37 (1956).
34. L. Mandelkern, *Crystallization of Polymers*, McGraw-Hill, New York, 1964.
35. J. L. Vodonik, E. I. du Pont de Nemours and Co., U.S. Patent 2,727,882 (Dec. 20, 1955).
36. E. I. du Pont de Nemours and Co., unpublished.
37. J. L. Vodonik, E. I. du Pont de Nemours and Co., U.S. Patent 2,758,915 (Aug. 14, 1956).
38. W. Hippe et al., Glanzstoffe A.G., U.S. Patent 2,973,341 (Feb. 28, 1961).
39. H. Scheller, *Chemiefasern*, **15**, 923 (1965).
40. S. D. Ross, E. R. Cobrun, W. A. Leach, and M. B. Robinson, *J. Polymer Sci.*, **13**, 406 (1954).
41. J. Goodman and B. F. Nesbitt, *Polymer*, **1**, 384 (1960).
42. J. Goodman and B. F. Nesbitt, *J. Polymer Sci.*, **48**, 423 (1960).
43. W. Stockmayer and H. Jacobsen, *J. Chem. Phys.*, **18**, 1600 (1950).
44. H. Zahn, P. Rathgeber, E. Rexroth, and R. Krzikalla, *Angew. Chemie*, **68**, 229 (1956).
45. B. Seidel, *Z. Phys. Chem.*, **19**, 254 (1959).
46. P. H. Hermans, *Rec. Trav. Chim.*, **72**, 798 (1953).
47. H. A. Pohl, *J. Am. Chem. Soc.*, **73**, 5660 (1951).
48. I. Marshall and A. Todd, *Trans. Faraday Soc.*, **49**, 67 (1953).
49. R. J. P. Allen and P. D. Ritchie, *Chem. Ind. (London)*, **1953**, 747.
50. A. Conix, *Makromol. Chem.*, **25**, 226 (1957).
51. B. Samways, British Cellophane Ltd., British Patent 744,977 (Feb. 15, 1956).
52. K. G. Gerber, Imperial Chemical Industries, Ltd., British Patent 787,479 (Dec. 11, 1957).

53. J. Browning and G. Lindsay, Imperial Chemical Industries, British Patent 732,894 (June 29, 1955).
54. W. Seifried and L. Klenk, Kalle A.G., German Patent 1,168,059 (Apr. 6, 1964).
55. E. Ferraris, *Materie Plastiche*, **24,** 775 (1960).
56. K. L. Knox, E. I. du Pont de Nemours and Co., U.S. Patent 2,686,931 (Aug. 24, 1954).
57. E. I. du Pont de Nemours and Co., British Patent 947,677 (Jan. 29, 1964).
58. M. Miller and D. McGregor, E. I. du Pont de Nemours and Co., U.S. Patent 2,920,352 (Jan. 12, 1960).
59. W. A. Chren and C. H. Hofrichter, E. I. du Pont de Nemours and Co., U.S. Patent 2,736,066 (Feb. 28, 1956).
60. J. E. Owens, E. I. du Pont de Nemours and Co., U.S. Patent 3,068,528 (Dec. 18, 1962).
61. W. Vieth and J. E. Owens, E. I. du Pont de Nemours and Co., U.S. Patent 3,223,757 (Dec. 14, 1965).
62. H. O. Corbett, National Distillers and Chemical Co., U.S. Patent 3,090,076 (May 21, 1963).
63. F. P. Alles, E. I. du Pont de Nemours and Co., U.S. Patent 2,779,684 (Jan. 29, 1957).
64. Minnesota Mining and Manufacturing Co., British Patent 922,481 (Apr. 3, 1963).
65. C. L. Long, E. I. du Pont de Nemours and Co., U.S. Patent 2,968,067 (Jan. 17, 1961).
66. E. I. du Pont de Nemours and Co., British Patent 850,993 (Oct. 12, 1960).
67. J. W. Cornforth and C. H. Crooks, Imperial Chemical Industries, U.S. Patent 3,107,139 (Oct. 15, 1963).
68. Minnesota Mining and Manufacturing Co., British Patent 922,091 (Mar. 27 1963).
69. W. H. Ryan, Polaroid Corp., U.S. Patent 2,854,697 (Oct. 7, 1968).
70. Agfa-Wolfen A.G., Belgian Patent 578,264 (May 15, 1959).
71. P. Hold, Farrel-Birmingham Co., U.S. Patent 2,875,985 (Mar. 3, 1959).
72. J. C. Nash, Marshall and Williams Corp., U.S. Patent 3,023,479 (Mar. 6, 1962).
73. K. L. Knox, E. I. du Pont de Nemours and Co., U.S. Patent 2,718,666 (Sept. 27, 1955).
74. F. P. Alles and K. A. Heilman, E. I. du Pont de Nemours and Co., U.S. Patent 2,627,088 (Feb. 3, 1953).
75. F. P. Alles and W. R. Saner, E. I. du Pont de Nemours and Co., U.S. Patent 2,627,088 (Feb. 3, 1953).
76. A. C. Scarlett, E. I. du Pont de Nemours and Co., U.S. Patent 2,823,421 (Feb. 18, 1958).
77. C. J. Heffelfinger, E. I. du Pont de Nemours and Co., U.S. Patent 3,256,379 (June 14, 1966).
78. C. J. Heffelfinger, E. I. du Pont de Nemours and Co., U.S. Patent 3,257,489 (June 21, 1966).
79. R. L. Richards, E. I. du Pont de Nemours and Co., U.S. Patent 2,928,132 (Mar. 15, 1960).

80. F. R. Winter, E. I. du Pont de Nemours and Co., U.S. Patent 2,995,779 (Aug. 15, 1961).
81. F. P. Alles, E. I. du Pont de Nemours and Co., U.S. Patent 2,884,663 (May 5, 1959).
82. F. P. Alles, E. I. du Pont de Nemours and Co., U.S. Patent 3,165,499 (Jan. 12, 1965).
83. E. I. du Pont de Nemours and Co., British Patent 861,636 (Feb. 22, 1961).
84. B. D. Stead, Imperial Chemical Industries, British Patent 930,962 (July 10, 1963).
85. H. P. Koppehele, American Viscose Corp., U.S. Patent 3,055,048 (Sept. 25, 1962).
86. D. F. Miller, E. I. du Pont de Nemours and Co., U.S. Patent 2,755,533 (July 24, 1956).
87. F. P. Alles, E. I. du Pont de Nemours and Co., U.S. Patent 2,767,435 (Oct. 23, 1956).
88. R. H. B. Buteux, Imperial Chemical Industries, British Patent 887,342 (Jan. 17, 1962).
89. C. A. Van Dierendonck, Gevaert Photo Producten N.V., Belgium Patent 594,085 (Aug. 14, 1959).
90. C. A. Van Dierendonck, Gevaert Photo Producten N.V., Belgium Patent 594,086 (Aug. 14, 1959).
91. W. R. Tooke and E. G. Lodge, American Viscose Corp., U.S. Patent 2,923,966 (Feb. 9, 1960).
92. H. Kurzke and H. Sattler, Farbwerke Hoechst A.G., U.S. Patent 3,030,173 (Apr. 17, 1962).
93. E. I. du Pont de Nemours and Co., British Patent 807,258 (Jan. 14, 1959).
94. A. F. O'Hanlon and J. W. Cornforth, Imperial Chemical Industries, U.S. Patent 2,728,951 (Jan. 3, 1956).
95. C. H. Crooks, Imperial Chemical Industries, U.S. Patent 2,728,944 (Jan. 3, 1956).
96. C. E. Slaughter, Extruded Plastics, U.S. Patent 2,451,986 (Oct. 19, 1948).
97. International Development Co., British Patent 914,100 (Dec. 28, 1962).
98. J. C. Swallow and D. K. Baird, Imperial Chemical Industries, U.S. Patent 2,497,376 (Feb. 14, 1950).
99. J. Pitat, C. Holcik, and M. Bacok, British Patent 822,834 (Nov. 4, 1959).
100. J. R. Whinfield and J. T. Dickson, E. I. du Pont de Nemours & Co., U.S. Patent 2,465,319 (Sept. 24, 1945).
101. E. I. du Pont de Nemours and Co., British Patent 870,639 (June 14, 1961).
102. R. F. Chambret, Societe Rhodiaceta, U.S. Patent 3,120,561 (Feb. 4, 1964).
103. E. Heisenberg, E. Siggel, and R. Lotz, Vereinigte Glanzstoffe-Fabriken A.G., U.S. Patent 3,037,050 (May 29, 1962).
104. E. Katzschmann, Chemische Werke Witten G.m.b.H., U.S. Patent 3,008,980 (Nov. 14, 1961).
105. E. Siggel, L. Riehl, R. Lotz, and G. Wick, Vereinigte Glanzstoffe-Fabriken A.G., U.S. Patent 3,148,208 (Sept. 8, 1964).
106. Kodak Ltd., British Patent 905,562 (Sept. 12, 1962).

107. D. L. Johnson, Minnesota Mining and Manufacturing Co., U.S. Patent 3,154,461 (Oct. 29, 1964).
108. W. D. Bills, E. I. du Pont de Nemours and Co., U.S. Patent 3,201,506 (Aug. 17, 1965).
109. V. M. Grabovez, E. I. du Pont de Nemours and Co., U.S. Patent 3,271,229 (Sept. 6, 1966).
110. P. H. Pelley and L. A. Paul, Wheelabrator Corp., U.S. Patent 3,067,021 (Dec. 4, 1962).
111. J. Browning, Imperial Chemical Industries, U.S. Patent 2,678,285 (May 11, 1954).
112. A. F. Chapman, E. I. du Pont de Nemours and Co., U.S. Patent 2,968,538 (Jan. 17, 1961).
113. R. H. Michel, E. I. du Pont de Nemours and Co., U.S. Patent 2,762,720 (Sept. 11, 1956).
114. N. G. Gaylord, E. I. du Pont de Nemours and Co., U.S. Patent 2,805,963 (Sept. 10, 1957).
115. D. I. Sapper, E. I. du Pont de Nemours and Co., U.S. Patent 2,898,237 (Aug. 4, 1959).
116. D. M. Brewis, A. C. Eagles, and N. R. Harworth, *J. Mater. Sci.*, **2**, 435 (1967).
117. Minnesota Mining and Manufacturing Co., British Patent 891,469 (Mar. 14, 1962).
118. L. E. Wolinski, E. I. du Pont de Nemours & Co., U.S. Patent 3,274,089 (Sept. 20, 1966).
119. R. O. Osborn, E. I. du Pont de Nemours and Co., U.S. Patent 2,829,070 (Apr. 1, 1958).
120. W. E. McIntyre, E. I. du Pont de Nemours and Co., U.S. Patent 2,824,025 (Feb. 18, 1958).
121. R. R. Charbonneau and J. F. Abere, Minnesota Mining and Manufacturing Co., U.S. Patent 3,188,265 (June 8, 1965).
122. A. F. Chapman, E. I. du Pont de Nemours and Co., U.S. Patent 2,824,024 (Feb. 18, 1958).
123. J. J. Stewart, E. I. du Pont de Nemours and Co., U.S. Patent 2,829,068 (Apr. 1, 1959).
124. N. M. Haynes, *Elements of Magnetic Tape Recording*, Prentice Hall, Englewood Cliffs, N.J., 1957.
125. S. J. Begun, *Magnetic Recording*, Rinehart, New York, 1943.
126. L. E. Amborski, E. I. du Pont de Nemours and Co., U.S. Patent 3,043,709 (July 10, 1962).
127. R. H. Doremus, B. W. Roberts, and D. Turnbull, *Growth and Perfection of Crystals*, Wiley, New York, 1958, Chapter 6.
128. A. J. Keller, *J. Polymer Sci.*, **17**, 351 (1955).
129. J. D. Ferry, *Viscoelastic Properties of Polymers*, Wiley, New York, Chapter 11, 1961.
130. A. B. Thompson, *J. Polymer Sci.*, **34**, 741 (1959).
131. V. A. Kargin, *J. Polymer Sci.*, **30**, 247 (1958).
132. G. L. Berestneva, V. A. Berestnev, T. V. Gatovskaya, V. A. Kargin, P. V. Kozlov, *Vysokomolekul. Soedin.*, **3**, 801 (1961).

133. G. S. Y. Yeh, Ph.D. Thesis, Case Institute of Technology, Cleveland, November 1966.
134. W. O. Statton, Am. Chem. Soc. Symposium on the "Meaning of Crystallinity in Polymers," Phoenix, Ariz., 1966.
135. W. Ruland, *Faserforsch Textiltech.*, **15**, 533 (1964).
136. C. J. Heffelfinger and R. L. Burton, *J. Polymer Sci.*, **47**, 289 (1960).
137. P. H. Geil, *Polymer Single Crystals*, Wiley-Interscience, New York, 1963.
138. P. H. Hermans, *Contributions to the Physics of Cellulose Fibers*, Elsevier, New York, 1946; *Physics and Chemistry of Cellulose Fibers*, Elsevier, New York, 1949.
139. O. Kratky, *Kolloid–Z.*, **64**, 213 (1933).
140. Z. W. Wilchinsky, *J. Appl. Physics*, **30**, 792 (1959).
141. R. A. Sack, *J. Polymer Sci.*, **54**, 543 (1961).
142. Z. W. Wilchinsky, *J. Appl. Physics*, **31**, 1969 (1960).
143. Z. W. Wilchinsky, *J. Polymer Sci.*, **7**, 923 (1963).
144. W. R. Krigbaum and R. J. Roe, *J. Chem. Phys.*, **40**, 2608 (1964).
145. W. R. Krigbaum and R. J. Roe, *J. Chem. Phys.*, **41**, 737 (1964).
146. R. de P. Daubeny, C. W. Bunn, and C. J. Brown, *Proc. Roy. Soc. (London)*, **A226**, 531 (1954).
147. W. J. Dulmage and A. L. Geddes, *J. Polymer Sci.*, **31**, 499 (1958).
148. C. J. Heffelfinger and P. G. Schmidt, *J. Appl. Polymer Sci.*, **9**, 2261 (1965).
149. W. O. Statton and G. M. Godard, *J. Appl. Phys.*, **28**, 1111 (1957).
150. W. P. Baker, *J. Polymer Sci.*, **57**, 993 (1962).
151. M. L. Wallach, *J. Polymer Sci.*, C, No. 13, 69 (1966).
152. Y. I. Balta, Ph.D. Thesis, Department of Chemistry, Duke University, Durham, N.C., 1966.
153. P. G. Schmidt, *J. Polymer Sci.*, **141**, 1271 (1963).
154. K. Z. Gumargalieva, E. M. Bebavtseva, M. R. Kiselev, E. I. Evko, B. M. Lukyanovich, *Vysokomolekul. Soedin.*, **8**, 1741 (1966).
155. N. Veda, S. Nishiumi, *Chem. High Polymers (Tokyo)*, **21**, 166, 337 (1964).
156. R. Bonart, *Kolloid-Z.*, **199**, 136 (1964).
157. G. Farrow, J. McIntosh, and I. M. Ward, *Makromol. Chem.*, **38**, 147 (1960).
158. H. Tadohorn, K. Tatsuka, and S. Murahashi, *J. Polymer Sci.*, **59**, 413 (1962).
159. K. H. Illers and H. Breuer, *J. Colloid Sci.*, **18**, 1 (1963).
160. J. Johnson, *J. Appl. Polymer Sci.*, **2**, 205 (1959).
161. D. Grime and I. M. Ward, *Trans. Faraday Soc.*, **54**, 959 (1958).
162. D. Patterson and I. M. Ward, *Trans. Faraday Soc.*, **53**, 291 (1957).
163. K. G. Mayhan, W. J. James, and W. Bosch, *J. Appl. Polymer Sci.*, **9**, 3605 (1965).
164. C. Y. Liang and S. Krimm, *J. Chem. Phys.*, **27**, 327 (1957).
165. P. R. Blakey and R. O. Sheldon, *J. Polymer Sci.*, A–2, 1043 (1964).
166. L. G. Kazaryan and Ya. G. Urman, *Zh. Strukt. Khim.*, **5**, 543 (1964).
167. Y. Mitsuishi and M. Ikeda, *J. Polymer Sci.*, A–2, **4**, 283 (1966).
168. L. G. Kazaryan and D. Ya. Tsvankin, *Vysokomolekul. Soedin.*, **7**, 80 (1965).
169. P. V. Kozlov and G. L. Berestneva, *Vysokomolekul. Soedin.*, **2**, 590 (1960).
170. L. Z. Rogovina and G. L. Slonimskii, *Vysokomolekul. Soedin.*, **8**, 219 (1966).
171. J. H. Dumbleton and B. B. Bowles, *J. Polymer Sci.*, A–2, **4**, 951 (1966).
172. G. Farrow and J. Bagley, *Textile Res. J.*, **32**, 587 (1962).

173. I. Kuriyama, *J. Soc. Fiber Sci. Technol., Japan*, **20**, 431 (1964).
174. G. Farrow and I. M. Ward, *Brit. J. Appl. Phys.*, **11**, 543 (1960).
175. G. Farrow and I. M. Ward, *Polymer*, **1**, 330 (1960).
176. W. O. Statton, *J. Appl. Polymer Sci.*, **1**, 803 (1963).
177. D. E. Bosley, *J. Appl. Polymer Sci.*, **8**, 1521 (1964).
178. G. Oster and M. Yamamoto, *Chem. Rev.*, **63**, 257 (1963).

CHAPTER 15

POLYVINYL FLUORIDE

ORVILLE J. SWEETING*

Yale University, New Haven, Connecticut

I. Manufacture of Vinyl Fluoride

Vinyl fluoride was known for a long time after its discovery in 1901 [1] as a curious analog of the chloride but of not much use, until chemists found commercial methods for synthesis and polymerization. The latter is considerably more difficult to accomplish than the polymerization of vinyl chloride.

Two methods of making the monomer are used commercially: (a) a two-step process in which acetylene is converted quantitatively to 1,1-

* Formerly Associate Director of Research and Development, Olin Film Division, Olin Corp., New Haven, Connecticut. Present address: Quinnipiac College, New Haven 06518.

difluoroethane which is then pyrolyzed [2, 3]; (b) a direct one-step catalytic synthesis from acetylene and hydrogen fluoride [4–6].

(a) $\qquad$ $CH{\equiv}CH + 2HF \longrightarrow CH_3{-}CHF_2$

$\qquad\qquad CH_3{-}CHF_2 \longrightarrow CH_2{=}CHF + HF$

(b) $\qquad$ $CH{\equiv}CH + HF \xrightarrow{\text{Hg}} CH_2{=}CHF$

The monomer is a colorless gas at the usual ambient conditions with a normal boiling point of $-72.2°C$. The critical temperature and pressure are $54.7°C$ and 760 psi (53.44 kg/cm^2), respectively. The liquid density is 0.7753 g/ml at $-30°C$ and 0.6808 at $10°C$. Vinyl fluoride is slightly soluble in water: 0.94 g/100 g at $80°C$ and 500 psi.

For successful polymerization the monomer must be carefully purified. Distillation separates 1,1-difluoroethane, hydrogen fluoride, and small amounts of other contaminants. The final traces of hydrogen fluoride are removed by passage through soda-lime towers and traces of acetylene by treatment with diamminocopper(I) chloride solution. A final fractionation at -50 to $-25°C$ and 40 to 100 psi removes traces of oxygen. The monomer for polymerization normally contains less than 5 ppm acetylene, less than 20 ppm oxygen, and no other detectable impurities [7].

II. Polymerization of Vinyl Fluoride

A. Polymerization

Vinyl fluoride does not polymerize readily. The first report of successful work was made by Starkweather in 1934, who noted "some polymerization" in a toluene solution saturated with monomer and held at $-35°C$ and 6000 atm at $67°C$ for 16 hr [8]. Claims of polymerization are also contained in a British patent issued in 1937 [9] which emphasized the difficulty involved.

Vinyl fluoride polymerizes in a typical free-radical process [4, 10–13]. The most thorough investigation yet made reported that the most useful polymers from the standpoint of properties and processability were those from thermally generated free-radicals [7].

Commercial polymerizations are usually done in stainless-steel autoclaves, but aqueous dispersions are preferred for some applications, and as with vinyl chloride or vinylidene chloride polymerizations

(Chapters 5 and 6) the latter method affords comparatively simple continuous operation and better control of polymer properties.

Table 1 [7] indicates the range of initiators that can be used and summarizes conditions and conversions.

TABLE 1

Initiators for Vinyl Fluoride Polymerization (35–100°C)

Initiator	Temperature (°C)	Pressure (atm)	Extent of conversion	Conversion time (hr)
α,α'-Azobis-α,γ,γ-trimethylvaleronitrile	37	1000	High	4
α,α'-Azobis-α,γ-dimethylvaleronitrile	50	500	Moderate	8
α,α'-Azobisisobutyronitrile	70	70	90%	19
α,α'-Azobisisobutyramidine hydrochloride	70	100	40%	4
Benzoyl peroxide	80	250	37%	13
Ammonium persulfate	85	250	33%	2
α,α'-Azodicyclohexane carbonitrile	115	175	32%	1.5
Diethyl peroxide	122	600	Good	14.5
tert-Butyl peroxide	140	500	Good	0.5
2(2'Hydroxyethylazo)-2-ethylbutyronitrile	145	250	Good	

B. Polymer Properties

Polyvinyl fluoride is a chemically inert substance, which melts at 198 to 200°C and has a density of 1.39 g/cm².

1. Molecular Weight

Conditions of polymerization have an enormous effect on molecular weights of the polymer, as expected. The most significant are type and amount of initiator, temperature, pressure, and impurities [7].

The type of initiator may have an inordinate effect on thermal stability and wettability, associated with the solubility of the initiator and the nature of the unit incorporated into the polymer. In general molecular weights are decreased by increased initiator concentrations.

The effect of temperature is a complex result of the well-known effect of temperature on rate of free-radical formation, of the more rapid propagation rate at higher temperatures, and also of the more rapid termination and increased chain branching at higher temperatures.

Over-all, increasing the temperature of polymerization reduces the molecular weight. Typical results are shown in Table 2 [7].

TABLE 2

Molecular Weight of Polyvinyl Fluoride[a]

Temperature (°C)	Intrinsic viscosity[b] $[(g/dl)^{-1}]$	Molecular weight[c]
85	4.40	180,000
95	3.50	169,000
110	1.47	58,000

[a] Initiator: 0.2% benzoyl peroxide; pressure, 250 atm.
[b] Measured in dimethylformamide at 110°C.
[c] Calculated by assuming each molecule to contain one initiator fragment (determined by counting carbon–14 tagged initiator).

Higher pressures resulted in increased polymerization rates and higher molecular weights.

The polymerization is sensitive to the presence of acetylene in the monomer, almost unaffected by oxygen (from air), and indifferent to the presence of 1,1-difluoroethane. Acetylene in 1000 ppm greatly accelerated the reaction and resulted in quantitative conversion to a crosslinked insoluble material; 2% acetylene stopped polymerization almost completely. Oxygen at 135 ppm increased the rate of polymerization and 500 ppm decreased the rate, both slightly.

Methanol, 2-propanol, and 2,3-dioxalane were found to be effective chain stoppers.

2. Solubility

Polyvinyl fluoride of molecular weight > 60,000 is insoluble in solvents below 110°C. Above 110°C dimethylformamide, tetramethylurea, and tetramethylene sulfone dissolve the polymer; the first two can be used for solvent-casting of films.

Plasticization is difficult. Dibutyl phthalate and tricresyl phosphate, however, are retained by the polymer at room temperature to the extent of 5 to 10% and can be used as aids in extrusion without producing rubbery films.

Further data appear below in a discussion of film properties.

III. Polyvinyl Fluoride Film

A. FILM FORMATION

Films may be made from the polymer by extrusion, but only with great difficulty because of decomposition at the temperatures required (225 to 250°C) and high pressures, though some heavily plasticized samples may be handled in this way.

Solvent casting of 8 % solutions in dimethylformamide at 125 to 130°C has been used to prepare film in gauges of 0.5 to 4 mils. The films are clear (especially in the quenched thin gauges) and momentary heating to 250°C and quenching improve clarity.

B. FILM PROPERTIES

High-molecular-weight films have high tensile strength, resistance to abrasion, good toughness, and excellent clarity. The high dielectric strength and dielectric constant are outstanding among transparent self-supporting films. Films are inert toward most solvents, up to 100°C at least, and film properties are not affected after 60-hr exposure to steam at 85 psi (325°F).

Properties are little affected by ultraviolet radiation during years of exposure [14, 15], a dominant factor in determining markets.

Chemical inertness and resistance to physical abrasion have led in recent years to important applications of polyvinyl fluoride films containing ultraviolet absorbers as laminates to protect photodegradable plastics, particularly those such as polystyrene and polymethyl methacrylate which are widely used as fluorescent-lamp diffusers [16].

1. Physical and Thermal Properties

The general physical and thermal properties of polyvinyl fluoride films are shown in Table 3 [17].

2. Electrical Properties

Polyvinyl fluoride films possess outstandingly high dielectric constant and dielectric strength, as shown in Table 4 [18]. The low rate of thermal aging and the high chemical resistance (cf 3 below) offer functional advantages over other materials in electrical insulation and cable fabrication.

TABLE 3

Properties of Tedlar[a] Polyvinyl Fluoride Film

Property	Transparent, 0.5 mil Type 20	Glossy white, 1.0 mil Type 30	Medium gloss white, 2.0 mil Type 30	Transparent, 2.0 mil Type 40	Low gloss white, 1.5 mil Type 30	Method
Area factor, in.²/lb	40,000	17,200	9100	10,000	11,100	Mullen ASTM D-774
Bursting strength, psi/mil	70	37	26	19	27	
Coefficient of friction, film-metal	0.16	0.15	0.24	0.16	0.30	
Density at 72°F, g/cc	1.38	1.53	1.57	1.38	1.55	ASTM D 1505 (Density gradient tube)
Impact strength, kg-cm/mil	5.3	4.6	4.8	2.7	3.1	Du Pont pneumatic tester
Water absorption, %	0.5	0.5	0.5	0.5	0.5	Water immersion
Water vapor transmission, g/100 m²-hr-mil-53 mm Hg	157	190	205	190	200	Du Pont test method
Water vapor transmission, perms/mil	1.5	1.5	1.5	1.5	1.5	ASTM E 96–58T modified
Refractive index, n_D 86°F (30°C)	1.46	1.46	1.46	1.46	1.46	ASTM D 542 (Abbé)
Tear strength (Elmendorf), g/mil	12	22	40	40	31	

Property						Test method
Tear strength initial (Graves), g/mil	450	542	512	620		ASTM 1004
Tensile modulus, psi	260,000	310,000	280,000	290,000	250,000	Instron ASTM D882, Method A–100%/min
Ultimate elongation, %	115	115	150	150–250	120	Instron ASTM D882, Method A–100%/min
Ultimate tensile strength, psi	18,000	16,000	11,000	7000	8500	Instron ASTM D882, Method A–100%/min
Ultimate yield strength, psi	6000	5700	4700	4900	4800	Instron ASTM D882, Method A–100%/min
Aging, hr to embrittlement	3000	3000	3000	3000	3000	Oven at 300°F
Flammability	Slow-burning to self-extinguishing	Slow-burning to self-extinguishing	Slow-burning to self-extinguishing	Slow-burning to self-extinguishing	Slow-burning to self-extinguishing	
Linear coefficient of expansion, in./in.-°F	2.8×10^{-5}	2.8×10^{-5}	2.8×10^{-5}	2.8×10^{-5}	2.8×10^{-5}	
Shrinkage, max, transversely, % at °F	4/266°	4/338°	4/338°	4/338°	4/338°	Oven, 30 min
Zero strength, °C	300	300	300	300	300	
Zero strength, °F	570	570	570	570	570	

[a] Registered Du Pont trademark.   Type 20 is highly oriented biaxially, and types 30 and 40 are much less oriented.

647

TABLE 4

Electrical Properties of Tedlar Polyvinyl Fluoride Film

	Transparent, 2.0 mil Type 40	Medium gloss white, 2.0 mil Type 30	Test method
Dielectric constant, 72°F, 1 kc	8.5	9.9	ASTM D 150
Dissipation factor, %			
1000 cycles, 23°C	1.6	1.4	ASTM D 150
1000 cycles, 70°C	2.7	1.7	
10 kc, 23°C	4.2	3.4	
10 kc, 70°C	2.1	1.6	
Volume resistivity			
Ω-cm, 23°C	4×10^{13}	7×10^{14}	ASTM D 257
Ω-cm, 100°C	2×10^{10}	1.5×10^{11}	
Dielectric strength, 60 cps, kV/mil	3.4	3.5	ASTM D 150
Corona endurance, hr at 60 cps,			
1000 V/mil	2.5	6.2	ASTM D 2275

3. Chemical Resistance

The film form and properties of polyvinyl fluoride are unaffected after many hours (sometimes weeks) of exposure to a range of common solvents, including hydrocarbons and chlorinated hydrocarbons, at boiling temperatures. Films of polyvinyl fluoride are impermeable to greases, fats, and oils.

As mentioned in II-B-2, a few ketones, dinitriles, N-substituted ureas and amides, and sulfones will dissolve polyvinyl fluoride at elevated temperatures.

A summary of the chemical resistance of these films is shown in Table 5 [19].

4. Optical Properties

Films can be made in a range of optical clarity, depending on the process. The clear films are transparent to both visible and ultraviolet radiation, and absorb strongly in the infrared [18, 20]. These properties make polyvinyl fluoride film useful in solar stills.

5. Permeability

Table 6 summarizes the permeability of polyvinyl fluoride to various gases [14, 21] (see also Chapter 1).

TABLE 5

Resistance to Attack by Chemical Agents[a]

Acetic acid (glacial)	1 year
Acetic acid (glacial)	31 days at 75°C
Acetic acid (4%)	168 hours at bp
Acetone	1 year at room temperature
Acetone	2 hours at bp
Ammonium hydroxide (12 and 39%)	1 year at room temperature
Ammonium hydroxide (10%)	31 days at 75°C
Benzene	1 year at room temperature
Benzene	2 hours at bp
Benzyl alcohol	31 days at 75°C
1,4-Dioxane	31 days at 75°C
Ethyl acetate	31 days at 75°C
Ethyl alcohol	31 days at 75°C
n-Heptane	1 year at room temperature
Hydrochloric acid (10%)	1 year at room temperature
Hydrochloric acid (10 and 30%)	31 days at 75°C
Hydrochloric acid (10%)	2 hours at bp
Hydrochloric acid (10%)	Suspension in vapors at 105°C for 1 week
Kerosene	1 year at room temperature
Methyl ethyl ketone	31 days at 75°C
Nitric acid (20%)	1 year at room temperature
Nitric acid (10 and 40%)	31 days at 75°C
Perchloric acid (60%)	25 days at room temperature
Phenol	1 year at room temperature
Phenol (5%)	31 days at 75°C
Phosphoric acid (20%)	1 year at room temperature
Sodium chloride (10%)	1 year at room temperature
Sodium hydroxide (10%)	1 year at room temperature
Sodium hydroxide (10 and 54%)	31 days at 75°C
Sodium hydroxide (10%)	2 hours at bp
Sodium sulfide (9%)	31 days at 75°C
Sulfuric acid (20%)	1 year at room temperature
Sulfuric acid (30%)	31 days at 75°C
Toluene	31 days at 75°C
Trichloroethylene	31 days at 75°C
Tricresyl phosphate	31 days at 75°C

[a] After exposure no change could be measured in tensile strength, impact strength, or elongation.

TABLE 6

Gas Permeability of Polyvinyl Fluoride Film
(1-mil film at 23.5°C)

Gas	Permeability[a] (ml/24 hr-100 in.^{2})	Method
Water (at 39.5°C)	2.8	Cf. Vol. I, p. 603ff
Oxygen	3.2	ASTM D 1434
Nitrogen	0.25	ASTM D 1434
Carbon dioxide	11.0	ASTM D 1434
Hydrogen	58.1	ASTM D 1434
Helium	150	ASTM D 1434
Ethanol	0.54	ASTM E 96
Benzene	1.4	ASTM E 96
Acetic acid	0.7	ASTM E 96
Ethyl acetate	15	ASTM E 96
Acetone	310	ASTM E 96
Hexane	0.85	ASTM E 96
Carbon tetrachloride	0.75	ASTM E 96

[a] See also Chapter 1.

6. Heat Sealability

The less oriented types of polyvinyl fluoride films (such as Du Pont's Tedlar Types 30 and 40) present no serious problems for sealing by heated jaws, bars, or rollers, but the more highly oriented products (e.g., Tedlar Type 20) will shrink to an objectionable degree unless suitable methods are used.

In general, impulse and dielectric sealers are recommended for Types 20 and 30, though some rotary band sealers may be used. Because of the low shrinkage, Type 40 can be sealed also with heated bars.

The heat-sealing temperature is 400 to 425°F; recommended jaw pressure is 20 psi. The dwell time at this temperature is approximately 1 to 2 sec, and a cooling period of 3 to 4 sec should be sufficient. These conditions vary slightly with different film thicknesses.

Typical seal strengths obtained with an impulse sealer are 16.0 lb/in. in shear for 2-mil film, 8.5 lb/in. for peel seals. The corresponding values for 1-mil film are 6.5 and 6.0 lb/in. respectively.

Dielectric heat sealing has been demonstrated at the common frequency of 27.18 Mc. For optimum performance, however, higher frequencies are recommended. Studies conducted with a commercial

40-Mc, kilowatt dielectric sealing unit resulted in excellent film-to-film bonds. Peel and shear strengths were approximately 80 and 50 % higher, respectively, than results obtained with impulse sealers.

The thermoplastic nature of the polymer will also give a weld seal under the proper conditions [22].

IV. Markets and Applications

The high cost of polyvinyl fluoride film ($3-5.50 has seriously limited its use as a large-volume film (cf. p. 203), but consumption is growing in special applications, notably outdoor exposure.

In Section III we mentioned its electrical, thermal, and corona stability and its good electrical properties.

An outstanding property of the unmodified polymer is its nonadherence to liquids or to solid surfaces, but special treatments can render one surface adhesive while the other is non-adhesive and retains all its resistant properties toward scuffing and other forms of wear as well as to chemical attack by solvents such as oils, waxes, greases, acids, alkalis, water, alcohols, and a host of other materials that stain and deteriorate wood and metallic surfaces. This capability has opened a very large laminating field for less resistant surfaces. At the present time thin polyvinyl fluoride film is used for the protection of house siding, prefinished trim, movable partitions, furniture, awnings, aluminum siding, architectural building panels, doors, stadium seating, and pre-finished metal sheet and coil stock in great variety.

The tendency of reinforced polyesters exposed to the outdoor elements to suffer yellowing, erosion, and fiber bloom can be prevented by the application of polyvinyl fluoride film to the surface [23]. The exceptional resistance to weathering will probably extend the application of both polyesters and polyvinyl fluoride into new fields.

References

1. J. Swarts, *Bull. Acad. Roy. Med. Belg.*, **7**, 383 (1901).
2. R. E. Burk, D. D. Coffman, and G. H. Kalb, U.S. Patent 2,425,991 (Aug. 19, 1947).
3. J. Harmon, U.S. Patent 2,599,631 (June 10, 1952).
4. D. D. Coffman and T. A. Ford, U.S. Patent 2,419,010 (Apr. 15, 1947).
5. J. Söll, U.S. Patent 2,118,901 (May 31, 1938).

6. L. F. Salisbury, U.S. Patent 2,519,199 (Aug. 15, 1950).
7. G. H. Kalb, D. D. Coffman, T. A. Ford, and F. L. Johnston, *J. Appl. Polymer Sci.*, **4**, 55 (1960).
8. H. W. Starkweather, *J. Am. Chem. Soc.*, **56**, 1870 (1934).
9. F. Schloffer and O. Scherer, British Patent 465,520 (May 3, 1937); German Patent 677,071 (June 17, 1939).
10. C. A. Thomas, U.S. Patent 2,362,960 (Nov. 14, 1944).
11. A. E. Newkirk, *J. Am. Chem. Soc.*, **68**, 2467 (1946).
12. P. J. Manno, *Polymer Preprints*, Am. Chem. Soc., Div. of Polymer Chemistry, **4** (1), 79 (1963); *Nucleonics*, **22** (6), 64 (1964); *Nucleonics*, **22** (9), 72 (1964).
13. F. J. Welch, U.S. Patent 3,112,298 (Nov. 26, 1963).
14. V. L. Simril and B. A. Curry, *Mod. Plastics*, **36**, No. 11, 121 (July, 1959).
15. V. L. Simril and B. A. Curry, *J. Appl. Polymer Sci.*, **4**, 62 (1960).
16. R. F. Davis, *Illuminating Engineering*, **62**, No. 2, 61 (1968).
17. E. I. du Pont de Nemours and Co., Bulletin TD–2, "Physical-Thermal Properties of *Tedlar* Film," Wilmington, Del., 1969.
18. E. I. du Pont de Nemours and Co., Bulletin TD–4A, "Electrical Properties of *Tedlar* Polyvinyl Fluoride Film," Wilmington, Del., 1969.
19. E. I. du Pont de Nemours and Co., Bulletin TD–3, "Chemical Properties of *Tedlar* Polyvinyl Fluoride Film," Wilmington, Del., 1969.
20. E. I. du Pont de Nemours and Co., Bulletin TD–5, "Optical Properties of *Tedlar* Polyvinyl Fluoride Film," Wilmington, Del., 1969.
21. L. E. Wolinski in Kirk-Othmer *Encyclopedia of Chemical Technology*, 2nd ed., Wiley-Interscience, 1966, Vol. 9, p. 839.
22. E. I. du Pont de Nemours and Co., Bulletin TD–14, "Heat Sealing of *Tedlar* Polyvinyl Fluoride Film," Wilmington, Del., 1969.
23. C. Gumerman and G. R. McKay, Soc. Plastics Industry, 18th Annual Meeting at Chicago, 1963, Sect. 11–D, p. 1.

CHAPTER 16

OTHER SELF-SUPPORTED POLYMER FILMS

ORVILLE J. SWEETING*

Yale University, New Haven, Connecticut

In this final chapter we present brief summaries of other polymers that are used as self-supported materials. These polymers, and films made from them, are newcomers to the film field and a full assessment of potential cannot be made now.

* Formerly Associate Director of Research and Development, Olin Film Division, Olin Corp., New Haven, Connecticut. Present address: Quinnipiac College, New Haven, 06518.

I. Ionomers

A. Nature of the Polymer

The term *ionomer* was first introduced in 1965 to designate a new concept in polymers and referred to the binding of carboxylated polymer chains by various cations, notably NH_4^+, Na^+, Zn^{+2}, Ca^{+2}, Mg^{+2}. Essentially the polymer consists of an olefin copolymer backbone with dependent carboxyl groups partially or completely interchained with cations [1].

$$\sim CH_2-CH-CH_2\sim CH_2-\overset{\displaystyle CO_2H}{\underset{\displaystyle CO_2^-}{\overset{\displaystyle |}{\underset{\displaystyle |}{CH}}}}-CH_2\sim$$

$$M^+, M^{+2}, \text{etc.}$$

$$\sim CH_2-\overset{\displaystyle CO_2H}{\overset{\displaystyle |}{CH}}-CH_2\sim CH_2-\overset{\displaystyle CO_2^-}{\overset{\displaystyle |}{CH}}-CH_2\sim$$

Copolymers include those of ethylene with acrylic acid or methacrylic acid, or ethylene with methacrylic acid and vinyl acetate (2 to 10 mole-% acid). The polymerized acid may be 3 to 15% by weight of the total polymer and 50 to 70% ionized. The strong ionic interchain forces, which can be varied by controlling the number and distribution of dependent carboxyls, the extent of ionization, and the type of cation, are exploited for the control of polymer properties. Low x-ray crystallinity results, yet is accompanied by exceptionally high strength.

Ionomer resins (such as Surlyn A, made by Du Pont) are in general extrudable by conventional techniques. In contrast with radiation- or peroxide-crosslinked olefin polymers, the ionomers are thermoplastic. The ionic bonds, however, result in association in the melt and consequently high viscosities of the melt may be encountered; but, by application of higher temperatures and higher shear rates, flow occurs in most cases and films are readily extruded by high-speed extrusion.

In a majority of cases the polymers behave typically like thermoplastic materials of high molecular weight. They may be processed over a wide range of temperatures from 130 to 375°C. Compositions containing diprotic acids in conjunction with M^{2+} or M^{3+} ions may resemble crosslinked polyolefins, but those containing monoprotic acids ionized by Na^+ are thermoplastic and extrudable. Thus a thermoplastic ionomer is obtained when a copolymer of ethylene and maleic acid is partly ionized with Na^+, but the same polymer ionized with Zn^{2+} to the same extent is intractable.

B. Ionomer Films

1. Extrusion

As already mentioned, commercial ionomers can be extruded by conventional techniques (cf. Volume I, Chapter 8). Extrusion temperatures range from approximately 150 to 370°C and pressures are not exceptionally high.

The dependence of melt viscosity of ionomers on temperature is greater than for copolymers of olefins, one of which carries α-carboxylic acid groups, and much greater than for conventional polyolefins. Approximate values for the activation energy of viscous flow are 7 to 10 kcal for polyethylenes, 14 kcal for copolymers of ethylene and methacrylic acid, and 17 to 20 kcal for ionomers that incorporate methacrylic acid. These data are consistent with the theory that hydrogen bonding in the α-olefin carboxylic acid copolymers and ionic bonding in the ionomers play a similar role in determining melt viscosity and that these interchain forces are sensitive to temperature changes [1].

Films may be handled on chill-roll or blown-tubing equipment.

2. Film Properties

The properties of several ionomer films produced by Du Pont are summarized in Table 1 [2]. (*Iolon* is the trademark used by the Du Pont Company for their new family of ionomer films.)

As the table shows, these films are characterized by several useful properties that make the ionomers suitable as tough self-supporting packaging films, Among them are excellent optical properties, puncture resistance, high dart-drop toughness, high tensile strength, broad heat-seal range, superior forming characteristics, good abrasion resistance, good low-temperature flex life, and exceptional oil, grease, and chemical resistance. In addition, the film has a long shelf-life and will not embrittle with age [3, 4].

3. Applications

The earliest use of ionomer films was in skin packaging of meat products, hardware, small machine parts, toys, and similar items [5, 6],

TABLE 1

Properties of Iolon[a] Ionomer Films

Property[c]	XQ–121 (321[b])	XQ–123 (323[b])	XQ–128 (328 [b])
Gauge, mils	1.0	1.0	1.0
Density, g/cc	0.940	0.940	0.960
Yield, in.^{2}/lb	29,400	29,400	29,400
Dart drop, g/mil, 60 cm	300	300	300
Gloss, 20°	45	110	110
Haze, %	4.0	1.5	1.5
Modulus, psi	28,000	40,000	60,000
Tensile strength, psi	4000	5000	4000
O_2 transmission, cc/mil-m^2-24 hr at 23°C	450–500	450–500	450–500
Water vapor transmission, g/24 hr-m^2 at 100°F, 90% rh	15	15	15
Heat-seal strength, g/in.	1500	1500	1500
Coefficient of friction:			
Film to film	1.0	0.8–1.0	1.0
Film to metal	0.3–0.4	0.3–0.5	0.3–0.4
Melt index[d]	4.0	1.2	0.7
Chemical resistance			
Acids	Resistant	Resistant	Resistant
Bases	Very resistant	Very resistant	Very resistant
Hydrocarbons	Slow swell	Slow swell	Slow swell
Ketones and alcohols	Resistant	Resistant	Resistant
Mineral oil	Very resistant	Very resistant	Very resistant
Thermal data			
Maximum processing temperature, °F	600	600	600
Vicat softening temperature, °F	170	160	145
Brittleness temperature, °F	− 200	− 200	− 200
Suggested use	Where exceptional heat-seal characteristics are desirable and improved adhesion in the presence of water	Where a balanced combination of properties is desirable; high transparency	Where higher stiffness is desirable

[a] Du Pont trademark.
[b] Designates the untreated film, not recommended for lamination or printing.
[c] Volume I, Chapter 13.
[d] Chapter 2, p. 169.

which exploited clarity, machinability, adhesion to paperboard, heat-sealability, long shelf-life, indifference to oxygen and light, and (with food products) resistance to fats and oils.   These are still suitable applications.

Iolon ionomer film meets all the requirements for a food contact material under regulations of the U.S. Food and Drug Administration [7].   It has also demonstrated its value in a laminate for the vacuum packaging of processed meats.   Many types of food package may be prepared with this film, either in laminate or unsupported form.

The ionomers, with their high draw rate, high melt elasticity, and high melt cohesion, perform admirably as a substrate coating.   Tendency to pinhole formation is slight under multiple flexing compared with a polyethylene hot-melt coating [3].

Laminates of ionomers and other substrates are readily made with corona-treated ionomer film (Chapter 2) or other suitable primary treatment.   The resulting materials do not separate readily, even under rough conditions of moisture and temperature.   Thus for products that are susceptible to moisture (e.g., unit medicinal tablets) ionomers combined with a polyvinylidene-chloride-coated polyester and fluorocarbon-based film can provide an almost indefinitely watertight package with good handling characteristics.

On a coverage basis ionomer films cost approximately 3 cents/1000 in.^{2}-mil; that is, in the same price range as cellulosics and vinyls (cf. p. 203).

In the future film applications are likely to include thin-gauge (2 to 5 mils) heavy duty overwraps as well as bags and box windows.

References

1. R. W. Rees and D. J. Vaughan, Am. Chem. Soc., Div. Polymer Chem., *Polymer Preprints*, **6**, No. 1, 287, 296 (Apr. 1965 Detroit meeting).
2. Personal communication from I. Frank Peake, Film Department Research and Development, E. I. du Pont de Nemours and Co., Wilmington, Del.
3. J. P. Broussard, *Mod. Packaging*, **40**, No. 9, 157 (May 1967).
4. J. P. Broussard, *Mod. Packaging*, **40**, No. 10, 173 (June 1967).
5. Anon., *Mod. Packaging*, **38**, No. 7, 144 (Mar. 1965).
6. T. F. McLaughlin, *Package Engineering* (November **1968**).
7. Section 121.2582, Ethylene-methacrylic acid copolymers, ethylene-methacrylic acid-vinyl acetate copolymers, and their partial salts.   Amendment of June 10, 1967, 32 *F. R.* 8360 (Volume I, Chapter 16. Volume II, Appendix p. 729).

II. Polyimides

A. INTRODUCTION

For many years a search has been made for polymers with good thermal stability at temperatures from 500 to 1000°C. Much of this research was supported by federal grants from the military or the space programs and in the early days centered on inorganic polymers of silicon, boron, phosphorus, fluorine, etc., with oxygen, incorporating as little carbon and hydrogen as possible (Volume I, Chapter 5). With the single exception of the silicon polymers, most of this effort was disappointing, and gradually the search for such systems centered on carbon chains with unusual stiffness and strong bonds of carbon to other elements in the structure.

More recently the search for high-temperature stability has concentrated on heterocyclic compounds of carbon and nitrogen (or carbon and a small number of other atoms that readily form heterocycles), and from this work has come the interesting film-forming polyimides.

B. PREPARATION OF POLYIMIDES

1. Polyamic Acids

The direct formation of aliphatic polyimides was first described in 1955 [1]. They were made by fusion of salts of diamines and tetraprotic acids or diamines and diprotic acids or diesters.

After having been heated at 110 to 138°C for several hours the product at that stage of low molecular weight was heated in a second stage at high vacuum at 250 to 300°C to remove small molecules (e.g., H_2O, CH_3OH) eliminated in the final condensation to form high polymer; for example, a pyromellitimide polymer could be made.

These procedures were successful only for cases in which the polyimide is a melt at the final temperature and therefore useful polymers from pyromellitic acid and diamines required R to consist of at least a nine-carbon open chain or a seven-carbon branched chain [2, 3].

The more general method of preparing polyimides, however, involves a two-step process: formation of **I**, a soluble polyamic acid by condensation of a dianhydride and a diamine, and **II**, formation of the insoluble polyimide by condensation of the polyamic acid [4–9].

A wide variety of diamines and dianhydrides has been used for synthesis of polyamic acids. They are listed in Table 1 [10].

Polyamic acid preparation proceeds successfully at room temperature in any of several suitable solvents (Table 2 [10]).

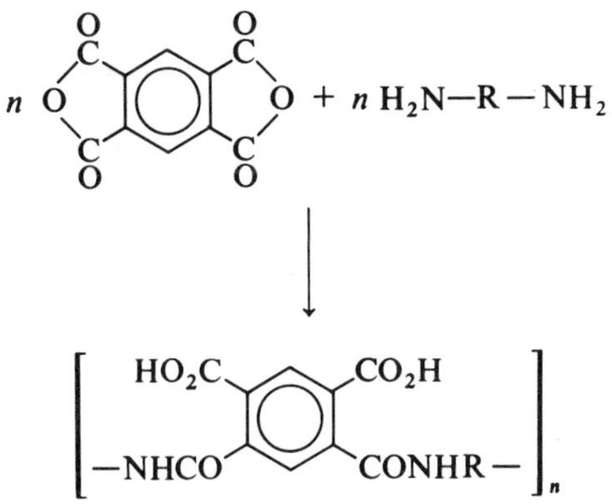

$$n \; \text{(pyromellitic dianhydride)} + n \, H_2N-R-NH_2$$

$$\downarrow$$

$$\left[\begin{array}{c} HO_2C \quad\quad CO_2H \\ -NHCO \quad\quad CONHR- \end{array} \right]_n$$

I

$$\mathbf{I} \longrightarrow \left[-N \underset{\text{(pyromellitic diimide)}}{} N-R- \right]_n + 2n \, H_2O$$

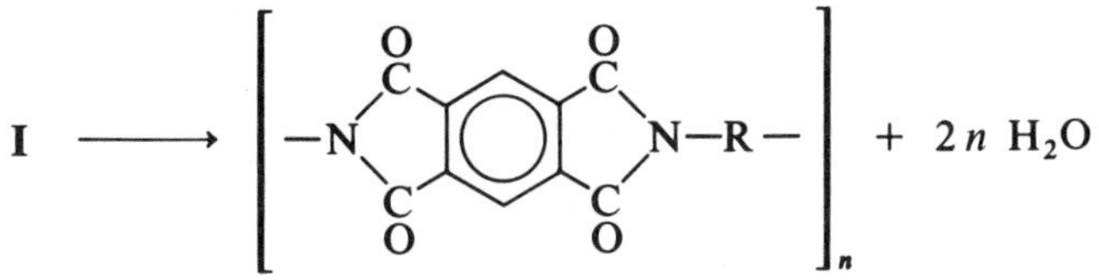

II

a. Properties of Polyamic Acids. Polyamic acids, as polyelectrolytes, are soluble in numerous polar solvents [11, 12]. Concentrated solutions undergo less degradation than dilute solutions, a fact that suggests a solvent role in cleavage. Solutions protected from moisture can be stored at temperatures below 0°C for long periods without serious change.

TABLE 1

Diamines and Dianhydrides for Polyamic Acid Synthesis

DIAMINES

m-Phenylenediamine	Decamethylenediamine
p-Phenylenediamine	Nonamethylenediamine
4,4'-Diaminodiphenylpropane	3-Methylheptamethylenediamine
4,4'-Diaminodiphenylmethane	4,4-Dimethylheptamethylene-
Benzidine	diamine
4,4'-Diaminodiphenyl sulfide	2,11-Diaminododecane
4,4'-Diaminodiphenyl sulfone	1,2-Bis(3-aminopropoxy)ethane
4,4'-Diaminodiphenyl ether	2,2-Dimethylpropylenediamine
1,5-Diaminonaphthalene	3-Methoxyhexamethylenediamine
3,3'-Dimethylbenzidine	2,5-Dimethylhexamethylene-
3,3'-Dimethoxybenzidine	diamine
2,4-Bis(β-amino-t-butyl)toluene	2,5-Dimethylheptamethylene-
Bis[p-(β-amino-t-butyl)phenyl] ether	diamine
Bis[p-(β-methyl-δ-aminopentyl)]benzene	3-Methylheptamethylenediamine
1-Isopropyl-2,4-m-phenylenediamine	5-Methylnonamethylenediamine
m-Xylylenediamine	Piperazine
p-Xylylenediamine	2,17-Diaminoeicosadecane
Di(p-aminocyclohexyl)methane	1,4-Diaminocyclohexane
Hexamethylenediamine	1,12-Diaminooctadecane
Heptamethylenediamine	$H_2N(CH_2)_3S(CH_2)_3NH_2$
Octamethylenediamine	$H_2N(CH_2)_3N(CH_3)(CH_2)_3NH_2$

DIANHYDRIDES

Pyromellitic dianhydride	3,4-Dicarboxyphenyl sulfone
Naphthalene-2,3,6,7-tetracarboxylic	dianhydride
dianhydride	Perylene-3,4,9,10-tetracarboxylic
Diphenyl-3,3',4, 4'-tetracarboxylic	acid dianhydride
dianhydride	Bis(3,4-dicarboxyphenyl)ether
Naphthalene-1,2,5,6-tetracarboxylic	dianhydride
dianhydride	Ethylene tetracarboxylic acid
Diphenyl-2,2',3,3'-tetracarboxylic	dianhydride
dianhydride	Benzophenone-3,4,3',4'-tetracar-
Thiophene-2,3,4,5-tetracarboxylic	boxylic dianhydride
dianhydride	
2,2-Bis(3,4-dicarboxyphenyl) propane	
dianhydride	

TABLE 2

Solvents for Polyamic Acid Synthesis

N,N-Dimethylformamide
N,N-Diethylformamide
N,N-Diethylacetamide
N,N-Dimethylmethoxyacetamide
N-Methylcaprolactam
Dimethyl sulfoxide
N-Methyl-2-pyrrolidone
Pyridine
Dimethyl sulfone
Hexamethylphosphoramide
Tetramethylene sulfone
N-Acetyl-2-pyrrolidone

Hydrolytic cleavage of the amide link is rapid, however, especially for the aromatic orthocarboxy amides [13, 14].

At room temperature cyclization (with elimination of water, which would, of course, accelerate cleavage) is very slow [15], amounting to 15 to 20% in a dilute solution that has stood at 35°C for 200 days.

A thorough study has been made by Wallach of the structure of the polymeric amide-acids obtained from pyromellitic dianhydride and 4,4'-diaminodiphenyl ether [16]. Number-average molecular weights from 13,000 to 55,000 reflected degrees of polymerization from 31 to 131. Weight-average molecular weights were 9900 to 266,000, giving ratios of $\overline{M}_w/\overline{M}_n$ from 2.2 to 4.8.

Use of very pure monomers was found important, for impurities in monomers or solvents gave low number-average molecular weights.

2. Dehydration of Polyamic Acids to Polyimides

The polyamic acids may be dehydrated (a) by heating them in a suitable solvent or as a cast form, or (b) by treatment with chemical dehydrating agents such as acid anhydrides or catalysts such as tertiary amines [12, 17].

Cyclization to polyimide may be done by heating a film of polyamic acid from 25 to 300°C at a carefully controlled rate [18]. Progress can be followed in the infrared as the N-H band at 3.08 μ disappears and the imide bands appear at 5.63 and 13.85 μ [12]. This is the process favored for preparing the most useful member of the imide polymers that derived from 4,4'diaminodiphenyl ether and pyromellitic dianhydride (Du Pont's Kapton Type H Film).

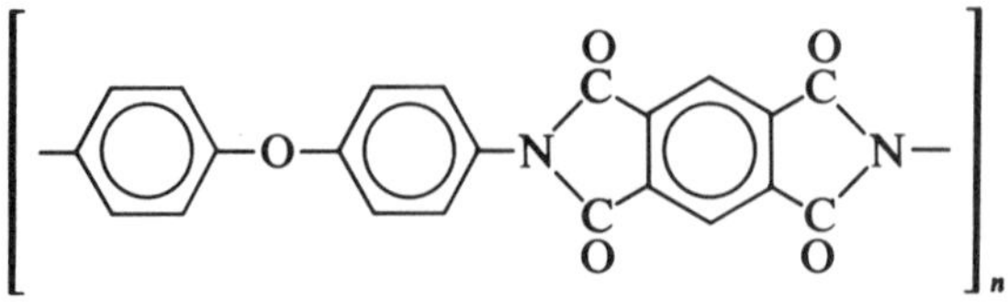

Chemical dehydrating agents that have been used include propionic anhydride, acetic anhydride, butyric anhydride, valeric anhydride, naphthoic anhydride, benzoic anhydride, aliphatic ketenes, pyridine, triethylamine, isoquinoline, morpholines, diethylcyclohexylamine, and other bases [19, 20].

C. PROPERTIES OF POLYIMIDES

The most useful polymers described so far are those prepared from pyromellitic dianhydride and aromatic amines. Substitution of aliphatic diamines gives polymers melting in general below 300°C; some of them degrade thermally below the melting point. Somewhat better results obtain if aliphatic diamines are condensed with dianhydrides of the following structure, $(R = -O-, -CH_2-, -CR_2-)$, but the

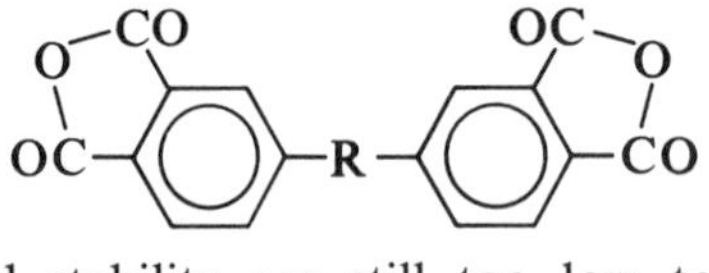

melting points and stability are still too low to permit successful molding or film casting [21, 22].

Table 3 summarizes the properties of polypyromellitimides in terms of the amine component. In general, the aromatic derivatives have extremely high melting points, are soluble only with difficulty in unusual solvents, and have exceptionally high resistance to degradation of properties by heat, oxygen, and radiation. The zero-strength temperatures† of films are higher than 700° (by comparison aluminum foil has a zero-strength temperature of 550°C).

† The zero-strength temperature is measured by holding a strip of film against a metal bar that has been preheated to a high temperature. The film is under a tensile load of 20 psi, and the test measures the temperature at which the film fails exactly 5 sec after contact with the bar [23].

TABLE 3

Properties of Polypyromellitimides
(R = diamine component)

R	Solubility	Crystallinity	Zero strength temperature (°C)	Thermal stability in air[a]	
				275°C	300°C
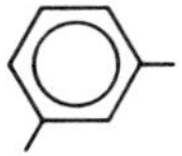 (m-phenylene ring)	Amorphous conc. H_2SO_4; crystalline insol.	Crystallizable	900	>1 yr	>1 mo
 (p-phenylene ring)	Amorphous conc. H_2SO_4; crystalline insol.	Crystallizes readily	900	>1 yr	
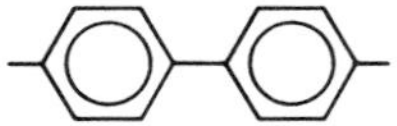 (biphenyl)	Fuming HNO_3	Highly crystalline	>900		1 mo
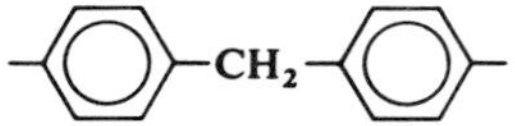 (diphenyl–CH_2)	Conc. H_2SO_4	Slightly crystalline	800		7–10 days
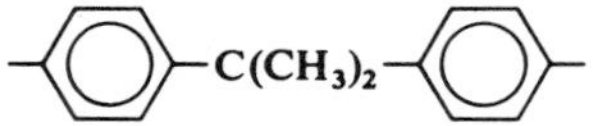 (diphenyl–$C(CH_3)_2$)	Conc. H_2SO_4	Crystallizable with difficulty	580		15–20 days
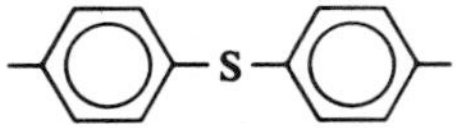 (diphenyl–S)	Fuming HNO_3	Crystallizable	800	10–12 mo (est.)	6 weeks
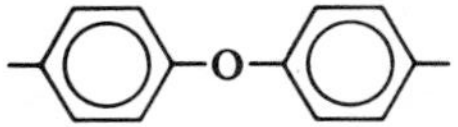 (diphenyl–O)	Fuming HNO_3	Crystallizable	850	>1 yr	>1 mo
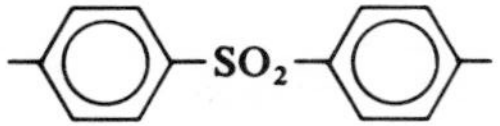 (diphenyl–SO_2, para)	Conc. H_2SO_4				>1 mo
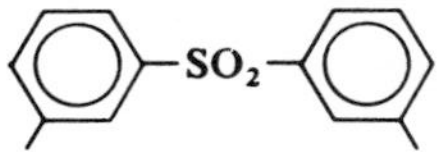 (diphenyl–SO_2, meta)	Conc. H_2SO_4				>1 mo

[a] As measured by retention of film creasability.

663

TABLE 4
Properties of Commercial Type H Kapton Film[a]
(1 mil)

Property	Value	Test method[b]
Density, g/cc	1.42	ASTM D 1505–63T
Yield, in.^{2}/lb	19,600	
Refractive index (Becke line)	1.78	
Melting point	Dec. above 900°C	
Zero strength temperature, 20 psi load for 5 sec	815°C	Hot bar (see text)
Cut-through temperature	435°C	Weighted probe on heated film (Du Pont test)
Coefficient of thermal expansion in./in.-°C, $-14°$ to $+38°$C	2.0×10^{-5}	ASTM D 696–44
Coefficient of thermal conductivity, cal-cm/cm^2-sec-°C		Model TC–1000 Twin heatmeter
25°C	3.72×10^{-4}	Comparative
75°C	3.89×10^{-4}	tester
200°C	4.26×10^{-4}	
300°C	4.51×10^{-4}	
Specific heat, cal/g-°C at 40°C	0.261	
Inflammability	Self-extinguishing	
Coefficient of friction, film-film	0.42	ASTM D 1894–63

	$-195°$C	25°C	200°C	
Ultimate tensile strength, psi (MD)	35,000	25,000	17,000	ASTM D 882–64T
Yield point, psi at 3% (MD)		10,000	6,000	ASTM D 882–64T
Stress to produce 5% elongation, psi (MD)		13,000	8,500	ASTM D 882–64T
Ultimate elongation, % (MD)	2	70	90	ASTM D 882–64T
Tensile modulus, psi (MD)	510,000	430,000	260,000	ASTM D 882–64T
Impact strength, kg-cm/mil		6		Du Pont Pneumatic Impact Test
Folding endurance (MIT), cycles		10,000		ASTM D 2176–63T
Tear strength—propagating (Elmendorf), g/mil		8		ASTM D 1922–61T
Tear strength—initial (Graves), g/mil		510		ASTM D 1004–61
Bursting test (Mullen), psi		75		ASTM D 774–63T
Permeability, ml/100 in.^{2}-24 hr-atm at 23°C				ASTM D 1434–63
CO_2		45		
H_2		250		
N_2		6		
O_2		25		
He		415		
H_2O (g/100 in.^{2}-24 hr)		5.4		ASTM E 96–63T

[a] [28].
[b] American Society for Testing and Materials, Philadelphia, Pa., except as noted.

The hydrolytic stability of these polymers is very high. The best retain film properties completely after a year's subjection to boiling water (*o*- and *p*-phenylenediamine polymers endure such treatment for one week only, however).

Aromatic pyromellitimides are resistant to thermal deomposition in air and are nearly inert in nonoxidizing atmospheres up to about 400°C [24–26]. A careful study of the pyrolysis of the polymer of pyromellitic dianhydride and 4,4'-diaminodiphenyl ether between 400 and 600°C at very low pressures (ca. 5×10^{-3} mm Hg) has demonstrated the thermal stability of this particular polymer and elucidated the mechanism of decomposition up to ca. 600°C [27].

D. PROPERTIES OF POLYIMIDE FILMS

The properties of commercial polyimide films are summarized in Table 4 for Du Pont's Type H Kapton (since Kapton does not melt, Du Pont offers also a Type F which is the base film coated on one or both sides with a heat-sealable Teflon FEP resin; see Chapter 13). The film is available in gauges from 0.5 to 5 mils.

Table 5 notes the effect on film properties caused by exposure to chemicals under a variety of conditions.

TABLE 5

Effect of Chemical Exposure upon Type H Film Properties
Typical Values for 1-mil Film

Property	Tensile retained (%)	Elongation retained (%)	Modulus retained (%)	Immersion test
Benzene	100	82	100	365 days at 20°C
Toluene	94	66	97	365 days at 20°C
Methanol	100	73	140	365 days at 20°C
Acetone	67	62	160	365 days at 20°C
10% sodium hydroxide		Degrades		5 days at 20°C
Glacial acetic acid	85	62	102	36 days at 110°C
p-Cresol	100	77	102	22 days at 200°C
Arochlor	100	53	142	365 days at 200°C
Transformer oil	100	100	100	180 days at 150°C
Aqueous *p*H 1	65	30	100	14 days at 100°C
*p*H 4.2	65	30	100	14 days at 100°C
*p*H 7.0	65	30	100	70 days at 100°C
*p*H 8.9	65	20	100	14 days at 100°C
*p*H 10.0	60	10	100	4 days at 100°C

Table 6 summarizes typical values for the electrical properties of
1-mil Type H Kapton polyimide film.

TABLE 6

Electrical Properties of Type H Kapton Film
(1 mil)

Property	−195°C	25°C	200°C	Test method
Dielectric strength, 60 cycles, V	10,800	7000	5600	ASTM D 149–64
Dielectric constant, 1 kc		3.5	3.0	ASTM D 150–64T
Dissipation factor, 1 kc		0.003	0.002	ASTM D 150–64T
Volume resistivity, Ω-cm		10^{18}	10^{14}	ASTM D 257–61
Surface resistivity, Ω, 50% rh		10^{16}		ASTM D 257–61
Corona start voltage, V, 50% rh		465		ASTM D 1868–61T
Insulation resistance, MΩ mfd		100,000		Based on 0.05 mfd. wound capacitor using 1-mil H film

E. Applications

Kapton Type H has been used in applications subject to temperatures
as low as −267°C and as high as 400°C. At 20°C the mechanical properties of Kapton and Mylar (Chapter 14) are similar, but at much higher
or lower temperatures, the properties of Kapton are less affected.

The polyimide cannot be heat-sealed (it begins to char above 800°C)
or solvent-sealed (there is no known organic solvent). A coated modification (Type F) is available for applications that require sealing, as
mentioned above.

Electrical properties (Table 6) do not vary greatly over a wide range
of temperatures, and therefore the film finds uses in wrapping wire or
cable, for flexible printed circuits, as motor slot-liners, in building
transformers, capacitors, and motors, and as a magnetic tape. These
films have been applied in the aerospace industry to effect savings of as
much as 50% in volume and 25% in weight of electrical conductors.

Metallized films 0.25 and 0.5 mil thick are available. They exploit
the thermal and dimensional stability of the base film, its toughness,
dielectric properties, and flame- and solvent-resistance in thermal
insulation applications that require high flexibility, low weight, and total

resistance to degradation in atmospheres of various sorts and at temperatures from -275 to $400°C$ plus.   Specific applications include space vehicles, solar cells, capacitors, protective clothing, and cryogenic storage tanks for gases.

References

1. W. M. Edwards and I. M. Robinson, U.S. Patent 2,710,853 (June 14, 1955).
2. W. M. Edwards and I. M. Robinson, U.S. Patent 2,867,609 (Jan. 6, 1959).
3. W. M. Edwards and I. M. Robinson, U.S. Patent 2,880,230 (Mar. 31, 1959).
4. W. M. Edwards, British Patent 898,651 (Sept. 18, 1959).
5. W. M. Edwards, U.S. Patent 3,179,614 (Apr. 20, 1965).
6. W. M. Edwards, U.S. Patent 3,179,634 (Apr. 20, 1965).
7. A. L. Endrey, Canadian Patent 659,328 (Mar. 12, 1963).
8. A. L. Endrey, U.S. Patent 3,179,631 (Apr. 20, 1965).
9. A. L. Endrey, U.S. Patent 3,179,633 (Apr. 20, 1965).
10. C. E. Sroog, *J. Polymer Sci.*, **C**, No. 16, 1191 (1967).
11. G. M. Bower and L. W. Frost, *J. Polymer Sci.*, **A1**, 3135 (1963).
12. C. E. Sroog, A. L. Endrey, S. V. Abrams, C. E. Berr, W. M. Edwards, and K. L. Olivier, *J. Polymer Sci.*, **A3**, 1373 (1965).
13. M. L. Bender, *J. Am. Chem. Soc.*, **79**, 1258 (1957).
14. M. L. Bender, *J. Am. Chem. Soc.*, **80**, 5380 (1958).
15. L. W. Frost and I. Kesse, *J. Appl. Polymer Sci.*, **8**, 1039 (1964).
16. M. L. Wallach, *J. Polymer Sci.*, **A-2, 5**, 653 (1967).
17. J. J. Jones, F. W. Ochynski, and F. A. Rackley, *Chem. Ind.* (*London*), **1962,** 1686.
18. J. A. Kreuz, A. L. Endrey, E. P. Gay, and C. E. Sroog, *J. Polymer Sci.*, **A-1**, 4, 2607 (1966).
19. A. L. Endrey, U.S. Patent 3,179,630 (Apr. 20, 1965).
20. A. K. Bose, F. Greer, and C. C. Price, *J. Org. Chem.*, **23**, 1335 (1958).
21. W. F. Gresham and M. A. Naylor, U.S. Patent 2,731,447 (Jan. 17, 1956).
22. M. L. Wallach, *J. Polymer Sci.*, **A-2, 6**, 953 (1968).
23. C. E. Sroog and F. P. Gay, *Discovery*, July **1966**, 13.
24. J. F. Heacock and C. E. Berr, *SPE Trans.*, **5**, 105 (Apr. 1965).
25. S. D. Bruck, *Polymer*, **5**, 435 (1964).
26. S. D. Bruck, *Polymer*, **6**, 49 (1965).
27. F. P. Gay and C. E. Berr, *J. Polymer Sci.*, **A-1, 6**, 1935 (1968).
28. Data from Du Pont Kapton Bulletins H–1A and H–2, E. I. du Pont de Nemours and Co., Film Department, Wilmington, Del. 19898 (1969).

III. Polyamides

A. Introduction

The basic chemistry of the polyamides of high molecular weight is outlined in Carothers' first patent on high polymers [1] and the prepara-

tion of such materials from suitable monomers is discussed in Volume I†
of this work.

The polyamides are usually called nylons and are identified by one or
two numerical coefficients that signify the number of carbons in the
repeat unit from the amine and acid, respectively. Thus 6,6-nylon
refers to the polymer formed by condensation of the C_6 diamine,
hexamethylenediamine, and C_6 diprotic acid, adipic acid, and 6,10-
nylon refers similarly to the corresponding polymer from sebacic acid.

$$[-NH-CH_2-CH_2-CH_2-CH_2-CH_2-CH_2-NH-CO-$$
$$-CH_2-CH_2-CH_2-CH_2-CO-]_n$$

6,6-Nylon

$$[-NH-(CH_2)_6-NH-CO-(CH_2)_8-CO-]_n$$

6,10-Nylon

If the nylon is derived from an ω-amino acid, or from a lactam, it is
evident that only one coefficient is required. Thus 6-nylon results from
either ε-aminocaproic acid or caprolactam.

$$[-NH-CH_2-CH_2-CH_2-CH_2-CH_2-CO-]_n$$

6-Nylon

6-Nylon and 11-nylon account for most of the nylon films market,
though a small amount of 6,6- and 6,10-nylons have been sold for film
from time to time.

As a class polyamide films are characterized by outstanding toughness,
excellent resistance to abrasion, good thermal stability, low gas permeability, and high water-vapor-transmission rates.

B. Manufacture of the Monomers

The C_6 monomers can be synthesized conveniently from phenol,
cyclohexane, 1,3-butadiene, or tetrahydrofuran. 11-Aminoundecanoic
acid is made from castor oil. Sebacic acid, the C_{10} diprotic acid, is also
obtained from castor oil. These processes are outlined below.

† Chapter 2, p. 47 ff.

Products from Phenol

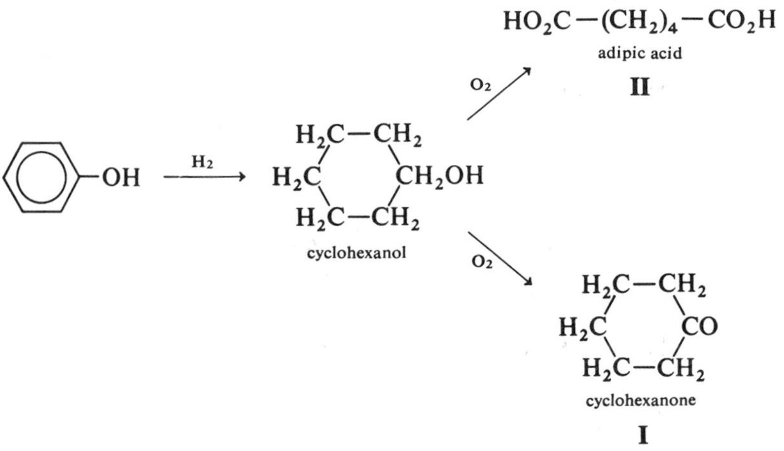

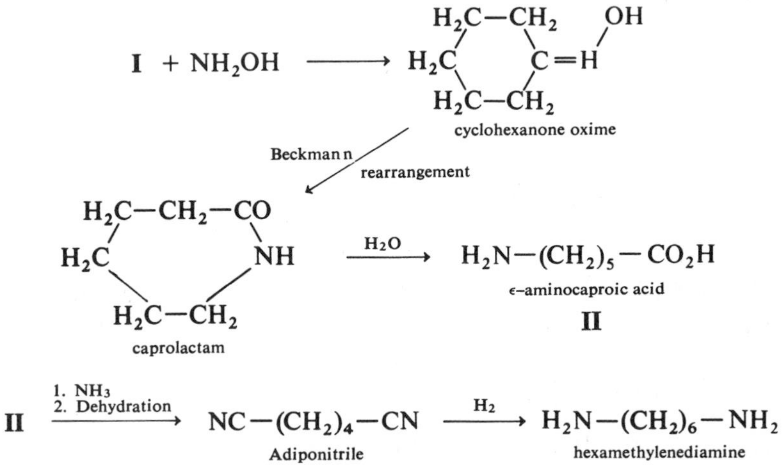

$$HO_2C-(CH_2)_4-CO_2H$$

adipic acid

II

$$\text{phenol} \xrightarrow{H_2} \text{cyclohexanol}$$

cyclohexanol $\xrightarrow{O_2}$ (adipic acid)

cyclohexanol $\xrightarrow{O_2}$ cyclohexanone

cyclohexanone

I

$$\text{I} + NH_2OH \longrightarrow \text{cyclohexanone oxime}$$

cyclohexanone oxime

Beckmann rearrangement

$$\text{caprolactam} \xrightarrow{H_2O} H_2N-(CH_2)_5-CO_2H$$

ε-aminocaproic acid

II

caprolactam

$$\text{II} \xrightarrow[\text{2. Dehydration}]{\text{1. } NH_3} NC-(CH_2)_4-CN \xrightarrow{H_2} H_2N-(CH_2)_6-NH_2$$

Adiponitrile   hexamethylenediamine

Product from Cyclohexane

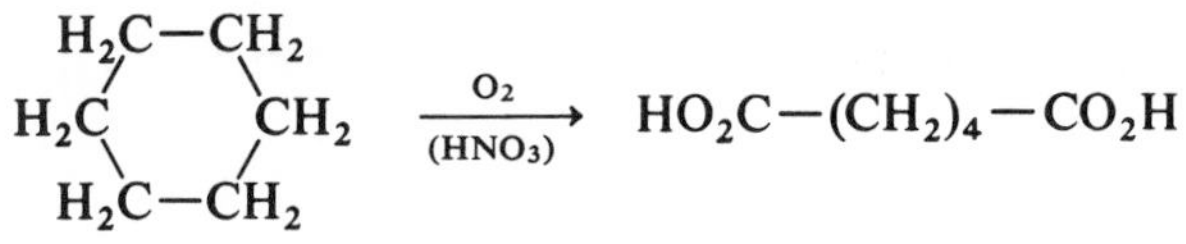

$$\text{cyclohexane} \xrightarrow[(HNO_3)]{O_2} HO_2C-(CH_2)_4-CO_2H$$

Hexamethylenediamine is made in large quantities from butadiene or from tetrahydrofuran, as indicated [2].

Products from 1,3-Butadiene

$$CH_2=CH-CH=CH_2 \xrightarrow{Cl_2}$$

$$ClCH_2-CH=CH-CH_2Cl \xrightarrow[\text{2. }H_2]{\text{1. NaCN}}$$

$$H_2N(CH_2)_6NH_2$$

Products from Tetrahydrofuran

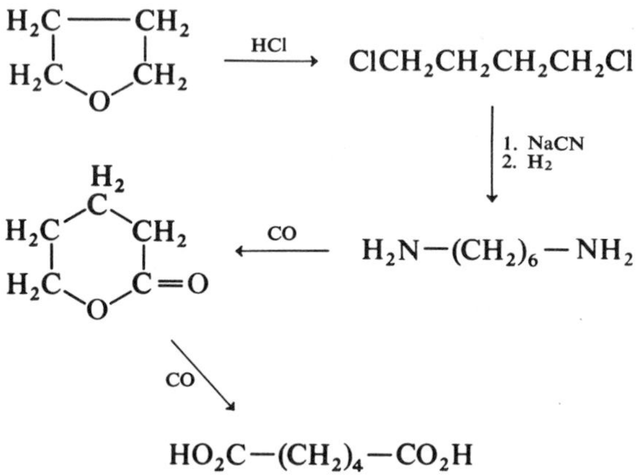

$$HO_2C-(CH_2)_4-CO_2H$$

Products from Castor Oil

Glyceryl triricinoleate is transesterified with methanol to yield the methyl ester and glycerol and the former is cracked thermally to methyl undecylenate and heptaldehyde [3, 4].

$$CH_3(CH_2)_5CH(OH)CH_2CH=CH(CH_2)_7CO_2CH_2$$
$$CH_3(CH_2)_5CH(OH)CH_2CH=CH(CH_2)_7CO_2\overset{|}{C}H+3CH_3OH \longrightarrow$$
$$CH_3(CH_2)_5CH(OH)CH_2CH=CH(CH_2)_7CO_2CH_2$$

glyceryl triricinoleate

$$3CH_3(CH_2)_5CH(OH)CH_2CH=CH(CH_2)_7CO_2CH_3$$

methyl ricinoleate

$$+ CH_2OHCH(OH)CH_2OH$$

glycerol

(heat)

$$CH_2=CH(CH_2)_8CO_2CH_3 + CH_3(CH_2)_5CHO$$

methyl undecylenate   heptaldehyde

Methyl undecylenate is hydrolyzed; the resulting acid adds hydrogen bromide in a reverse-Markownikoff reaction, under the influence of air, to yield largely 11-bromoundecanoic acid along with a small proportion of the 10-isomer. The bromo acid is aminated and the amino acid is polymerized in the usual way.

$$CH_2{=}CH(CH_2)_8CO_2CH_3 + H_2O$$

$$\searrow \text{NaOH}$$

$$CH_2{=}CH(CH_2)_8CO_2H + CH_3OH$$
$$\text{undecylenic acid}$$

$$\nearrow \underset{(O_2)}{HBr}$$

$$BrCH_2(CH_2)_9CO_2H \xrightarrow{+\ NH_3} H_2NCH_2(CH_2)_9CO_2H$$
$$\text{11-bromoundecanoic acid} \qquad\qquad \text{11-aminoundecanoic acid}$$

$$n\ H_2NCH_2(CH_2)_9CO_2H \xrightarrow{(heat)} [-NH(CH_2)_{10}CO_2-]_n + n\ H_2O$$
$$\text{11-nylon}$$

The still residue from the cracking step can be cleaved by fusion with alkali to give 2-octanol and sebacic acid.

$$CH_3(CH_2)_5CH(OH)CH_2CH{=}CH(CH_2)_7CO_2H$$
$$\text{ricinoleic acid}$$

$$\xrightarrow[\text{(fuse)}]{\text{NaOH}} CH_3(CH_2)_5CHOHCH_3 + HO_2C(CH_2)_8CO_2H$$
$$\text{2-octanol} \qquad\qquad\qquad \text{sebacic acid}$$

C. Manufacture of Nylons

1. Polymer Preparation

The manufacture of nylons from a diamine and a diprotic acid is done essentially by Carothers' original method. To ensure high polymer a nylon salt is prepared by the addition of equivalent amounts of acid and amine to boiling methanol from which the salt precipitates in rather pure form.

The salt is separated, purified if necessary, and a 60 to 75% aqueous solution or slurry is autoclaved at approximately 200 to 225°C and 250 psi. The temperature is finally raised to 275 to 285°C to remove traces of water, and the polymer is unloaded through the bottom of the reactor and extruded under nitrogen onto a chill-roll and pelletized.

The polymerization of caprolactam can be accomplished either hydrolytically or catalytically. The latter is preferable, by the use of metal alkoxides (e.g., $CH_3O^-Na^+$, $C_2H_5O^-K^+$), metal hydrides (NaH), or alkali metal lactams (e.g., sodium butyrolactam) [5–7]. Polymerization is rapid, requiring only a few minutes after introduction of the catalyst to the stirred molten lactam at 250 to 260 °C under a nitrogen blanket. The process is suited to batch or continuous operation.

Polymerization of caprolactam appears to consist of three steps: (1) formation of lactam ion, (2) ring cleavage, and (3) chain propagation [5].

1. Formation of Alkali Metal Derivative

$$M^+B^- + CH_2\text{—}(CH_2)_4\text{—}CO\text{—}NH$$
$$\underline{\qquad\qquad\qquad\qquad}$$

$$\longrightarrow \quad CH_2\text{—}(CH_2)_4\text{—}CO\text{—}N^-\ M^+ + HB$$
$$\underline{\qquad\qquad\qquad\qquad}$$

$(B^- = H^-, OH^-, OR^-, \text{etc.})$

2. Ring Cleavage

$$CH_2\text{—}(CH_2)_4\text{—}CO\text{—}N^- + CH_2\text{—}(CH_2)_4\text{—}CO\text{—}NH$$
$$\underline{\qquad\qquad\qquad}\quad\underline{\qquad\qquad\qquad}$$

$$\longrightarrow \quad CH_2\text{—}(CH_2)_4\text{—}CO\text{—}N\text{—}CO\text{—}CH_2\text{—}(CH_2)_4\text{—}NH^-$$
$$\underline{\qquad\qquad\qquad\qquad}$$

3. Chain Propagation

$$CH_2\text{—}(CH_2)_4\text{—}CO\text{—}N\text{—}CO\text{—}(CH_2)_5\text{—}NH^-$$
$$\underline{\qquad\qquad\qquad}$$

$$+\ n\ CH_2\text{—}(CH_2)_4\text{—}CO\text{—}NH \quad \longrightarrow$$
$$\underline{\qquad\qquad\qquad}$$

$$CH_2\text{—}(CH_2)_4\text{—}CO\text{—}N\text{—}[CO\text{—}(CH_2)_5\text{—}NH\text{—}]_nCO\text{—}(CH_2)_5\text{—}NH^-$$
$$\underline{\qquad\qquad\qquad}$$

It is also possible that ring cleavage may occur by direct attack on the monomer by such a strong Lewis base as hydride ion (this particular polymerization reaction is extremely rapid with 0.4–0.6 mole% of sodium hydride).

2'. *Ring cleavage*

$$H^- + CH_2-(CH_2)_4-CO-NH \longrightarrow H-CO-(CH_2)_5-NH^-$$

3'. *Chain Propagation*

$$H-CO-(CH_2)_5-NH^- + n\ CH_2-(CH_2)_4-CO-NH$$

$$\longrightarrow H[-CO-(CH_2)_5-NH-]_nCO-(CH_2)_5-NH^-$$

The growing polymer chain can easily be stopped by any of several agents that quench the anion, such as H^+ and O_2.

2. Polymer Properties and Processing

Table 1 presents some of the important properties of nylon resins used

TABLE 1

Properties of Nylon Polymers

Property	Method of measurement[a]	6-Nylon	11-Nylon	6,6-Nylon	6,10-Nylon
Specific gravity	D 792	1.12–1.14	1.4	1.13–1.15	1.07–1.09
Mp, °C		220	185	225	215
Heat of fusion, cal/g		ca. 30	ca. 20	ca. 30	
Water absorption, %	D 570	9.5		8.0	3.5
Tensile strength, psi	D 638	6700–12,500	6800–8500	8200–12,600	7100–8500
Elongation to break, %	D 638	30–300	100–300	60–300	85–220
Impact strength, Izod, ft-lb/in. at 25°C	D 256	0.7–4.0	3.5–4.8	0.9–2.1	0.6–1.6
Deflection temperature, °F	D 648				
66 psi stress		340–365		365–470	300–310
264 psi stress		140–167		155–220	135–140

[a] ASTM methods [10].

for film applications [8]. The range of values given depends on the methods of preparation (modifiers, chain stoppers, etc., added) and on the short- and long-range order of the polymer. Further data for some of the well-studied nylons such as 6,6- may be found in the hand-books [9].

Films can be made by extrusion or by casting a prepolymer and subsequent polymerization, but only the extrusion method is used commercially. Both tubular and flat-film techniques are practiced (cf. Volume I, Chapter 8). Film made by the flat die method is generally not transparent. Conventional converting methods, including the manufacture of extrusion and adhesion laminates, are usually employed. Coextrusion or extrusion coating of nylon onto other substrates are possible alternatives. Nylon films can also be modified by coatings and by orientation.

Extrusion temperatures range from 225 to 290°C for 6-nylon, from 190 to 250°C for 11-nylon, from 260 to 280°C for 6,6-nylon, and from 220 to 280°C for 6,10-nylon. At these temperatures the melt densities are 0.97, 0.85, 0.97, and 0.92, in the same order.

The so-called nylon screw, that is, one with a compression section length of 1 diameter or less, is in general not recommended for nylon extrusion. A gradual compression section followed by a long metering section is the preferred design (Volume I, Chapter 8 and [11]).

D. FILM PROPERTIES AND APPLICATIONS

Nylon films are very tough, as reflected in high tensile, tear, impact, and bursting strength. They possess high elongation and flexing life. The density is about 1.13, which results in a reasonably high coverage for film.

The nylons have high water-vapor permeability and do not protect well in such applications, but this is exploited in steam sterilization procedures, since after autoclaving the moisture that penetrated escapes rapidly. Permeability to other gases is, in general, much lower than found for polyethylenes, polystyrenes, polypropylene, or vinyls. It is approximately the same as for sarans and higher than for coated cellophanes.

Table 2 summarizes the common film properties.

Nylon films accept printing inks readily. A permanent bond is formed that survives stretching while being thermoformed and vacuum drawn, heat sealing through printed areas, boiling water for 30 min., steam sterilization, and flexing and abrasion at frozen-food temperatures.

TABLE 2

Properties of Nylon Films[a]
(1 mil)

Property	Method of measurement[b]	
Density, g/cc		1.13–1.14
Yield, in.^{2}/lb		23,500–24,500
Clarity		Transparent to translucent
Tensile strength, lb/in.^{2}	D 882	10,000–18,000
Stretch factor, %	D 882	250–500
Impact strength, kg-cm	Ref. 12	4–6
Tear strength, Elmendorf, g/mil	D 1922	50–150
Stiffness, Handle-O-Meter		5–40
Heat-seal range, °F		350–500
Maximum use temperature, °F		355–475
Minimum use temperature, °F		−100
Dimensional change at high rh, %		1–4
Water absorption, % in 24 hr		1
Flammability		Self-extinguishing
Water-vapor transmission, g/m^2-24 hr at 100°F and 90% rh	E 96 Method E 1	140
Gas transmission, cc/m^2-24 hr-atm at 73°F, 0% rh	D 1434	
O_2		30–110
CO_2		150–400
Dielectric strength, V/mil	D 149	350–450
Dielectric constant, 60 Hz	D 150	3.8–5.3
Dissipation factor, 60 Hz	D 150	0.03–0.04

[a] This table is a composite and represents averages for 6-, 11-, 6,6-, and 6,10-nylons. They are not greatly different, one from another.
[b] ASTM methods [10].

Their dimensional stability and high melting point recommend nylon films for such applications as goods boiled in the shipping bag (especially frozen foods), meats roasted in the container to avoid shrinkage during cooking, and doughs to be baked in the container. A growing use is for meat casings.

The effective grease resistance of polyamide films makes them useful for wrapping butter, lard, shortening, and foods containing them. Rancidity is much delayed and permeation by odoriferous foreign materials is hindered. For the same reason pharmaceuticals, herbs, flavors, lobster, and shrimp are being packaged in polyamide films.

Several types of nylon film are acceptable for food packaging under regulations of the Food and Drug Administration (cf. Appendix, p. 696) [13].

References

1. W. H. Carothers, U.S. Patent 2,071,250 (Feb. 16, 1937).
2. H. Hopff, *Die Polyamide*, Springer, Berlin, 1954.
3. M. Genas, U.S. Patent 2,462,855 (Mar. 1, 1949).
4. R. Aelion, U.S. Patent 2,600,953 (June 17, 1952).
5. W. E. Hanford and R. M. Joyce, *J. Polymer Sci.*, **3**, 171 (1948).
6. H. R. Mighton, U.S. Patent 2,647,105 (July 28, 1953).
7. G. H. Berthold, U.S. Patent 2,727,017 (Dec. 13, 1955).
8. Personal communication from A. H. Steinberg, Director of Research and Development, Allied Chemical Corp., and other sources.
9. J. Brandrup and E. H. Immergut, *Polymer Handbook*, Wiley-Interscience, New York, 1966.
10. *1968 Standards*, American Society for Testing and Materials, Philadelphia, Pa.
11. J. M. McKelvey, *Polymer Processing*, Wiley, New York, 1962, Chapters 10, 11.
12. K. W. Ninnemann, *Mod. Packaging*, **30**, No. 3, 163 (Nov. 1956).
13. Section 121.2502 (see Appendix).

APPENDIX I

GLOSSARY OF

TRADE NAMES FOR FILMS[a]

Name	Composition	Manufacturer
Acetophane	Cellulose acetate	UCB-Sidac
Aclar	Polytrifluorochloroethylene	Allied Chemical Corp.
Algoflon	Polytetrafluoroethylene	Montecatini Edison
Avisco	Cellophane	FMC Corp. (American Viscose Div.)
Avistar	Polyethylene terephthalate	FMC Corp. (American Viscose Div.)
Cadco	Cast vinyl	Cadillac Plastic and Chem. Co.
Capran	Polycaprolactam (6-Nylon)	Allied Chemical Corp.
Celanar	Polyethylene terephthalate	Celanese Plastics Co.
Cello	Cellulose	British Cellophane Ltd.
Cellophane[b]	Cellulose	British Cellophane Ltd.
Cellothene	Polyethylene-coated cellulose	UCB-Sidac, British Cellophane Ltd.
Conolene	Low-density polyethylene	Continental Can Co.
Conolex	High-density polyethylene	Continental Can Co.
Conolon	6-Nylon	Continental Can Co.
Cronar (photographic)	Polyethylene terephthalate	E. I. du Pont de Nemours and Co.
Cuprophan	Cellophane	J. P. Bemberg A.-G.
Diafoil	Polyethylene terephthalate	Mitsubishi Plastics Industries Ltd.
Durethene	Polyethylene	Sinclair-Koppers Co.
Duvalon	Cellophane	Wolff and Co. A.-G., Walsrode
Dylan	Polyethylene	Sinclair-Koppers Co.
Estar (photographic)	Polyethylene terephthalate	Eastman Kodak Co.
Estar	Polyethylene terephthalate	Kodak Ltd. (Great Britain)
Estar	Polyethylene terephthalate	Kodak-Pathé (France)
Europhan	Polyvinyl chloride	Polytherm-Kassel Corp.
Folex (photographic)	Polethylene terephthalate	Celfa A.-G.
Fresh-Pak	Polyolefins	UCB-Sidac
Fuji film	Polyethylene terephthalate	Fuji Photo Film Co.

(Continued)

Name	Composition	Manufacturer
Garmil	Polyethylene	Canadian Industries Ltd.
Genotherm	Polyvinyl chloride	American Hoechst Corp.
Gevar	Polyethylene terephthalate	Gevaert
Hostaphan	Polyethylene terephthalate	Kalle A.-G.
Iolon	Ionomers (cf. Surlyn A)	E. I. du Pont de Nemours and Co.
Irrathene	Irradiated polyethylene	General Electric Co.
Kapton	Polyimide	E. I. du Pont de Nemours and Co.
Kardel	Polystyrene	Union Carbide Corp.
Kodacel	Cellulose	Eastman Chemical Products, Inc.
Koni (photographic)	Polyethylene terephthalate	Teijin-Konishiroku Film Co. Ltd.
Korad	Acrylic	Rohm and Haas Co.
Kordite	Polyolefins	Mobil Chemical Co.
Koroseal	Polyvinyl chloride	B. F. Goodrich Chemical Co.
Krene	Plasticized polyvinyl chloride	Union Carbide Corp.
Lactophane	Cellulose	British Cellophane Ltd.
Lexan	Polycarbonate	General Electric Co.
Lumirror	Polyethylene terephthalate	Toyo Rayon Co. Ltd.
Melinex	Polyethylene terephthalate	Imperial Chemical Industries Ltd.
Metalumy	Metallized polyethylene terephthalate	Toyo Rayon Co.
Milpack (also Milrol and Milwrap)	Polyethylene	Canadian Industries Ltd.
Montivel	Polyester	Montecatini Edison
Moplefan	Polypropylene	Montecatini Edison
Mylar	Polyethylene terephthalate	E. I. du Pont de Nemours and Co.
Myraclear (also Myradur, Myraform, Myralon and Myraplast)	Vinyl	Kunstoffwerk Staufen
Olefane	Polypropylene	Avisun Corp.
Optithene	Polyethylene	Gulf Oil Corp.
Orgavyl	Shrinkable polyvinyl chloride	Aquitaine-Organico
Orlex	Polyvinyl chloride	Tenneco Chemicals Inc.
Pentaphane	Poly-3,3-bis(chloromethyl) oxetane (chlorinated Hercules Penton)	British Cellophane Ltd.

(Continued)

Name	Composition	Manufacturer
Pliofilm	Polyisoprene	Goodyear Tire and Rubber Co.
Poly Eth	Polyethylene	Gulf Oil Corp.
Poly-Fresh	Polyethylene	Crown-Zellerbach Corp.
Polyphane	Polyethylene	W. Ralston and Co.
Polythene	Polyethylene	British Cellophane Ltd. Imperial Chemical Industries Ltd.
Prime Wrap	Polyvinyl chloride	Goodyear Tire and Rubber Co.
Propathene	Polypropylene	Imperial Chemical Industries Ltd.
Propophane	Polypropylene	British Cellophane Ltd.
Rampak	Polyvinyl chloride	Ramp Plastic Co.
Rayophane	Cellulose	British Sidac Ltd.
Resinite	Polyvinyl chloride	Borden Chemical Co.
Reynolon	Shrinkable polyvinyl chloride	Reynolds Metals Co.
Rilsan	11-Nylon	American Soplaril Co.
Robex	Polystyrene	Plastica Caleppio s.r.l.
Rucoam	Vinyl	Ruco Div., Hooker Chemical Corp.
Sakura (photographic)	Polyethylene terephthalate	Teijin-Konishiroku Film Co. Ltd.
Saran Wrap[c] (also Saranex)	Polyvinylidene chloride copolymer	The Dow Chemical Co.
Sarophan	Saran-coated cellophane	J. P. Bemberg A.-G.
Scotchpar	Polyethylene terephthalate	3 M Co.
Sidac	Cellulose	UCB-Sidac
Sidavine	Vinyl	UCB-Sidac
Sumilite VSS	Polyvinyl chloride	Sumitomo Bakelite Co.
Surlyn A	Ethylene/methacrylic acid ionized copolymers	E. I. du Pont de Nemours and Co.
Talovin	Polyvinyl chloride	W. R. Grace and Co.
Tedlar	Polyvinyl fluoride	E. I. du Pont de Nemours and Co.
Terphane	Polyethylene terephthalate	La Cellophane S.A.
Teteron	Polyethylene terephthalate	Teijin-Konishiroku Film Co. Ltd.
Thermolene	Medium-density polyethylene	Continental Can Co.
Tolgen	Vinyl	General Tire and Rubber Co.
Torayfan	Polypropylene	Toyo Rayon Co.
Transothen	Polyethylene-polypropylene	J. P. Bemberg, A.-G.

(Continued)

Name	Composition	Manufacturer
Transparit	Cellophane	Wolff and Co., A.-G. Walsrode
Trycite	Polystyrene	The Dow Chemical Co.
Tuftane	Polyurethane	B. F. Goodrich Chemical Co.
Udel	Polypropylene	Union Carbide Corp.
Ultron	Vinyl	Monsanto Co.
Velon	Polyvinyl chloride	Firestone Plastics Co.
Videne	Polyethylene terephthalate	Goodyear Tire and Rubber Co.
Vinophane	Polyvinyl chloride	British Cellophane Ltd.
Vinyfoil	Polyvinyl chloride	Mitsubishi Plastics Industries Ltd.
Viplatherm	Shrinkable polyvinyl chloride	Montecatini Edison
Viscacelle	Cellulose	British Cellophane Ltd.
Viscophane	Cellulose	British Cellophane Ltd.
Viskseal	Shrinkable polyvinyl chloride	Viscose Development Co.
Vis Queen	Polyethylene	Visking Corp.
Vistal	Polyethylene	UCB-Sidac
Vitrose	Cellophane	Technical Packaging N.V.
Vondafol	Cellulose acetate	Fabriek van Chemische Producten Vondelingenplaat N.V.
Vuepak	Cellulose acetate	Monsanto Co.
Vylene	Vinyl	W. R. Grace and Co.
Vynaloy	Vinyl	B. F. Goodrich Chemical Co.
Vynatherm	Polyvinyl chloride	Tenneco Chemicals, Inc.
Vynawrap	Shrinkable polyvinyl chloride	Polytherm-Kassel Corp.
Vypro	Polypropylene	W. R. Grace and Co.
Waloplast	Polyethylene	Wolff and Co. A.-G., Walsrode
Walotherm	Shrinkable polyethylene	Wolff and Co., A.-G., Walsrode
Zendel	Polyethylene	Union Carbide Corp.

[a] Trademarks of converters are not included. Also omitted are trademarks of resins from which films are manufactured; these are legion and are constantly changing.

[b] Registered trademark in Great Britain.

[c] Saran is a trademark abroad only.

APPENDIX II

FOOD ADDITIVE REGULATIONS OF
THE FOOD AND DRUG ADMINISTRATION

(See also Volume I, Chapter 16)

I. Amendments to Subpart B and Subpart E Previously Published

The status of polymeric and other chemical substances that might become food additives within the meaning of the 1958 Amendment of the Food, Drug, and Cosmetic Act was reviewed in Volume I, Chapter 16.

We intend to cover here the relatively minor changes that have been made in Subparts B and E since publication of Volume I and present the pertinent sections of subparts D and F that regulate the use of polymers discussed in Volume II that are now used, or may be used, as food containers or as food wrappings.

Material presented previously (Volume I, pp. 829–846) is still applicable except as noted.

Food and Drug Administration

Food Additive Regulations

SUBPART B. EXEMPTION OF CERTAIN FOOD ADDITIVES
FROM THE REQUIREMENTS OF TOLERANCES

*§121.101.   Substances that are Generally Recognized as Safe
(d)(2).   Chemical Preservatives*

Nordihydroguaiaretic acid (p. 831) was deleted from the list on April 11, 1968.*

The following, inadvertently omitted, should be added (Volume I, p. 840).

(e)(1) Spices and Other Natural Seasonings and Flavorings . . .

Common name	Botanical name of plant source
Chervil	*Anthriscus cerefolium* (L.) Hoffm.
Chives	*Allium schoenoprasum* L.
Cinnamon, Ceylon	*Cinnamomum zeylanicum* Nees.
Cinnamon, Chinese	*Cinnamomum cassia* Blume.
Cinnamon, Saigon	*Cinnamomum loureirli* Nees.
Clary (clary sage)	*Salvia sclarea* L.
Clover	*Trifolium* spp.
Cloves	*Eugenia caryophylata* Thunb.
Coriander	*Coriandrum sativum* L.
Cumin (cummin)	*Cuminum cyminum* L.
Cumin, black (black caraway)	*Nigella sativa* L.
Dill	*Anethum graveolens* L.
Elder flowers	*Sambucus canadensis* L.
Fennel, common	*Foeniculum vulgare* Mill.
Fennel, sweet (finocchio, Florence fennel)	*Foeniculum vulgare* Mill. var. *dulce* (DC) Alef.

* **33** F.R. 5619.

Common name	Botanical name of plant source
Fenugreek	*Trigonella foenum-graecum* L.
Galanga (galangal)	*Alpinia officinarum* Hance.
Garlic	*Allium sativum* L.
Geranium	*Pelargonium* spp.
Ginger	*Zingiber officinale* Rosc.
Glycyrrhiza	*Glycyrrhiza glabra* L. and other spp. of *Glycyrrhiza*.
Grains of paradise	*Amomum melegueta* Rosc.
Horehound (hoarhound)	*Marrubium vulgare* L.
Horseradish	*Armoracia lapathifolia* Gilib.
Hyssop	*Hyssopus officinalis* L.
Lavender	*Lavandula officinalis* Chaix.
Licorice	*Glycyrrhiza glabra* L. and other spp. of *Glycyrrhiza*.
Linden flowers	*Tilia* spp.
Mace	*Myristica fragrans* Houtt.
Marigold, pot	*Calendula officinalis* L.
Marjoram, pot	*Majorana onites* (L.) Benth.
Marjoram, sweet	*Majorana hortensis* Moench.
Mustard, black or brown	*Brassica nigra* (L.) Koch.
Mustard, brown	*Brassica juncea* (L.) Coss.
Mustard, white or yellow	*Brassica hirta* Moench.
Nutmeg	*Myristica fragrans* Houtt.
Oregano (oreganum, Mexican oregano, Mexican sage, origan)	*Lippia* spp.
Paprika	*Capsicum annuum* L.
Parsley	*Petroselinum crispum* (Mill.) Mansf.
Pepper, black	*Piper nigrum* L.
Pepper, cayenne	*Capsicum frutescens* L. or *Capsicum annuum* L.
Pepper, red	*Capsicum frutescens* L. or *Capsicum annuum* L.
Pepper, white	*Piper nigrum* L.
Peppermint	*Mentha piperita* L.
Poppy seed	*Papaver somniferum* L.
Pot marigold	*Calendula officinalis* L.
Pot marjoram	*Majorana onites* (L.) Benth.
Rosemary	*Rosmarinus officinalis* L.
Rue	*Ruta graveolens* L.
Saffron	*Crocus sativus* L.
Sage	*Salvia officinalis* L.
Sage, Greek	*Salvia triloba* L.
Savory, summer	*Saturela hortensis* L. (Satureja).
Savory, winter	*Saturela montana* L. (Satureja).

Common name	Botanical name of plant source
Sesame	*Sesamum indicum* L.
Spearmint	*Mentha spicata* L.
Star anise	*Illicium verum* Hook. F.
Tarragon	*Artemisia dracunculus* L.
Thyme	*Thymus vulgaris* L.
Thyme, wild or creeping	*Thymus serpyllum* L.
Tumeric	*Curcuma longa* L.
Vanilla	*Vanilla planifolia* Andr. or *Vanilla tahitensis* J. W. Moore
Zedoary	*Curcuma zedoaria* Rosc.

(2) *Essential Oils, Oleoresins* (*Solvent-Free*)*, and Natural Extractives* (*including distillates*)

Common name	Botanical name of plant source
Alfalfa	*Medicago sativa* L.
Allspice	*Pimenta officinalis* Lindl.
Almond, bitter (free from prussic acid)	*Prunus amygdalus* Batsch, *Prunus armeniaca* L. or *Prunus persica* (L.) Batsch.
Ambrette (seed)	*Hibiscus moschatus* Moench.
Angelica root	*Angelica archangelica* L.
Angelica seed	*Angelica archangelica* L.
Angelica stem	*Angelica archangelica* L.
Angostura (cusparia bark)	*Gallpea officinalis* Hancock.
Anise	*Pimpinella anisum* L.
Asafetida	*Ferula asafoetida* L. and related spp. of *Ferula*.
Balm (lemon balm)	*Melissa officinalis* L.
Balsam of Peru	*Myroxylon perelrae* Klotzch.
Basil	*Ocimum basilicum* L.
Bay leaves	*Laurus nobilis* L.
Bay (myrcia oil)	*Pimenta racemosa* (Mill.) J. W. Moore.
Bergamot (bergamot orange)	*Citrus aurantium* L. subsp. *bergamia* Wright et Arn.
Bitter almond (free from prussic acid)	*Prunus amygdalus* Batsch, *Prunus armeniaca* L., or *Prunus persica* (L.) Batsch.
Bois de rose	*Aniba rosaeodora* Ducke.
Cacao	*Theobroma cacao* L.
Camomile (chamomile) flowers, Hungarian	*Matricaria chamomilla* L.

Common name	Botanical name of plant source
Camomile (chamomile) flowers, Roman or English	*Anthemis nobilis* L.
Cananga	*Cananage odorata* Hook. f. and Thomas.
Capsicum	*Capsicum frutescens* L. and *Capsicum annuum* L.
Caraway	*Cerum carvi* L.
Cardamom seed (cardamon)	*Elettaria cardamomum* Maton.
Carob bean	*Ceratonia siliqua* L.
Carrot	*Daucus carota* L.
Cascarilla bark	*Croton eluteria* Benn.
Cassia bark, Chinese	*Cinnamomum cassia* Blume.
Cassia bark, Padang or Batavia	*Cinnamomum Burmanni* Blume.
Cassia bark, Saigon	*Cinnamomum loureirli* Nees.
Celery seed	*Apium graveolens* L.
Cherry, wild, bark	*Prunus serotina* Ehrh.
Chervil	*Anthriscus cerefolium* (L.) Hoffm.
Chicory	*Cichorium intybus* L.
Cinnamon bark, Ceylon	*Cinnamomum zeylanicum* Nees.
Cinnamon bark, Chinese	*Cinnamomum cassia* Blume.
Cinnamon bark, Saigon	*Cinnamomum loureirli* Nees.
Cinnamon leaf, Ceylon	*Cinnamomum zeylanicum* Nees.
Cinnamon leaf, Chinese	*Cinnamomum cassia* Blume.
Cinnamon leaf, Saigon	*Cinnamomum loureirli* Nees.
Citronella	*Cymbopogon nardus* Rendle.
Citrus peels	*Citrus* spp.
Clary (clary sage)	*Salvia sclarea* L.
Clove bud	*Eugenia caryophyllata* Thunb.
Clove leaf	*Eugenia caryophyllata* Thunb.
Clove stem	*Eugenia caryophyllata* Thunb.
Clover	*Trifolium* spp.
Coca (decocainized)	*Erthroxylum coca* Lam. and other spp. of *Erythroxylum.*
Coffee	*Coffea* spp.
Cola nut	*Cola acuminata* Schott and Endl. and other spp. of *Cola.*
Coriander	*Coriandrum sativum* L.
Corn silk	*Zea mays* L.
Cumin (cummin)	*Cuminum cyminum* L.
Curacao orange peel (orange, bitter peel)	*Citrus aurantium* L.
Cusparia bark	*Galipea officinalis* Hancock.
Dandelion	*Taraxacum officinale* Weber and *T. laevigatum* DC.
Dandelion root	*Taraxacum officinale* Weber and *T. laevigatum* DC.

Common name	Botanical name of plant source
Dill	*Anethum graveolens* L.
Dog grass (quackgrass, triticum)	*Agropyron repens* (L.) Beauv.
Elder flowers	*Sambucus canadensis* L. and *S. nigra* L.
Estragole (esdragol, esdragon, tarragon)	*Artemisia dracunculus* L.
Estragon (tarragon)	*Artemisia dracunculus* L.
Fennel, sweet	*Foeniculum vulgare* Mill.
Fenugreek	*Trigonella foenum-graecum* L.
Galanga (galangal)	*Alpinia officinarum* Hance.
Garlic	*Allium sativum* L.
Geranium	*Pelargonium* spp.
Geranium, East Indian	*Cymbopogon martini* Stapf.
Geranium, rose	*Pelargonium graveolens* L'Her.
Ginger	*Zingiber officinale* Rosc.
Glycyrrhiza	*Glycyrrhiza glabra* L. and other spp. of *Glycyrrhiza.*
Glycyrrhizin, ammoniated	*Glycyrrhiza glabra* L. and other spp. of *Glycyrrhiza.*
Grapefruit	*Citrus paradisi* Macf.
Guava	*Psidium* spp.
Hickory bark	*Carya* spp.
Horehound (hoarhound)	*Marrubium vulgare* L.
Hops	*Humulus lupulus* L.
Horsemint	*Monarda punctata* L.
Hyssop	*Hyssopus officinalis* L.
Immortelle	*Helichrysum augustifolium* DC.
Jasmine	*Jasminum officinale* L. and other spp. *Jasminum.*
Juniper (berries)	*Juniperus communis* L.
Kola nut	*Cola acuminata* Schott and Endl. and other spp. of *Cola.*
Laurel berries	*Laurus nobilis* L.
Laurel leaves	*Laurus nobilis* L.
Lavender	*Lavandula officinalis* Chaix.
Lavender, spike	*Lavandula latifolia* Vill.
Lavandin	*Hybrids* between *Lavandula officinalis* Chaix and *Lavandula latifolia* Vill.
Lemon	*Citrus limon* (L.) Burm. f.
Lemon balm (see Balm)	
Lemon grass	*Cymbopogon citratus* DC. and *Cymbopogon flexuosus* Stapf.
Lemon peel	*Citrus limon* (L.) Burm. f.
Licorice	*Glycyrrhiza glabra* L. and other spp. of *Glycyrrhiza.*
Lime	*Citrus aurantifolia* Swingle.

Common name	Botanical name of plant source
Linden flowers	*Tilia* spp.
Locust bean	*Ceratonia siliqua* L.
Lupulin	*Humulus lupulus* L.
Mace	*Myristica fragrans* Houtt.
Malt (extract)	*Hordeum vulgare* L. or other grains.
Mandarin	*Citrus reticulata* Blanco.
Marjoram, sweet	*Majorana hortensis* Moench.
Maté	*Illex paraguariensis* St. Hil.
Melissa (see Balm)	
Menthol	*Mentha* spp.
Menthyl acetate	*Mentha* spp.
Molasses (extract)	*Saccharum officinarum* L.
Mustard	*Brassica* spp.
Naringin	*Citrus paradisi* Macf.
Neroli, bigarade	*Citrus aurantium* L.
Nutmeg	*Myristica fragrans* Houtt.
Onion	*Allium cepa* L.
Orange, bitter, flowers	*Citrus aurantium* L.
Orange, bitter, peel	*Citrus aurantium* L.
Orange leaf	*Citrus sinensis* (L.) Osbeck.
Orange, sweet	*Citrus sinensis* (L.) Osbeck.
Orange, sweet, flowers	*Citrus sinensis* (L.) Osbeck.
Orange, sweet, peel	*Citrus sinensis* (L.) Osbeck.
Origanum	*Origanum* spp.
Palmarosa	*Cymbopogon martini* Stapf.
Paprika	*Capsicum annuum* L.
Parsley	*Petroselinum crispum* (Mill.) Mansf.
Pepper, black	*Piper nigrum* L.
Pepper, white	*Piper nigrum* L.
Peppermint	*Mentha piperita* L.
Peruvian balsam	*Myroxylon peretrae* Klotzsch.
Petitgrain	*Citrus aurantium* L.
Petitgrain lemon	*Citrus limon* (L.) Burm. f.
Petitgrain mandarin or tangerine	*Citrus reticulata* Blanco.
Pimenta	*Pimenta officinalis* Lindl.
Pimenta leaf	*Pimenta officinalis* Lindl.
Pipsissewa leaves	*Chimaphila umbellata* Nutt.
Pomegranate	*Punica granatum* L.
Prickly ash bark	*Xanthoxylum* (or *Zanthoxylum*) *Americanum* Mill. or *Xanthoxylum clava-herculis* L.
Rose absolute	*Rosa alba* L., *Rosa centifolia* L., *Rosa damascena* Mill., *Rosa gallica* L., and vars. of these spp.

Common name	Botanical name of plant source
Rose (otto of roses, attar of roses)	*Rosa alba* L., *Rosa centifolia* L., *Rosa damascena* Mill., *Rosa gallica* L., and vars. of these spp.
Rose buds	*Rosa* spp.
Rose flowers	*Rosa* spp.
Rose fruit (hips)	*Rosa* spp.
Rose geranium	*Pelargonium graveolens* L'Her.
Rose leaves	*Rosa* spp.
Rosemary	*Rosmarinus officinalis* L.
Rue	*Ruta graveolens* L.
Saffron	*Crocus sativus* L.
Sage	*Salvia officinalis* L.
Sage, Greek	*Salvia triloba* L.
Sage, Spanish	*Salvia lavandulaefolia* Vahl.
St. John's bread	*Ceratonia siliqua* L.
Savory, summer	*Saturela hortensis* L.
Savory, winter	*Saturela montana* L.
Schinus molle	*Schinus molle* L.
Sloe berries (blackthorn berries)	*Prunus spinosa* L.
Spearmint	*Mentha spicata* L.
Spike lavender	*Lavandula latifolia* Vill.
Tamarind	*Tamarindus indica* L.
Tangerine	*Citrus reticulata* Blanco.
Tannic acid	Nutgalls of *Quercus infectoria* Oliver and related spp. of *Quercus*. Also in many other plants.
Tarragon	*Artemisia dracunculus* L.
Tea	*Thea sinensis* L.
Thyme	*Thymus vulgaris* L. and *Thymus zygis* var. *gracilis* Boiss.
Thyme, white	*Thymus vulgaris* L. and *Thymus zygis* var. *gracilis* Boiss.
Thyme, wild or creeping	*Thymus serpyllum* L.
Triticum (see Dog grass)	
Tuberose	*Polianthes tuberosa* L.
Turmeric	*Curcuma longa* L.
Vanilla	*Vanilla planifolia* Andr. or *Vanilla tahitensis* J. W. Moore.
Violet flowers	*Viola odorata* L.
Violet leaves	*Viola odorata* L.
Violet leaves absolute	*Viola odorata* L.
Wild cherry bark	*Prunus serotina* Ehrh.
Ylang-Ylang	*Cananga odorata* Hook. f. and Thomas.
Zedoary bark	*Curcuma zedoaria* Rosc.

(3) Natural Substances Used in Conjunction with Spices and Other Natural Seasonings and Flavorings

Common name	Botanical name of plant source
Algae, brown (kelp)	*Laminaria* spp. and *Nereocystis* spp.
Algae, red	*Porphyra* spp. and *Rhodymenia palmata* L. Grev.
Dulse	*Rhodymenia palmata* L.

(4) Natural Extractives (Solvent-Free) Used in Conjunction with Spices, Seasonings and Flavorings

Common name	Botanical name of plant source
Algae, brown	*Laminaria* spp. and *Nereocystis* spp.
Algae, red	*Porphyra* spp. and *Rhodymenia palmata* L. Grev.
Apricot kernel (persic oil)	*Prunus armeniaca* L.
Dulse	*Rhodymenia palmata* L. Grev.
Kelp (see Algae, brown)	
Peach kernel (persic oil)	*Prunus persica* Sieb. et Zucc.
Peanut stearine	*Arachis hypogaea* L.
Persic oil (see Apricot kernel and Peach kernel)	
Quince seed	*Cydonia oblonga* Miller.

(5) Miscellaneous

Common name	Derivation
Ambergris	*Physeter macrocephalus* L.
Castoreum	*Castor fiber* L. and *C. canadensis* Kuhl.
Civet (zibeth, zibet, zibetum)	Civet cats, *Viverra civetta* Schreber and *Viverra zibetha* Schreber.
Cognac oil, white and green	Ethyl cenanthate, so-called
Musk (Tonquin musk)	Musk deer, *Moschus moschiferus* L.

SUBPART E. SUBSTANCES FOR WHICH PRIOR SANCTIONS HAVE BEEN GRANTED

§121.2001. Substances Employed in the Manufacture of Food Packaging Materials

The following substances were delisted on October 15, 1968.*

(f) Release agents

Polyethylene glycol 400
Polyethylene glycol 1500
Polyethylene glycol 4000

(g) Substances Used in the Manufacture of Paper and Paperboard Products in Food Packaging

Pyrethrins in combination with piperonyl butoxide in outside plies of multiwall bags were delisted on September 2, 1966.†

II. Sections of Food, Drug, and Cosmetic Act Related to Polymer Food Wraps

SUBPART D. FOOD ADDITIVES PERMITTED IN FOOD FOR HUMAN CONSUMPTION

§121.1029. Sorbitan Monostearate

The food additive, sorbitan monostearate, which is a mixture of partial stearic and palmitic acid esters of sorbitol anhydrides, may be safely used in or on food in accordance with the following prescribed conditions:

(a) The food additive is manufactured by reacting stearic acid (usually containing associate fatty acids, chiefly palmitic) with sorbitol to yield essentially a mixture of esters.

(b) The food additive meets the following specifications:

Saponification number	147–157
Acid number	5– 10
Hydroxyl number	235–260

* **33** F.R. 15281.
† **31** F.R. 11608.

(c) It is used, or intended for use, alone or in combination with polyoxyethylene (20) sorbitan monostearate, as follows:

(1) As an emulsifier in whipped vegetable oil topping with or without one or a combination of the following:

(i) Polysorbate 60 [polyoxyethylene (20) sorbitan monostearate];
(ii) Polyoxyethylene (20) sorbitan tristearate;
(iii) Polysorbate 80;

whereby the maximum amount of the additive or additives used does not exceed 0.4% of the weight of the finished whipped vegetable oil topping, except that a combination of the additive with polysorbate 60 [polyoxyethylene (20) sorbitan monostearate] may be used in excess of 0.4%, provided that the amount of the additive does not exceed 0.27% and the amount of polysorbate 60 [polyoxyethylene (20) sorbitan monostearate] does not exceed 0.77% of the weight of the finished whipped vegetable oil topping.

(2) As an emulsifier . . . .

§121.1030.   Polysorbate 60

The food additive, polysorbate 60 [polyoxyethylene (20) sorbitan monostearate], which is a mixture of polyoxyethylene ethers of mixed partial stearic and palmitic acid esters of sorbitol anhydrides and related compounds, may be safely used in food in accordance with the following prescribed conditions:

(a) The food additive is manufactured by reacting stearic acid (usually containing associated fatty acids, chiefly palmitic) with sorbitol to yield a product with a maximum acid number of 10 and a maximum water content of 0.2%, which is then reacted with ethylene oxide. . . .

(c) It is used or intended for use as follows:

(1) As an emulsifier . . . .
(10) As a foaming agent in nonalcoholic mixes to be added to alcoholic beverages in the preparation of mixed alcoholic drinks at a level not to exceed 4.5% by weight of the nonalcoholic mix.
*(11) As a dough conditioner in yeast-leavened bakery products in an amount not to exceed 0.5% by weight of the flour used.
*(12) As an emulsifier, alone or in combination with sorbitan monostearate, in the minimum quantity required to accomplish the intended effect, in formulations of white mineral oil conforming with §121.1146 and/or petroleum wax conforming with §121.1156 for use as protective coatings on raw fruits and vegetables.
† (13) As a dispersing agent in artificially sweetened gelatin desserts and in artificially sweetened gelatin dessert mixes whereby the amount of the additive does not exceed 0.5% on a dry-weight basis.

* Amendments published in the *Federal Register*, March 27, 1969.   **34** F.R. 5720, 5721.
† Amendment published in the *Federal Register*, April 19, 1969.   **34** F.R. 6684.

§121.1160. Hydroxypropyl Cellulose

* The food additive hydroxypropyl cellulose may be safely used in food, except standardized foods that do not provide for such use, in accordance with the following prescribed conditions:

(a) The additive is a cellulose ether containing propylene glycol groups attached by an ether linkage and contains, on an anhydrous basis, not more than 4.6 hydroxypropyl groups per anhydroglucose unit. The additive has a minimum viscosity of 75 cp for 5% by weight aqueous solution at 25°C.

(b) The additive is used or intended for use as an emulsifier, film former, protective colloid, stabilizer, suspending agent, or thickener, in accordance with good manufacturing practice.

SUBPART F. FOOD ADDITIVES RESULTING FROM CONTACT WITH CONTAINERS OR EQUIPMENT, AND FOOD ADDITIVES OTHERWISE AFFECTING FOOD

†§121.2500. General Provisions Applicable to Subpart F

(a) Regulations prescribing conditions under which food additive substances may be safely used predicate usage under conditions of good manufacturing practice. For the purpose of this Subpart F good manufacturing practice shall be defined to include the following restrictions:

(1) The quantity of any food additive substance that may be added to food as a result of use in articles that contact food shall not exceed, where no limits are specified, that which results from use of the subtance in an amount not more than reasonably required to accomplish the intended physical or technical effect in the food-contact article; shall not exceed any prescribed limitations; and shall not be intended to accomplish any physical or technical effect in the food itself, except as such may be permitted by regulations in this Part 121.

(2) Any substance used as a component of articles that contact food shall be of a purity suitable for its intended use.

(b) The existence in this Subpart F of a regulation prescribing safe conditions for the use of a substance as an article or component of articles that contact food shall not be construed to relieve such use of the substance or article from compliance with any other provision of the Federal Food, Drug, and Cosmetic Act. For example, if a regulated food-packaging material were found on appropriate test to impart odor or taste to a specific food product such as to render it unfit within the meaning of section 402 (a)(3) of the act, the regulation would not be construed to relieve such use from compliance with section 402 (a)(3).

(c) The existence in this Subpart F of a regulation prescribing safe conditions for the use of a substance as an article or component of articles that contact food shall not be construed as implying that such substance may be safely used as a direct additive in food.

* Amendment published in the *Federal Register*, December 23, 1965. **30** F.R. 15922.

† Amendment published in the *Federal Register*, March 19, 1964. **29** F.R. 3523.

(d) Substances that under conditions of good manufacturing practice may be safely used as components of articles that contact food include the following, subject to any prescribed limitations:

(1) Substances generally recognized as safe in or on food.

(2) Substances generally recognized as safe for their intended use in food packaging.

(3) Substances used in accordance with a prior sanction or approval.

(4) Substances permitted for use by regulations in this Subpart F.

§121.2501.   Olefin Polymers

The olefin polymers listed in paragraph (a) of this section may be safely used as articles or components of articles intended for use in contact with food, subject to the provisions of this section.

(a) For the purpose of this section, olefin polymers are basic polymers manufactured as described in this paragraph, so as to meet the specifications prescribed in paragraph (c) of this section, when tested by the methods described in paragraph (d) of this section.

(1) Polypropylene . . .

(2) Polyethylene . . .

(3) Olefin basic copolymers consist of basic copolymers manufactured by the catalytic copolymerization of:

(i) Two or more of the 1-alkenes having 2 to 8 carbon atoms.   Such olefin basic copolymers contain not less than 96 weight-percent of polymer units derived from ethylene and/or propylene, except that olefin basic copolymers manufactured by the catalytic copolymerization of two or more of the monomers, ethylene, propylene, butene-1,2-methylpropene-1, and 2,4,4-trimethylpentene-1 shall contain not less than 85 weight-percent of polymer units derived from ethylene and/or propylene; or

*(ii) 4-Methylpentene-1 and 1-alkenes having 6 to 10 carbon atoms.   Such olefin basic copolymers shall contain not less than 95 mole-percent of polymer units derived from 4-methylpentene-1; or

*(iii) Ethylene and propylene that may contain as modifiers not more than 5 weight-percent of total polymer units derived by copolymerization with one or more of the following monomers: 5-ethylidine-2-norbornene, 5-methylene-2-norbornene.

(4) Polymethylpentene consists of basic polymers manufactured by the catalytic polymerization of 4-methylpentene-1.

(b) The basic olefin polymers identified in paragraph (a) of this section may contain optional adjuvant substances required in the production of such basic olefin polymers. The optional adjuvant substances required in the production of the basic olefin polymers may include substances permitted for such use by applicable regulations in this Part 121, substances generally recognized as safe in food and food packaging, and substances used in accordance with a prior sanction or approval.

(c) Specifications: [opposite page]

(d) [Analytical procedures omitted.]

* Amendment published in the *Federal Register*, August 31, 1968.   **33** F.R. 12309.

Olefin polymers	Density	Melting or softening temperature (°C)	Maximum extractable fraction (expressed as percent by weight of polymer) in n-hexane at specified temperatures	Maximum soluble fraction (expressed as percent by weight of polymer) in xylene at specified temperatures
1.1 Polypropylene	0.880–0.913	m, 160–180	6.4% at reflux temperature	9.8% at 25°C
1.2 Polypropylene, noncrystalline . . .	0.80–0.88			
1.3 Polypropylene, noncrystalline, for use only: To plasticize polypropylene . . .	0.80–0.88	s, 115–138		
2.1 Polyethylene for use in articles that contact food . . .	0.85–1.00		5.5% at 50°C	11.3% at 25°C
2.2 Polyethylene for use in articles used for packing or holding food during cooking	0.85–1.00		2.6% at 50°C	11.3° at 25°C
2.3 Polyethylene for use only as component of food-contact coatings . . .	0.85–1.00		53.0% at 50°C	75.0% at 25°C
3.1 Olefin copolymers described in paragraph (a) (3) (i) of this section for use in articles that contact food . . .	0.85–1.00		5.5% at 50°C	30.0% at 25°C
3.2 Olefin copolymers described in (a) (3) (i) of this section for use in articles used for packing or holding food during cooking	0.85–1.00		2.6% at 50°C	30.0% at 25°C
3.3 Olefin copolymers described in paragraph (a) (3) (ii) of this section	0.82–0.85	m, 235–250	6.6% at reflux temperature	7.5% at 25°C
*3.4 Olefin copolymers, primarily non-crystalline described in paragraph (a) (3) (iii) of this section . . .	0.85–0.90			
*4 Polymethylpentene	0.82–0.85	m, 235–250	6.6% at reflux temperature	*7.5% at 25°C

* Amendment published in the *Federal Register*, August 31, 1968.   **33** F.R. 12309.

§121.2502. *Nylon Resins*

The nylon resins listed in paragraph (a) of this section may be safely used to produce articles intended for use in processing, handling, and packaging food, subject to the provisions of this section:

(a) The nylon resins are manufactured as described in this paragraph so as to meet the specifications prescribed in paragraph (b) of this section when tested by the methods described in paragraph (c) of this section.

(1) Nylon-6,6 resins are manufactured by the condensation of hexamethylenediamine and adipic acid.

(2) Nylon-6,10 resins are manufactured by the condensation of hexamethylenediamine and sebacic acid.

(3) Nylon-6,6/6,10 resins are manufactured by the condensation of equal-weight mixtures of nylon-6,6 salts and nylon-6,10 salts.

(4) Nylon-6/6,6 resins are manufactured by the condensation and polymerization of mixtures of nylon-6,6 salts and ε-caprolactam under such conditions that the ε-caprolactam monomer content does not exceed 0.7% by weight of the finished nylon-6/6,6 resins.

(5) Nylon-11 resins are manufactured by the condensation of 11-aminoundecanoic acid.

(6) Nylon 6 resins are manufactured by the polymerization of ε-caprolactam.

(b) Specifications:

Nylon resins	Specific gravity	Melting point (°F)	Solubility in boiling 4.2N HCl	Maximum extractable fraction in selected solvents (expressed as percent by weight of resin)			
				Water	95% ethyl alcohol	Ethyl acetate	Benzene
Nylon-6,6 resins	1.14 ± 0.015	475–495	Dissolves in 1 hr	1.5	1.5	0.2	0.2
Nylon-6,10 resins	1.09 ± 0.015	405–425	Insoluble after 1 hr	1.0	2.0	1.0	1.0
Nylon-6,6/6,10 resins	1.10 ± 0.015	375–395	Dissolves in 1 hr	1.5	2.0	1.0	1.0
Nylon-6/6,6 resins	1.13 ± 0.015	440–460	Dissolves in 1 hr	2.0	2.0	1.5	1.5
Nylon-11 resins for use in articles intended for one-time use or repeated use in contact with food	1.04 ± 0.015	355–375	Insoluble after 1 hr	*0.30	0.35	0.25	0.30
Nylon-11 resins for use only:							
a. In articles intended for repeated use in contact with food	1.04 ± 0.015	355–375	Insoluble after 1 hr	*0.35	1.60	0.35	0.40
b. In side-seam cements for articles intended for one-time use in contact with food and which are in compliance with 121.2514							
Nylon-6 resins	1.15 ± 0.015	392–446	Dissolves in 1 hr	1.0	2.0	1.0	1.0
Nylon-6 resins for use only in food-contact films having an average thickness not to exceed 0.001 in.	1.15 ± 0.015	392–446	Dissolves in 1 hr	1.5	2.0	1.0	1.0

* Amendment published in the *Federal Register*, September 2, 1967. **32** F.R. 12716.

(c) [Analytical methods omitted.]

§121.2507.   Cellophane

Cellophane may be safely used for packaging food in accordance with the following prescribed conditions:

(a) Cellophane consists of a base sheet made from regenerated cellulose to which have been added certain optional substances of a grade of purity suitable for use in food packaging as constituents of the base sheet or as coatings applied to impart desired technological properties.

(b) Subject to any limitations prescribed in this part, the optional substances used in the base sheet and coating may include:

(1) Substances generally recognized as safe in food.

(2) Substances for which prior approval or sanctions permit their use in cellophane, under conditions specified in such sanctions and substances listed in Subpart E, §121.2001 [Volume I, p. 843].

(3) Substances that by any regulation promulgated under section 409 of the act may be safely used as components of cellophane.

(4) Substances named in this section and further identified as required.

(c) List of substances:

Substance	Limitations (residue and limits of addition expressed as percent by weight of finished packaging cellophane)
Acrylonitrile-butadiene copolymer resins	As the basic polymer
Acrylonitrile-butadiene-styrene copolymer resins	As the basic polymer
Acrylonitrile-styrene copolymer resins	As the basic polymer
Acrylonitrile-vinyl chloride copolymer resins	As the basic polymer
[a]N-Acylsarcosines where the acyl group is lauroyl or stearoyl	For use only as release agents in coatings at levels not to exceed a total of 0.3% by weight of the finished packaging cellophane
Alkyl ketene dimers identified in §121.2538	
Aluminum hydroxide	
Aluminum silicate	
Ammonium persulfate	
Ammonium sulfate	
Behenamide	
Butadiene-styrene copolymer	As the basic polymer
1,3-Butanediol	
n-Butyl acetate	0.1%
n-Butyl alcohol	0.1%
Calcium ethyl acetoacetate	
Carboxymethyl hydroxyethylcellulose polymer	
Castor oil, hydrogenated	
Castor oil phthalate with adipic acid and fumaric acid-diethylene glycol polyester	As the basic polymer
Castor oil phthalate, hydrogenated	Alone or in combination with other phthalates where total phthalates do not exceed 5%
Castor oil, sulfonated, sodium salt	

[a] Amendment published in the *Federal Register*, March 15, 1968. **33** F.R. 4575.

700

Substance	Limitations (residue and limits of addition expressed as percent by weight of finished packaging cellophane)
Cellulose acetate butyrate	
Cellulose acetate propionate	
Cetyl alcohol	
Clay, natural	
Coconut oil fatty acid (C_{12}–C_{18}) diethanolamide coconut oil fatty acid (C_{12}–C_{18}) diethanolamine soap, and diethanolamine mixture having total alkali (calculated as potassium hydroxide) of 16–18% and having an acid number of 25–35	For use only as an adjuvant employed during the processing of cellulose pulp used in the manufacture of cellophane base sheet
Copal resin, heat processed	As basic resin
Damar resin	
Defoaming agents identified in §121.2557	
Dialkyl ketones where the alkyl groups are lauryl or stearyl	Not to exceed a total of 0.35%
Dibutyl phthalate	Alone or in combination with other phthalates where total phthalates do not exceed 5%
Dicyclohexyl phthalate	Alone or in combination with other phthalates where total phthalates do not exceed 5%
Diethylene glycol ester of the adduct of terpene and maleic anhydride	
Di(2-ethylhexyl) adipate	
Di(2-ethylhexyl) phthalate	Alone or in combination with other phthalates where total phthalates do not exceed 5%
[b]Diisobutyl phthalate	Alone or in combination with other phthalates where total phthalates do not exceed 5%
Dimethylcyclohexyl phthalate	Alone or in combination with other phthalates where total phthalates do not exceed 5%

[b] Amendment published in the *Federal Register*, November 7, 1968.   **33** F.R. 16334.

Substance	Limitations (residue and limits of addition expressed as percent by weight of finished packaging cellophane)
Dimethyldialkyl (C$_8$–C$_{18}$) ammonium chloride	0.005 % for use only as a flocculant for slip agents
Di-n-octyltin-bis(2-ethylhexyl maleate)	For use only as a stabilizer at a level not to exceed 0.55 % by weight of the coating solids in vinylidene chloride copolymer waterproof coatings prepared from vinylidene chloride copolymers identified in this paragraph, provided that such vinylidene chloride copolymers contain not less than 90 % by weight of polymer units derived from vinylidene chloride
N,N′-Dioleoylethylenediamine, N,N′-dilinoleoylethylenediamine, and N-oleoyl-N′-linoleoylethylenediamine mixture produced when tall oil fatty acids are made to react with ethylenediamine such that the finished mixture has a melting point of 212–228°F, as determined by ASTM Method D 127–60, and an acid value of 10 maximum	0.5 %
N,N′-Dioleoyl ethylenediamine[N,N′-ethylene-bis(oleamide)]	
Disodium salt of ethylenediaminetetraacetic acid	
Distearic acid ester of di(hydroxyethyl)diethylenetriamine mono-acetate	0.06 %
N,N′-Distearylethylenediamine (N,N′-ethylene-bis(stearamide)	
Epoxidized polybutadiene	For use only as a primer subcoat to anchor surface coatings to the base sheet
Erucamide	
Ethyl acetate	
Ethylene-vinyl acetate copolymers complying with §121.2570	
2-Ethylhexyl alcohol	0.1 %, for use only as a lubricant
Fatty acids derived from animal and vegetable fats and oils, and the following salts of such acids, single or mixed: aluminum, ammonium, calcium, magnesium, potassium, sodium	

701

Substance	Limitations (residue and limits of addition expressed as percent by weight of finished packaging cellophane)
Ferrous ammonium sulfate	
Fumaric acid	
°Glycerol-maleic anhydride	As the basic polymer
Glycerol diacetate	
Glycerol monoacetate	
Hydroxyethyl cellulose, water-insoluble	
Hydroxypropyl cellulose identified in §121.1160	
Isopropyl acetate	Residue limit 0.1%
Isopropyl alcohol	Residue limit 0.1%
Itaconic acid	
Lanolin	
Lauryl alcohol	
Lauryl sulfate salts: ammonium, magnesium, potassium, sodium	
Maleic acid adduct of butadiene-styrene copolymer	
Melamine-formaldehyde	As the basic polymer
Melamine-formaldehyde modified with one or more of the following: butyl alcohol, diaminopropane, diethylenetriamine, ethyl alcohol, guanidine, iminobis-propylamine, methyl alcohol, polyamines made by reacting ethylenediamine or trimethylene-diamine with dichloroethane or dichloropropane, sulfanilic acid, tetraethylenepentamine, triethanolamine, triethylene-tetramine, iminobis-ethylamine	As the basic polymer and used as a resin to anchor coatings to substrate
Methyl ethyl ketone	Residue limit 0.1%
Methyl hydrogen siloxane	0.1% as the basic polymer

° Items, *glycerin ester of dimerized rosin* and *maleic acid* were deleted by amendment published in the *Federal Register*, December 23, 1964. **29** F.R. 1817.

702

Substance	Limitations (residue and limits of addition expressed as percent by weight of finished packaging cellophane)
Mineral oil, white	
Naphthalenesulfonic acid-formaldehyde condensate, sodium salt	0.1%, for use only as an emulsifier
Nitrocellulose, 10.9–12.2% nitrogen	
Nylon resins complying with §121.2502	
n-Octyl alcohol	For use only as a defoaming agent in the manufacture of cellophane base sheet
Olefin copolymers complying with §121.2501	
Oleic acid reacted with N-alkyl trimethylenediamine (alkyl C_{16} to C_{18})	
Oleic acid, sulfonated, sodium salt	
Oleyl palmitamide	
N,N'-Oleoyl-stearylethylenediamine (N-(2-stearoylaminoethyl)-oleamide)	
Paraffin, synthetic, complying with §121.2575	
Pentaerythritol tetrastearate	0.1%
Polyamide resins derived from dimerized vegetable oil acids (containing not more than 20% of monomer acids) and ethylenediamine as the basic resin	For use only in cellophane coatings that contact food at temperatures not to exceed room temperature
dPolyamide resins having a maximum acid value of 5 and a maximum amine value of 8.5 derived from dimerized vegetable oil acids (containing not more than 10% of monomer acids), ethylenediamine, and 4,4-bis(4-hydroxyphenyl) pentanoic acid (in an amount not to exceed 10% by weight of said polyamide resins)	As the basic resin for use only in coatings that contact food at temperatures not to exceed room temperature, provided that the concentration of the polyamide resins in the finished food-contact coating does not exceed 5 mg/in.^{2} of food-contact surface

d Amendment published in the *Federal Register*, October 23, 1968.   **33** F.R. 15654.

Substance	Limitations (residue and limits of addition expressed as percent by weight of finished packaging cellophane)
Polybutadiene resin (molecular weight range 2,000–10,200; bromine number range 210–320)	For use only as an adjuvant in vinylidene chloride copolymer coatings
Polycarbonate resins complying with §121.2574	
Polyester resin formed by the reaction of the methyl ester of rosin, phthalic anhydride, maleic anhydride, and ethylene glycol, such that the polyester resin has an acid number of 4 to 11, a drop-softening point of 70–92°C, and a color of K or paler	
Polyethylene	
Polyethyleneaminostearamide ethyl sulfate produced when stearic acid is made to react with equal parts of diethylene-triamine and triethylenetetramine and the reaction product is quaternized with diethyl sulfate	0.1%
°Polyethylene glycol (400) monolaurate	
Polyethylene glycol (600) monolaurate	
Polyethylene glycol (400) monooleate	
Polyethylene glycol (600) monooleate	
Polyethylene glycol (400) monostearate	
Polyethylene glycol (600) monostearate	
Polyethylene, oxidized; complying with the identity prescribed in §121.2517 (a)	
Polyethylenimine	As the basic polymer, for use as a resin to anchor coatings to the substrate and for use as an impregnant in the food-contact surface of regenerated cellulose sheet, in an amount not to exceed that required to improve heat-sealable bonding between coated and uncoated sides of cellophane

° Polyethylene glycol (molecular weight greater than 300) was deleted by amendment published in the Federal Register, October 15, 1968. **33** F.R. 15281.

704

Substance	Limitations (residue and limits of addition expressed as percent by weight of finished packaging cellophane)
Polyisobutylene complying with §121.2590	
Polyoxypropylene-polyoxyethylene block polymers (molecular weight 1900–9000)	For use as an adjuvant employed during the processing of cellulose pulp used in the manufacture of cellophane base sheet
Polypropylene complying with §121.2501	
Polysorbate 60 [polyoxyethylene (20) sorbitan monostearate] conforming to the identity prescribed in 121.1030	
Polystyrene	As the basic polymer
Polyvinyl acetate	As the basic polymer
Polyvinyl alcohol (minimum viscosity of 4% aqueous solution at 20°C of 4 cp)	
Polyvinyl chloride	As the basic polymer
Polyvinyl stearate	As the basic polymer
n-Propyl acetate	Residue limit 0.1%
n-Propyl alcohol	Residue limit 0.1%
Rapeseed oil, blown	
Rosins and rosin derivatives as provided in §121.2592	
Rubber, natural (natural latex solids)	
Silica	
Silicic acid	
Sodium metabisulfite	
Sodium dioctyl sulfosuccinate	
Sodium dodecylbenzenesulfonate	
Sodium lauroyl sarcosinate	0.35%; for use only in vinylidene chloride copolymer coatings
Sodium oleyl sulfate-sodium cetyl sulfate mixture	For use only as an emulsifier for coatings; limit 0.005% where coating is applied to one side only and 0.01% where coating is applied to both sides

Substance	Limitations (residue and limits of addition expressed as percent by weight of finished packaging cellophane)
Sodium silicate	
Sodium sulfate	
Sodium sulfite	
Sorbitan monopalmitate	
Sorbitan monostearate conforming to the identity prescribed in §121.1029	
fSorbitan trioleate	
Sorbitan tristearate	
Spermaceti wax	
Stannous oleate	
Stearyl alcohol	
Styrene-maleic anhydride resins	As the basic polymer
Terpene resins identified in §121.1059	
Tetrahydrofuran	Residue limit of 0.1%
Titanium dioxide	
Toluene	Residue limit of 0.1%
Toluene sulfonamide formaldehyde	0.6% as the basic polymer
Triethylene glycol	
Triethylene glycol diacetate, prepared from triethylene glycol containing not more than 0.1% of diethylene glycol	
g2,2,4-Trimethyl-1,3-pentanediol diisobutyrate	For use only in cellophane coatings and limited to use at a level not to exceed 10% by weight of the coating solids except when used as provided in §121.2511
Urea (carbamide)	

---

f *Sorbitan monooleate* deleted by amendment published in the *Federal Register*, June 25, 1968. **33** F.R. 9288.

g Amendment published in the *Federal Register*, November 7, 1968. **33** F.R. 16334.

Substance	Limitations (residue and limits of addition expressed as percent by weight of finished packaging cellophane)
Urea-formaldehyde	As the basic polymer
Urea-formaldhyde modified with methanol, ethanol, butanol, diethylenetriamine, triethylenetetramine, tetraethylenepentamine, guanidine, sodium sulfite, sulfanilic acid, imino-bis-ethylamine, imino-bis-propylamine, imino-bis-butylamine, diaminopropane, diaminobutane, aminomethylsulfonic acid, polyamines made by reacting ethylenediamine or trimethylenediamine with dichloroethane or dichloropropane	As the basic polymer, and used as a resin to ancvor coatings to the subtrate
Vinyl acetate-vinyl chloride copolymer resins	As the basic polymer
Vinyl acetate/vinyl chloride/maleic acid copolymer resins	As the basic polymer
Vinylidene chloride copolymerized with one or more of the following: acrylic acid, acrylonitrile, butyl acrylate, butyl methacrylate, ethyl acrylate, 2-ethylhexyl acrylate, 2-ethylhexyl methacrylate, ethyl methacrylate, itaconic acid, methacrylic acid, methyl acrylate, methyl methacrylate, propyl acrylate, propyl methacrylate, vinyl chloride	As the basic polymer
Vinylidene chloride-methacrylate decyloctyl copolymer	As the basic polymer
Wax, petroleum, complying with §121.2586	

* (d) Any optional component listed in this section covered by a specific food additive regulation must meet any specifications in that regulation.

* Amendment published in the *Federal Register*, December 10, 1966. **31** F.R. 15570.

§121.2510.   *Polystyrene and Rubber-Modified Polystyrene*

Polystyrene and rubber-modified polystyrene identified in this section may be safely used as components of articles intended for use in contact with food, subject to the provisions of this section.

(a) Identity.   For the purposes of this section, polystyrene and rubber-modified polystyrene are basic polymers manufactured as described in this paragraph . . .

(1) Polystyrene consists of basic polymers produced by the polymerization of styrene.

(2) Rubber-modified polystyrene consists of basic polymers produced by combining styrene-butadiene copolymers and/or polybutadiene with polystyrene, either during or after polymerization of the polystyrene, such that the finished basic polymers contain not less than 75 % of total polymer units derived from styrene monomer.

(b) [Optional adjuvants omitted.]
(c) [Specifications omitted.]
(d) [Analytical methods omitted.]

§121.251$\overline{0}$.   *Polyethylene*

Polyethylene may be safely used to produce packaging materials, containers, and equipment intended for use in producing, manufacturing, packing, processing, preparing, treating, packaging, transporting, or holding food, in accordance with the following prescribed conditions:

(a) Polyethylene consists of a basic resin produced by the catalytic polymerization of ethylene to which may have been added certain optional substances of a grade of purity suitable for use in contact with food to impart desired technological properties. Subject to any limitation prescribed in this section, the optional substances may include:

(1) Substances generally recognized as safe in food and food packaging.

(2) Substances the use of which is permitted under regulations in this part, prior sanctions, or approvals.

[Specifications and analytical methods omitted.]

§121.2511.   *Plasticizers in Polymeric Substances*

Subject to the provisions of this regulation, the substances listed in paragraph (b) of this section may be safely used as plasticizers in polymeric substances used in the manufacture of articles or components of articles intended for use in producing, manufacturing, packing, processing, preparing, treating, packaging, transporting, or holding food.

(a) The quantity used shall not exceed the amount reasonably required to accomplish the intended technical effect.

(b) List of substances:

Substance	Limitations
Butyl benzyl phthalate	For use only: 1. As provided in §§121.2520 and 121.2571 2. In polymeric substances used in food-contact articles complying with §§121.2514, 121.2526, or 121.2569: Provided, That the butyl benzyl phthalate contain not more than 1 % by weight of dibenzyl phthalate 3. In polymeric substances used in other permitted food-contact articles: Provided, That the butyl benzyl phthalate contain not more than 1 % by weight of dibenzyl phthalate; and Provided further, That the finished food-contact article, when extracted with the solvent or solvents characterizing the type of food and under the conditions of time and temperature characterizing the conditions of its intended use as determined from tables 1 and 2 of §121.2514(d), shall yield net chloroform-soluble extractives not to exceed 0.5 mg/in.^{2}, as determined by the methods prescribed in §121.2514(e)
Dicyclohexyl phthalate	For use only: 1. As provided in §§121.2507, 121.2520, 121.2526, and 121.2571 2. Alone or in combination with other phthalates, in polyvinyl chloride and polyvinyl acetate film and sheet that contact food at temperatures not to exceed room temperature provided that total phthalates, calculated as phthalic acid, do not exceed 10% by weight of the finished film or sheet
Di-2-ethylhexyl adipate Di-n-hexyl azelate	Not to exceed 15% by weight of finished article
Dihexyl phthalate	For use only: 1. As provided in §121.2520 2. In articles that contact food only of the types identified in §121.2526 (c), table 1, under categories I, II, IV-B, VI B, and VIII
Diphenyl phthalate	For use only: 1. As provided in §121.2520 2. Alone or in combination with other phthalates, in polyvinyl chloride and polyvinyl acetate film and sheet that contact food at temperatures not to exceed room temperature, provided that total phthalates, calculated as phthalic acid, do not exceed 10% by weight of the finished film or sheet

Substance	Limitations
Epoxidized butyl esters of linseed oil fatty acids	Iodine number, maximum 5; oxirane oxygen, minimum 7.5%
Epoxidized linseed oil	Iodine number, maximum 5; oxirane oxygen, minimum 9%
Mineral oil, white	
*Polybutene, hydrogenated (minimum viscosity at 210°F, 39 Saybolt Universal seconds, as determined by ASTM Methods D 445 and D 2161; and bromine number of 3 or less as determined by ASTM Method D 1492)	For use only: 1. In polymeric substances used in contact with nonfatty food 2. In polyethylene complying with §121.2501 and used in contact with fatty food, provided that the hydrogenated polybutene is added in an amount not to exceed 0.5% by weight of the polyethylene, and further provided that such plasticized polyethylene shall not be used as a component of articles intended for packing or holding food during cooking 3. In polystyrene complying with §121.2510 and used in contact with fatty food, provided that the hydrogenated polybutene is added in an amount not to exceed 5% by weight of the polystyrene, and further provided that such plasticized polystyrene shall not be used as a component of articles intended for packing or holding food during cooking
Polyisobutylene (mol. wt. 300–5,000)	For use in polyethylene complying with §121.2501 provided that the polyisobutylene is added in an amount not exceeding 0.5% weight of the polyethylene, and further provided that such plasticized polyethylene shall not be used as a component of articles intended for packing or holding food during cooking
Polyisobutylene complying with §121.2590	
Triethylene glycol	Diethylene glycol content not to exceed 0.1%
2,2,4-Trimethyl-1, 3-pentanediol diisobutyrate	For use only in cellulosic plastics in an amount not to exceed 15% by weight of the finished food-contact article, provided that the finished plastic article contacts food only of the types identified in §121.2526 (c), table 1, under categories I, II, VI-B, VII-B, and VIII

* Amendment published in the *Federal Register*, January 17, 1968. **33** F.R. 567.

(c) The use of the plasticizers in any polymeric substance or article subject to any regulation in this Subpart F must comply with any specifications and limitations prescribed by such regulation for the finished form of the substance or article.

§121.2512.   Acrylamide-Acrylic Acid Resins

Acrylamide-acrylic acid resins may be safely used as components of articles intended for use in producing, manufacturing, packing, processing, preparing, treating, packaging, transporting, or holding food, subject to the provisions of this section.

(a) Acrylamide-acrylic acid resins are produced by the polymerization of acrylamide with partial hydrolysis or by the copolymerization of acrylamide and acrylic acid.

(b) The acrylamide-acrylic acid resins contain less than 0.2% monomer.

(c) The resins are used as adjuvants in the manufacture of paper and paperboard in amounts not to exceed that necessary to accomplish the technical effect and not to exceed 2% by weight of the paper or paperboard.

§121.2514.   Resinous and Polymeric Coatings

Resinous and polymeric coatings may be safely used as the food-contact surface of articles intended for use in producing, manufacturing, packing, processing, preparing, treating, packaging, transporting, or holding food, in accordance with the following prescribed conditions:

(a) The coating is applied as a continuous film or enamel over a metal substrate, or the coating is intended for repeated food-contact use and is applied to any suitable substrate as a continuous film or enamel that serves as a functional barrier between the food and the substrate.   The coating is characterized by one or more of the following descriptions:

(1) Coatings cured by oxidation.

(2) Coatings cured by polymerization, condensation, and/or cross-linking without oxidation.

(3) Coatings prepared from prepolymerized substances.

(b) The coatings are formulated from optional substances that may include:

(1) Substances generally recognized as safe in food.

(2) Substances the use of which is permitted by regulations in this part or which are permitted by prior sanction or approval and employed under the specific conditions, if any, of the prior sanction or approval.

*(3) Any substance employed in the production of resinous and polymeric coatings that is the subject of a regulation in this Subpart F and conforms with any specification in such regulation.   Substances named in this subparagraph and further identified as required . . .

[Extensive tables omitted.]

* Amendment published in the *Federal Register*, December 10, 1966.   **31** F.R. 15570.

(c) The coating in the finished form in which it is to contact food, when extracted with the solvent or solvents characterizing the type of food, and under conditions of time and temperature characterizing the conditions of its intended use as determined from Tables 1 and 2 of paragraph (d) of this section, shall yield chloroform-soluble extractives, corrected for zinc extractives as zinc oleate, not to exceed the following:

(1) From a coating intended for or employed as a component of a container not to exceed 1 gal. and intended for one-time use, not to exceed 0.5 mg/in.^{2} nor to exceed that amount as milligrams per square inch that would equal 0.005% of the water capacity of the container, in milligrams, divided by the area of the food-contact surface of the container in square inches.   From a fabricated container conforming with the description in this subparagraph, the extractives shall not exceed 0.5 mg/in.^{2} of food-contact surface nor exceed 50 ppm of the water capacity of the container as determined by the methods provided in paragraph (e) of this section.

(2) From a coating intended for or employed as a component of a container having a capacity in excess of 1 gal. and intended for one-time use, not to exceed 1.8 mg/in.^{2} nor to exceed that amount as milligrams per square inch that would equal 0.005% of the water capacity of the container in milligrams, divided by the area of the food-contact surface of the container in square inches.

(3) From a coating intended for or employed as a component of a container for repeated use, not to exceed 18 mg/in.^{2} nor to exceed that amount as milligrams per square inch that would equal 0.005% of the water capacity of the container in milligrams, divided by the area of the food-contact surface of the container in square inches.

(4) From coating intended for repeated use, and employed other than as a component of a container, not to exceed 18 mg/in.^{2} surface.

(d) Tables:

TABLE 1

Types of Food

I. Nonacid (pH above 5.0) aqueous products; may contain salt or sugar or both, and including oil-in-water emulsions of low- or high-fat content.

II. Acidic (pH 5.0 or below) aqueous products; may contain salt or sugar or both, and including oil-in-water emulsions of low- or high-fat content.

III. Aqueous, acid, or nonacid products containing free oil or fat; may contain salt, and including water-in-oil emulsions of low- or high-fat content.

IV. Dairy products and modifications:
   A. Water-in-oil emulsion, high or low fat.
   B. Oil-in-water emulsion, high or low fat.

V. Low-moisture fats and oils.

VI. Beverages:
   A. Containing alcohol
   B. Nonalcoholic

VII. Bakery products

VIII. Dry solids (no end-test required).

TABLE 2

Test Procedures for Determining the Amount of Extractives from Resinous or Polymeric Coatings, Using Solvents Simulating Types of Foods and Beverages

Condition of use	Types of food (see table 1)	Extractant		
		Water	Heptane[a]	8% Alcohol
		Time and temperature	Time and temperature	Time and temperature
A. High temperature heat-sterilized (e.g., over 212°F)	I, IV-B	250°F, 2 hr		
	III, IV-A, VII	250°F, 2 hr	150°F, 2 hr	
B. Boiling water sterilized	II	212°F, 30 min		
	III, VII	212°F, 30 min	120°F, 30 min	
C. Hot filled or pasteurized above 150°F	II, IV-B	Fill boiling, cool to 100°F		
	III, IV-A	Fill boiling, cool to 100°F	120°F, 15 min	
	V		120°F, 15 min	
D. Hot filled or pasteurized below 150°F	II, IV-B, VI-B	150°F, 2 hr		
	III, IV-A	150°F, 2 hr	100°F, 30 min	
	V		100°F, 30 min	
	VI-A			150°F, 2 hr

714

Storage condition	Material			
E. Room temperature filled and stored (no thermal treatment in the container)	I, II, IV-B, VI-B	120°F, 24 hr	70°F, 30 min	
	III, IV-A	120°F, 24 hr	70°F, 30 min	
	V, VII			120°F, 24 hr
	VI-A			
F. Refrigerated storage (no thermal treatment in the container)	I, II, III, IV-A, IV-B, VI-B, VII	70°F, 48 hr		
	VI-A			70°F, 48 hr
G. Frozen storage (no thermal treatment in the container)	I, II, III, IV-B, VII	70°F, 24 hr		
H. Frozen storage: Ready-prepared foods intended to be reheated in container at time of use:				
1. Aqueous or oil in water emulsion of high or low fat	I, II, IV-B	212°F, 30 min		
2. Aqueous, high or low free oil or fat	III, IV-A, VII	212°F, 30 min	120°F, 30 min	

---

[a] Heptane extractant not to be used on wax-lined containers. Heptane extractivity results must be divided by a factor of five in arriving at the extractivity for a food product.

[b] Amendment published in the *Federal Register*, January 25, 1964. **29** F.R. 1318.

(e) [Analytical methods omitted.]

*§121.2517. *Polyethylene, Oxidized*

Oxidized polyethylene identified in paragraph (a) of this section may be safely used as a component of food-contact articles, in accordance with the following prescribed conditions:

(a) Oxidized polyethylene is the basic resin produced by the mild air oxidation of polyethylene conforming to the density, maximum *n*-hexane extractable fraction, and maximum xylene soluble fraction specifications prescribed under item 2.3 of the table in §121.2501 (c). Such oxidized polyethylene has a minimum number-average molecular weight of 1200, as determined by high temperature vapor pressure osmometry, contains a maximum of 5% by weight of total oxygen, and has an acid value of 9 to 19. . . .

§121.2520. *Adhesives*

(a) Adhesives may be safely used as components of articles intended for use in packaging, transporting, or holding food in accordance with the following prescribed conditions:

(1) The adhesive is prepared from one or more of the optional substances named in paragraph (c) of this section, subject to any prescribed limitations.

(2) The adhesive is either separated from the food by a functional barrier or used subject to the following additional limitations:

(i) *In dry foods.* The quantity of adhesive that contacts packaged dry food shall not exceed the limits of good manufacturing practice.

(ii) *In fatty and aqueous foods.* The quantity of adhesive that contacts packaged fatty and aqueous foods shall not exceed the trace amount at seams and at the edge exposure between packaging laminates that may occur within the limits of good manufacturing practice.

(a) Under normal conditions of use the packaging seams or laminates will remain firmly bonded without visible separation.

(b) To assure safe usage of adhesives, the label of the finished adhesive container shall bear the statement "food-packaging adhesive."

(c) Subject to any limitations prescribed in this section and in any other regulation promulgated under section 409 of the act which prescribes safe conditions of use for substances that may be employed as constituents of adhesives, the optional substances used in the formulation of adhesives may include the following:

(1) Substances generally recognized as safe for use in food or food packaging.

(2) Substances permitted for use in adhesives by prior sanction or approval and employed under the specific conditions of use prescribed by such sanction or approval.

(3) Flavoring substances permitted for use in food by regulations in this part, provided that such flavoring substances are volatilized from the adhesives during the packaging fabrication process.

(4) Color additives approved for use in food.

* Amendment published in the *Federal Register*, April 11, 1967. **32** F.R. 5773.

(5) Substances permitted for use in adhesives by other regulations in this subpart and substances named in this subparagraph: Provided however, that any substance named in this subparagraph and covered by a specific regulation in this subpart must meet any specifications in such regulation.

Components of Adhesives . . .

[Table of more than 830 substances omitted.]

§121.2521. Vinyl Chloride-Propylene Copolymers

*The vinyl chloride-propylene copolymers identified in paragraph (a) of this section may be safely used as components of articles intended for contact with food, under condition of use D, E, F, or G described in table 2 of §121.2526 (c), subject to the provisions of this section.

(a) For the purpose of this section, vinyl chloride-propylene copolymers consist of basic copolymers produced by the copolymerization of vinyl chloride and propylene such that the finished basic copolymers meet the specifications and extractives limitations prescribed in paragraph (c) of this section, when tested by the methods described in paragraph (d) of this section.

(b) The basic vinyl chloride-propylene copolymers identified in paragraph (a) of this section may contain optional adjuvant substances required in the production of such basic polymers. The optional adjuvant substances required in the production of the basic vinyl chloride-propylene copolymers may include substances permitted for such use by regulations in this Part 121, substances generally recognized as safe in food, and substances used in accordance with a prior sanction or approval.

(c) The vinyl chloride-propylene basic copolymers meet the following specifications and extractives limitations:

(1) Specifications. (i) Total chlorine content is in the range of 53 to 56%, as determined by any suitable analytical procedure of generally accepted applicability. . . .

(d) [Analytical methods omitted.]

§121.2522. Polyurethane Resins

The polyurethane resins identified in paragraph (a) of this section may be safely used as the food-contact surface of articles intended for use in contact with bulk quantities of dry food of the type identified in §121.2525 (c), table 1, under type VIII, in accordance with the following prescribed conditions:

(a) For the purpose of this section polyurethane resins are those produced when one or more of the isocyanates listed in subparagraph (1) of this paragraph is made to react with one or more of the substances listed in subparagraph (2) of this paragraph:

(1) Isocyanates:
4,4′-Diisocyanato-3,3′-dimethylbiphenyl(bitolylene diisocyanate)
Diphenylmethane diisocyanate

* Amendment published in the *Federal Register*, April 28, 1967. **32** F.R. 6569.

Hexamethylene diisocyanate
Toluene diisocyanate
   (2) List of substances:
Adipic acid
1,4-Butanediol
1,3-Butylene glycol
2,2-Dimethyl-1,3-propanediol
Ethylene glycol
Maleic anhydride
$\alpha,\alpha',\alpha'',\alpha'''$-Neopentanetetrayltetrakis[ω-hydroxypoly(oxypropylene)]   (1–2 moles),
   average molecular weight 400
Pentaerythritol-linseed oil alcoholysis product
Phthalic anhydride
Polyethyleneadipate modified with ethanolamine with the molar ratio of the amine
   to the adipic acid less than 0.1 to 1.
Polyoxypropylene ethers of 4,4'-isopropylidenediphenol (containing an average of
   2–4 moles of propylene oxide)
Polypropylene glycol
Propylene glycol
Trimethylolpropane

   (b) Optional adjuvant substances employed in the production of the polyurethane
resins or added thereto to impart desired technical or physical properties may include
the following substances:

Substances	Limitations
N,N-Dimethyldodecylamine	As a catalyst
N-Dodecylmorpholine	As a catalyst
4,4'-Methylenebis(2-chloroaniline)	As a curing agent
*Phthalocyanine blue (C.I. pigment blue 15, C.I. No. 74160)	As a pigment ("Copper phthalocyanine" is deleted.)
Polyvinyl isobutyl ether	
Polyvinyl methyl ether	
Soya alkyd resin	Conforming in composition with §121.2514 and containing litharge not to exceed that residual from its use as the reaction catalyst and cresol not to exceed that required as an antioxidant
N,N,N',N'-Tetrakis(2-hydroxypropyl)-ethylenediamine	As a curing agent
Triethanolamine	As a curing agent
Ultramarine blue	As a pigment

* Amendment published in the *Federal Register*, March 28, 1969. **34** F.R. 5837.

(c) An appropriate sample of the finished resin in the form in which it contacts food, when subjected to Method 6191 in Federal Test Method Standard No. 141, using No. 50 Emery abrasive in lieu of Ottawa sand, shall exhibit an abrasion coefficient of not less than 20 liters per mil of film thickness.

*§121.2523.   Fluorocarbon Resins

Fluorocarbon resins may be safely used as articles or components of articles intended for use in contact with food, in accordance with the following prescribed conditions:

(a) For the purpose of this section, fluorocarbon resins consist of basic resins produced as follows:

(1) Chlorotrifluoroethylene resins. . . .
(2) Chlorotrifluoroethylene-1,1-difluoroethylene copolymer resins. . . .
(3) Chlorotrifluoroethylene-1,1-difluoroethylene-tetrafluoroethylene   copolymer resins. . . .

§121.2524.   Polyethylene Terephthalate Film

Polyethylene terephthalate film may be safely used as an article or component of articles used in producing, manufacturing, packing, processing, preparing, treating, packaging, transporting, or holding food subject to the provisions of this section.

(a) Polyethylene terephthalate film consists of a base sheet of ethylene terephthalate polymer, to which have been added optional substances, either as constituents of the base sheet or as constituents of coatings applied to the base sheet.
(b) The quantity of any optional substance employed in the production of polyethylene terephthalate film does not exceed the amount reasonably required to accomplish the intended physical or technical effect or any limitation further provided.
†(c) Any substance employed in the production of polyethylene terephthalate film that is the subject of a regulation in Subpart F of this part conforms with any specification in such regulation.
(d) Substances employed in the production of polyethylene terephthalate film include:

(1) Substances generally recognized as safe in food.
(2) Substances subject to prior sanction or approval for use in polyethylene terephthalate film and used in accordance with such sanction or approval.
(3) Substances which by regulation in this Subpart F may be safely used as components of resinous or polymeric coatings and film used as food-contact surfaces subject to the provisions of such regulation.
(4) Substances identified in this subparagraph and subject to such limitations as are provided.

* Amendment published in the *Federal Register*, June 15, 1965.   **30** F.R. 7706.
† Amendment published in the *Federal Register*, December 10, 1966.   **31** F.R. 15570.

List of Substances and Limitations

(i) *Base sheet*:

Ethylene terephthalate copolymers: prepared by the copolymerization of dimethyl terephthalate or terephthalic acid with ethylene glycol, modified with one or more of the following: azelaic acid, dimethyl azelate, dimethyl sebacate, sebacic acid

Ethylene terephthalate polymer: prepared by the condensation of dimethyl terephthalate and ethylene glycol

Ethylene terephthalate polymer: prepared by the copolymerization of terephthalic acid and ethylene glycol

(ii) *Coatings*:

2-Ethylhexyl acrylate copolymerized with one or more of the following:

Acrylonitrile
Methacrylonitrile
Methyl acrylate
Methyl methacrylate
Itaconic acid

Vinylidene chloride copolymerized with one or more of the following:

Methacrylic acid and its methyl, ethyl, propyl, butyl, or octyl esters
Acrylic acid and its methyl, ethyl, propyl, butyl, or octyl esters
Acrylonitrile
Methacrylonitrile
Vinyl chloride
Itaconic acid

(iii) *Emulsifiers*:

Sodium dodecylbenzenesulfonate: as an adjuvant in the application of coatings to the base sheet

Sodium laurylsulfate: as an adjuvant in the application of coatings to the base sheet

---

(e) . . .

§121.2526.   *Components of Paper and Paperboard in Contact with Aqueous and Fatty Foods*

Substances identified in this section may be safely used as components of the uncoated or coated food-contact surface of paper and paperboard intended for use in producing, manufacturing, packing, processing, preparing, treating, packaging, transporting, or holding aqueous and fatty foods, subject to the provisions of this section.   Components of paper and paperboard in contact with dry food of the type identified under type VIII of table 1 in paragraph (c) of this section are subject to the provisions of §121.2571.

(a) Substances identified in subparagraphs (1) through (5) of this paragraph may be used as components of the food-contact surface of paper and paperboard.   Paper and paperboard products shall be exempted from compliance with the extractives limitations prescribed in paragraph (c) of this section: *Provided*, That the components of the food-contact surface consist entirely of one or more of the substances identified in this paragraph: *And provided further*, That if the paper or paperboard when extracted under the conditions prescribed in paragraph (c) of this section exceeds the limitations on extractives contained in paragraph (c) of this section, information shall be available from manufacturing records from which it is possible to determine that only substances identified in this paragraph (a) are present in the food-contact surface of such paper or paperboard.

(1) Substances generally recognized as safe in food.

(2) Substances generally recognized as safe for their intended use in paper and paperboard products used in food packaging.

(3) Substances used in accordance with a prior sanction or approval.

(4) Substances that by regulation in this Part 121 may be safely used without extractive limitations as components of the uncoated or coated food-contact surface of paper and paperboard in contact with aqueous or fatty food subject to the provisions of such regulation.

(5) Substances identified in this subparagraph, as follows:

[Table listing approximately 125 substances omitted.]

TABLE 1

Types of Raw and Processed Foods

I. Nonacid, aqueous products; may contain salt or sugar or both (pH above 5.0)

II. Acid aqueous products; may contain salt or sugar or both, and including oil-in-water emulsions of low- or high-fat content

III. Aqueous, acid, or nonacid products containing free oil or fat; may contain salt, and including water-in-oil emulsions of low- or high-fat content

IV. Dairy products and modifications:
    A. Water-in-oil emulsions, high- or low-fat
    B. Oil-in-water emulsions, high- or low-fat

V. Low-moisture fats and oils

VI. Beverages:
    A. Containing up to 8% of alcohol
    B. Nonalcoholic
    C. Containing more than 8% of alcohol

VII. Bakery products other than those included under types VIII or IX of this table:
    A. Moist bakery products with surface containing free fat or oil
    B. Moist bakery products with surface containing no free fat or oil

VIII. Dry solids with the surface containing no free fat or oil (no end-test required)

IX. Dry solids with the surface containing free fat or oil

TABLE 2

Test Procedures with Time-Temperature Conditions for Determining Amount of Extractives from the Food-Contact Surface of Uncoated or Coated Paper and Paperboard, Using Solvents Simulating Types of Foods and Beverages

Condition of use	Types of food (see table 1)	Food-simulating solvents			
		Water	Heptane[a]	8% Alcohol	50% Alcohol
		Time and temperature	Time and temperature	Time and temperature	Time and temperature
A. High temperature heat-sterilized (e.g., over 212°F)	I, IV-B, VII-B	250°F, 2 hr			
	III, IV-A, VII-A	250°F, 2 hr	150°F, 2 hr		
B. Boiling water sterilized	II, VII-B	212°F, 30 min			
	III, VII-A	212°F, 30 min	120°F, 30 min		
C. Hot filled or pasteurized above 150°F	II, IV-B	Fill boiling, cool to 100°F			
	III, IV-A	Fill boiling, cool to 100°F	120°F, 15 min		
	V		120°F, 15 min		
D. Hot filled or pasteurized below 150°F	II, IV-B, VI-B	150°F, 2 hr			
	III, IV-A	150°F, 2 hr			
	V		100°F, 30 min		
	VI-A		100°F, 30 min	150°F, 2 hr	
	VI-C				150°F, 2 hr

Category	Packaging types	Condition 1	Condition 2	Condition 3
E. Room-temperature filled and stored (no thermal treatment in the container)	I, II, IV-B, VI-B, VII-B	120°F, 24 hr		
	III, IV-A, VII-A	120°F, 24 hr		
	V, IX		70°F, 30 min	
	VI-A		70°F, 30 min	120°F, 24 hr
	VI-C			120°F, 24 hr
F. Refrigerated storage (no thermal treatment in the container)	III, IV-A, VII-A	70°F, 48 hr		
	I, II, IV-B, VI-B, VII-B	70°F, 48 hr		
	VI-A		70°F, 30 min	70°F, 48 hr
	VI-C			70°F, 48 hr
G. Frozen storage (no thermal treatment in the container)	I, II, IV-B, VII-B	70°F, 24 hr		
	III, VII-A	70°F, 24 hr	70°F, 30 min	
H. Frozen or refrigerated storage: Ready-prepared foods intended to be reheated in container at time of use:				
1. Aqueous or oil-in-water emulsion of high- or low-fat	I, II, IV-B, VII-B	212°F, 30 min		
2. Aqueous, high- or low-free oil or fat	III, IV-A, VII-A	212°F, 30 min		120°F, 30 min

a Heptane extractability results must be divided by a factor of five in arriving at the extractability for a food product having water-in-oil emulsion or free oil or fat. Heptane food-simulating solvent is not required in the case of wax-polymer blend coatings for corrugated paperboard containers intended for use in bulk packaging of iced meat, iced fish, and iced poultry. Amendment published in the *Federal Register*, January 28, 1966. **31** F.R. 1149.

(b) Substances identified in subparagraphs (1) and (2) of this paragraph may be used as components of the food-contact surface of paper and paperboard, provided that the food-contact surface of the paper or paperboard complies with the extractive limitations prescribed in paragraph (c) of this section.

[Table of about 200 substances omitted.]

(c) The food-contact surface of the paper and paperboard in the finished form in which it is to contact food, when extracted with the solvent or solvents characterizing the type of food, and under conditions of time and temperature characterizing the conditions of its intended use as determined from tables 1 and 2 of this paragraph, shall yield net chloroform-soluble extractives (corrected for wax, petrolatum, mineral oil, and zinc extractives as zinc oleate) not to exceed 0.5 mg/in.^{2} of food contact surface as determined by the methods described in paragraph (d) of this section. [Tables 1 and 2 appear on pp. 721-723.]

(d) [Analytical methods omitted.]

*§121.2528.   Ethylene-Methyl Acrylate Copolymer Resins

Ethylene-methyl acrylate copolymer resins may be safely used as articles or components of articles intended for use in contact with food, in accordance with the following prescribed conditions:

(a) For the purpose of this section, the ethylene-methyl acrylate copolymer resins consist of basic copolymers produced by the copolymerization of ethylene and methyl acrylate such that the copolymers contain no more than 25 weight percent of polymer units derived from methyl acrylate.

(b) The finished food-contact article, when extracted with the solvent or solvents characterizing the type of food and under the conditions of time and temperature characterizing the conditions of its intended use as determined from tables 1 and 2 of §121.2526 (c). . . .

§121.2538.   Alkyl Ketene Dimers

Alkyl ketene dimers may be safely used as a component of articles intended for use in producing, manufacturing, packing, processing, preparing, treating, packaging, transporting, or holding food, subject to the provisions of this section.

(a) The alkyl ketene dimers are manufactured by the dehydrohalogenation of the acyl halides derived from the fatty acids of animal or vegetable fats and oils.

(b) The akyl ketene dimers are used as an adjuvant in the manufacture of paper and paperboard under such conditions that the alkyl ketene dimers and their hydrolysis products dialkyl ketones do not exceed 0.4% by weight of the paper or paperboard . . .

* Issued September 4, 1968.

*§121.2543. *Packaging Materials for Use During the Irradiation of Prepackaged Foods*

The packaging materials identified in this section may be safely subjected to irradiation incidental to the radiation treatment and processing of prepackaged foods, subject to the provisions of this section and to the requirement that no induced radioactivity is detectable in the packaging material itself:

(a) The radiation of the food itself shall comply with regulations in Subpart G of this Part 121.

(b) The following packaging materials may be subjected to a dose of radiation, not to exceed 1 megarad, unless otherwise indicated, incidental to the use of gamma radiation in the radiation treatment of prepackaged foods:

(1) Nitrocellulose-coated or vinylidene chloride copolymer-cellophane complying with §121.2507.

(2) Glassine paper complying with §121.2526.

(3) Wax-coated paperboard complying with §121.2526.

(4) Polyolefin film prepared from one or more of the basic olefin polymers complying with §121.2501. . . .

(5) Kraft paper prepared from unbleached sulfate pulp to which rosin complying with §121.2592, and alum may be added. The kraft paper is used only as a container for flour and is irradiated with a dose not exceeding 50,000 rads.

(6) Polyethylene terephthalate film prepared from the basic polymer as described in §121.2524 (d) (4) (i). . . .

(7) Polystyrene film prepared from styrene basic polymer. . . .

(8) Rubber hydrochloride film prepared from rubber hydrochloride basic polymer having a chlorine content of 30–32 weight percent and having a maximum extractable fraction of 2 weight percent when extracted with n-hexane at reflux temperature for 2 hr. . . .

(9) Vinylidene chloride-vinyl chloride copolymer film prepared from vinylidene chloride-vinyl chloride basic copolymers containing not less than 70 weight percent of vinylidene chloride. . . .

(10) Nylon-11 conforming to §121.2502.

(c) The following packaging materials may be subjected to a dose of radiation, not to exceed 6 megarads incidental to the use of gamma or x-radiation in the radiation processing of prepackaged foods.

(1) Vegetable parchments. . . .

(2) Films prepared from basic polymers and with or without adjuvants, as follows:

(i) Polyethylene film prepared from the basic polymer as described in §121.2501 (a). . . .

(ii) Polyethylene terephthalate film prepared from the basic polymer as described in §121.2524 (d) (4) (i). . . .

(iii) Nylon-6 films prepared from the nylon-6 basic polymer as described in §121.2502 (a) (6). . . .

* Reissued March 16, 1968.

(iv) Vinyl chloride-vinyl acetate copolymer film prepared from the basic copolymer containing 38.5 to 90.0 weight percent of vinyl chloride with 10.0 to 11.5 weight percent of vinyl acetate and having a maximum volatility of not over 3.0% (1 hr at 105°C) and viscosity not less than 0.30 determined by ASTM D 1243–60, Method A....

*§121.2555. Perfluorocarbon Resins

Perfluorocarbon resins may be safely used as articles or components of articles used in producing, manufacturing, packing, processing, preparing, treating, packaging, transporting, or holding food, in accordance with the following prescribed conditions:

(a) Perfluorocarbon resins are produced by the homopolymerization and/or copolymerization of hexafluoropropylene and tetrafluoroethylene, to which may have been added certain optional substances to impart desired technological properties to the resins. Subject to any limitations prescribed in this section, the optional substances may include:

(1) Substances generally recognized as safe in food and food packaging.
(2) Substances, the use of which is permitted under applicable regulations in this part, prior sanctions, or approvals....

(b) [Specifications omitted.]

§121.2557. Defoaming Agents Used in Coatings

The defoaming agents described in this section may be safely used as components of articles intended for use in producing, manufacturing, packing, processing, preparing, treating, packaging, transporting, or holding food, subject to the provisions of this section.

(a) The defoaming agents are prepared as mixtures of substances described in paragraph (d) of this section.
(b) The quantity of any substance employed in the formulation of defoaming agents does not exceed the amount reasonably required to accomplish the intended physical or technical effect in the defoaming agents or any limitation further provided.
(c) Any substance employed in the production of defoaming agents and which is the subject of a regulation in this Subpart F conforms with any specifications in such regulation.

(d) Substances employed in the formulation of defoaming agents include:

(1) Substances generally recognized as safe in food.
(2) Substances subject to prior sanction or approval for use in defoaming agents and used in accordance with such sanction or approval.
(3) Substances identified in this subparagraph and subject to such limitations as are provided:

* Amendment published in the *Federal Register*, October 13, 1962.   **27** F.R. 10098.

[Table of more than 90 substances omitted.]

*§121.2564.   Ethylene-Acrylic Acid Copolymers

The ethylene-acrylic acid copolymers identified in paragraph (a) of this section may be safely used as components of articles intended for use in contact with food subject to the provisions of this section.

(a) For the purpose of this section, ethylene-acrylic acid copolymers consist of basic copolymers produced by the copolymerization of ethylene and acrylic acid such that the finished basic copolymers contain no more than 10 weight-percent of total polymer units derived from acrylic acid. . . .

§121.2570.   Ethylene-Vinyl Acetate Copolymers

Ethylene-vinyl acetate copolymers may be safely used as articles or components of articles intended for use in producing, manufacturing, packing, processing, preparing, treating, packaging, transporting, or holding food in accordance with the following prescribed conditions:

(a) Ethylene-vinyl acetate copolymers consist of basic resins produced by the catalytic copolymerization of ethylene and vinyl acetate to which may have been added certain optional substances to impart desired technological or physical properties to the resin. . . .

§121.2574.   Polycarbonate Resins

Polycarbonate resins may be safely used as articles or components of articles intended for use in producing, manufacturing, packing, processing, preparing, treating, packaging, transporting, or holding food, in accordance with the following prescribed conditions:

(a) Polycarbonate resins are polyesters produced by:

(1) The condensation of 4,4'-isopropylidenediphenol and carbonyl chloride to which may have been added certain optional adjuvant substances required in the production of the resins; or by

(2) The reaction of molten 4,4'-isopropylidenediphenol with molten diphenyl carbonate in the presence of the disodium salt of 4,4'-isopropylidenediphenol. . . .

(b) (c) [Adjuvants and specifications omitted.]

§121.2575.   Paraffin, Synthetic

Synthetic paraffin may be safely used as an impregnant in, coating on, or component of coatings on articles used in producing, manufacturing, packing, processing, preparing, treating, packaging, transporting, or holding food. . . .

* Amendment published in the *Federal Register*, April 7, 1967.   **32** F.R. 5675.

§121.2576.   Crosslinked Polyester Resins

Crosslinked polyester resins may be safely used as articles or components of articles intended for repeated use in contact with food, in accordance with the following prescribed conditions:

*(a) The crosslinked polyester resins are produced by the condensation of one or more of the acids listed in subparagraph (1) of this paragraph with one or more of the alcohols or epoxides listed in subparagraph (2) of this paragraph, followed by copolymerization with one or more of the crosslinking agents listed in subparagraph (3) of this paragraph:

(1) *Acids*:
Adipic
Fatty acids, and dimers thereof, from natural sources
Fumaric
Isophthalic
Maleic
*Methacrylic
Orthophthalic
Sebacic
Terephthalic
Trimellitic
(2) *Polyols and polyepoxides*:
Butylene glycol
Diethylene glycol
2,2-Dimethyl-1,3-propanediol
Dipropylene glycol
Ethylene glycol
Glycerol
*4,4'-Isopropylidenediphenol-epichlorohydrin
Mannitol
α-Methyl glucoside
Pentaerythritol
Polyoxypropylene ethers of 4,4'-isopropylidenediphenol (containing an average of
    2–7.5 moles of propylene oxide)
Propylene glycol
Sorbitol
Trimethylolethane
Trimethylolpropane
2,2,4-Trimethyl-1,3-pentanediol
(3) *Crosslinking agents*:
Butyl acrylate
Butyl methacrylate
Ethyl acrylate
2-Ethylhexyl acrylate

* Amendments published in the *Federal Register*, January 11, 1966.   **31** F.R. 290.

Methyl acrylate
Methyl methacrylate
Styrene
Vinyltoluene

*§121.2580.   Polyethylene Resins, Carboxyl-Modified

Carboxyl-modified polyethylene resins may be safely used as the food-contact surface of articles intended for use in contact with food in accordance with the following prescribed conditions:

(a) For the purpose of this section, carboxyl-modified polyethylene resins consist of basic polymers produced when ethylene-methyl acrylate basic copolymers, containing no more than 25 weight percent of polymer units derived from methyl acrylate, are made to react in an aqueous medium with one or more of the following substances: ammonium hydroxide, calcium carbonate, potassium hydroxide, sodium hydroxide.

. . . .

§121.2581.   Chlorinated Polyether Resins

Chlorinated polyether resins may be safely used as articles or components of articles intended for repeated use in producing, manufacturing, packing, processing, preparing, treating, packaging, transporting, or holding food, in accordance with the following prescribed conditions:

(a) The chlorinated polyether resins are produced by the catalytic polymerization of 3,3-bis(chloromethyl)oxetane, and shall contain not more than 2% residual monomer. . . .

§121.2582.   Ethylene-Methacrylic Acid Copolymers, Ethylene-Methacrylic Acid-Vinyl Acetate Copolymers, and Their Partial Salts

†Ethylene-methacrylic acid copolymers, ethylene-methacrylic acid-vinyl acetate copolymers, and/or their ammonium, calcium, magnesium, sodium, and/or zinc partial salts may be safely used as articles or components of articles intended for use in contact with food, in accordance with the following prescribed conditions:

†(a) For the purpose of this section, the ethylene-methacrylic acid copolymers consist of basic copolymers produced by the copolymerization of ethylene and methacrylic acid such that the copolymers contain no more than 20 weight percent of polymer units derived from methacrylic acid, and the ethylene-methacrylic acid-vinyl acetate copolymers consist of basic copolymers produced by the copolymerization of ethylene, methacrylic acid, and vinyl acetate such that the copolymers contain no more than 15 weight percent of polymer units derived from methacrylic acid.

* Amendment published in the *Federal Register*, July 31, 1965.   **30** F.R. 9575.
† Amendments published in the *Federal Register*, June 10, 1967.   **32** F.R. 8360.

(b) The finished food-contact article, when extracted with the solvent or solvents characterizing the type of food and under the conditions of time and temperature characterizing the conditions of its intended use as determined from tables 1 and 2 of §121.2526 (c), yields net acidified chloroform-soluble extractives in each extracting solvent not to exceed 0.5 mg/in.[2] of food-contact surface when tested by the methods described in paragraph (c) of this section; and if the finished food-contact article is itself the subject of a regulation in this Subpart F, it shall also comply with any specifications and limitations prescribed for it by that regulation. . . .

(c) [Analytical methods omitted.]

§121.2583.   Adjuvant Substances Used in the Manufacture of Foamed Polystyrene

The following substances may be safely used as adjuvants in the manufacture of foamed polystyrene intended for use in contact with food, subject to any prescribed limitations:

Substances	Limitations
Isopentane	For use as blowing agent
n-Pentane	For use as blowing agent
1,1,2,2-Tetrachloethylene	For use only as a blowing agent adjuvant at a level not to exceed 0.3% by weight of finished foamed polystyrene intended for use in contact with food only of the types identified in §121.2526 (c), table 1, under categories I, II, VI and VIII
Toluene	For use only as a blowing agent adjuvant at a level not to exceed 0.35% by weight of finished foamed polystyrene

§121.2586.   Petroleum Wax

Petroleum wax may be safely used as a component of nonfood articles in contact with food, in accordance with the following conditions:

(a) Petroleum wax is a mixture of solid hydrocarbons, paraffinic in nature, derived from petroleum, and refined to meet the specifications prescribed in this section.

(b) The petroleum wax meets the following ultraviolet absorbance limits when subjected to the analytical procedure described in §121.1156 (b).

Ultraviolet absorbance per centimeter path length:

280–289 millimicrons	0.15 maximum
290–299 millimicrons	0.12 maximum
300–359 millimicrons	0.08 maximum
360–400 millimicrons	0.02 maximum

(c) Petroleum wax may contain any antioxidant permitted in food by regulations issued in accordance with section 409 of the act, in an amount not greater than that required to produce its intended effect. . . .

§121.2590. Isobutylene Polymers

Isobutylene polymers may be safely used as components of articles intended for use in producing, manufacturing, packing, processing, preparing, treating, packaging, transporting, or holding food, in accordance with the following prescribed conditions:

(a) For the purpose of this section, isobutylene polymers are those produced as follows:

(1) Polyisobutylene produced by the homopolymerization of isobutylene such that the finished polymers have a molecular weight of 750,000 (Flory) or higher.

(2) Isobutylene-isoprene copolymers produced by the copolymerization of isobutylene with not more than 3 mole percent of isoprene such that the finished polymers have a molecular weight of 300,000 (Flory) or higher.

(3) Chlorinated isobutylene-isoprene copolymers produced when isobutylene-isoprene copolymers [molecular weight 300,000 (Flory) or higher] are modified by chlorination with not more than 1.3 weight percent of chlorine. . . .

§121.2591. Semirigid and Rigid Acrylic and Modified Acrylic Plastics

Semirigid and rigid acrylic and modified acrylic plastics may be safely used as articles intended for use in contact with food, in accordance with the following prescribed conditions:

(a) The optional substances used in the formulation of the semirigid and rigid acrylic and modified acrylic plastics include substances generally recognized as safe in food, substances used in accordance with a prior sanction or approval, substances permitted for use in such plastics by regulations in this Part 121, and substances identified in this paragraph. At least 50 weight percent of the polymer content of the finished plastics shall consist of polymer units derived from one or more of the acrylic or methacrylic monomers listed in subparagraph (1) of this paragraph.

(1) Homopolymers and copolymers of the following monomers:

n-Butyl acrylate
n-Butyl methacrylate
Ethyl acrylate
2-Ethylhexyl acrylate
Ethyl methacrylate
Methyl acrylate
Methyl methacrylate

(2) Copolymers produced by copolymerizing one or more of the monomers listed in subparagraph (1) of this paragraph with one or more of the following monomers:

Acrylonitrile
Methacrylonitrile
α-Methylstyrene
Styrene
Vinyl chloride
Vinylidene chloride

(3) Polymers identified in subparagraphs (1) and (2) of this paragraph containing no more than 5 weight percent of total polymer units derived by copolymerization with one or more of the monomers listed in subdivisions (i) and (ii) of this subparagraph. Monomers listed in subdivision (ii) of this subparagraph are limited to use only in plastic articles intended for repeated use in contact with food.

(i) List of minor monomers:

Acrylamide
Acrylic acid
1,3-Butylene glycol dimethacrylate
1,4-Butylene glycol dimethacrylate
Diethylene glycol dimethacrylate
Dipropylene glycol dimethacrylate
Divinylbenzene
Ethylene glycol dimethacrylate
Itaconic acid
Methacrylic acid
N-Methylolacrylamide
N-Methylolmethacrylamide
4-Methyl-1,4-pentanediol dimethacrylate
Propylene glycol dimethacrylate
Trivinylbenzene

(ii) List of minor monomers limited to use only in plastic articles intended for repeated use in contact with food:

tert-Butyl acrylate
tert-Butylaminoethyl methacrylate
sec-Butyl methacrylate
tert-Butyl methacrylate
Cyclohexyl methacrylate
Dimethylaminoethyl methacrylate
2-Ethylhexyl methacrylate
Hydroxyethyl methacrylate
Hydroxyethyl vinyl sulfide
Hydroxypropyl methacrylate
Isobornyl methacrylate
Isobutyl methacrylate
Isopropyl acrylate
Isopropyl methacrylate
Methacrylamide
Methacrylamidoethylene urea
Methacryloxyacetamidoethylethylene urea
Methacryloxyacetic acid
n-Propyl methacrylate
3,5,5-Trimethylcyclohexyl methacrylate

(4) Polymers identified in subparagraphs (1), (2), and (3) of this paragraph mixed together and/or with the following polymers, provided that no chemical reactions, other than addition reactions, occur when they are mixed:

Butadiene-acrylonitrile copolymers
Butadiene-acrylonitrile-styrene copolymers
Butadiene-acrylonitrile-styrene-methyl methacrylate copolymers
Butadiene-styrene copolymers
Butyl rubber
Natural rubber
Polybutadiene
Poly-3-chloro-1,3-butadiene
Polyesters identified in §121.2514 (b) (3) (vii)
Polyvinyl chloride
Vinyl chloride-vinyl acetate copolymers

*(5) Antioxidants and stabilizers identified in §121.2514 (b) (3) (xxx) and the following:

Di-*tert*-butyl-*p*-cresol
2-Hydroxy-4-methoxybenzophenone
2-Hydroxy-4-methoxy-2-carboxybenzophenone
3-Hydroxyphenyl benzoate
p-Methoxyphenol
Methyl salicylate
Phenyl salicylate

(6) Release agents: Fatty acids derived from animal and vegetable fats and oils, and fatty alcohols derived from such acids.

(7) Surface active agent: Sodium dodecylbenzenesulfonate.

(8) Miscellaneous materials:

Dimethyl phthalate
Oxalic acid, for use only as a polymerization catalyst aid
Tetraethylenepentamine, for use only as a catalyst activator at a level not to exceed 0.5 weight percent based on the monomers
Toluene
Xylene

[Analytical method omitted.]

§121.2592. *Rosins and Rosin Derivatives*

The rosins and rosin derivatives identified in paragraph (a) of this section may safely be used in the manufacture of articles or components of articles intended for

* 2(2′-Hydroxy-5′-methylphenyl)benzotriazole was stricken from the list March 15, 1969. **34** F.R. 5292.

use in producing, manufacturing, packing, processing, preparing, treating, packaging, transporting, or holding food, subject to the provisions of this section.

(a) The rosins and rosin derivatives are identified as follows:

(1) Rosins:

(i) Gum rosin, refined to color grade of K or paler.
(ii) Wood rosin, refined to color grade of K or paler.
(iii) Tall oil rosin, refined to color grade of K or paler.
(iv) Dark tall oil rosin, a fraction resulting from the refining of tall oil rosin produced by multicolumn distillation of crude tall oil to effect removal of fatty acids and pitch components and having a saponification number of 110–135 and 32–44% rosin acids.
(v) Dark wood rosin, all or part of the residue after the volatile terpene oils are distilled from the oleoresin extracted from pine wood.

(2) Modified rosins manufactured from rosins identified in subparagraph (1) of this paragraph . . .

(3) Rosin esters manufactured from rosins and modified rosins identified in sub-paragraphs (1) and (2) of this paragraph. . . .

(4) Rosin salts and sizes. Ammonium, calcium, potassium, sodium, or zinc salts of rosin manufactured by the partial or complete saponification of gum rosin, wood rosin, tall oil rosin, dark wood rosin, partially hydrogenated rosin, fully hydrogenated rosin, or disproportionated rosin as identified in subparagraphs (1) and (2) of this paragraph, or blends thereof, and with or without modification by reaction with one or more of the following: formaldehyde, fumaric acid, maleic anhydride, saligenin.
   . . . .

*§121.2593.   Polyvinylidene Fluoride Resins

Polyvinylidene fluoride resins may be safely used as articles or components of articles intended for repeated use in contact with food, in accordance with the following prescribed conditions:

(a) For the purpose of this section, the polyvinylidene fluoride resins consist of basic resins produced by the polymerization of vinylidene fluoride. . . .

†§121.2594.   Odorless Light Petroleum Hydrocarbons

Odorless light petroleum hydrocarbons may be safely used, as a component of nonfood articles intended for use in contact with food, in accordance with the following prescribed conditions:

* Amendment published in the *Federal Register*, April 17, 1965. **30** F.R. 5510.
† Amendment published in the *Federal Register*, June 22, 1965. **30** F.R. 7997.

(a) The additive is a mixture of liquid hydrocarbons derived from petroleum or synthesized from petroleum gases. The additive is chiefly paraffinic, isoparaffinic, or naphthenic in nature.

. . . .

§121.2597.   Polymer Modifiers in Semirigid and Rigid Vinyl Chloride Plastics

*The polymers identified in paragraph (a) of this section may be safely admixed, alone or in mixture with other permitted polymers, as modifiers in semirigid and rigid vinyl chloride plastic food-contact articles prepared from vinyl chloride homopolymers and/or from vinyl chloride copolymers complying with §121.2608, in accordance with the following prescribed conditions:

(a) For the purpose of this section, the polymer modifiers are identified as follows:

(1) Acrylic polymers identified in this subparagraph provided that such polymers contain at least 50 weight-percent of polymer units derived from one or more of the monomers listed in subdivision (i) of this subparagraph.

(i) Homopolymers and copolymers of the following monomers: n-butyl acrylate, n-butyl methacrylate, ethyl acrylate, methyl methacrylate.
(ii) Copolymers produced by copolymerizing one or more of the monomers listed in subdivision (i) of this subparagraph with one or more of the following monomers: acrylonitrile, butadiene, styrene, vinylidene chloride.
(iii) Polymers identified in subdivisions (i) and (ii) of this subparagraph containing no more than 5 weight-percent of total polymer units derived by copolymerization with one or more of the following monomers: acrylic acid, 1,3-butylene glycol dimethacrylate, divinylbenzene, methacrylic acid.

. . . .

§121.2598.   Polyvinyl Alcohol Film

Polyvinyl alcohol film may be safely used in contact with food of the types identified in §121.2526 (c), table 1, under types V, VIII, and IX, in accordance with the following prescribed conditions:

(a) The polyvinyl alcohol film is produced from polyvinyl alcohol having a minimum viscosity of 4 cp when a 4% aqueous solution is tested at 20°C.

. . . .

†§121.2608.   Vinyl Chloride-Lauryl Vinyl Ether Copolymers

The vinyl chloride-lauryl vinyl ether copolymers identified in paragraph (a) of this section may be used as an article or as a component of an article intended for use in contact with food subject to the provisions of this section.

* Amendment published in the *Federal Register* April 8, 1969.   **34** F.R. 6240.
† Amendment published in the *Federal Register*, November 7, 1968.   **33** F.R. 16335.

(a) Identity.   For the purposes of this section vinyl chloride-lauryl vinyl ether copolymers consist of basic copolymers produced by the copolymerization of vinyl chloride and lauryl vinyl ether such that the finished copolymers contain not more than 3 weight-percent of polymer units derived from lauryl vinyl ether and meet the specifications and extractives limitations prescribed in paragraph (c) of this section.

(b), (c), (d), (e) [Adjuvants, tests, and specifications omitted.]

*§121.2609.   Vinyl Chloride-Ethylene Copolymers

The vinyl chloride-ethylene copolymers identified in paragraph (a) of this section may be safely used as components of articles intended for contact with food, under conditions of use D, E, F, or G described in table 2 of §121.2526 (c), subject to the provisions of this section.

(a) For the purpose of this section, vinyl chloride-ethylene copolymers consist of basic copolymers produced by the copolymerization of vinyl chloride and ethylene such that the finished basic copolymers meet the specifications and extractives limitations prescribed in paragraph (c) of this section, when tested by the methods described in paragraph (d) of this section.

(b), (c), (d), (e) [Specifications and analytical methods omitted.]

* Amendment published in the *Federal Register*, March 12, 1969.   **34** F.R. 5101.

INDEX